Conceptual Academy Biology

Jennifer Yeh

Conceptual Academy

Conceptual Academy

Editor-in-Chief: John Suchocki
Operations Specialist: John Wright
Senior Marketing Manager: Graham Orr
Composition: Tracy Suchocki, Ian Suchocki
Indexing: Ian Suchocki
Photo Research: Conceptual Productions, LLC
Consulting Editor: Suzanne Lyons
Cover Image Credit: Ondrej Prosicky/Shutterstock

Credits and acknowledgments appear with images.

Library of Congress Cataloging-in-Publication Data available upon request from Publisher.

Printed in the United States
3 4 5 6 7 8 9 10

Conceptual Academy

Conceptual Academy, PBC www.ConceptualAcademy.com

ISBN: 979-8-9862706-4-7

About Conceptual Academy

Conceptual Academy is a public benefit company of science teachers, media producers, and authors known for their conceptual science textbooks, such as *Conceptual Physics*, *Conceptual Chemistry*, *Conceptual Biology*, and others. Conceptual Academy provides a comprehensive library of video lessons that, altogether, create the "video textbook"—just as comprehensive and cohesive as the traditional textbook but featuring video. Conceptual Academy also produces traditional textbooks, such as the one you are now holding, as well as many other learning resources, such as lab activities, worksheets, podcasts, test banks, an automated encouragement-based assessment system, and resources for an instructor's professsional development all now hosted by the PocketLab Notebook learning and lesson plan system. Variety is the key to serving the varying needs of our diverse students and helping to keep each classroom truly interactive and learner-centered. Toward this, our programs follow a double helical model as described here:

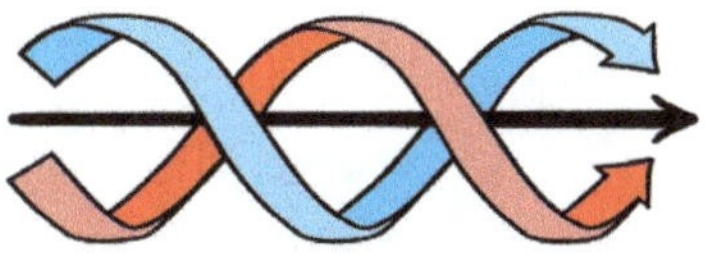

About PocketLab

PocketLab is a hands-on, classroom-ready, learning and lesson plan system that brings STEM labs and lessons to life. This learning system features hand-held data sensors and digital lab software where students collect, visualize, and analyze real-time experimental data. PocketLab hosts hundreds of comprehensive lesson plans available for instructors to customize for their own classroom needs. Lesson plan developers include PhET Interactive Simulations, Kesler Science, TeacherGeek, Green Ninja, Open SciEd, Hooked on Science with Jason Lindsay, StarTalk with Neil deGrasse Tyson, and Conceptual Academy with Paul Hewitt. PocketLab also sponsors the community-building "Science is Cool" virtual unconferences for science teachers across the world. Known for a relaxed and engaging atmosphere, each webinar unconference allows teachers to connect with science celebrities and, just as celebrated, science educators doing amazing things in their classroom.

TEXAS EDITIONS

About the Conceptual Academy Science Series

SBOE approved

The Conceptual Academy authors have revived their physical science and integrated science "Explorations" titles once published for middle and secondary level students. The content has been 1) updated and reorganized into distinct module lessons, 2) integrated with the Conceptual Academy video library and other resources including PocketLab sensor activities, 3) made available through the dynamic Pocket Lab Notebook platform that allows each instructor to organize and implement this library of content according to their classroom needs, and 4) aligned to the TEKS and ELPS Texas state standards. To both teach and reach our modern students effectively, we need not just curricula of high quality but also of great variety and adaptability. Combine this with affordability, then we have the means to pay it forward to a sustainable future built by compassionate, curious, and scientifically literate citizens who love and respect the natural universe for what it truly is. That's good energy!

To the Student

Dear Student,

Welcome to biology—the study of life. This program maps out the overarching and interconnected concepts of biology at an introductory level. We begin with a review of the chemistry needed to understand and appreciate the molecular nature of life. From this foundation, we explore the nature of cells, genetics, and inheritance, which sets the stage for undestanding evolution and the diversity of life. From there the journey continues into human anatomy and physiology culminating in an exploration of populations, communities, and ecosystems of which we humans are an integral part. Throughout, the concepts of biology are connected to relevant examples from medicine, nutrition, health, and biotechnology. Our main goal: nurturing curiosity and respect for the wonders of our living world.

To Life!

Table of Contents

Chapter 1 Science and Biology 1

1.1 What Is Life? 1
1.2 The Scientific Method 3
    The Scientific Experiment
    Example: Sea Butterflies
1.3 Science and Technology 8
1.4 Facts, Laws, and Theories 9
1.5 Working with Numbers 10
    The Metric System
Concept Review 13
Living Earth Essay: Biogeology—An Introduction 16

Chapter 2 The Chemistry of Life 17

2.1 Atoms and Molecules 17
2.2 Chemical Compounds 20
2.3 Mixtures 24
    A Closer Look at Sharing Electrons
    Some Definitions
2.4 Chemical Reactions 27
    Acids and Bases
    Oxidations and Reductions
2.5 Types of Reactions 30
2.6 Organic Molecules 34
2.7 Macromolecules Needed for Life 37
Concept Review 40
Living Earth Essay: Isotopic Dating 47

Chapter 3 The Cell 51

3.1 What is a Cell? 51
    The Two Types of Cells
3.2 Cell Theory 53
3.3 Looking at Cells 54
3.4 Eukaryotic Cell 57
3.5 The Cell Membrane 58
3.6 Cell Organelles 60
Concept Review 61
Living Earth Essay: Geologic Time 64

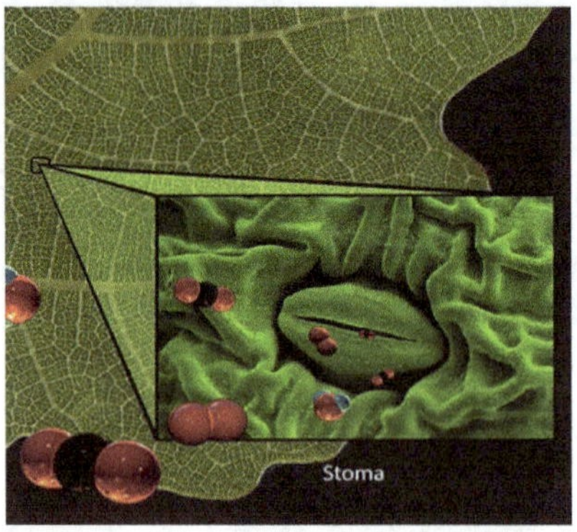

Chapter 4 How Cells Work 69

4.1 Cellular Transport 69
Simple Diffusion
Facilitated Diffusion
Endocytosis and Exocytosis
4.2 Cell Communication 73
4.3 ATP and Chemical Reactions 74
4.4 Enzymes 77
4.5 Photosynthesis 78
4.6 Cellular Respiration 81
Glycolysis
The Krebs Cycles and Electron Transport
Fermentation
Concept Review 84
Living Earth Essay: Plate Tectonics 89

Chapter 5 DNA and Genes 93

5.1 What is a Gene? 93
5.2 Chromosomes 94
5.3 The Structure of DNA 96
5.4 How DNA Is Copied 97
5.5 How Proteins Are Built 98
RNA
Transcription
Translation
5.6 Genetic Mutations 102
Concept Review 104
Living Earth Essay: The Great Oxygenation 108

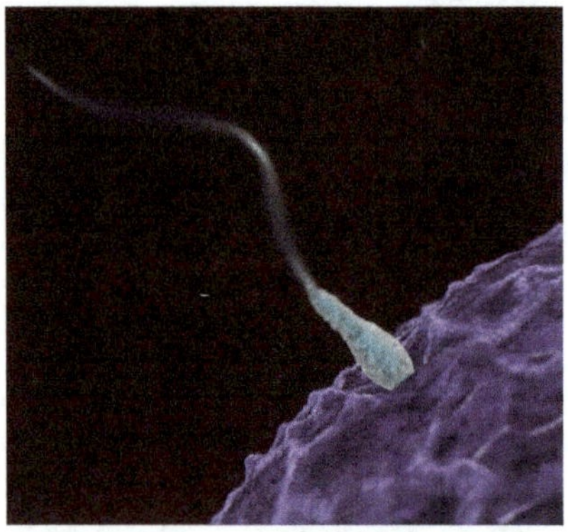

Chapter 6 Inheritance 111

6.1 How Cells Reproduce 111
6.2 Cell Division and Genetic Diversity 114
Meiosis and Sexual Reproduction Produce Genetic Diversity
Comparing Mitosis and Meiosis
6.3 Traits and Inheritance 116
6.4 First Law of Inheritance 118
6.5 Second Law of Inheritance 120
6.6 Beyond Mendel 121
Incomplete Dominance
Codominance
Polygenic Traits
Pleiotropy
Linked Genes
Sex-Linked Traits
Horizontal Gene Transfer
Concept Review 124
Living Earth Essay: Life in the Trenches 128

Chapter 7 Genetic Technologies 131

7.1 The Human Genome 131
7.2 Genetic Testing 132
7.3 Cancer 133
Environmental Causes of Cancer
7.4 DNA Technology 134
Genetically Modified Organisms
Cloning
7.5 Genetically Engineered Mosquitoes 138
7.6 Genome Editing with CRISPR-Cas9 139
7.7 Concerns about DNA Technology 140
Environmental Concern
Contamination
Cost
Concept Review 142
Living Earth Essay: Supervolcanoes 146

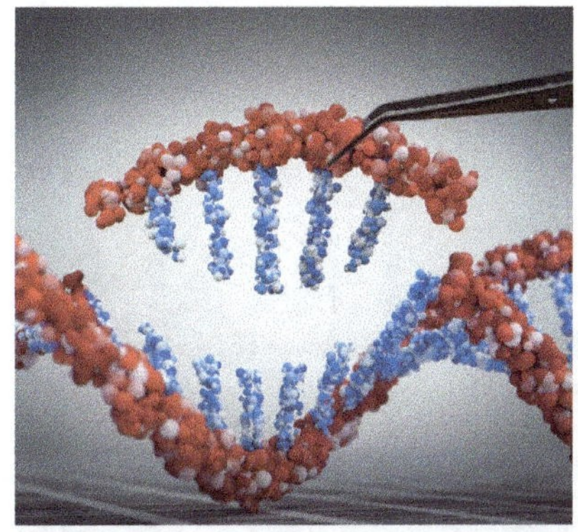

Chapter 8 Natural Selection 149

8.1 The Origin of Life 149
8.2 Is there life on Mars or Venus? 152
8.3 Charles Darwin 154
8.4 Natural Selection 157
8.5 Examples of Natural Selection 159
The Peppered Moth
Antibiotic-Resistant Bacteria
8.6 Adaptation 162
Staying Warm and Keeping Cool
Concept Review 166
Living Earth Essay: Magnetoreception 172

Chapter 9 Evidence of Evolution 175

9.1 Mechanisms of Evolution 175
Evolution and Genetics
Mutation Pressure
Genetic Drift
Gene Flow
Where Genetic Variation Comes From
9.2 How New Species Form 179
9.3 Natural Selection in Action 182
Observations of Natural Selection
9.4 Fossils 184
9.5 Body Structures and Genetics 186
Vestigial Organs
Patterns of Development
DNA and Molecular Evidence
Hierarchical Organization of Living Things
9.6 Biogeography 188
9.7 The Evolution of Humans 190
Concept Review 193
Living Earth Essay: Living Under the Ice 198

Chapter 10 Diversity of Life 1 201

10.1 Classifying Life 201
10.2 Evolutionary Trees 203
10.3 Three Domains of Life 205
10.4 Bacteria 206
10.5 Archaea 207
10.6 Protists 208
10.7 Plants 210
Mosses
Ferns
Seed Plants
Concept Review 215
Living Earth Essay: The Wonder of Wind 220

Chapter 11 Diversity of Life 2 223

11.1 Fungi 223
11.2 Animals Part 1 225
Sponges
Cnidarians
11.3 Animals Part 2 229
Roundworms
Arthropods
Mollusks
Annelids
11.4 Animals Part 3 233
Chordates
11.5 Viruses and Prions 238
11.6 Life is Interconnected 239
Concept Review 240
Living Earth Essay: Interconnectedness 244

Chapter 12 The Nervous System 247

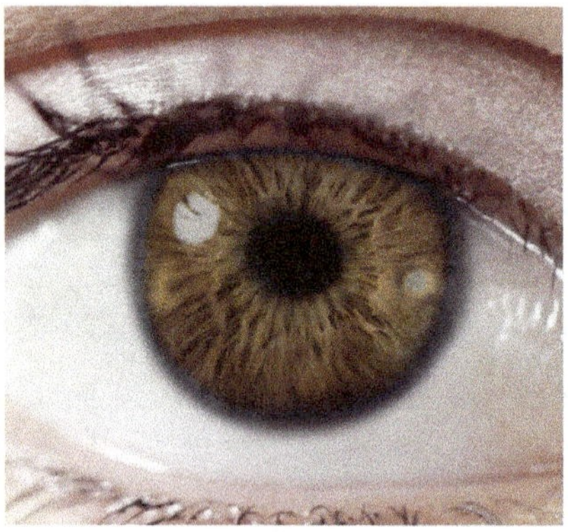

12.1 Organization of the Human Body 247
12.2 Homeostasis 249
12.3 The Brain 250
12.4 The Nervous System 252
12.5 How Neurons Fire 254
The Action Potential
How an Action Potential is Propagated
12.6 How Neurons Communicate 257
12.7 The Senses 259
Vision
Hearing
Smell and Taste
Touch
Other Senses
Concept Review 262
Living Earth Essay: Plant Nutrients 268

Chapter 13 Control and Development 271

13.1 Hormones 271

Endocrine Organs and Their Hormones

13.2 Reproduction 275

13.3 Development 277

13.4 The Skeleton 278

13.5 Muscles 279

Muscle Contraction

Concept Review 282

Living Earth Essay: Soil Fertility 286

Chapter 14 Maintaining the Body 289

14.1 The Circulatory System 289

Blood Vessels

14.2 The Path of Blood Flow 291

Blood Pressure

14.3 Blood 293

Hemoglobin

14.4 Respiration 294

The Path of Air

Inhaling and Exhaling

14.5 Digestion 296

Concept Review 299

Living Earth Essay: Pesticide Bioaccumulation 303

Chapter 15 Protecting Health 307

15.1 Nutrition, Exercise, and Health 307

15.2 The Excretory System 309

15.3 The Innate Immune System 311

Components of the Innate Immune System

Mucous Membranes

Innate Immune Cells

15.4 The Acquired Immune System 314

Components of the Acquired Immune System

B Cells

T Cells

Vaccines

Diseases of the Immune System

Concept Review 318

Living Earth Essay: Topsoil in Trouble 322

Chapter 16 Populations 325

16.1 Organisms and Their Environments 325
16.2 Population Growth 327
Exponential Growth
Logistic Growth
16.3 Life History 329
16.4 Human Population Growth 330
Ecological Footprint
Concept Review 332
Living Earth Essay: Topsoil Regenerated 338

Chapter 17 Communities 343

17.1 Food Webs 343
17.2 Competition 345
17.3 Symbiosis 346
17.4 Invasive Species 347
Concept Review 349
Living Earth Essay: Ocean Acidification 352

Chapter 18 Ecosystems 355

18.1 Terrestrial Biomes 355
18.2 Aquatic Biomes 359
Freshwater Habitats
Saltwater Habitats
18.3 Biogeochemical Cycles 362
Water
Carbon
Nitrogen
18.4 Energy Flow in Ecosystems 365
18.5 Ecological Succession 367
Concept Review 370
Living Earth Essay: Global Climate Change 374

Index 381

Conceptual Academy

PocketLab
SCIENCE. EVERYWHERE.

Chapter 1

Science and Biology

THE MAIN IDEA

Biology is the study of life.

1.1 What Is Life?
1.2 The Scientific Method
1.3 Science and Technology
1.4 Facts, Laws, and Theories
1.5 Working with Numbers

Humans, dogs, germs, giraffes, big trees—these are all living things. Clouds, fire, rock, and puddles—these are not living. But what exactly are living things? Do all living things move? Do all living things reproduce? What are living things made of? In this chapter, you will begin your study of biology, the science of life. You will also learn what science is and how science works. Science has given us much. Our modern world is built on it. Nearly all forms of technology—from medicine to space travel—are applications of science. But what exactly is this amazing thing called science? Science is a body of knowledge, but is it also a process for finding the connections in nature? Does science give us a more insightful way to see the world around us? How should science be used? Where did science come from? And what would the modern world be like without it?

1.1 What Is Life?

Biology is the study of life and living organisms. But what is a living organism? What distinguishes living things from nonliving things?

Living things share certain characteristics. For one thing, they use energy. Living things, such as the sunflowers and lions in **Figure 1.1**, take energy from the environment and convert it into other forms of energy for their own use. Plants take radiant energy from sunlight and convert it into the chemical energy of molecules, which they can use to build their stems and leaves or fuel their activities. Animals eat, converting the energy they get from food into chemical energy, which they store in their bodies. This chemical energy is eventually converted into kinetic energy (energy of motion) and heat (thermal energy).

> READING CHECK
>
> What do living things do to with energy they take from their environment?

Another characteristic of living things is that they develop and grow. When chicks hatch, they are small and covered with downy yellow feathers. Over time, they grow bigger, and their downy feathers are replaced by stiff adult feathers (**Figure 1.2**).

Living things maintain themselves. They generate structures, such as stems and leaves or skin and bones, and they repair damage done to those structures. When you scrape your knee, your blood clots to stop the bleeding, and the wounded skin scabs over and heals. Living things also maintain their internal environment, keeping it stable despite changing external conditions. Whether it is freezing cold or blisteringly hot, your body temperature stays right around 37°C (98.6°F).

(a) (b)

Figure 1.1

Living things take energy from the environment and convert it into other forms of energy. (a) Sunflowers convert energy from sunlight into chemical energy, which can be used to build tissues or fuel activity. (b) Lions convert the chemical energy stored in food into motion or other activity, or use it for growth and reproduction.

Figure 1.2

Living things develop and grow over time.

Living things have the capacity to reproduce. They make offspring that are exact or inexact copies of themselves. **Figure 1.3** shows the two ways living things reproduce, asexually and sexually. In *asexual reproduction*, a living organism reproduces all by itself, such as by dividing into two. Bacteria typically reproduce asexually. In *sexual reproduction*, organisms form special sex cells, such as sperm and eggs, that join to develop into new individuals. Humans, penguins, beetles, and oak trees reproduce sexually.

Finally, living things are parts of populations that evolve. Populations do not remain constant from one generation to the next but change over time. Often, populations change in response to their environments. During the Industrial Revolution, when cities became polluted and blackened with soot, peppered moth populations evolved so that camouflaged dark-winged moths became more common than light-winged moths. After antipollution laws were passed and cities were cleaned up, light-winged moths again became more common.

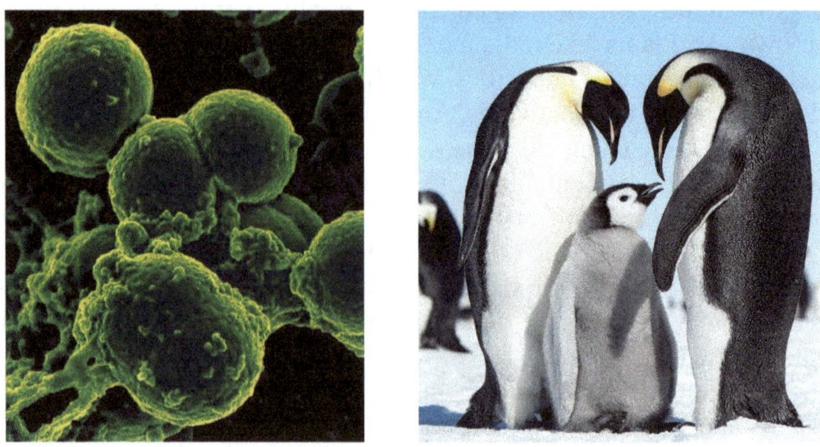

(a) (b)

Figure 1.3

Living things have the capacity to reproduce. (a) Bacteria reproduce asexually by dividing. (b) Penguins reproduce sexually.

CONCEPT **CHECK**

1. Cars are not living things. Which characteristics of living organisms do cars have, and which are they lacking?

2. Do all living things have all the characteristics of life we have listed?

CHECK YOUR ANSWER

1. Cars use energy, converting the energy in gasoline into motion. We might be able to argue that cars "develop" over time, acquiring nicks and dents and wearing down the treads on their tires. However, cars do not maintain themselves, do not reproduce, and do not evolve.

2. There are some exceptions—for example, mules are sterile and unable to reproduce, but they are certainly alive.

More to Explore

For more on the characteristics of life, check out the website below:

https://sciencing.com/ten-characteristics-living-organisms-8119158.html

1.2 The Scientific Method

Science is a process of gathering and organizing human knowledge about nature. The practice of science, including biology, involves keen observations, rational analysis, and experimentation. In the 17th century, Galileo and the English philosopher Francis Bacon were the first to formalize a particular method for doing science. What they outlined has come to be known as the classic **scientific method**.

It essentially includes the following steps:

1. **Observe**
   Closely observe the physical world around you.
2. **Question**
   Recognize a question or a problem.
3. **Hypothesize**
   Make an educated guess—a *hypothesis*—to answer the question.
4. **Predict**
   Predict consequences that can be observed if the hypothesis is correct. The consequences should be *absent* if the hypothesis is not correct.
5. **Test predictions**
   Do experiments to see if the consequences you predicted are present.
6. **Draw a conclusion**
   Formulate the simplest general rule that organizes the hypotesis, predicted effects, and experimental findings.

Although the traditional scientific method is powerful, good science is often done differently, in a less systematic way. In the example later in this section, you will see a recent application of the classic scientific method. However, many scientific advances involve trial and error, experimenting without guessing, or just plain accidental discovery. Besides the methods that are employed, success in science has much to do with an attitude common to scientists. This attitude is one of inquiry, experimentation, and humility before the facts.

Let's now take a closer look at some key components of the scientific method. It is a reliable way for you to investigate the natural world and generate scientific findings.

The Scientific Hypothesis

A scientific **hypothesis** is an educated guess that tentatively answers a question or solves a problem in regard to the natural world. Typically, experiments are done to test hypotheses.

The cardinal rule in science is that all hypotheses must be testable—in other words, they must, at least in principle, be capable of being shown wrong.

In science, it is more important that there be a means of proving an idea wrong than that there be a means of proving it right. This is a major feature that distinguishes science from nonscience. The idea that scientific hypotheses must be capable of being proven wrong is a pillar of the philosophy of science, and it is stated formally as the **principle of falsifiability**:

> **For a hypothesis to be considered scientific it must be testable—it must, in principle, be capable of being proven wrong.**

At first this principle may seem strange, for when we wonder about most things, we concern ourselves with ways of finding out whether they are true. Scientific hypotheses are different. In fact, if you want to determine whether a hypothesis is scientific or not, look to see whether there is a test for proving it wrong. If there is no test for possible wrongness, then the hypothesis is not scientific. One famous scientist, the physicist Albert Einstein, put it well: "No number of experiments can prove me right; a single experiment can prove me wrong."

Here are some examples of hypotheses that are testable and capable of being proven wrong: "When it rains and the sun is out, there will be a visible rainbow in the sky." "Robins and house finches eat some of the same foods, and they compete for these foods." "If opossums 'play dead' upon encountering a predator, they are less likely to be attacked." "After a fire, new species of plants, not seen prior to the fire, will germinate."

On the other hand, consider this hypothesis: "Finding a four-leaf clover means you will have good luck." Many people believe it, but this hypothesis is not scientific. It cannot be proven wrong. It is speculation.

Likewise, the hypothesis "Intelligent life exists on planets somewhere in the universe besides Earth" is not scientific. Although it can be proven correct by the verification of a single instance of life existing elsewhere in the universe, there is no way to prove it wrong if no life is ever found. If we searched the far reaches of the universe for eons and found no life, we would not prove that it doesn't exist "around the next corner." A hypothesis that is capable of being proven right but not capable of being proven wrong is not a scientific hypothesis. That does not necessarily mean that it is of no

READING CHECK

Suppose a hypothesis is capable of being proven right but not capable of being proven wrong. Is it a scientific hypothesis?

interest, however. For example, scientists are still interested in searching for intelligent life elsewhere in the universe—in fact, some scientists are working on just that right now, and it will be a memorable day if they are ever successful.

The Scientific Experiment

Once scientists have a hypothesis, it can be tested. Scientists do this by making predictions based on their hypothesis and then performing experiments to see if the expected results are observed.

Scientists must accept their experimental findings even when they would like them to be different. They must strive to distinguish between the results they see and those they wish to see. This is not easy. Scientists, like most people, are capable of fooling themselves. People have always tended to adopt general rules, beliefs, creeds, ideas, and hypotheses without thoroughly questioning their validity. And sometimes we retain these ideas long after they have been shown to be meaningless, false, or at least questionable. The most widespread assumptions are often the least questioned. Too often, when an idea is adopted, great attention is given to the instances that support it. Contrary evidence is often distorted, belittled, or ignored.

The fact that scientific statements will be thoroughly tested before they are believed helps to keep science honest. Sooner or later, mistakes (or deceptions) are found out. A scientist exposed for cheating doesn't get a second chance in the community of scientists. Honesty, so important to the progress of science, thus becomes a matter of self-interest to scientists. There is relatively little bluffing in a game where all bets are called.

Example: Sea Butterflies

Let's consider an example of a recent scientific research project that shows how the scientific method can be put to work. Along the way, we'll get a taste of how biology and chemistry are integrated in the physical world.

The Antarctic research team headed by James McClintock, Professor of Biology at the University of Alabama at Birmingham, and Bill Baker, Professor of Chemistry at the University of South Florida, was studying the toxic chemicals Antarctic marine organisms secrete to defend themselves against predators.

McClintock and Baker observed an unusual relationship between two animal species, a sea butterfly and an amphipod— a relationship that led to a question, a scientific hypothesis, a prediction, tests concerning the chemicals involved in the relationship, and finally a conclusion. The research generally proceeded according to the steps of the classic scientific method.

1. Observe The sea butterfly *Clione antarctica* is a brightly colored, shell-less snail with winglike extensions used in swimming (**Figure 1.4a**), and the amphipod *Hyperiella dilatata* resembles a small shrimp. McClintock and Baker observed a large percentage of amphipods carrying sea butterflies on their backs, with the sea butterflies held tightly by the legs of the amphipods (**Figure 1.4b**). Any amphipod that lost its sea butterfly would quickly seek another—the amphipods were actively abducting the sea butterflies!

2. Question McClintock and Baker noted that amphipods carrying butterflies were slowed considerably. This seemed like it would make the amphipods more vulnerable to predators and less adept at catching prey. Why then did the amphipods abduct and hang on to the sea butterflies?

(a)

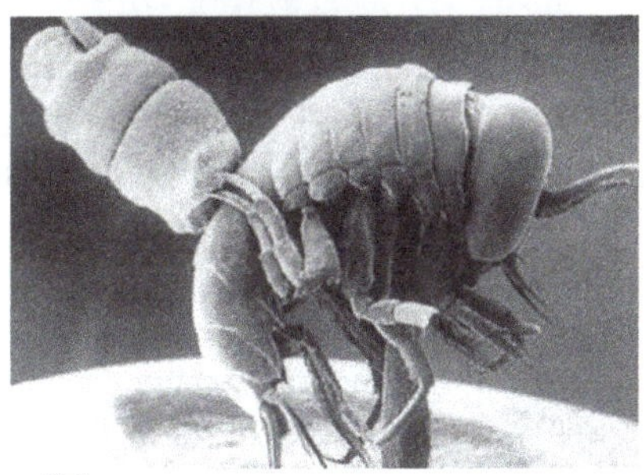

(b)

Figure 1.4
(a) The graceful Antarctic sea butterfly is a species of snail that does not have a shell. (b) The shrimp-like amphipod attaches a sea butterfly to its back even though doing so limits the amphipod's mobility.

3. Hypothesize Given their experience with the chemical defense systems of various sea organisms, the research team hypothesized that amphipods carry sea butterflies to take advantage of a chemical that deters a predator of the amphipod.

4. Predict Based on their hypothesis, they predicted (a) that they would be able to isolate this chemical and (b) that an amphipod predator would be deterred by it.

5. Test predictions To test their hypothesis and predictions, the researchers captured several predator fish species and conducted the test shown in **Figure 1.5**. The fish were presented with solitary sea butterflies, which they took into their mouths but promptly spat back out. The fish readily ate uncoupled amphipods but spit out any amphipod coupled with a sea butterfly. These are the results expected if the sea butterfly was secreting some sort of chemical deterrent. The same results would be obtained, however, if a predator fish simply didn't like the feel of a sea butterfly in its mouth. The results of this simple test were therefore ambiguous. A conclusion could not yet be drawn.

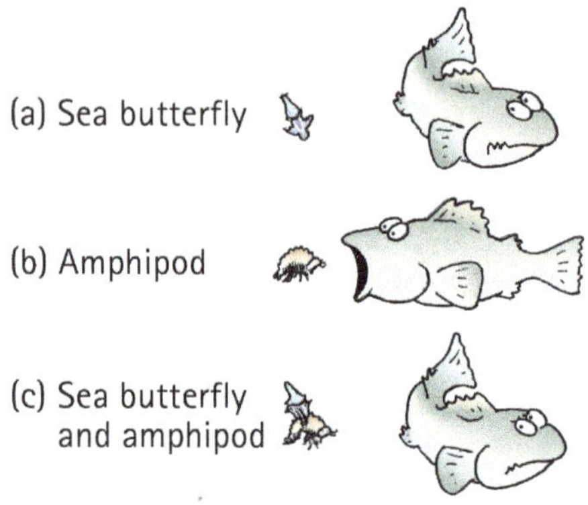

Figure 1.5
In McClintock and Baker's initial experiment, a predatory fish (a) rejected the sea butterfly, (b) ate the free-swimming amphipod, and (c) rejected the amphipod coupled with a sea butterfly.

This experiment highlights that all scientific tests need to minimize the number of possible conclusions. Often this is done by running an experimental test along with a *control*. Ideally, the experimental test and the control should differ by only one variable. Any differences in results can then be attributed to how the experimental test differed from the control.

To confirm that the deterrent was chemical and not physical, the researchers made one set of food pellets containing both fish meal and sea butterfly extract (the experimental pellets). For their control test, they made a physically identical set containing only fish meal (the control pellets). As shown in **Figure 1.6**, the predator fish readily ate the control pellets but not the experimental ones. These results strongly supported the chemical hypothesis.

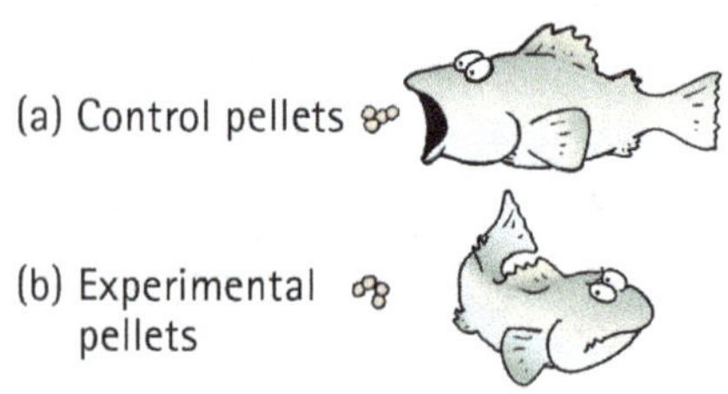

Figure 1.6
The predator fish (a) ate the control pellets but (b) rejected the experimental pellets, which contained sea butterfly extract.

Further processing of the sea butterfly extract yielded five major chemical compounds, only one of which deterred the predator fish from eating the pellets. Chemical analysis of this compound revealed it to be the previously unknown molecule that the scientists named pteroenone.

6. Draw a conclusion In addition to running control tests, scientists confirm experimental results by repeated testing. In this case, the Antarctic researchers made many food pellets, both experimental and control, so that each test could be repeated many times. Only after obtaining consistent results in repeated tests can a scientist draw a conclusion. McClintock and Baker were thus able to conclude that amphipods abduct sea butterflies in order to use the sea butterflies' secretion of pteroenone as a defense against predator fish.

Yet, this conclusion might still be regarded with skepticism in the scientific community. Why? There is a great potential for unseen error in any experiment. A laboratory may have faulty equipment that leads to consistently wrong results, for example. Because of the potential for unseen error from any particular research group, experimental results must be *reproducible* to be considered valid. This means that other scientists must be able to reproduce the same experimental findings in separate experiments. Thus, you can see that it is a long road from a bright idea to accepted scientific finding! The plodding, painstaking nature of this process is beneficial, though—it is the very reason that scientific knowledge is highly trustworthy.

As frequently happens in science, McClintock and Baker's results led to new questions. What are the properties of pteroenone? Does this substance have applications—for example, can it be used as a pest repellent? Could it be useful for treating human disease? In fact, a majority of the chemicals we use were originally discovered in natural sources. This illustrates that there is an important reason for preserving marine habitats, tropical rainforests, and the other diverse natural environments on Earth—they are storehouses of countless yet-to-be-discovered substances.

CONCEPT CHECK

Which of the statements below are scientific hypotheses? Explain your reasoning.

a. The stock market has good days when the planets Venus, Earth, and Mars are aligned.

b. Atoms are the smallest particles of matter that exist.

c. The Moon is made of Swiss cheese.

d. Outer space contains a kind of matter whose existence can't be detected or tested.

e. Charles Darwin was the greatest biologist of the 19th century.

CHECK YOUR ANSWERS

All these statements are hypotheses, but only statements a, b, and c are scientific hypotheses because they are testable statements.

a. This statement can be tested (and proven wrong) by researching the performance of the stock market during times when these planets were aligned.

b. Not only can this statement be tested, it has been tested. Although the statement has been found to be untrue (many particles smaller than atoms have been discovered), the statement is nevertheless a testable scientific hypothesis.

c. Visits to the Moon have proven that the statement is wrong. It is definitely not made of Swiss cheese!

d. This statement is easily seen to be unscientific because it cannot be tested.

e. Finally, this is an assertion that has no test. What possible test, beyond collective opinion, could prove that Darwin was the greatest biologist? How could we know? Greatness is a quality that cannot be objectively known.

More to Explore

To read more about the scientific method, check out the websites below:

https://www.livescience.com/21490-what-is-a-scientific-hypothesis-definition-of-hypothesis.html

https://www.livescience.com/20896-science-scientific-method.html

1.3 Science and Technology

READING CHECK

From what does technology arise?

Science is concerned with gathering knowledge about the natural world. When we apply this knowledge for practical purposes, we have what we call **technology.** New technologies often arise from scientific discoveries. In biology, scientific discoveries have led to advances in genetic and medical technologies, among other areas. New technologies, in turn, can be of assistance to scientists in conducting their research. So technology arises from science, but technology also supports progress in science—the two are different from each other but very much related.

Technology is a double-edged sword that can be both helpful and harmful. We have the technology, for example, to extract fossil fuels from the ground and then to burn the fossil fuels for the production of energy. Energy production from fossil fuels has benefited our society in countless ways. On the flip side, the burning of fossil fuels warms the planet and endangers the environment. It is tempting to blame technology itself for problems such as pollution, resource depletion, and even overpopulation. But ultimately, it is humans who are developing the technology. Humans must constantly think about how to use technology responsibly.

Remarkably, we already possess the technology to solve many environmental problems. In this 21st century, we are seeing a switch from fossil fuels to more sustainable energy sources, such as photovoltaics, hydroelectric, wind, solar thermal electric generation, and biomass conversion. Whereas most paper printed comes from trees, there is also paper available from fast-growing weeds, and less of even these materials will be needed as e-books gain popularity. In some parts of the world, progress is being made to stem the rapid growth of our human population, which is an issue that aggravates almost every problem faced by humans today. We live on a finite planet, and Earth's population carrying capacity is being acknowledged. The greatest obstacle to solving today's problems lies more with social inertia than with a lack of technology. Technology is our tool. What we do with this tool is up to us.

CONCEPT **CHECK**

Is technology helpful or harmful?

CHECK YOUR ANSWER

Technology can be either helpful or harmful depending on how it is used. Using technology responsibly, and in helpful ways, should be one of the goals of society.

More to Explore

You will read more about biology and technoogy later in this book, but here's a website with more information if you would like to learn more now:

https://www.pnas.org/content/98/18/10051

1.4 Facts, Laws, and Theories

When a scientific hypothesis has been tested over and over again and has not been contradicted, it may become known as a **law** or *principle*. A scientific fact, on the other hand, is generally something that competent observers can observe and agree to be true. For example, it is a fact that a pill bug, commonly known as a roly-poly, is able to roll into a ball when it is disturbed (**Figure 1.7**). Anyone can watch it happen.

Scientists use the word *theory* in a way that differs from its use in everyday speech. In everyday speech, a theory is similar to a hypothesis—a statement that hasn't been tested. But in science, a **theory** is a synthesis of facts and well-tested hypotheses. The scientific theory tends to be overarching, explaining a wide range of observations.

Physicists use quantum theory to explain the behavior of light. Chemists have theories about how atoms bond to form molecules. Astronomers speak of the Big Bang theory to explain the origin of our Universe. And, in biology, the theory of evolution explains how the Earth's diversity of living things have come to be the way they are today.

Theories are a foundation of science, but they are not fixed. Rather, theories can evolve and change over time. Often, they pass through stages of refinement. For example, in chemistry, since the theory of the atom was proposed 200 years ago, it has been refined many times in light of new evidence. Theories in biology have also changed over time. The theory of evolution, for example, underwent important changes once scientists had learned more about how DNA, genes, and heredity worked.

Those who know only a little about science may argue that scientific theories can't be taken seriously because they are always changing. Those who understand science, however, see it differently: Theories grow stronger and more precise as they evolve to include new information.

READING CHECK

What is the definition of a scientific fact?

Figure 1.7

A scientific fact: healthy pill bugs are able to roll into a ball when disturbed.

CONCEPT **CHECK**

How is the word "theory" used differently in everyday speech compared to in science?

CHECK YOUR ANSWER

In everyday speech, most people use the word "theory" in a way that is similar to a hypothesis—to describe an idea that is uncertain, that has not been tested. In science, however, a theory is a synthesis of facts and well-tested hypotheses explaining a broad range of observations. Theories in science are well supported..

More to Explore

You can read more about scientific theories here:

https://www.livescience.com/21491-what-is-a-scientific-theory-definition-of-theory.html

1.5 Working With Numbers

Science starts with observations. When possible, it is helpful to quantify observations by taking measurements. By quantifying observations, we are able to make objective comparisons, share accurate information with others, or look for trends that might reveal some inner workings of nature.

Scientific Notation

Scientific notation is a different way of writing numbers. You know that one million is 1,000,000 and that one billion is 1,000,000,000. Pretty quickly, it becomes cumbersome to write out all those digits. This is where scientific notation comes in. Scientific notation is especially useful for very big numbers and very small numbers, but it can be used for just regular numbers as well.

In *scientific notation*, you use digits, with a decimal point after the first digit, multiplied by a power of 10. For example, remember that 10^3 is 10 x 10 x 10, and so one thousand in scientific notation is:

$1{,}000 = 1 \times 10^3$

One million is: $1{,}000{,}000 = 1 \times 10^6$

Three million is: $3{,}000{,}000 = 3 \times 10^6$

And the number 4,800,000 is: $4{,}800{,}000 = 4.8 \times 10^6$

You can also write small numbers in scientific notation by using a negative exponent:

$0.001 = 1 \times 10^{-3}$

$0.00001 = 1 \times 10^{-5}$

$0.00000000007 = 7 \times 10^{-11}$

READING CHECK
Name six physical quantities that biologists often measure.

The Metric System

Scientists measure physical quantities. Some physical quantities commonly measured in biology and other sciences include: length, time, weight, volume, energy and temperature.

It would be meaningless, for example, to say that your dog weighs 40, because without a specific unit, no one would know what that meant: 40 ounces, 40 pounds, 40 kilograms? A dog that weighed 40 kg would be more than 35 times heavier than one that weighed 40 oz. Units such as ounces, pounds, and kilograms, or feet, yards, and kilometers allow us to make meaningful comparisons when we measure physical quantities. Therefore, they must be included to complete the description. Again for emphasis: Any quantity must be accompanied by a unit.

There are two major unit systems used in the world today. One is the United States Customary System (USCS, formerly called the British System of Units), used in the United States, primarily for nonscientific purposes. The other is the Système International (SI), which is used in most other nations. This system is also known as the International System of Units or as the *metric system*. The orderliness of this system makes it useful for scientific work, so it is used by scientists all over the world, including those in the United States. (And the International System is beginning to be used for nonscientific work in the United States, as **Figure 1.8** shows.

This book uses the SI units given in **Table 1.1**. On occasion, USCS units are also used to help you make comparisons. One major advantage of the metric system is that it uses a decimal system, which means all units are related to the next smaller or larger units by a factor of 10. Some of the more commonly used prefixes, along with their decimal equivalents, are shown in **Table 1.2**. From this table, you can see that 1 kilometer is equal to 1000 meters, where the prefix *kilo-* indicates 1000. Likewise, 1 millimeter is equal to 0.001 meter, where the prefix *milli-* indicates 1/1000. You need not memorize this table, but you will find it a useful reference when you come across these prefixes in your course of study.

^ Figure 1.8
The metric system is finally making some headway in the United States, where various commercial goods, such as Evan's favorite soda, are now sold in metric quantities.

Table 1.1 Metric Units for Physical Quantities and Their USCS Equivalents

PHYSICAL QUANTITY	METRIC UNIT	ABBREVIATION	USCS EQUIVALENT
length	kilometer	km	1 km = 0.621 miles (mi)
	meter	m	1 m = 3.285 feet (ft)
	centimeter	cm	1 cm = 0.3937 inches (in.)
			1 in = 2.54 cm
	millimeter	mm	none commonly used
time	second	s	second also used in USCS
mass	kilogram	kg	1 kg = 2.205 pounds (lb)
	gram	g	1 g = 0.03528 ounces (oz)
			1 oz = 28.345 g
	milligram	mg	none commonly used
volume	liter	L	1 L = 1.057 quarts (qt)
	milliliter	mL	1 mL = 0.0339 fl oz
	cubic centimeter	cm^3	1 cm^3 = 0.0339 fl oz
energy	kilojoule	kJ	1 kJ = 0.239 kilocalories (kcal)
	joule	J	1 J = 0.239 calories (cal)
			1 cal = 4.184 J
temperature	degree Celsius	°C	(°C * 1.8) + 32 = degrees Fahrenheit, °F
	kelvin	K	°C + 273 = K

Table 1.2 Metric Prefixes

PREFIX	SYMBOL	DECIMAL EQUIVALENT	EXPONENTIAL FORM	EXAMPLE
tera-	T	1,000,000,000,000.	10^{12}	1 **tera**meter (Tm) 1 trillion meters
giga-	G	1,000,000,000.	10^{9}	1 **giga**meter (Gm) 1 billion meters
mega-	M	1,000,000.	10^{6}	1 **mega**meter (Mm) 1 million meters
kilo-	k	1000.	10^{3}	1 **kilo**meter (km) 1 thousand meters
hecto-	h	100.	10^{2}	1 **hecto**meter (hm) 1 hundred meters
deka-	da	10.	10^{1}	1 **deka**meter (dam) ten meters
no prefix	–	1.	10^{0}	1 meter (m) 1 meter
deci-	d	0.1	10^{-1}	1 **deci**meter (dm) 1 tenth of a meter
centi-	c	0.01	10^{-2}	1 **centi**meter (cm) 1 hundredth of a meter
milli-	m	0.001	10^{-3}	1 **milli**meter (mm) 1 thousandth of a meter
micro-	µ	0.000 001	10^{-6}	1 **micro**meter (µm) 1 millionth of a meter
nano-	n	0.000 000 001	10^{-9}	1 **nano**meter (nm) 1 billionth of a meter
pico-	p	0.000 000 000 001	10^{-12}	1 **pico**meter (pm) 1 trillionth of a meter

CONCEPT CHECK

1. What is the number 56,780,000 in scientific notation?
2. A cell is 30 micrometers in diameter. How many meters is this?

CHECK YOUR ANSWER

1. In scientific notation, 56,780,000 = 5.678×10^{7}
2. This cell's diameter is 30 micrometers = 30×10^{-6} meters = 30 x 0.000001 meters = 0.00003 meters.

More to Explore

Read about scientific notation and practice with numbers:

https://www.mathsisfun.com/numbers/scientific-notation.html

Learn about the metric system and compare it to the USCS system:

https://www.mathsisfun.com/measure/metric-system.html

https://www.interexchange.org/articles/career-training-usa/2012/05/24/imperial-vs-metric-system/

Concept Review

Chapter 1

Detailed Chapter Summary

Biology is the study of life and living organisms. Living organisms share many key characteristics. For example, they use energy, develop and grow, maintain themselves, and have the capacity to reproduce. Living things are also a part of populations that evolve—that is, populations change over time, often in response to their environments.

The classic scientific method describes how much scientific research is conducted, including in biology. The key steps in the scientific method are:

1. Observe: Closely observe the world around you.
2. Question: Recognize a question or a problem.
3. Hypothesize: Make an educated guess—a *hypothesis*—to answer the question.
4. Predict: Predict consequences that can be observed if the hypothesis is correct. The consequences should be *absent* if the hypothesis is not correct.
5. Test predictions: Do experiments to see if the consequences you predicted are present.
6. Draw a conclusion: Formulate the simplest general rule that organizes the hypothesis, predicted effects, and experimental findings.

To be valid, a scientific hypothesis must be *falsifiable*—that is, it must be able to be proven wrong. Scientists conduct carefully designed experiments to test their hypotheses.

The application of scientific knowledge for practical purposes is known as *technology*. New technologies often arise from scientific discoveries. In biology, scientific discoveries have led to technological advances in genetics and medicine, among other areas. Technology is a double-edged sword that can be both helpful and harmful. It is society's responsibility to make sure that technology is used in helpful, rather than harmful, ways.

In science, a *fact* is something that competent observers can observe and agree to be true. A *scientific law* or *principle* is a hypothesis that has been tested over and over again and that has not been contradicted. A *scientific theory* is a synthesis of facts and well-tested hypotheses. The value of a scientific theory is in its ability to explain a wide range of observations. Scientific theories are not fixed, but evolve and change over time as they are refined in light of new evidence.

Numbers are important in many scientific observations and experiments. In *scientific notation*, numbers are expressed using digits, with a decimal point after the first digit, multiplied by a power of 10. For measurements, scientists generally use the units of the *metric system*.

Key Terms

Biology The study of life and living organisms.

Fact A phenomenon about which competent observers can agree.

Hypothesis An educated guess or a reasonable explanation. When the hypothesis can be tested and is falsifiable, it qualifies as a *scientific hypothesis*.

Law A statement about the relationship of natural quantities that has been tested repeatedly and has not been contradicted; also known as a *principle*.

Principle of falsifiability A scientific hypothesis must be testable or "falsifiable"—it must, in principle be capable of being proven wrong.

Science The collective findings about nature, and the process of gathering and organizing knowledge about nature.

Scientific method An orderly method for gaining and organizing new knowledge about nature.

Technology The means of solving practical problems by applying the findings of science.

Theory A synthesis of a large body of information that encompasses well-tested hypotheses about certain aspects of the natural world with great explanatory powers.

Review Questions

Simple Review Questions

1.1 What Is Life?

1. What are some of the characteristics of living organisms?
2. Describe what it means to say that living things are parts of populations that evolve.

1.2 The Scientific Method

3. What do we mean when we say that a scientific hypothesis must be testable?
4. Is any hypothesis that is not scientific necessarily unreasonable? Explain.

1.3 Science and Technology

5. Clearly distinguish between science and technology.
6. Is technology beneficial to society, harmful to society, or both?

1.4 Facts, Laws, and Theories

7. Distinguish among a scientific fact, a hypothesis, a law, and a theory.
8. How does the definition of the word theory differ in science versus in everyday life?

1.5 Working with Numbers

9. What is scientific notation, and why do people use this notation?
10. What is the metric system?

Challenging Review Questions

1.1 What Is Life?

11. How can you tell a plant is alive even though it doesn't talk or run around?
12. What are some features of living organisms? Describe how human beings show each of these features.
13. Bacteria reproduce by dividing in two. Is this an example of asexual reproduction or sexual reproduction? Defend your answer.

1.2 The Scientific Method

14. Which of the following are scientific hypotheses? (a) Chlorophyll makes grass green. (b) Pandas are the cutest mammals. (c) The leaves of maple trees will turn red in the autumn even if temperatures stay high.
15. The Texas horned lizard has sharp spikes on its head and body, as you can see in the photo. Can you formulate a hypothesis for why horned lizards have their horns? How would you test your hypothesis?

1.3 Science and Technology

16. Forensic scientists collect skin and hair samples from crime scenes and use these samples to collect information about the DNA of the people at the scene. They can then compare this DNA to the DNA of potential suspects. Explain how this is an example of technology.
17. Is a vaccine, such as the flu vaccine or Covid-19 vaccine, an example of technology? Why or why not?

1.4 Facts, Laws, and Theories

18. Zebras have black and white stripes on their coats. Is this a fact, a law, or a theory?
19. "Living things can reproduce." Is this a scientific law?

1.5 Working with Numbers

20. The typical adult human body has about 3×10^{10} cells. How would you write out this number, currently in scientific notation, with normal notation?
21. How many centimeters are there in a meter? How many millimeters are there in a meter?
22. The human egg is 0.1 millimeters in diameter. How many meters is this? Express your answer in regular notation as well as scientific notation.

Apply & Discuss Questions

23. Discuss a technology based on advances in biology. How has this technology been helpful? Has it also been harmful in any way? You may choose one of the following topics, or you may come up with your own: lab-grown "meat," the Covid-19 vaccine, cancer screening techniques, the antibiotic penicillin, animal cloning.
24. Consider again the topic you discussed in the previous question. Can you think of any guidelines or laws that would help prevent the harmful effects you identified?
25. In what sense is science grand and breathtaking? In what sense is it dull and painstaking?

Solutions to Odd-Numbered Questions

Simple Review Solutions

1. Living things share certain characteristics. For one thing, they use energy. Living things take energy from the environment and convert it to other forms of energy for their own use. Another characteristic of living things is that they develop and grow. Living things maintain themselves. They generate structures, such as stems and leaves or skin and bones, and they repair damage done to those structures. Living things have the capacity to reproduce. Finally, living things are parts of populations that evolve. Populations do not remain constant from one generation to the next but change over time.

3. The cardinal rule in science is that all hypotheses must be testable—in other words, they must, at least in principle, be capable of being shown wrong.

5. Science is concerned with gathering knowledge about the natural world. When we apply this knowledge for practical purposes, we have what we call technology.

7. When a scientific hypothesis has been tested over and over again and has not been contradicted, it may become known as a law or principle. A scientific fact, on the other hand, is generally something that competent observers can observe and agree to be true. In science, a theory is a synthesis of facts and well-tested hypotheses.

9. Scientific notation is a different way of writing numbers. You know that one million is 1,000,000 and that one billion is 1,000,000,000. Pretty quickly, it becomes cumbersome to write out all those digits. This is where scientific notation comes in. Scientific notation is especially useful for very big numbers and very small numbers, but it can be used for just regular numbers as well.

Challenging Review Solutions

11. Plants show all the characteristics of life—they use energy, which they obtain from the Sun. They develop and grow, maintain their bodies. Plants also reproduce, for example when they produce seeds that grow into new plants. Finally, plant populations evolve over time.

13. This is an example of asexual reproduction, because a bacteria that divides in two is reproducing all by itself.

15. One hypothesis is that the horns of these lizards help protect them from predators. One way a scientist might test this hypothesis is by observing interactions between lizards and potential predators. A possible prediction is that lizards might actively use their horns in defense, by attacking potential predators. Or, predators might be unable to grab the lizards because of their horns. (In fact, biologists have observed lizards jerking their heads at potential predators in order to stab them with their horns.) Another way to test this hypothesis is by comparing lizards that were caught by predators to ones that had not, in order to see if there is any difference between the horns of these two groups. Scientists might predict that horned lizards that escape predation could have more horns or larger horns. (In fact, this has been done, and scientists did find that horned lizards who were caught by bird predators had shorter horns than those who were not caught by birds.)

17. Yes, a vaccine uses scientific knowledge to address the practical problem of reducing the impact of epidemics such as the flu and COVID epidemics.

19. A scientific law is a hypothesis that has been tested over and over again and that has not been contradicted. Although the statement above describes a general characteristic of living things, it falls short of being a law because not every living thing can reproduce. Some organisms might be too young or too old to reproduce, for example. We also saw that the mule, though clearly living, is unable to reproduce.

21. There are 100 centimeters in a meter. There are 1000 millimeters in a meter.

Apply & Discuss Solutions

23. Answers will vary!

25. Science is grand and breathtaking in its ability to tell us about the natural world around us. The process of building knowledge in science can be dull and painstaking, though, as scientists test their hypotheses repeatedly and with utmost care.

Living Earth Essay

Life on Earth Connections

Biogeology—An Introduction

Planet Earth can be classified into a number of domains, such as the **geosphere**, the **hydrosphere**, and the **atmosphere**. The geosphere is the rocky planet itself and is, by far, the most massive of these three domains. The hydrosphere is all the water of planet Earth, which is found mostly in the oceans, but also in the polar ice caps, lakes, and rivers. The atmosphere consists of all the gaseous materials, such as nitrogen and oxygen, that are held down by gravity close to the Earth's surface.

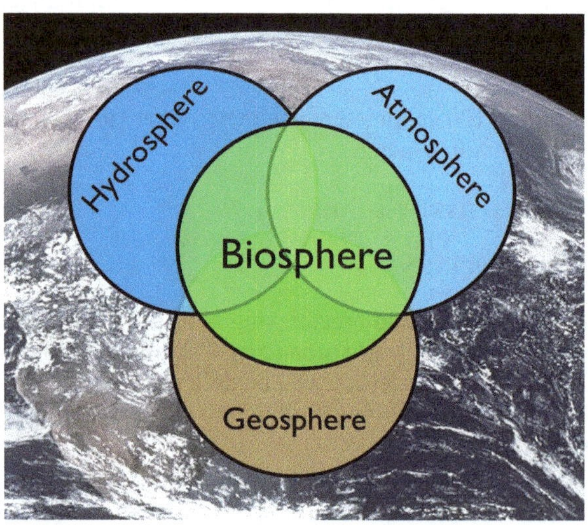

A fourth domain is the **biosphere**, which consists of all the living things here on planet Earth. These living things rely on the resources of the geosphere, hydrosphere, and atmosphere in order to survive. In many ways, the biosphere arises where the geosphere, hydrosphere, and atmosphere intersect.

Changes to the geosphere, hydrosphere, or atmosphere can greatly impact living things. In response to these changes, the living things adapt, or evolve, as we'll be discussing in Chapters 8 - 9. But might it be possible for the biosphere to have an impact on the geosphere, hydrosphere, or atmosphere? The answer is a resounding yes.

A good example is with the development of photosynthetic organisms (Chapter 4), which in absorbing the energy of sunlight generate molecular oxygen. Most alll the oxygen of the atmosphere is, in fact, a by-product of these living things. Thus, much of the atmosphere itself was formed by the biosphere. Further, molecular oxygen is a rather reactive material, which means it can react with minerals within the geosphere, such as iron, thus changing the chemical composition of parts of that geosphere.

While biology is the study of life, **biogeology** is the study of the many interactions between life and geology. The impacts of these interactions are most significant and will be featured within these special Living Earth "LE" sections of *Conceptual Academy Biology*. We'll be exploring topics such as the history of life on this planet, the availability of natural resources, climate change, the importance of biodiversity for sustainability, and more.

Planet Earth is our home. It supports us, but, in return, we need to support planet Earth. We've no other choice but to go extinct.

To Life!

Chapter 2

The Chemistry of Life

THE MAIN IDEA

Atoms are the fundamental building blocks of matter and living organisms.

2.1 Atoms and Molecules
2.2 Chemical Compounds
2.3 Mixtures
2.4 Chemical Reactions
2.5 Types of Reactions
2.6 Organic Molecules
2.7 Macromolecules Needed for Life

All the stuff around us is made of atoms. A group of atoms connected together is what we call a molecule. A water molecule, for example, consists of two hydrogen atoms together with one oxygen atom, which you know as H_2O. Add another oxygen atom, and you no longer have water. You instead have hydrogen peroxide, H_2O_2, which is a common disinfectant. There are only about 100 different kinds of atoms, but there are a limitless number of ways in which they can connect together. This is why we have so many different kinds of materials around us. Chemistry is the study of how atoms connect to form new materials, including the tissues that make living organisms.

2.1 Atoms and Molecules

Knowledge of chemistry lays the foundation for knowledge of biology. Chemistry is, in many ways, the physics of atoms and molecules. Biology is, in many ways, the *chemistry* of life. This chapter focuses on the concepts of chemistry that are key to understanding biology.

The science of chemistry is concerned with the behavior of atoms and molecules. If you have already taken a chemistry course, you understand much about how atoms and molecules behave. This chapter will be a helpful review for you. However, if you are coming into biology with a minimal background in chemistry, then this chapter is essential.

An atom is a fundamental unit of matter. Hydrogen is the simplest, lightest, and most common atom. It makes up more than 90% of the atoms in the known universe—most of these hydrogen atoms formed at the beginning of our universe about 13.8 billion years ago. Heavier atoms are produced in stars, which are massive collections of hydrogen atoms pulled together by gravitational forces. The immense pressures and temperatures deep in a star's interior cause hydrogen atoms to combine into heavier atoms. With the exception of hydrogen, and the next lightest atom, helium, all the atoms that occur naturally on Earth—including those in your body—are the products of stars. You are made of stardust, as is everything that surrounds you.

READING CHECK

What is the most common kind of atom in the universe?

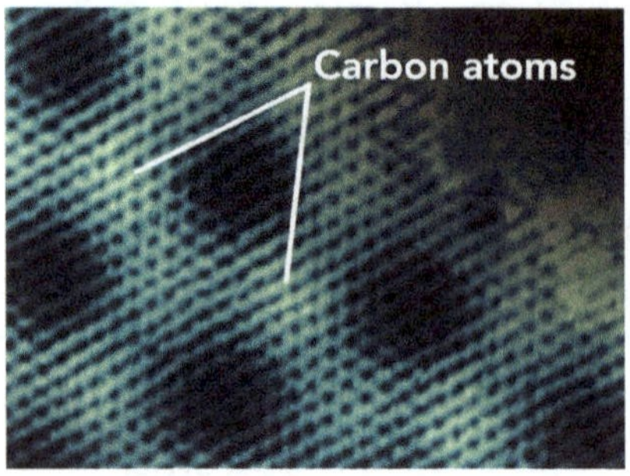

Figure 2.1

An image of carbon atoms obtained with a scanning microscopy.

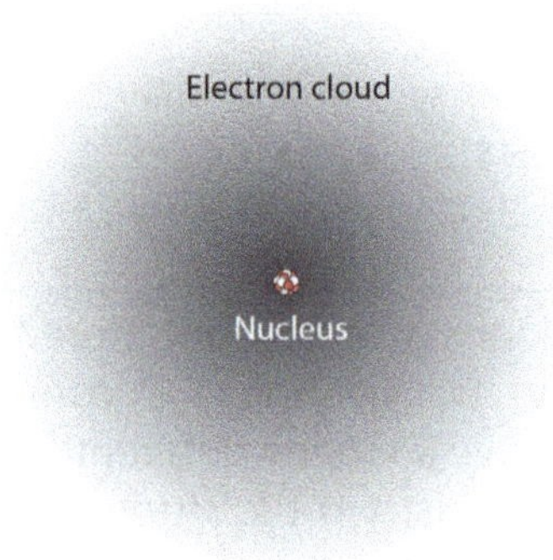

Figure 2.2

Electrons whiz around the atomic nucleus, forming what can be described as a cloud that is more dense where the electrons tend to spend most of their time. Furthermore, if this illustration were drawn to scale, the atomic nucleus with its protons and neutrons would be too small to be seen.

Atoms are tiny. In fact, they are so minuscule that each breath you exhale contains more than 10 billion trillion of them. Atoms are so small that they can't be seen with visible light. This is because they are even smaller than the wavelengths of visible light. We could stack microscope on top of microscope and never "see" an atom. Photographs of atoms, as in Figure 2.1, however, can be obtained with the scanning probe microscope, an imaging device that bypasses light and optics altogether.

An atom is made of smaller "subatomic" particles—*electrons, protons*, and *neutrons*. We know that atoms differ from one another only in the number of subatomic particles they contain. **Protons** and **neutrons** are subatomic particles bound together at the atom's center to form the **atomic nucleus.** The atomic nucleus is a dense, positively charged center of every atom. This nucleus is a relatively heavy particle that makes up most of an atom's mass. Each proton carries a +1 charge, while each neutron has a neutral charge. Surrounding the nucleus are the ultra-tiny **electrons**, which swarm to create an electron cloud, as shown in Figure 2.2. Each electron carries a -1 charge. Within a neutral atom, the number of protons in the nucleus equals the number of surrounding electrons.

In the outer regions of any atom are electrons, which repel the electrons of neighboring atoms. Two atoms, therefore, can get only so close to each other before they start repelling (provided they don't join in a chemical bond, as will be discussed in a bit). When the atoms of your hand push against the atoms of a wall, electrical repulsions between electrons in your hand and electrons in the wall prevent your hand from passing through the wall. These same electrical repulsions prevent us from falling through the solid floor. They also allow us the sense of touch. Interestingly, when you touch someone, your atoms and those of the other person do not meet. Instead, atoms from the two of you get close enough so that you sense an electrical repulsion. A tiny, though imperceptible, gap still exists between the two of you (Figure 2.3).

Given all the types of materials, you might think there must be many kinds of atoms. But the number of different kinds of atoms is surprisingly small. The great variety of substances results from the many ways a few kinds of atoms can be combined. Just as the three colors red, green, and blue can be combined to form any color on a computer screen or the 26 letters of the alphabet make up all the words in a dictionary, only a few kinds of atoms combine in different ways to produce all substances. To date, we know of about 118 distinct atoms. Of these, about 90 are found in nature. The remaining atoms have been created in the laboratory.

Any material made of only one type of atom is classified as an **element.** Three examples are shown in Figure 2.4. Pure gold, for example, is an element— it contains only gold atoms. Nitrogen gas is an element because it contains only nitrogen atoms. Likewise, the graphite in your pencil is an element—carbon. Graphite is made up solely of carbon atoms. All of the elements are listed in a chart called the **periodic table**, Figure 2.5.

Figure 2.3

When we physically touch, we feel an electric force of repulsion between our atoms that prevents us from getting even closer. An emotional touch, by contrast, holds no such boundaries.

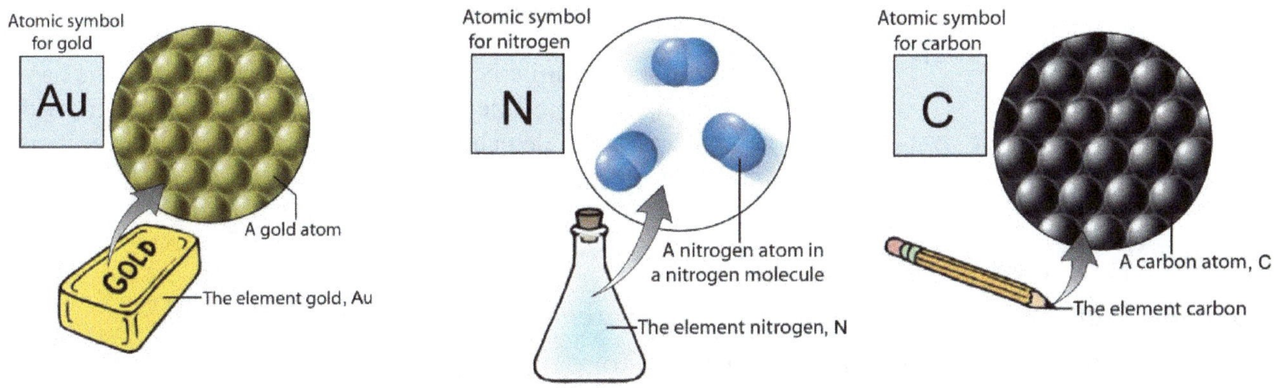

Figure 2.4

Any element consists of only one kind of atom.

As you can see from the periodic table, each element is designated by its **atomic symbol**, which comes from the letters of the element's name. For example, the atomic symbol for carbon is C, and the symbol for chlorine is Cl. In many cases, the atomic symbol is derived from the element's Latin name. Gold has the atomic symbol Au after its Latin name, *aurum*. Lead has the atomic symbol Pb after its Latin name, *plumbum*. Elements with symbols derived from Latin names are usually those that were discovered earliest.

Note that only the first letter of an atomic symbol is capitalized. For instance, the symbol for the element cobalt is Co, but CO is a combination of two elements: carbon, C, and oxygen, O.

The terms *element* and *atom* are often used in a similar context. You might hear, for example, that gold is an element made of gold atoms. Generally, *element* is used in reference to an entire macroscopic or microscopic sample, and *atom* is used when speaking of the submicroscopic particles in the sample. The important distinction is that elements are made of atoms and not the other way around.

Atoms can combine to form a **molecule**, which is a group of atoms held tightly together by chemical bonds. Two oxygen atoms, for example, form an oxygen molecule. We use the elemental formula to indicate multiple atoms of the same element bonded into a molecule. An **elemental formula** shows

1 H																	2 He
3 Li	4 Be											5 B	6 C	7 N	8 O	9 F	10 Ne
11 Na	12 Mg											13 Al	14 Si	15 P	16 S	17 Cl	18 Ar
19 K	20 Ca	21 Sc	22 Ti	23 V	24 Cr	25 Mn	26 Fe	27 Co	28 Ni	29 Cu	30 Zn	31 Ga	32 Ge	33 As	34 Se	35 Br	36 Kr
37 Rb	38 Sr	39 Y	40 Zr	41 Nb	42 Mo	43 Tc	44 Ru	45 Rh	46 Pd	47 Ag	48 Cd	49 In	50 Sn	51 Sb	52 Te	53 I	54 Xe
55 Cs	56 Ba	57 La	72 Hf	73 Ta	74 W	75 Re	76 Os	77 Ir	78 Pt	79 Au	80 Hg	81 Tl	82 Pb	83 Bi	84 Po	85 At	86 Rn
87 Fr	88 Ra	89 Ac	104 Rf	105 Db	106 Sg	107 Bh	108 Hs	109 Mt	110 Ds	111 Rg	112 Cn	113 Nh	114 Fl	115 Mc	116 Lv	117 Ts	118 Og

58 Ce	59 Pr	60 Nd	61 Pm	62 Sm	63 Eu	64 Gd	65 Tb	66 Dy	67 Ho	68 Er	69 Tm	70 Yb	71 Lu
90 Th	91 Pa	92 U	93 Np	94 Pu	95 Am	96 Cm	97 Bk	98 Cf	99 Es	100 Fm	101 Md	102 No	103 Lr

Figure 2.5

The periodic table lists all the known elements.

the atomic symbol followed by a numeral subscript indicating the number of atoms in each molecule. Thus, a molecule of two oxygen atoms has the elemental formula O_2. Similarly, a molecule of eight sulfur atoms has the elemental formula S_8. For elements in which the basic units are individual atoms (no molecules), the elemental formula is simply the chemical symbol. This is the case for most elements.

CONCEPT CHECK

1. Are atoms made of molecules, or are molecules made of atoms?
2. Are atoms made of elements, or are elements made of atoms?

CHECK YOUR ANSWERS

1. A molecule is a group of atoms held together by chemical bonds. Thus, molecules are made of atoms.
2. An element can be considered a macroscopic sample of matter containing only one type of atom. Thus, elements are made of atoms.

2.2 Chemical Compounds

READING CHECK

Do chemical compounds have similar properties to the elements that they are made up of?

When atoms of different elements bond to one another, they make a **chemical compound**. Sodium atoms and chlorine atoms, for example, bond to make the compound sodium chloride, commonly called "table salt." Likewise, nitrogen and hydrogen atoms can join to make the compound ammonia, which is a common household cleaner.

A chemical compound is represented by its **chemical formula**. The chemical formula for sodium chloride is NaCl, and the formula for ammonia is NH_3. Numerical subscripts indicate the ratio of atoms combining. By convention, the subscript 1 is understood and omitted. So, the chemical formula NaCl tells us that the compound sodium chloride has one sodium atom for every chlorine atom; the chemical formula NH_3 tells us that the compound ammonia has one nitrogen atom for every three hydrogen atoms, as Figure 2.6 shows.

Compounds have physical and chemical properties completely different from the properties of their elemental components. The compound sodium chloride, NaCl (shown in Figure 2.7), is very different from elemental sodium and elemental chlorine. Elemental sodium, Na, consists of nothing but sodium atoms, which form a soft, silvery metal that can be cut easily with a knife. Its melting point is 97.5°C, and it reacts violently with water. Elemental chlorine, Cl_2, consists of chlorine molecules. This material, a yellow-green gas at room temperature, is very toxic; it was used as a chemical warfare agent during World War I. Its boiling point is -34°C. The compound

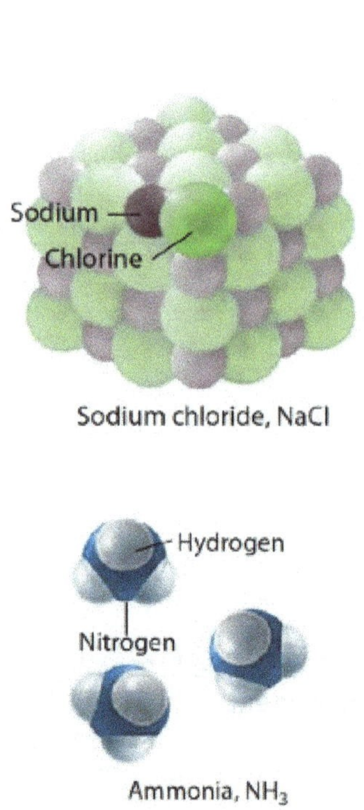

< Figure 2.6

The compounds sodium chloride and ammonia are represented by their chemical formulas, NaCl and NH_3. A chemical formula shows the ratio of atoms that make up the compound.

Figure 2.7

Sodium metal and chlorine gas react together to form sodium chloride. Although the compound sodium chloride is composed of sodium and chlorine, the physical and chemical properties of sodium chloride are very different from the physical and chemical properties of either sodium metal or chlorine gas.

sodium chloride, NaCl, is a translucent, brittle, colorless crystal with a melting point of 800°C. Sodium chloride does not react chemically with water and is nontoxic. It is an essential component of all living organisms. Sodium chloride is not sodium, nor is it chlorine; it is uniquely sodium chloride, a tasty chemical compound we lightly sprinkle over popcorn.

CONCEPT CHECK

Hydrogen sulfide H_2S, is one of the smelliest compounds. Rotten eggs get their characteristic bad smell from the hydrogen sulfide they release. Can you infer from this information that elemental sulfur, S_8, is just as smelly?

CHECK YOUR ANSWER

No. In fact, the odor of elemental sulfur is negligible compared with that of hydrogen sulfide. Compounds are different from the elements from which they are formed. Hydrogen sulfide, H_2S, is as different from elemental sulfur, S_8, as water, H_2O, is from elemental oxygen, O_2.

How are atoms held together to form a chemical compound? The answer is through the *chemical bond*. There are three major types of chemical bonds: the ionic bond, the covalent bond, and the metallic bond. For the purpose of learning biology, we can focus only on the ionic and covalent bonds.

Atoms can lose or gain electrons. Upon losing an electron, the atom will have more protons than electrons. Recall that electrons are negatively charged, and protons are positively charged. This means that an atom that loses one or more electrons will have a net positive charge. For example, the calcium atom easily loses two electrons. What remains is an atom with 20 positively charged protons and only 18 negatively charged electrons. The net charge of this calcium atom thus becomes 2+.

Any atom with a different number of protons versus electrons is called an **ion**. When calcium loses two electrons it transforms into the calcium ion, which we represent as Ca^{2+}. As another example, the fluorine atom tends to gain one electron forming the fluoride ion represented as F^{1-}.

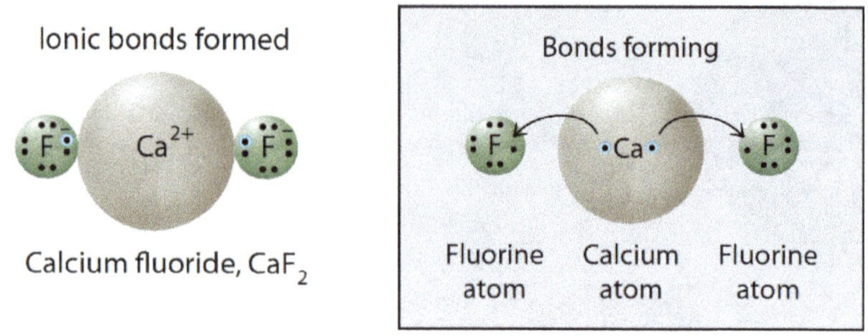

Figure 2.8 >

A calcium atom loses two electrons to become a positively charged calcium ion, Ca^{2+}. These two electrons can be captured by two fluorine atoms, which thus become fluoride ions, F^{1-}. All three ions then bind together to form the compound calcium fluoride, CaF_2.

Now consider what might happen when a calcium atom that tends to lose two electrons moves adjacent to two fluorine atoms, which each tend to gain one electron? Answer: There is an exchange of electrons. The electrons from the calcium atom will jump to each of the fluorine atoms. The result is a calcium ion, Ca^{2+}, and two fluoride ions, F^{1-}. Because opposite signs attract, the ions become held together as the chemical compound calcium fluoride, CaF_2, as shown in **Figure 2.8**. What we have just described is called the **ionic bond**, which is the electrical attraction between oppositely charged ions.

Ionic bonds tend to form between atoms of elements found on opposite sides of the periodic table. The hydrogen atom plus atoms of elements found to the upper right side of the periodic table are able to connect with each other through a different kind of bond called the *covalent bond.*

A **covalent bond** occurs when two atoms become attracted to each other's electrons. But rather than the electrons fully jumping from one atom to the other (as in an ionic bond), the electrons merely congregate in between the two atoms. By analogy, consider two kids playing with two toys that they share. You'll find the toys located between the two kids. You'll find that each child is attracted to those two toys. You'll also find that the kids tend to stay connected because of this attraction to the toys. Similarly, two atoms can be attracted to electrons that they share. This is the nature of the covalent bond. It is the covalent bond that holds two or more atoms together in a molecule, as shown in **Figure 2.9**. A covalent bond is often indicated merely by drawing a line between the two covalently bonded atoms, as shown in **Figure 2.9c**.

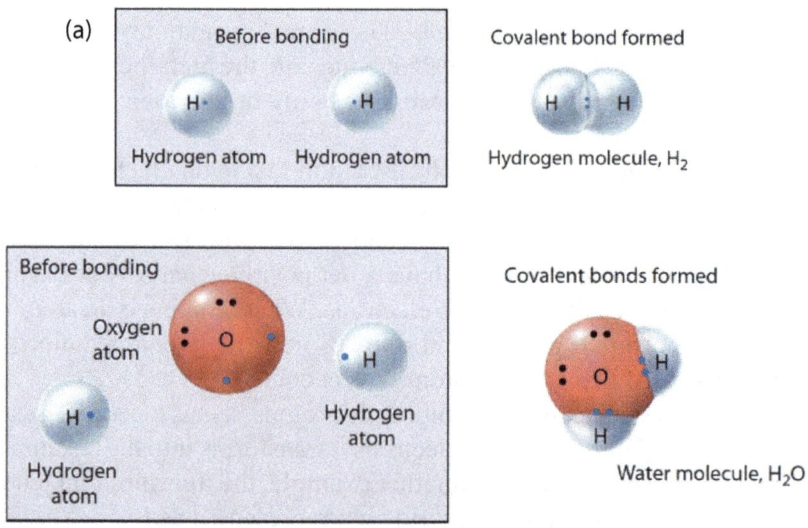

Figure 2.9 >

(a) Two hydrogen atoms can share their electrons resulting in a covalent bond.

(b) The oxygen atom is able to share electrons with two adjacent hydrogen atoms to form the compound water, H_2O.

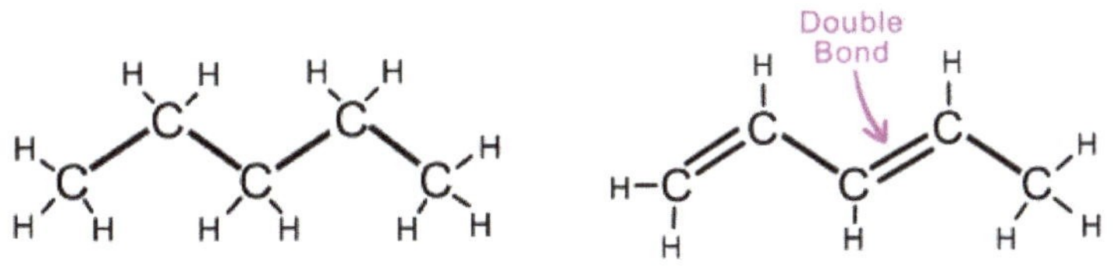

(a) (b)

< Figure 2.10

(a) All the bonds within the pentane molecule, C_5H_{12}, are single covalent bonds. (b) The pentadiene molecule, C_5H_{10}, has two double bonds.

When each atom shares one of its electrons, it is called a *single covalent bond*. When each atom shares two of its electrons, it forms a stronger bond called the *double covalent bond*, which can be represented as a double line, as shown in **Figure 2.10**.

Interestingly, the single covalent bond can rotate. This means that molecules made of single bonds can have all sorts of different shapes, which are called *conformations*, as shown in **Figure 2.11**. The double bond, by contrast, cannot rotate. Therefore, any double bond within a molecule restricts the number of different potential shapes.

(a)

(b)

< Figure 2.11

(a) These photos show the same arm in three different conformations. (b) Similarly, these molecular models represent the same molecule, pentane, in three different conformations. Carbon atoms (shown in black) held together by single covalent bonds are able to rotate relative to one another. This means the molecule as a whole can twist and turn into different conformations. This explains how larger organic biomolecules, such as proteins and nucleic acids, are also able to twist and turn into different shapes.

CONCEPT CHECK

1. Magnesium tends to lose two electrons while oxygen tends to gain two electrons. What is the chemical formula for the ionic compound magnesium oxide?
2. How many electrons make up a single covalent bond?

CHECK YOUR ANSWERS

1. A magnesium atom loses two electrons to form a Mg^{2+} ion. An oxygen atom gains two electrons to form an O^{2-} ion. These charges balance in a one-to-one ratio, and so the formula for magnesium oxide is MgO.
2. Two—one from each participating atom.

2.3 Mixtures

Λ Figure 12.12

The sharing of the two bonding electrons between a hydrogen and oxygen atom is less than even because the oxygen atom has a greater affinity for these electrons. As a result, the bonding electrons spend the majority of their time closer to the oxygen side of the bond. This makes the oxygen side of the bond slightly negative in charge, indicated by the δ- symbol, and the hydrogen side slightly positive, indicated by the δ+ symbol.

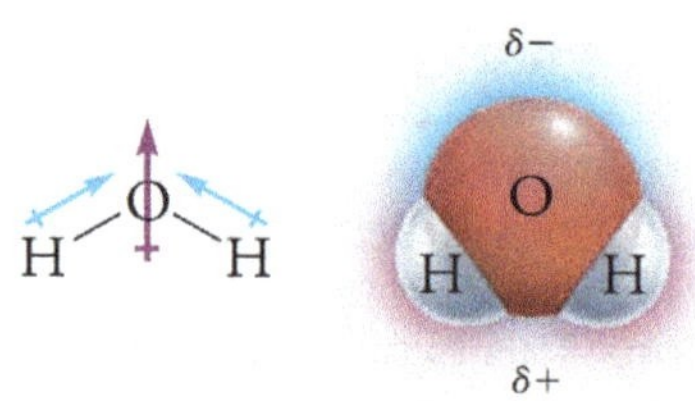

Λ Figure 12.13

Water is a polar molecule. The orientation of this polarity can be represented using a crossed arrow as shown on the left. As shown in the structure to the right, one side of the water molecule is thus slightly negative while the opposite side is slightly positive.

READING CHECK

What do you create when you stir salt into water so thoroughly you no longer see the salt crystals?

In this section, we look at what happens at the level of atoms and molecules when materials come together in the same place. Will the materials mix? Will one dissolve in the other? Let's explore.

A Closer Look at Sharing Electrons

As you have learned, electrons are shared between two atoms in a single covalent bond. However, the sharing may not be "fair." This is because some atoms are "greedy" for electrons. In other words, some atoms have a greater *electron affinity* than others. For example, consider the covalent bond between a hydrogen atom and an oxygen atom, as shown in **Figure 2.12**. Oxygen has a greater electron affinity, or the ability to attract the shared electrons, than hydrogen has. The result is a covalent bond where the bonding electrons spend more time closer to the oxygen atom than the hydrogen atom.

Of course, electrons are negatively charged. This means that the oxygen side of the bond is somewhat negative in character. At the same time, the hydrogen side of the bond is slightly positive because its positively charged nucleus is more exposed. This type of bond is what we call a *polar covalent bond.* A water molecule has two polar covalent bonds, as you can see in **Figure 2.13**. The oxygen side of the molecule is slightly negative, while the hydrogen side is slightly positive.

Molecules such as water, where there is an uneven distribution of electrons, are called **polar** molecules. A polar molecule is also sometimes called a *dipole* (pronounced die-pole). "Di" is Latin for two; thus, the word translates to "two poles." A dipole is represented with a crossed arrow where the arrow points to the negative side of the dipole. The greater the dipole's strength, the larger the crossed arrow, as shown in Figure 2.13.

Now, how do you suppose two water molecules might feel about each other? Bring them together and they'll be attracted to each other such that each positive side becomes aligned to each negative side, as shown in **Figure 2.14**. In this sense, water molecules are said to be "sticky." They stick to one another quite well. This explains why water has such a high boiling point—it takes a lot of thermal energy to pull them away from each other, which is what happens when going from a liquid to a gaseous phase.

What do you suppose happens when the ionic compound sodium chloride, NaCl, is added to water? Each sodium ion is positively charged. Therefore each sodium ion will attract and be surrounded by the negative sides of the water molecules. Meanwhile, each chloride ion is negatively charged. Therefore each of these ions will be surrounded by the positive sides of the water molecules. Mixing has occurred at the level of atoms and molecules. This is what happens when you stir table salt into water so thoroughly that you no longer see the salt particles. Sodium chloride has *dissolved* in the liquid water. You have made a *solution* of saltwater. You were able to do this because of the affinity between polar water molecules and the electrically charged sodium and chloride ions, as shown in **Figure 2.15**.

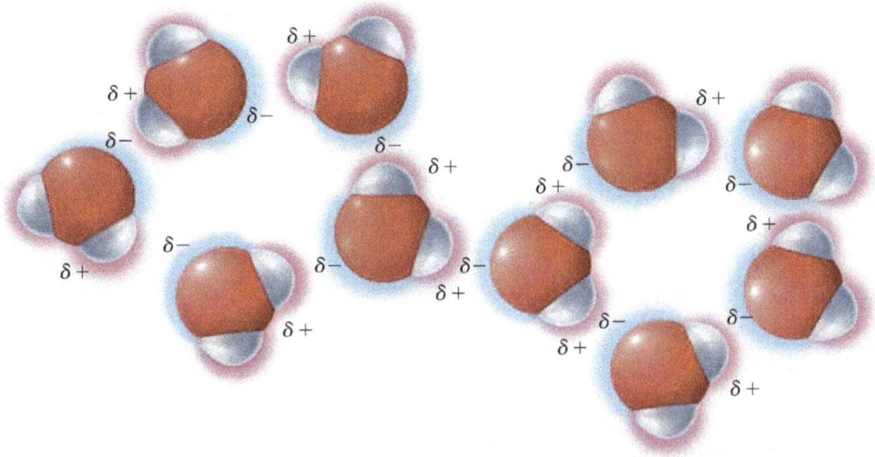

< Figure 2.14

Water molecules attract one another because each contains a slightly positive side and a slightly negative side. The molecules position themselves such that the positive side of one faces the negative side of a neighbor. Because of these attractions, water molecules are "sticky."

Some Definitions

Let's formalize some definitions now that we understand a bit about solutions at the submicroscopic level. First, if a material is **pure**, it consists of only a single element or a single chemical compound. Pure gold, for example, contains nothing but gold atoms. Pure table salt contains nothing but the chemical compound sodium chloride.

On the other hand, a **mixture** is a collection of two or more pure substances that are physically mixed. Within a mixture, each pure substance retains its own properties.

A **solution** is a type of mixture where all the components are of the same phase, as shown in **Figure 2.16**. Saltwater is a liquid solution because both the water and the dissolved sodium chloride are found in a single liquid phase. The clearest air, air high up in the atmosphere, consisting just of the elements nitrogen and oxygen plus minor amounts of other gaseous materials, would be considered a solution. An example of a solid solution is the alloy white gold. It is a mixture of the elements gold and palladium or sometimes platinum.

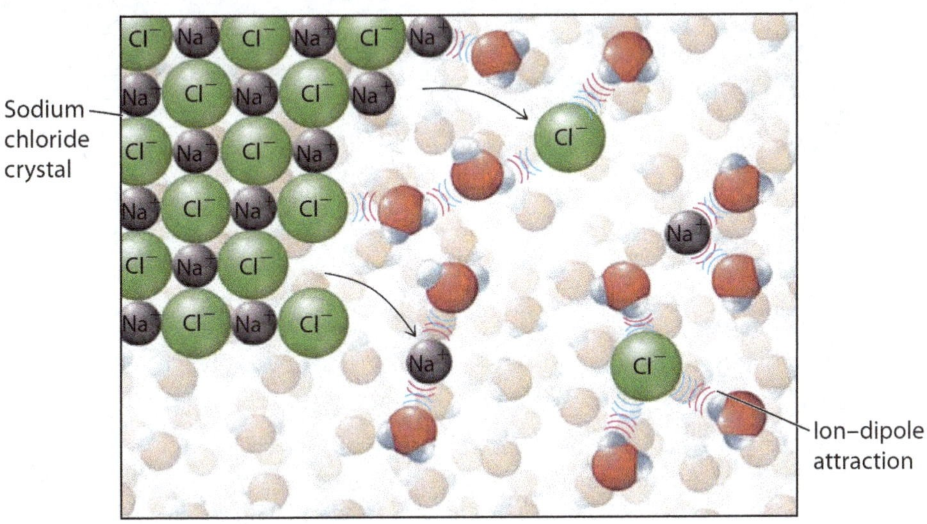

Aqueous solution of sodium chloride

< Figure 2.15

Water molecules and sodium and chloride ions in a crystal lattice are separated from one another by the collective attraction exerted by many water molecules.

Air

Clear seawater

White gold

Figure 2.16 >

Examples of solutions.

A **suspension** is a homogeneous mixture in which the different components are in different phases, such as solids in liquids or liquids in gases. In a suspension, the mixing is so thorough that the different phases cannot be readily distinguished. Milk is a suspension because it is a homogeneous mixture of solid proteins and fats finely dispersed in liquid water. Blood is a suspension composed of finely dispersed blood cells in water. Another example of a suspension is clouds, which are mixtures of tiny water droplets suspended in air. Shining a light through a suspension, as in **Figure 2.17**, results in a visible cone as the light is reflected by the suspended components.

In describing solutions, the component present in the largest amount is the **solvent**; other components are called **solutes**. For example, when a teaspoon of table sugar is mixed with 1 L of water, sugar is the solute and water is the solvent. **Figure 2.18** shows sugar dissolving in water at a molecular level.

The process of mixing a solute with a solvent is called **dissolving**. To make a solution, a solute must dissolve in a solvent—that is, the solute and solvent must form a homogeneous mixture. Whether one material dissolves in another is a function of their electrical attractions for each other. Generally speaking, the stronger the attractions, the better the mixing.

Figure 2.17 >

The path of light becomes visible when the light passes through a suspension.

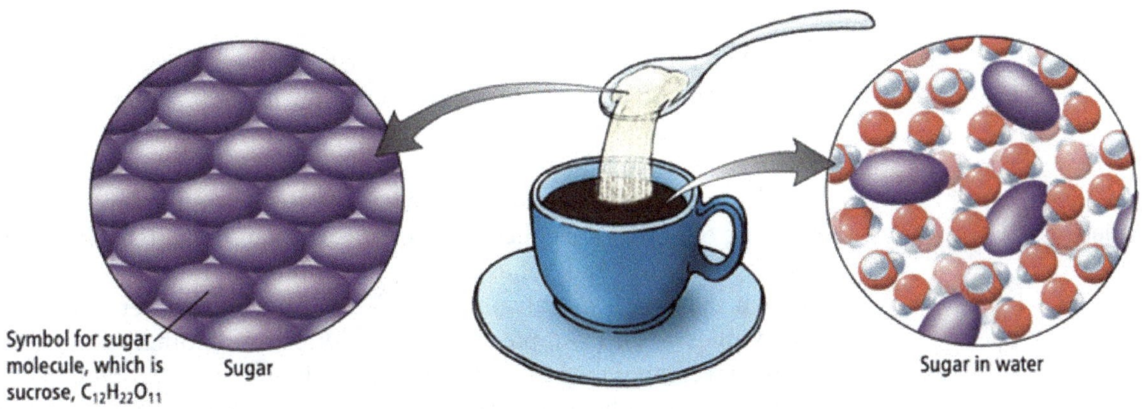

Figure 2.18

Water molecules pull the sucrose molecules in a sucrose crystal away from one another. This pulling away does not, however, affect the covalent bonds within each sucrose molecule, which is why each dissolved sucrose molecule remains intact as a single molecule.

CONCEPT CHECK

Impure water can be purified by

a. removing the impure water molecules.
b. removing everything that is not water.
c. breaking down the water into its simplest components.
d. adding some disinfectant such as chlorine.

CHECK YOUR ANSWER

The answer is (b) when something other than water molecules is found in the water, we say that the water is impure.

2.4 Chemical Reactions

Sometimes different chemicals mix together without changing one another. On the other hand, some chemicals are not at all stable when they come into contact They *react*! A good example is what happens when hydrogen gas, H_2, and oxygen gas, O_2 are mixed. These two chemicals have a tendency to react with one another, which means that their atoms change partners and new types of molecules are formed.

Imagine a dance floor where people dancing in couples suddenly step back from their partner to find a new partner or multiple partners. The initial couples are gone and new ones have been created. Similarly, after a **chemical reaction**, the atoms have bonded in new ways to different atoms in new arrangements. For example, every two hydrogen molecules, H_2, will react with one oxygen molecule, O_2, to form two water molecules, H_2O, as shown in **Figure 2.19**.

$$2\,H_2\,(g) + 1\,O_2\,(g) \longrightarrow 2\,H_2O\,(g) \quad \text{(balanced)}$$

Figure 2.19

In hydrogen gas, H_2, hydrogen atoms are connected to hydrogen atoms. In oxygen gas, O_2 oxygen atoms are connected to oxygen atoms. These two chemicals can react to form water, H_2O, in which two hydrogen atoms are connected to a single oxygen atom. Shown below the molecules is the chemical equation that represents this chemical change.

READING CHECK

How do scientists use chemistry to make new materials?

The result of any chemical reaction is that new materials are formed. Water, H_2O, is different from the hydrogen, H_2, and oxygen, O_2, from which it was formed. So how do chemists make new materials? Answer: through chemical reactions. And how are new materials made in nature? Same answer: through chemical reactions, as shown in **Figure 2.20**.

Chemical reactions also involve energy. Reactions in which there is a net release of energy are called **exothermic**. Rocket ships lift off into space, and campfires glow red hot as a result of exothermic reactions. Reactions in which there is a net absorption of energy are called **endothermic**. Photosynthesis, for example, involves a series of endothermic reactions that are driven by the energy of sunlight, as shown in **Figure 2.21**.

The speed of chemical reactions varies. Some chemical reactions, such as the rusting of iron, happen very slowly, perhaps even taking years. Other chemical reactions, such as the burning of gasoline, happen in minutes or seconds. The speed of any reaction is known as its *reaction rate*. This is the time it takes for reactants to be transformed into products.

The flask shown in **Figure 2.22** contains only reactant molecules initially. Over time, these reactants form product molecules so that the concentration of product molecules increases. This gives us a useful way to define the reaction rate in terms of concentration. We say **reaction rate** is defined as how quickly the concentration of products increases or how quickly the concentration of reactants decreases.

The speed of most chemical reactions is influenced by temperature. This includes reactions that occur in living bodies. The body temperature of some animals, such as humans, is fairly constant. However, the body temperature of other animals, such as the alligator shown in **Figure 2.23**, may vary a bit depending on the temperature of the environment. On a warm day, the chemical reactions occurring in an alligator are "up to speed," and the animal may be more active. On a chilly day, however, the chemical reactions proceed at a lower rate and, as a consequence, the alligator's movements are more sluggish.

Sekundemal

Figure 2.20 >

The heat of a lightning bolt causes atmospheric nitrogen and oxygen to react, leading to the formation of nitric acid, HNO_3, and nitrous acid, HNO_2. As part of the nitrogen cycle, these acids are carried by rain into the ground. There, they are transformed by microorganisms into nitrate ions that plants use for growing. We, in turn, eat the plants, or plant-eating animals, to support life-sustaining chemical reactions within ourselves.

< Figure 2.21

Chemical reactions that occur in a photosynthetic plant have a net absorption of energy. These are endothermic reactions. Chemical reactions that occur when wood is burning have a net release of energy. These are exothermic reactions.

The rate of a chemical reaction can be sped up by using a catalyst. A **catalyst** is any substance that increases the rate of a chemical reaction but is not itself consumed by the chemical reaction. The chemical industry depends on catalysts in many ways. They reduce manufacturing costs by lowering required temperatures and by providing greater product yields without being consumed (the catalysts can be recycled). Indeed, more than 90% of all manufactured goods are produced with the assistance of catalysts. Without catalysts, the price of gasoline would be much higher, as would the prices of consumer goods such as rubber, plastics, pharmaceuticals, automobile parts, clothing, and food grown with chemical fertilizers.

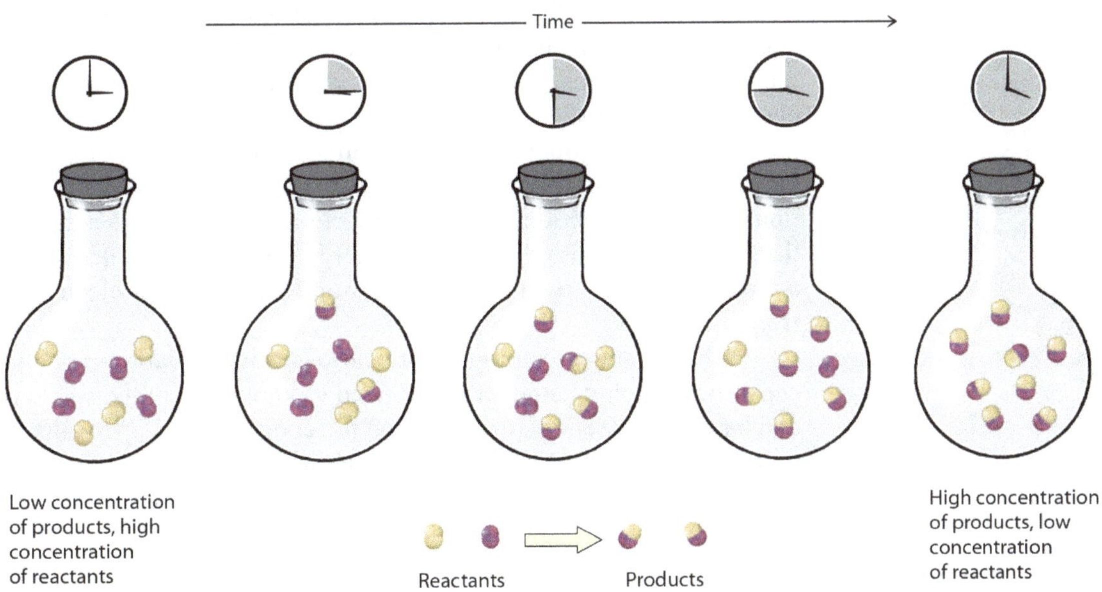

Λ Figure 2.22

Over time, the reactants in this reaction flask may transform into products. If this happens quickly, the reaction rate is high. If this happens slowly, the reaction rate is low.

Figure 2.23 >

Why did the alligator cross the road on a chilly day? The alligator was attracted to the daytime warmth of the pavement because the heat speeds up the chemical reactions going on inside its body.

Living organisms rely on special types of catalysts in their cells known as *enzymes*. Enzymes allow complex biochemical reactions to occur with ease. As we'll be exploring, this includes the digestion of food and the activity of DNA.

CONCEPT CHECK

What happens to atoms during a chemical reaction?

CHECK YOUR ANSWER

During a chemical reaction there is no change in the number or types of atoms. Rather, the change involves how those atoms are connected. As those connections change, a new material is created.

2.5 Types of Chemical Reactions

Two important classes of chemical reactions are *acid–base* reactions and *oxidation–reduction* reactions. Acid–base reactions involve the transfer of hydrogen ions, H^+, from one reactant to another. These sorts of reactions occur in your stomach to help you digest food. They also play a key role in global climate. Most consumer goods can trace their origins to acid–base chemical reactions.

Oxidation–reduction reactions involve the transfer of one or more *electrons* from one reactant to another. The burning of wood is an oxidation–reduction reaction, as are the reactions your body uses to transform the food you eat into biochemical energy. Oxidation–reduction reactions are responsible for the rusting of a car and the release of electrical energy from a battery.

What are the most important acid-base and oxidation-reduction concepts that a biology student might need to know? From the world of acids and bases, it's essential to understand the concept of pH. From the world of oxidations and reductions, it's important to understand how oxygen atoms have such a strong affinity for electrons. Let's take a closer look at these main ideas.

Acids and Bases

We can define an **acid** as any chemical that donates a hydrogen ion, H^+, and a **base** as any chemical that accepts a hydrogen ion. Because a hydrogen atom consists of one electron surrounding a one-proton nucleus, a hydrogen ion formed from the loss of an electron is nothing more than a lone proton. Thus, it is also sometimes said that an acid is a chemical that donates a proton, and a base is a chemical that accepts a proton. Consider what happens when hydrogen chloride is mixed into water:

$$HCl + H_2O \longrightarrow Cl^- + H_3O^+$$

H^+ donor (acid) — H^+ acceptor (base)

< Figure 2.24

Acid-base reaction.

Hydrogen chloride donates a hydrogen ion to the water molecule, resulting in a third hydrogen bonded to the oxygen. In this case, hydrogen chloride behaves as an acid (proton donor), and water behaves as a base (proton acceptor). The products of this reaction are a chloride ion, Cl^-, and a **hydronium ion**, H_3O^+, which is a water molecule with an extra proton.

READING CHECK

What is a hydronium ion?

When added to water, ammonia behaves as a base as it accepts a hydrogen ion from water, which, in this case, behaves as an acid:

$$H_2O + NH_3 \longrightarrow OH^- + NH_4^+$$

H^+ donor (acid) — H^+ acceptor (base)

< Figure 2.25

Ammonia added to water.

This reaction results in the formation of an ammonium ion, NH_4^+, and a **hydroxide ion**, OH^-, which is a water molecule without the nucleus of one of the hydrogen atoms.

How might these concepts apply to biology? Consider what happens when the drug pseudoephedrine is exposed to the acid hydrogen chloride. In the presence of hydrogen chloride, the pseudoephedrine behaves as a base by accepting H^+ from a hydrogen chloride. The negative Cl^- then joins the pseudoephedrine-H^+ ion to form the salt pseudoephedrine hydrochloride, which is a decongestant, shown in **Figure 2.26**.

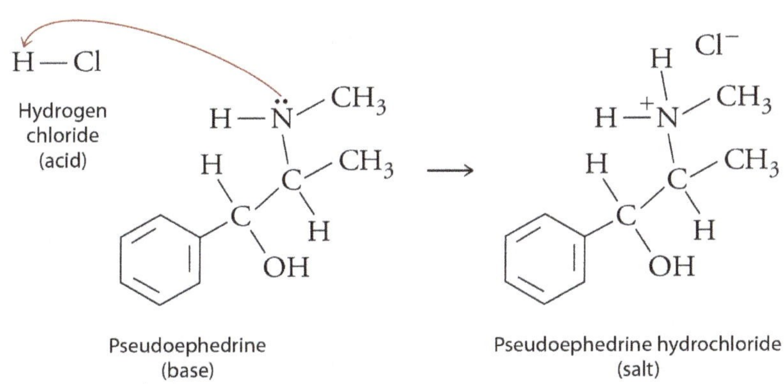

< Figure 2.26

Hydrogen chloride and pseudoephedrine react to form the salt pseudoephedrine hydrochloride, which, because of its solubility in water, is readily absorbed into the body. Many pharmaceuticals, especially those taken orally, are similarly provided in a salt form so that they are easily absorbed.

Notice that the pseudoephedrine hydrochloride salt has both negative and positive charges, much like table salt, NaCl. The pseudoephedrine molecule itself is an oil and doesn't mix very well in water. Its salt, by contrast, has these positive and negative charges, which means it can be dissolved in water, much like table salt. This means that the drug can be taken orally. It will dissolve in your stomach fluids and then be absorbed into your body (as a salt) through your intestines.

The quantity **pH** is a measure of the concentration of hydronium ions, H_3O^+, within a solution. In this sense, it's a measure of the solution's "acidity." Interestingly, the concentration of hydronium ions, H_3O^+, is intimately tied to the concentration of hydroxide ions, OH^-. As one goes up, the other necessarily goes down, much like two ends of a see-saw, as shown in **Figure 2.27**. At a pH of 7, the concentration of the two are equal, and the solution is said to be *neutral*. As the hydronium ion, H_3O^+, concentration increases, the pH goes down, and the solution is *acidic*. As the hydronium ion, H_3O^+, concentration decreases, the pH goes up, and the solution is *basic* (sometimes called *alkaline*). Your body works hard to maintain a very stable blood pH of about 7.40. Anything significant variation would be lethal. How so? The answer has to do with acid-base reactions. Notice the role of the nitrogen atom back in Figure 2.26. When exposed to an acid, it accepts the hydrogen ion to become ionized. The biomolecules within your body abound with nitrogen and other atoms that are susceptible to becoming ionized. In some cases, it's good that they are ionized. In other cases, it's important that they do *not* become ionized. Becoming ionized can impact the shape of the large biomolecule within which that atom is found. And if there's one major takeaway from the study of biochemistry, it is that the shape of a biomolecule plays a big role in determining which reactions are allowed and what the molecule's biological function can be. This includes biomolecules that act as enzymes, biomolecules that act as our genetic carriers, those that transport oxygen within our blood (**Figure 2.28**), and many other examples.

There are many other examples in biology where the concept of pH is key. Perhaps most significant, as discussed in Chapter 17, is that increasing levels of atmospheric carbon dioxide leads to a decrease in the pH of the oceans. This, in turn, results in the destruction of the calcium carbonate-based shells of marine organisms such as phytoplankton, which are part of the foundation of marine food chains.

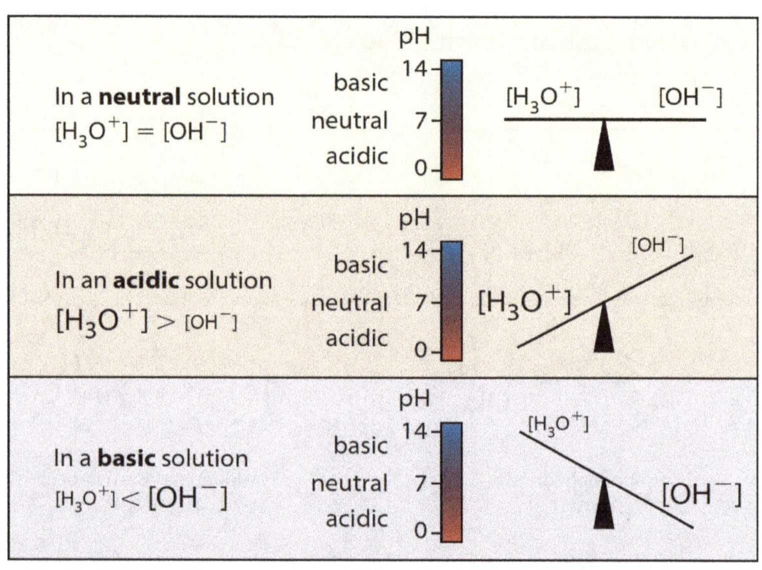

Figure 2.27 >

The relative concentrations of hydronium and hydroxide ions determine whether a solution is acidic, basic, or neutral.

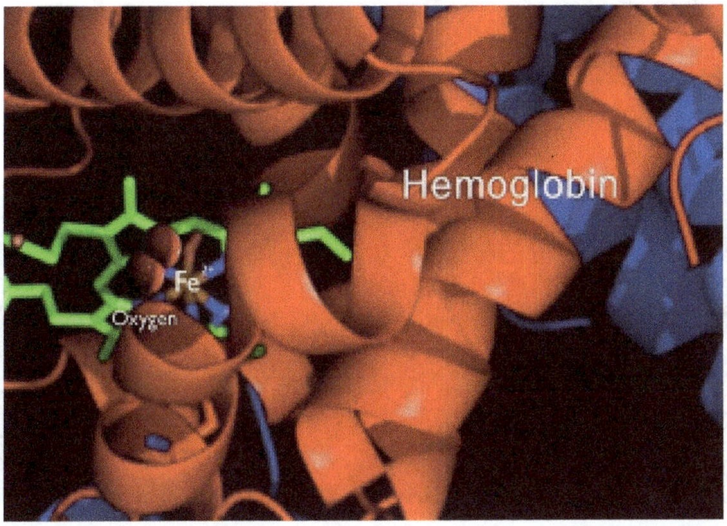

< **Figure 2.28**

The hemoglobin biomolecule is able to bind the oxygen molecule only within a very narrow pH range. Where there is much muscle activity, the pH tends to drop. This causes hemoglobin molecules within that vicinity to release their oxygen molecules, which helps to support the continued muscle activity. The role of the pH of body fluids is very important.

CONCEPT **CHECK**

How does adding ammonia, NH_3, to water make a basic solution?

CHECK YOUR ANSWER

Ammonia increases the hydroxide ion concentration by reacting with water:

$$NH_3 + H_2O \rightarrow NH_4^+ + OH^-$$

This reaction raises the hydroxide ion concentration, which has the effect of lowering the hydronium ion concentration. With the hydroxide ion concentration now higher than the hydronium ion concentration, the solution is basic.

Oxidations and Reductions

Oxidation is the process whereby a chemical loses one or more electrons. **Reduction** is the opposite process, whereby it gains one or more electrons. Oxidation and reduction are complementary processes that occur at the same time. They always occur together; you cannot have one without the other. How so? The electrons lost by one chemical in an oxidation reaction don't simply disappear; they are gained by another chemical in a reduction reaction.

Different atoms have different tendencies to gain or lose electrons. This is neatly spelled out by the periodic table, as shown in Figure 2.29. Briefly, atoms of elements residing near the upper right side of the table tend to have the best ability to gain electrons. Conversely, atoms of elements toward the lower left side tend to have the best ability to lose electrons. Thus, you can gauge the tendency of any atom to gain or lose electrons simply by finding its relative position within the periodic table.

An atom that is notorious for grabbing electrons is the oxygen atom. You'll note that oxygen is positioned in the far upper right side of the periodic table. In fact, this is where the term "oxidation" comes from. Almost any atom that gets close to an oxygen atom will be robbed of its electrons by that oxygen atom. We say that the atom has become *oxidized* (it has lost electrons). The oxygen atom, in turn, becomes "reduced," which means it has gained a negative charge. Wait a second. Gaining is a reduction? Think of it this way: What happens to your bank account when you add a *negative* amount of money? Answer: your bank account is "reduced".

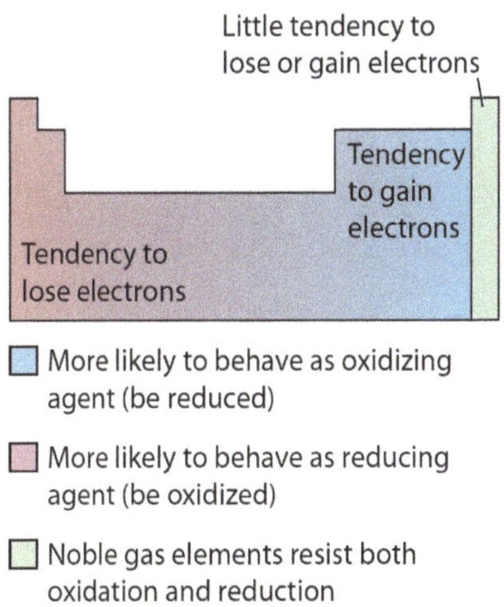

∧ **Figure 2.29**

The ability of an atom to gain or lose electrons is indicated by its position in the periodic table. Those at the upper right tend to gain electrons, and those at the lower left tend to lose them.

Figure 2.30 >

The oxygen, O_2, found in our atmosphere is vital for the burning of a campfire. It is also needed for the chemical reactions that produce energy within our cells.

A campfire, **Figure 2.30**, is an example of an oxidation-reduction reaction, as is cellular respiration, the biological process that gives cells energy. In both cases, oxygen grabs electrons from some source. In the campfire, oxygen grabs electrons from the burning wood. In cellular respiration, oxygen grabs electrons that come from food molecules. These chemical reactions are exothermic. They release energy as electrons jump over to oxygen atoms where they'll find themselves a bit closer to an atomic nucleus. Oxidation and reduction are at the heart of why campfires are warm and how we get energy from the food we eat.

CONCEPT CHECK

Oxygen, O_2, is a good oxidizer, but so is chlorine, Cl_2. What does this indicate about their relative positions in the periodic table?

CHECK YOUR ANSWER

Chlorine and oxygen must lie in the same area of the periodic table. Both are strong oxidizers, which means they're good at pulling electrons away from other atoms.

2.6 Organic Molecules

Carbon atoms are perhaps the most versatile of all atoms in their ability to bond among themselves repeatedly. This enables them to form long carbon chains. Carbon atoms also have a great capacity to bond with atoms of other elements. Due to these two special abilities, a practically endless number of different carbon-based molecules is possible. Each molecule has its own unique set of physical, chemical, and biological properties. The flavor of vanilla, for example, is perceived when the compound vanillin lands upon sensory cells within the nose. The flavor of chocolate is generated when a selection of compounds, such as tetramethylpyrazine, are detected by the nose, as shown in **Figure 2.31**.

The study of carbon-containing compounds is known as **organic chemistry**. Because organic compounds are so closely tied to living organisms and because they have many applications—flavorings, fuels, polymers, medicines, agriculture, and more—it is important to have a basic understanding of them.

< Figure 2.31

Food and flavorings are made of organic molecules.

The most basic organic molecules are the hydrocarbons. As their name suggests, hydrocarbons are made only of hydrogen and carbon atoms. Three examples are shown in **Figure 2.32**. Here, as in all hydrocarbons, carbon atoms bond together to form a "backbone" or basic structure. Only carbon atoms are good at forming extended chains.

READING CHECK

What kinds of atoms do hydrocarbons contain?

Atoms other than carbon, such as hydrogen, nitrogen, and oxygen, can also attach themselves to chains of carbon atoms. This helps to give each organic molecule a unique character. Look back to Figure 2.32 to see that the vanillin organic molecule has oxygen atoms (shown in red), while the tetramethylpyrazine molecule found in chocolate features nitrogen atoms (shown in blue).

Oxygen and nitrogen can form special groups, called *functional groups*. Some of these are shown in **Table 2.1**. Organic molecules can be defined by the functional groups they contain. There is no need for you to memorize these functional groups. We present them here as an introduction. You will see that functional groups abound in the biomolecules you will study in biology.

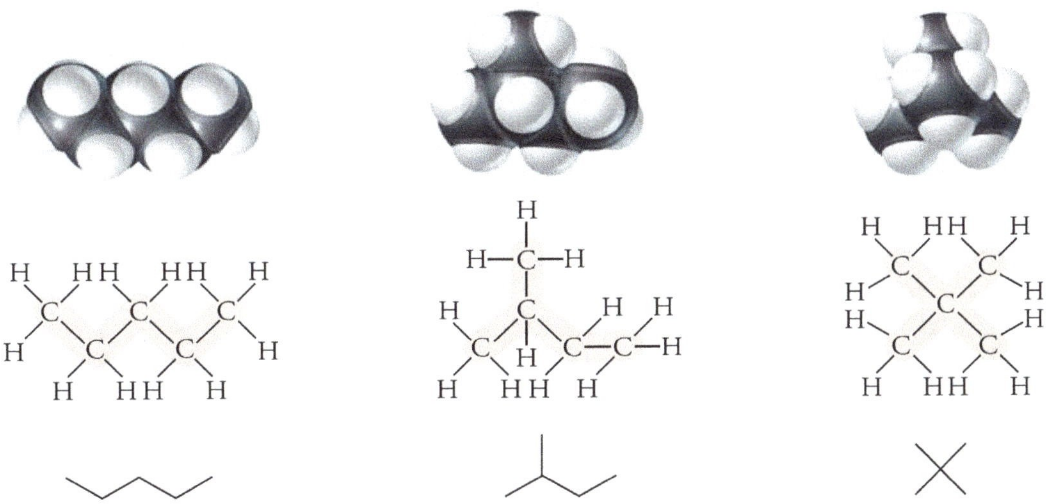

Pentane, C_5H_{12} Isopentane, C_5H_{12} Neopentane, C_5H_{12}

< Figure 2.32

These three hydrocarbons all have the same molecular formula. We can see their different structural features by highlighting the carbon framework in two dimensions. Easy-to-draw stick structures that use lines for all carbon–carbon covalent bonds are also useful.

TABLE 2.1 Functional groups in organic molecules.

GENERAL STRUCTURE	CLASS	GENERAL STRUCTURE	CLASS
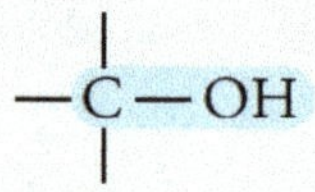	Alcohols	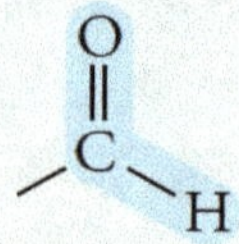	Aldehydes
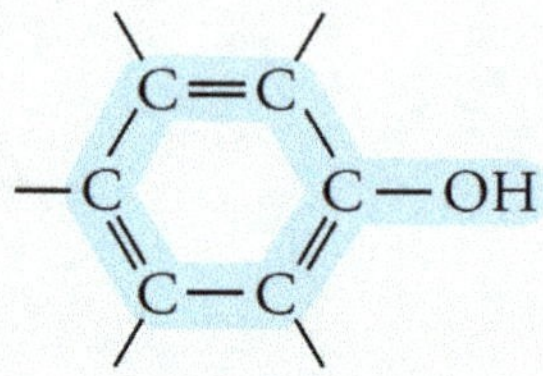	Phenols	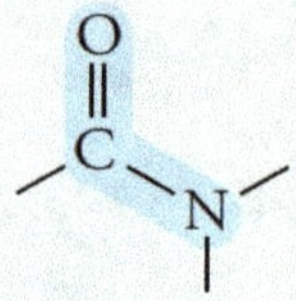	Amides
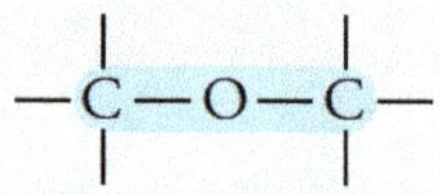	Ethers	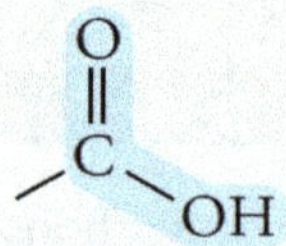	Carboxylic acid
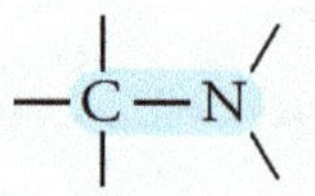	Amines	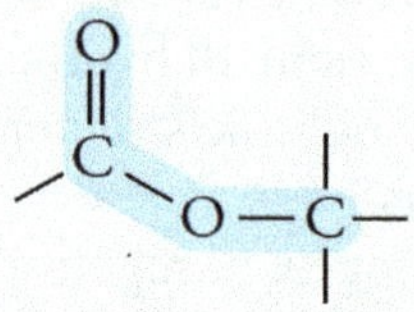	Esters
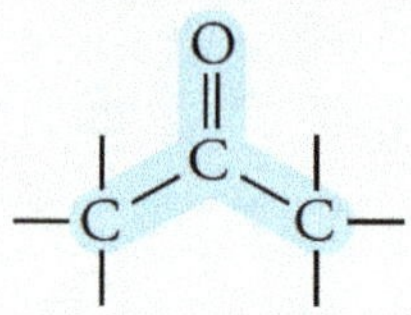	Ketones	**Functional Group:** A collection atoms within a molecule that behave as a single unit.	

CONCEPT CHECK

What is the significance of non-carbon atoms in an organic molecule?

CHECK YOUR ANSWER

Non-carbon atoms such as oxygen and nitrogen found within an organic molecule strongly impact the chemical and physical properties of that organic molecule.

2.7 Macromolecules Needed for Life

Living things are made up of four main types of organic *macromolecules*, or "big molecules." These macromolecules are: proteins, carbohydrates, lipids, and nucleic acids. They are shown in **Figure 2.33**. You will see that macromolecules are indeed very large molecules and that they consist of multiple smaller molecules joined together. You may already know quite a bit about macromolecules if you eat protein to grow, avoid foods that contain certain kinds of fat, or have ever "carb-loaded" before a race.

Carbohydrates store energy in living organisms. Simple sugars, such as glucose (**Figure 2.33b**) and fructose, are carbohydrates. More complex carbohydrates are made up of chains of simple sugars. Starch and glycogen—the primary energy-storage substances in plants and animals, respectively—consist of linked glucose molecules. Carbohydrates can also have structural functions. Cellulose, found in plant cell walls, is a structural carbohydrate built from glucose subunits—and it is the most abundant organic compound in the world (**Figure 2.34**). Carbohydrates are made up of carbon, hydrogen, and oxygen atoms in the form of $(CH_2O)_n$ where n is any number.

Lipids serve diverse functions in living organisms. As fats or oils, lipids are used by many living organisms to store energy. Lipids store energy much more efficiently than carbohydrates—that is, 1 gram of fat or oil contains a lot more energy than 1 gram of carbohydrate. For this reason, lipids are used for long-term energy storage by many organisms, including humans. Lipids can serve structural purposes as well; for example, phospholipids are an essential component of cell membranes. One of the most familiar lipids is cholesterol, which the body uses to make such hormones as estrogen and testosterone. Lipids have a variety of chemical structures, but many include fatty acids—strings of hydrocarbons (carbon and hydrogen atoms) with a carboxyl (COOH) group at one end—as a major component (**Figure 2.33c**). Lipids are hydrophobic; that is, they are not soluble in water. Fat deposits made of lipids also work well to insulate against cold (**Figure 2.35**).

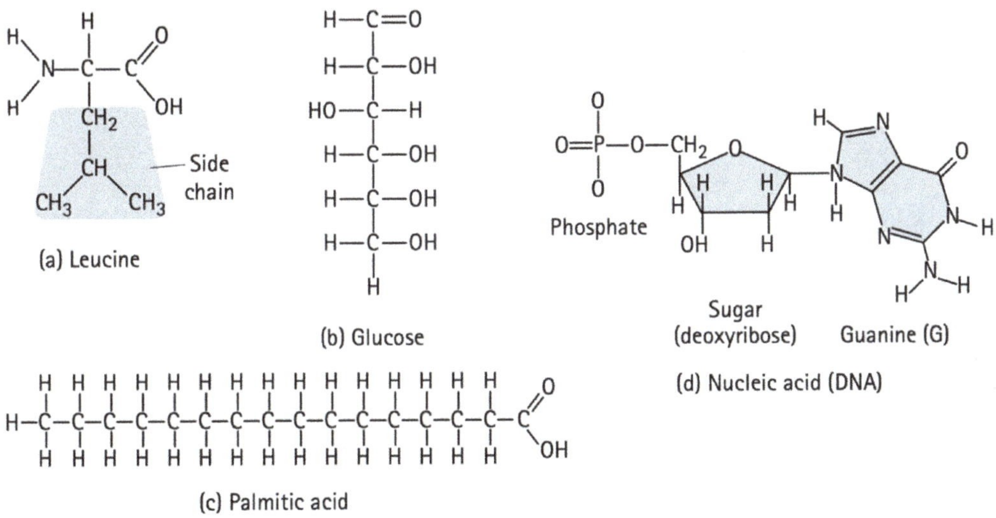

< Figure 2.33

(a) Proteins are made up of amino acids. Leucine is one of the 20 amino acids found in living organisms. (b) Carbohydrates are made up of one or more simple sugars. This structure is glucose, a primary energy source for many living things. (c) Lipids have a variety of chemical structures, but many include fatty acids. Palmitic acid is a fatty acid found in lard and butter. (d) Nucleic acids are made up of nucleotides. This guanine nucleotide is one of the four nucleotides that make up DNA.

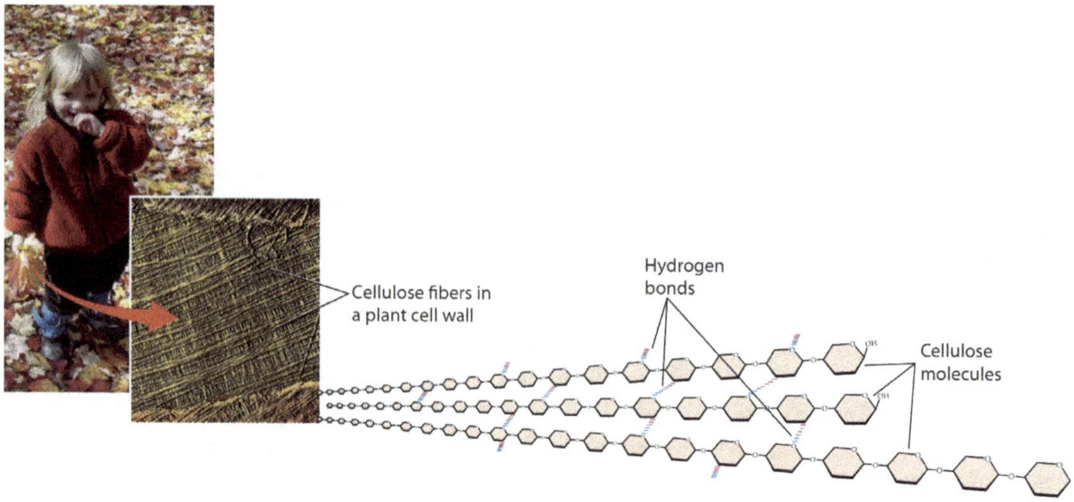

Figure 2.34 >

Strands of cellulose in a plant, including the leaves held by Maitreya, are joined by hydrogen bonds. These microscopic fibers are laid down in a crisscross pattern to give strength in many directions.

Proteins perform a wide range of functions in living organisms (**Figure 2.36**). The protein keratin provides structure—it is a major component of skin, hair, and feathers. Insulin is a protein that acts as a hormone, enabling one type of cell in the body to communicate with other types. Actin and myosin are proteins that allow muscles to contract. Hemoglobin, a protein found in red blood cells, transports oxygen to body tissues. Proteins called antibodies protect the body from disease. And proteins known as digestive enzymes break down food during digestion.

What are proteins? Proteins are folded chains of organic molecules called *amino acids*. All amino acids include a central carbon (C) atom bonded to an amino group (NH_2), a carboxyl group (COOH), a hydrogen atom (H), and a side chain (called R) that varies from one amino acid to another. The amino acid leucine is shown in **Figure 2.33a**. Although only 20 amino acids are found in living organisms, they can be strung together and folded in countless ways to create proteins with unique three-dimensional structures. This is why proteins are able to perform such a wide variety of functions.

The role of the fourth type of macromolecule, **nucleic acids**, is to store genetic information in living organisms. All of our genes are made up of nucleic acids. Genes provide the information that cells need to build proteins. Nucleic acids are made up of strands of smaller units called nucleotides. A *nucleotide* includes a sugar molecule, a phosphate group, and a nitrogenous base. The two kinds of nucleic acids found in living organisms are *deoxyribonucleic acid* (DNA) and *ribonucleic acid* (RNA).

< Figure 2.35

The walrus and other polar species are insulated from the cold by a think layer of fat beneath their skin.

The DNA molecule consists of two nucleic acid strands twisted into a spiral, or " double helix" (**Figure 2.37**). DNA has four kinds of nucleotides—adenine, cytosine, guanine, and thymine. These are called A, C, G, and T for short. All the genetic information in living organisms is expressed using this four-letter alphabet.

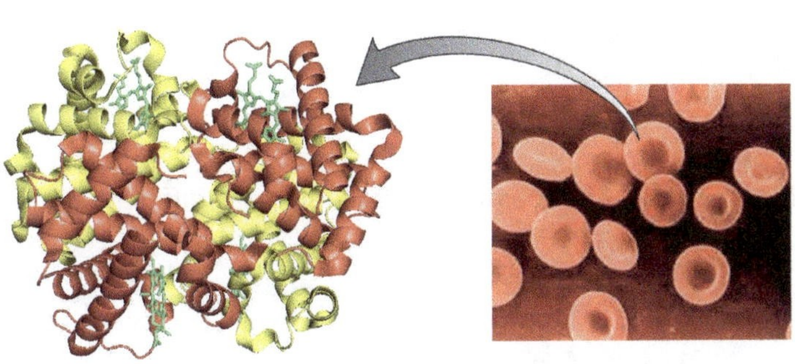

Hemoglobin, a **transport protein**, is the component of blood that takes the oxygen we breathe to our cells.

Contractile proteins in muscles allow us to move.

Structural proteins are found in skin, hair, and bones.

Storage proteins serve as a source of amino acids in milk.

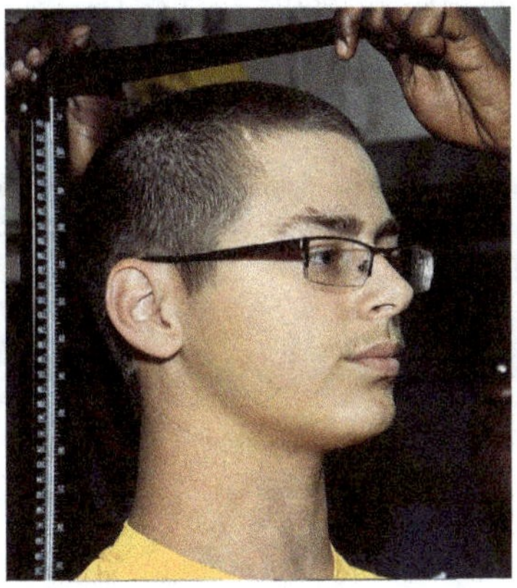

Some proteins are **hormones** that regulate body metabolism and growth.

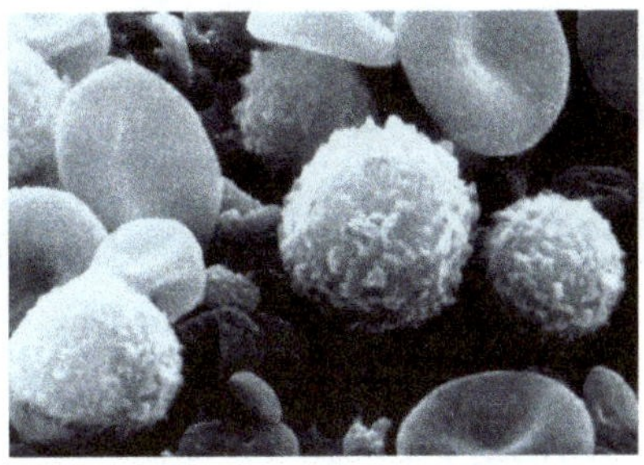

White blood cells produce **antibodies**, which are proteins that fight infections.

Enzymes are proteins that act as catalysts for reaction in the body, including the digestion of food.

^ Figure 2.36

A variety of proteins in the human body.

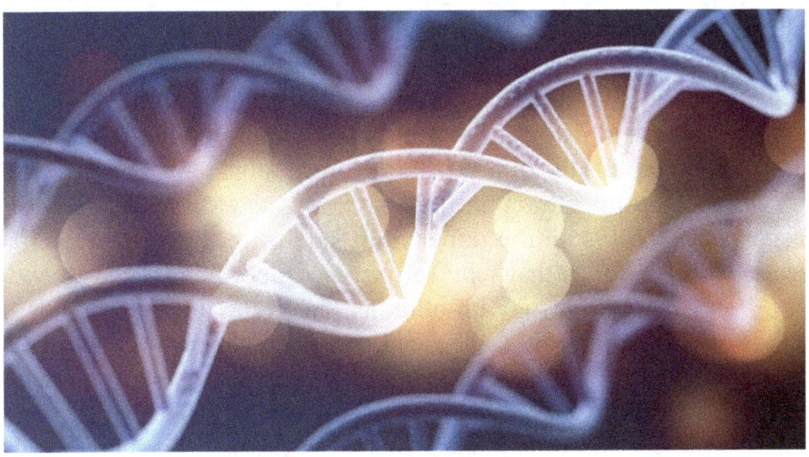

^ Figure 2.37

This double helix is the nucleic acid, deoxyribonucleic acid, or DNA.

CONCEPT CHECK

1. Gecko lizards store fat in their tails, using this energy supply to survive during lean periods. What is the advantage of storing energy as fat rather than as carbohydrate?

2. Living organisms contain thousands of different kinds of proteins. How is it possible to make so many different kinds of proteins from only 20 different amino acids?

CHECK YOUR ANSWERS

1. Because a gram of fat contains more energy than a gram of carbohydrate, storing the energy as fat allows geckos to store the energy they need without weighing themselves down.

2. The 20 amino acids that make up proteins can be strung together and then folded in a huge number of different ways, thus creating many different proteins.

Concept Review

Chapter 2

Detailed Chapter Summary

The study of atoms and molecules is **chemistry**. This chapter is a review of the basic chemistry concepts underlying biology, which is the study of life.

Atoms are a fundamental building block of matter. Atoms are incredibly small—smaller than the wavelengths of visible light. Atoms are made of even smaller subatomic particles, including the central **atomic nucleus** surrounding a cloud of tiny **electrons**. While the atomic nucleus is positively charged, electrons are negatively charged. Atoms can bind together to form a **molecule**. Thus, molecules are made of atoms. A material made of only one type of atom is called an **element**. The **periodic table** shows a listing of all known elements. Each element is represented by its **atomic symbol**.

Different types of atoms can join together to form a **chemical compound**, such as sodium chloride, NaCl, or ammonia, NH_3. A chemical compound is represented by its **chemical formula**, which shows the atomic symbols for the elements that combine to make that compound. A chemical compound is uniquely different from the elements used to make that compound.

Atoms can become attached to adjacent atoms by what is known as a **chemical bond**. The **ionic bond** is the attraction between oppositely charged ions. An **ion** is an atom that has either lost or gained electrons. Atoms of elements to the lower left of the periodic table tend to lose electrons, while atoms to the upper right tend to gain electrons. Thus, ionic bonds tend to occur between atoms found on opposite sides of the periodic table. An example is sodium chloride, NaCl, which is common table salt. In a **covalent bond**, atoms are held together by the attraction they have for the electrons they share. Atoms on the upper right side of the periodic table are the ones that tend to form covalent bonds, and they do so primarily with other atoms on the upper right side of the periodic table. An example is carbon dioxide, CO_2.

Within a covalent bond, the electrons might not be shared evenly because one of the two atoms has a greater affinity for the electrons. This results in one side of the covalent bond becoming slightly negative in charge while the opposite side becomes slightly positive. Such a covalent bond is described as a **polar** covalent bond. A water molecule has two polar covalent bonds that make the oxygen side of a water molecule slightly negative and the hydrogen sides slightly positive. This makes a water molecule electrically "sticky." In other words, it's easy for two water molecules to stick to each other, which explains why water has such a high boiling point. It also explains how ionic materials such as salt readily dissolve in water.

A **pure** substance is made of only one material. A **mixture** is a collection of two or more pure substances. A **solution** is an example of a mixture where all the components are in the same phase, such as a liquid, gas, or solid. A **suspension** is when the mixture consists of different phases. A cloud is an example of a suspension (liquid water droplets suspended in the air), as is blood (solid blood cells suspended in the blood plasma). In describing solutions, the component present in the largest amount is the **solvent**, and any other components are **solutes**.

In a **chemical reaction**, atoms change partners to form new materials. A chemical reaction that results in the net release of energy is **exothermic**. One that has a net absorption of energy is **endothermic**. How fast a chemical reaction occurs is described by its **reaction rate**. The faster the rate, the more quickly products are formed. Warmer temperatures tend to favor faster reactions, as do **catalysts**.

Two types of chemical reactions are acid-base reactions and oxidation-reduction reactions. An **acid-base** reaction involves the transfer of a hydrogen ion from one reactant to the other. An **oxidation-reduction** reaction involves the transfer of an electron. The pH of a solution is a measure of its concentration of **hydronium ions**, H_3O^+. The lower the pH, the greater this concentration, and the greater the acidity (H_3O^+ is a very strong acid made from water). The pH of the solution can determine whether an atom, such as nitrogen or oxygen, is ionized. Relative to biomolecules, this can have a major effect on the overall shape of the biomolecule, which profoundly impacts the functioning of that biomolecule.

A well-known behavior of oxygen atoms is that they like to attract electrons. Recall that this is why water is a polar molecule. It also explains how oxygen can cause other atoms or molecules to lose their electrons, a process called **oxidation**. These sorts of reactions tend to be exothermic. We see this in the burning of a campfire and in the breakdown of the food molecules we eat.

Carbon is unique in its ability to form repeated bonds with other carbon atoms as well as bonds with non-carbon atoms, such as hydrogen, nitrogen, and oxygen. The study of carbon-containing compounds is a special field known as **organic chemistry**.

There are four basic large organic molecules used by living things. These are **proteins**, **carbohydrates**, **lipids**, and **nucleic acids**. Proteins are folded chains of smaller units called amino acids. Proteins play a wide range of functions in living things. For example, they provide structure, such as in skin and hair, act as hormones that allow communication between body cells, allow for muscle contraction, transport oxygen to body tissues, break down food during digestion, and protect the body from disease. Carbohydrates store energy in organisms -- sugars such as glucose and fructose are examples. Lipids include fats and oils and are also used to store energy. Some carbohydrates and lipids play structural roles as well. Finally, nucleic acids, such as DNA and RNA, store genetic material. DNA is made up of four different nucleotides -- adenine, cytosine, guanine, and thymine.

Key Terms

Acid A chemical that donates hydrogen ions to other chemicals.

Atom An incredibly small fundamental unit of matter consisting of subatomic particles, such as electrons, protons, and neutrons.

Atomic nucleus The dense, positively charged center of an atom.

Atomic Number The number of protons in the nucleus of the atoms of an element.

Atomic symbol An abbreviation for an element or atom.

Base A chemical that accepts hydrogen ions from other chemicals.

Carbohydrate Organic molecules containing only carbon, hydrogen, and oxygen, including simple sugars and chains of sugars.

Catalyst Any substance that increases the rate of a chemical reaction without itself being consumed by the reaction.

Chemical compound A material in which atoms of different elements are bonded to one another.

Chemical formula A notation that indicates the composition of a compound, consisting of the atomic symbols for the different elements of the compound and numerical subscripts indicating the ratio in which the atoms combine.

Chemical reaction The transformation of one or more chemicals into different chemicals by a rearrangement of atoms.

Covalent compound A substance, such as an element or chemical compound, in which atoms are held together by covalent bonds.

Dissolving The process of mixing a solute in a solvent to produce a homogeneous mixture.

Electron An extremely small, negatively charged subatomic particle found in a cloud outside the atomic nucleus.

Element Any material that is made up of only one type of atom.

Elemental formula The chemical symbol of an element followed by a numeral subscript indicating the number of atoms in each molecule of the element.

Endothermic Description of a chemical reaction in which there is a net absorption of energy.

Exothermic Description of a chemical reaction in which there is a net release of energy.

Hydronium ion A polyatomic ion made by adding a proton (hydrogen ion) to a water molecule.

Hydroxide ion A polyatomic ion made by removing a proton (hydrogen ion) from a water molecule.

Ion An atom having a net electric charge because of either a loss or gain of electrons.

Ionic bond A chemical bond in which there is an electric force of attraction between two oppositely charged ions.

Lipid Hydrophobic organic molecules, many of which include fatty acids as a primary component.

Mixture A combination of two or more substances in which each substance retains its chemical properties.

Molecule A group of atoms held tightly together by chemical bonds.

Neutron A subatomic particle with no electric charge found within the atomic nucleus.

Nucleic acid Organic molecules that are composed of chains of nucleotides and store genetic information.

Organic chemistry The study of carbon-containing compounds.

Oxidation The process whereby a reactant loses one or more electrons.

Periodic table A chart in which all known elements are listed.

pH A measure of the acidity of a solution related to the hydronium-ion concentration.

Polar Description of a chemical bond or molecule that has a dipole. In a polar bond or molecule, electrons are congregated on one side. This makes that side slightly negative, while the opposite side (lacking electrons) becomes slightly positive.

Protein Organic molecules composed of folded chains of amino acids.

Proton A subatomic particle with a +1 electric charge charge found in an atom's nucleus.

Pure The state of a material that consists solely of a single element or compound.

Reaction rate A measure of how quickly the concentration of products in a chemical reaction increases or the concentration of reactants decreases.

Reduction The process whereby a reactant gains one or more electrons.

Solute Any component in a solution that is not the solvent.

Solution A homogeneous mixture in which all components are dissolved in the same phase.

Solvent The component in a solution that is present in the largest amount.

Suspension A homogeneous mixture in which the various components are finely mixed, but not dissolved.

Review Questions

Simple Review Questions

2.1 Atoms and Molecules

1. How many types of atoms can you expect to find in a pure sample of any element?
2. Distinguish between an atom and an element.
3. What is the atomic symbol for the element cobalt?
4. Is it possible to see an atom using visible light?
5. What is at the center of every atom?
6. Are elements made of atoms, or are atoms made of elements?

2.2 Chemical Compounds

7. How does an ion differ from an atom?
8. Does an atom lose or gain electrons to become a negative ion?
9. Which elements tend to form ionic bonds?
10. Which elements tend to form covalent bonds?

2.3 Mixtures

11. What is a dipole?
12. What defines a material as being a mixture?
13. Distinguish between a solute and a solvent.

2.4 Chemical Reactions

14. What generally happens to the rate of a chemical reaction with increasing temperature?
15. What net effect does a chemical reaction have on a catalyst?
16. Why are catalysts so important to our economy?
17. What is released by an exothermic reaction?

2.5 Two Types of Chemical Reactions

18. When an acid is dissolved in water, what ion does the water form?
19. When a chemical loses a hydrogen ion, is it behaving as an acid or a base?
20. What does the pH of a solution indicate?
21. Which elements have the greatest tendency to behave as oxidizing agents?

2.6 Organic Molecules

22. What is an organic molecule?
23. Why is the carbon atom so unique?
24. Why do non-carbon atoms make such a difference in the properties of an organic molecule?

2.7 Macromolecules Needed for Life

25. What are some of the different functions of proteins?
26. Give an example of (a) a carbohydrate that functions in energy storage and (b) a carbohydrate that has a structural function.
27. Describe the structure of DNA.
28. The macromolecules made from folded chains of amino acids are

    (a) proteins.
    (b) carbohydrates.
    (c) lipids.
    (d) nucleic acids.

Challenging Review Questions

2.1 Atoms and Molecules

29. A cat strolls across your backyard. An hour later, a dog with its nose to the ground follows the trail of the cat. Explain what is going on from a molecular point of view.

30. Where did the atoms that make up a newborn baby originate?
31. The atoms that compose your body are mostly empty space, and structures such as the chair you're sitting on are composed of atoms that are also mostly empty space. So why don't you fall through the chair?
32. Why isn't water an element?
33. Name some elements you have access to macroscopic samples of as a consumer here on the Earth.

2.2 Chemical Compounds

34. Which is more polar: a carbon-oxygen bond or a carbon-nitrogen bond?
35. Magnesium ions carry a 2+ charge, and chloride ions carry a 1- charge. What is the chemical formula for the ionic compound magnesium chloride?

36. Barium ions carry a 2+ charge, and nitrogen ions carry a 3- charge. What would be the chemical formula for the ionic compound barium nitride?

37. In each molecule, which atom carries the greater positive charge?

H-Cl Br-F C≡O Br-Br

38. The charges with sodium chloride are all balanced—for every positive sodium ion there is a corresponding negative chloride ion. Since its charges are balanced, how can sodium chloride be attracted to water and vice versa?

2.3 Mixtures

39. Chemists do not consider the air we breathe to be an element. Why not?

40. What happens to the volume of a sugar solution as more sugar is dissolved in it?

41. The below boxes depict a view at the level of atoms and molecules. Which best represents a suspension? Which best represents a solution? Which best represents a compound?

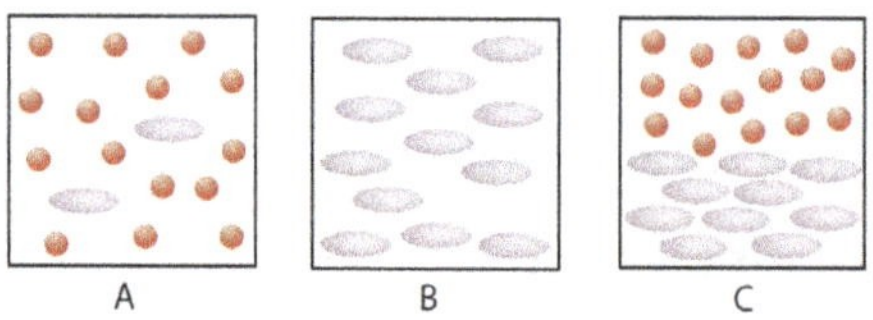

42. Why is the oxygen atom of a water molecule slightly negative in charge?

43. Why is it not practical to have a macroscopic sample that is 100 percent pure?

44. How might you separate a mixture of sand and salt? How about a mixture of iron and sand?

2.4 Chemical Reactions

45. How many oxygen atoms are indicated on the right side of this balanced chemical equation:

$$4\ Cr(s) + 3\ O_2(g) \rightarrow 2\ Cr_2O_3(g)$$

46. Are the chemical reactions that take place in a disposable battery exothermic or endothermic? What evidence supports your answer? Is the reaction going on in a rechargeable battery while it is recharging exothermic or endothermic?

2.5 Two Types of Chemical Reactions

47. What happens to the polarity of oxygen atoms as they transform from molecular oxygen, O_2, into water molecules, H_2O?

48. This reaction between two water molecules is endothermic, which means the reaction requires the input of heat energy in order to proceed:

$$\text{Energy} + H_2O + H_2O \rightarrow H_3O^+ + OH^-$$

The warmer the water, the more heat energy is available for this reaction, and the more hydronium and hydroxide ions are formed. So, which has a lower pH: pure water that is hot or pure water that is cold? Explain.

49. The general chemical equation for photosynthesis is shown below. Through this reaction are the oxygens of the water molecules, H_2O, oxidized or reduced? Explain.

$$6\ CO_2 + 6\ H_2O \rightarrow C_6H_{12}O_6 + 6\ O_2$$

50. Water is 88.88 percent oxygen by mass. Oxygen is exactly what a fire needs to grow brighter and stronger. So why doesn't a fire grow brighter and stronger when water is added to it?

51. Why is the air over an open flame always moist?

2.6 Organic Molecules

52. Organic molecules can be represented by "stick structures," as shown below. At each end and each corner, there is an implied carbon atom. For example, each of the three stick structures shown below consists of four carbon atoms. The stick structure is helpful in showing the basic structure or "framework" of a particular organic molecule. But of the three structures shown below, two are identical molecules. Which two are identical molecules?

53. Warning: This is a super hard question. Remember that carbon-carbon single bonds can rotate while carbon-carbon double bonds cannot rotate. How many different structures are shown below?

2.7 Macromolecules Needed for Life

54. Are you made up of the same macromolecules as a sea urchin? Explain your answer.

55. When birds migrate south for the winter, they may fly hundreds or even thousands of kilometers in a relatively short period of time. Why do birds put on a layer of fat before their annual migration? What would happen if they stored this energy in the form of carbohydrates?

56. Swimming mammals such as sea otters and beavers are often described as having "waterproof" fur that keeps them relatively dry as they swim. However, a close look at their fur coats shows that they have very dense fur. In addition, the fur is covered with an oil that helps repel water. Which macromolecules make up fur and oil and are involved in keeping swimming mammals dry?

57. DNA uses only four different kinds of nucleotides. How can only four nucleotides code for all the different kinds of genes that are found in different living things?

58. Why does an inflated, securely tied rubber balloon slowly deflate over several days?

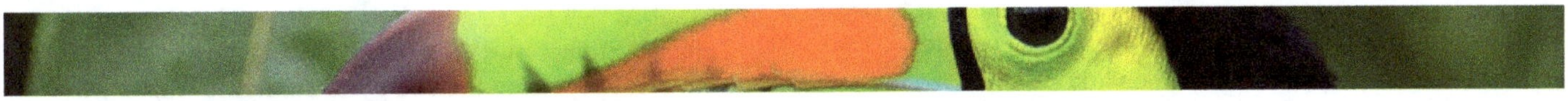

Solutions to Odd-Numbered Questions

Simple Review Solutions

1. An element consists of only one type of atom.

3. As can be found in the periodic table, the atomic symbol for the element cobalt is Co.

5. The atomic nucleus.

7. An ion has a charge and an atom does not.

9. Elements on opposite sides of the periodic table tend to form ionic bonds.

11. A dipole is an uneven distribution of electrons in a chemical bond caused by a difference in the attraction for electrons between two atoms.

13. A solute is the component of lesser quantity in a solution (that being dissolved; e.g., salt in water). A solvent is the component of greater quantity in a solution (e.g., water in salt water solution).

15. A catalyst is unchanged by a chemical reaction.

17. Energy is released by an exothermic reaction.

19. A chemical that loses a hydrogen ion is behaving as an acid.

21. The elements in the upper right of the periodic table have the greatest tendency to behave as oxidizing agents. Not discussed in this chapter is that the elements in the last vertical column (group 18) tend to be inert.

23. Unlike all other elements of the periodic table, carbon is able to form repeated covalent bonds with other carbon atoms. And it can do so in various directions. This makes the carbon atom a good "building block" allowing for complex chemical structures. That carbon is also able to bond with other non-carbon atoms provides for even greater versatility.

25. Proteins perform a wide range of functions in living organisms. The protein keratin provides structure in the form of skin, hair, and feathers. Insulin is a protein that acts as a hormone, allowing one type of cell in the body to communicate with other types. Actin and myosin are proteins that allow muscles to contract. Hemoglobin, a protein found in red blood cells, transports oxygen to body tissues. Antibodies are proteins that protect the body from disease. And proteins known as digestive enzymes break down food during digestion.

27. Nucleic acids are made up of strands of smaller units called nucleotides. A nucleotide includes a sugar molecule, a phosphate group, and a nitrogenous base. DNA consists of two nucleic acid strands twisted into a spiral, which is why it is sometimes called a double helix. Four kinds of nucleotides are found in DNA—adenine, cytosine, guanine, and thymine, or A, C, G, T for short.

Challenging Review Solutions

29. The cat leaves a trail of molecules across the yard. These molecules leave the ground and mix with the air, where they enter the dog's nose, activating the sense of smell.

31. The outsides of the atoms of the chair are made of negatively charged electrons, as are the outsides of the atoms that make up your body. Atoms don't pass through one another because of the repulsive forces that exist between these electrons. When you sit on the chair these repulsive forces hold you up against the force of gravity, which is pulling you downward.

33. Here is a list of sixteen. Aluminum (as in aluminum foil); tin (as in tin foil and tin cans); carbon (as in graphite and diamond); helium (as in a helium balloon); nitrogen (which comprises about 78% of the air we breathe); oxygen (which comprises about 21% of the air we breathe); argon (which comprises about 1% of the air we breathe); silicon (as in integrated circuits for computers and calculators); sulfur (a mineral used for many industrial processes); iron (as in most metal structures); chromium (as in chromium bumpers on cars); zinc (as in the coating of any galvanized nail or as the insides of any post 1982 copper penny); copper (as in copper pennies); nickel (as in nickel nickels); silver (as in jewelry and old silver coins); gold (as in jewelry); platinum (as in jewelry).

35. $MgCl_2$ (Two single negatively charged chlorine ions are needed to balance the one doubly positively charge magnesium ion). A shortcut way of solving these sorts of problems is to take the charge of one ion and make it the subscript of the opposite ion. For example, take the 2+ charge of the magnesium and make it the subscript on the chlorine. Then take the 1- charge on the chlorine and make that the subscript of the magnesium. Because

numeral 1 subscripts are implied when not written, we have not Mg_1Cl_2, but $MgCl_2$.

37. The atoms found closer to the lower left-hand corner of the periodic table are those that will bear the positive charge: a) hydrogen b) bromine c) carbon d) neither!

39. An element is a material made of only one type of atom. Air is a mixture of different kinds of atoms, mostly nitrogen and oxygen.

41. Box "c" best represents a suspension. What may be depicted in "c" is the interface between a solid and liquid phase within a suspension. Box "a" shows oval molecules dissolved within smaller circle molecules, which best illustrates a solution. Box "b" shows a pure material, which may be indicative of a compound.

43. Atoms and molecules are very small so if one atom or molecule out of a trillion is different, then the sample is no longer pure.

45. There are six oxygen atoms on the right hand side of this equation. Note the coefficient of 2 indicates that there are two chromium oxide, Cr_2O_3, molecules.

47. The oxygen atoms become more negatively charged, which is to say they are gaining electrons.

49. Within any water molecule, H_2O, the oxygen carries a slight negative charge. Why? Because this oxygen atom is able to pull the shared electrons away from the hydrogen atoms. Within any oxygen molecule, O_2, the bonding electrons between the two oxygen atoms are evenly shared. Why? Because the two atoms are equally strong at pulling on the electrons. Thus, an oxygen atom moving from being in a water molecule to an oxygen molecule is going to lose its negative charge. The loss of a negative charge is an example of oxidation. So, within photosynthesis, the oxygen is being oxidized. This is unusual. Why? Because oxygen prefers to have a negative charge and here we're talking that negative charge away. How so? By using the energy of sunlight. This is why photosynthesis is an endothermic reaction.

51. One of the products of combustion is water vapor.

53. The second and the fourth structures are the same. In all, there are three different structures shown.

55. Birds need a large energy supply to complete long, strenuous migratory flights. If they store the energy as fat rather than as carbohydrate, they won't weight themselves down as much, since 1 gram of fat contains a lot more energy than 1 gram of carbohydrate.

57. The four nucleotides in DNA can be strung together in a huge number of ways to code for many unique genes. It's a lot like the way you can make lots of different English sentences using only the 26 letters in the alphabet—you can string the letters together in different ways.

Living Earth Essay

Life on Earth Connections

Isotopic Dating

Atoms of the same element having a different number of neutrons are called **isotopes**. We can identify an isotope by the total number of protons and neutrons, which is the **atomic mass number**. For example, all carbon atoms have six protons in the nucleus. Those that also have 6 neutrons have an atomic mass number of 12, which is the sum of 6 protons and 6 neutrons. These carbon atoms are represented as carbon-12. A carbon atom with 8 neutrons, however, is carbon-14. Carbon-12 and carbon-14 are said to be isotopes of each other, as shown in **Figure A**.

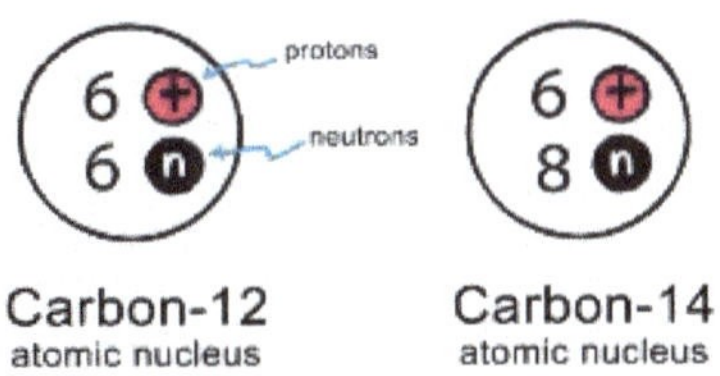

Figure A
The carbon-12 atomic nucleus has 6 protons and 6 neutrons while the carbon-14 nucleus has 6 protons and 8 neutrons. In nature, we find that about 99 percent of all carbon atoms are carbon-12. About 1 percent are the carbon-13 isotope. The carbon-14 isotope is quite rare. Collect about a trillion carbon atoms, and only one of them will be carbon-14.

For some isotopes, the atomic nucleus is unstable. While carbon-12 is quite stable, not so for carbon-14, which is unstable. The instability of carbon-14 results in one of the neutrons transforming into a proton, as shown in **Figure B**. This produces the stable nitrogen-14 isotope as well as an electron that shoots away from the nucleus at high velocities. Like a bullet, that fast-flying electron can be dangerous and is a form of *radioactivity* called beta radiation. For this reason, we say that carbon-14 is a *radioactive isotope*.

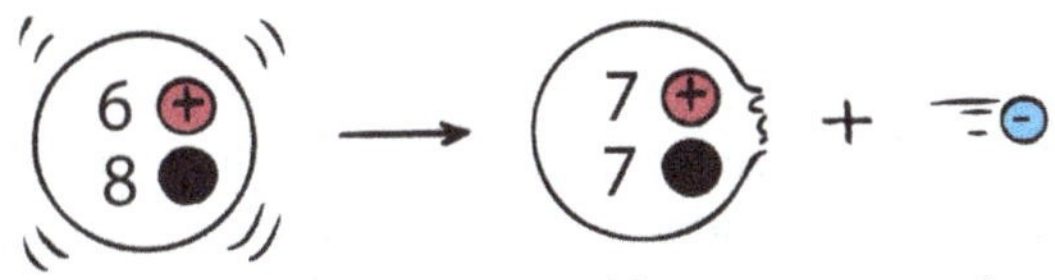

$$^{14}_{6}C \longrightarrow {}^{14}_{7}N + {}^{0}_{-1}e$$

Figure B
The carbon-14 nucleus is unstable and eventually transforms into the nitrogen-14 nucleus while also emitting a fast-flying electron.

There are many radioactive isotopes. Examples include hydrogen-3, technicium-99, and uranium-235. An amazing fact about radioactive isotopes is that when in bulk, the rate at which they release radioactivity is extremely consistent. Consider carbon-14. If you had 1 kilogram of carbon-14, given its rate of radioactivity, then after 5730 years, you would only have 0.5 kilograms of carbon-14 remaining. Why? Because half of that carbon-14 has been transformed into non-radioactive nitrogen-14.

The time it takes for half of a radioactive isotope to transform is called the radioactive half-life. **Table A** shows a variety of examples.

Table A

Isotope	Half-Life
Francium-209	0.05 seconds
Technetium-99m	6 hours
Polonium-210	138 days
Radium-226	1600 years
Carbon-14	5730 years
Uranium-238	4.5 billion years

If an isotope is super radioactive this means it will quickly disappear. And this means its half-life will be quite short. Thus, the shorter the half-life, the greater the radioactivity. In other words, it would be far safer to be exposed to a kilogram of uranium-235 than a kilogram of francium-209.

Radioactive isotopes have found many uses. Technicium-99, for example, is used for medical imaging. Within science, radioactive and non-radioactive isotopes are remarkable in helping us understand the past. Let's look at some examples.

Cosmic rays continually bombard Earth's atmosphere resulting in a "spraying out" of protons and neutrons into the environment. When one of the neutrons collides within a nitrogen atom, it can result in the radioactive carbon-14 isotope:

As a result, one-millionth of 1 percent of the carbon in the atmosphere is this radioactive carbon-14. Because carbon-12 and carbon-14 are both forms of carbon, they have the same chemical properties. Both of these isotopes, for example, chemically react with oxygen to form carbon dioxide, which is consumed by plants through the process of photosynthesis. This means that all plants contain a tiny quantity of radioactive carbon-14.

All animals eat either plants or plant-eating animals; therefore, all animals, including humans, have a little carbon-14 in them. So, we see why all living things on Earth contain some carbon-14.

Because plants absorb carbon dioxide while alive, any carbon-14 that decays is immediately replenished with fresh carbon-14 from the atmosphere. In this way, a radioactive equilibrium is reached in which there is a constant ratio of about one carbon-14 atom to every 100 billion carbon-12 atoms. When a plant dies, replenishment of carbon-14 ends. Then the percentage of carbon-14 decreases at a constant rate determined by its half-life. The amount of carbon-12, however, does not change because this isotope does not undergo radioactive decay. The longer a plant or other organism is dead, the less carbon-14 it contains relative to the constant amount of carbon-12.

Knowing that the half-life of carbon-14 is 5730 years, scientists are able to calculate the age of carbon-containing artifacts, such as wooden tools or the skeleton shown in **Figure C**. Their age is measured by their current level of radioactivity. This process, known as **carbon-14 dating**, enables investigators to probe as far as 50,000 years into the past. Beyond this time span, there is too little carbon-14 remaining for accurate analysis.

Because of fluctuations in cosmic ray bombardment rates over the centuries, carbon dating has an uncertainty of about 15 percent. This means, for example, that the straw of an old adobe brick dated to be 500 years old may really be only 425 years old on the low side or 575 years old on the high side. For many purposes, this is an acceptable level of uncertainty.

Figure D
Carbon-14 dating was developed by the American chemist Willard F. Libby (1908–1980) at the University of Chicago in the 1950s. For this work, he received the Nobel Prize in 1960.

READING CHECK

Plants contain a little carbon-14, but why do humans also contain carbon-14?

Figure C
The amount of carbon-14 in the skeleton is reduced by one-half every 5730 years. The result is that the same skeleton today contains only a trace amount of the original carbon-14. The blue arrows represent the relative amounts of carbon-14.

CONCEPT CHECK

Suppose that an archaeologist extracts 1 gram of carbon from an ancient ax handle and finds that it is one-fourth as radioactive as 1 gram of carbon extracted from a freshly cut tree branch. About how old is the ax handle?

CHECK YOUR ANSWER

Assuming the ratio of C-14/C-12 was the same when the ax was made, the ax handle is as old as two half-lives of C-14, or about 11,460 years old.

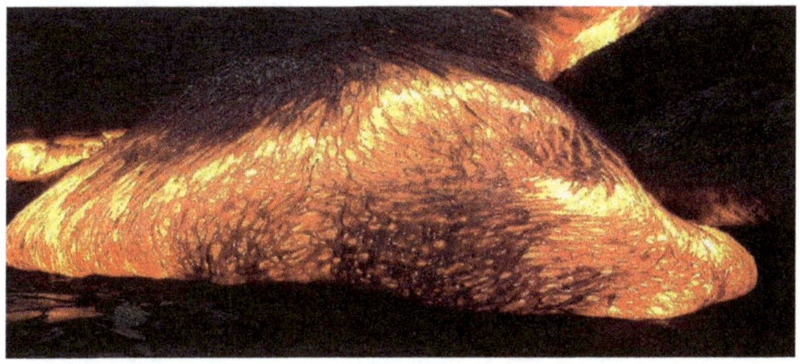

Figure E

When a rock forms, say from molten material, the proportion of isotopes it contains becomes set. Over time, this proportion of isotopes changes as radioactive isotopes decay. The current proportion of isotopes, therefore, is a function of how long ago the rock formed.

Artifacts derived from nonliving materials can also be dated based on the radioactive minerals they contain. The naturally occurring mineral isotopes uranium-238 and uranium-235, for example, decay very slowly and ultimately become lead—but not the common isotope lead-208. Instead, uranium-238 decays to the isotope lead-206. Uranium-235, on the other hand, decays to lead-207. Thus, the lead-206 and lead-207 that now exist in a uranium-bearing rock were at one-time uranium. Older rocks contain higher percentages of these trace isotopes.

If you know the half-lives of uranium isotopes and the percentage of lead isotopes in some uranium-bearing rock, you can calculate the date of the rock's formation. Rocks dated in this manner have been found to be as much as 3.7 billion years old. Samples from the Moon have been dated at 4.2 billion years, which is close to the estimated age of our solar system: 4.6 billion years.

Isotopes need not be radioactive to be useful in helping us to understand the past. Consider hydrogen atoms have 1 proton in the nucleus. Most isotopes of hydrogen, H-1, have zero neutrons. However, about 1 out of every 6400 hydrogen atoms is the deuterium isotope, H-2, which has 1 proton and 1 neutron. Neither of these isotopes is radioactive. However, it should be clear that the H-2 isotope is twice as heavy as the H-1 isotope.

Water molecules that contain H-2 rather than H-1 are heavier, making them more difficult to evaporate from a liquid to a gas. This difference provides us with a handle on past climates. When it's warmer, the fraction of "heavy water" that evaporates from the ocean is greater. The resulting precipitation, including snow, is thus also richer in heavy water. Ice caps, such as the one over Greenland, are the result of snow accumulations over millions of years. Studying the ratio of heavy water within ice cores thus provides a way to assess past global temperature changes and hence, past climates (**Figure F**).

CONCEPT CHECK

What role have isotopes played in helping us to understand the past.

CHECK YOUR ANSWER

Isotopes have played a central role in helping us to understand the past. Radioactive isotopes allow for radioactive dating of materials once-living and not once-living. Even non-radioactive isotopes provide us valuable clues to the past by the ratios in which they occur.

To Life!

a. h

Figure F

a) Greenland as seen from space is covered with an extensive ice sheet up to three kilometers thick. b) An ice core obtained by drilling deep into the Greenland ice sheet contains isotopic evidence from past climates.

Chapter 3

The Cell

THE MAIN IDEA

All living things are made of one or more units called cells.

3.1 What is a Cell?
3.2 Cell Theory
3.3 Looking at Cells
3.4 Eukaryotic Cell
3.5 The Cell Membrane
3.6 Cell Organelles

What is a cell? Are all cells similar, or are there differences between the cells of bacteria, animals, and plants? What are the two main types of cells? Do all cells come from other cells? Can we see cells with our eyes alone? What about with a microscope? What is inside a cell? In this chapter, we will explore the nature of life and the world of the cell.

3.1 What Is a Cell?

Cells are the basic units of life, the same way that atoms are the basic units of matter. All living organisms are made up of one or more cells. Organisms made up of just one cell are *unicellular*—bacteria are unicellular. Organisms made up of many cells are *multicellular*.

You, yourself, have about 37 trillion cells in your body, and the diversity of things they do is amazing. Right now, muscle cells are moving your eyeballs as you follow the text on this page, sensory cells in your eyes are taking in the shapes of the letters, cells in your ears are absorbing nearby sounds, red blood cells are carrying oxygen to all the other cells in your body, and digestive cells are making the enzymes that will break down your last snack. Furthermore, some extremely impressive cells—the neurons in your brain—are producing your thoughts about how amazing cells are.

The Two Types of Cells

Two distinct types of cells are found in different living organisms today: prokaryotic cells and eukaryotic cells. They are distinguished primarily by the presence or absence of a **nucleus**, a structure within the cell that contains the cell's DNA. *Prokaryotic cells* do not have a nucleus (*pro* means "before," and *karyote* refers to "nut" or "nucleus"). *Eukaryotic cells* ("true nucleus") have a nucleus as well as other structures not present in prokaryotic cells. Organisms with prokaryotic cells are called **prokaryotes**, and organisms with eukaryotic cells are called **eukaryotes**. **Figure 3.1** compares typical prokaryotic and eukaryotic cells.

READING CHECK

How do prokaryotic cells differ from eukaryotic cells?

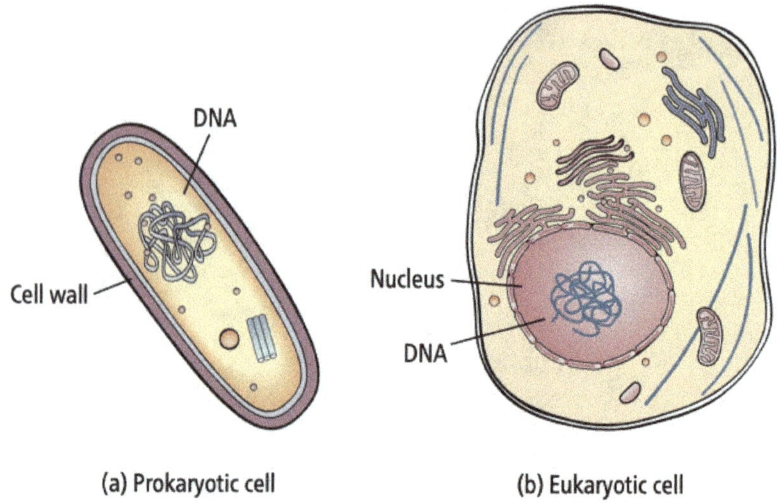

Figure 3.1 >

(a) Prokaryotic cells have no nucleus.
(b) Eukaryotic cells have a nucleus.

Prokaryotes have existed on Earth far longer than eukaryotes. Prokaryotes first evolved 3.5 to 4 billion years ago and were the only living things on Earth for several billion years. Prokaryotes now include two major lineages, the bacteria and the archaea.

Prokaryotes are single-celled organisms and are very small, ranging from about 0.1 to 10 micrometers (10^{-6} meter) in diameter. Their structure is simpler than that of eukaryotes. The DNA of prokaryotes is found in a single circular structure and is not contained within a nucleus. Most prokaryotes have an outer *cell wall* that helps protect the cell. The prokaryote *Escherichia coli*, which lives in the human digestive tract and is one of the best-studied organisms in the world, is shown in **Figure 3.2**.

Eukaryotes first appeared on Earth about 2 billion years ago. Eukaryotes can be single-celled, like prokaryotes, or they can be composed of many cells. The fungus known as baker's yeast, commonly used in baking and brewing, is a single-celled eukaryote (**Figure 3.3**). Humans are multicellular eukaryotes.

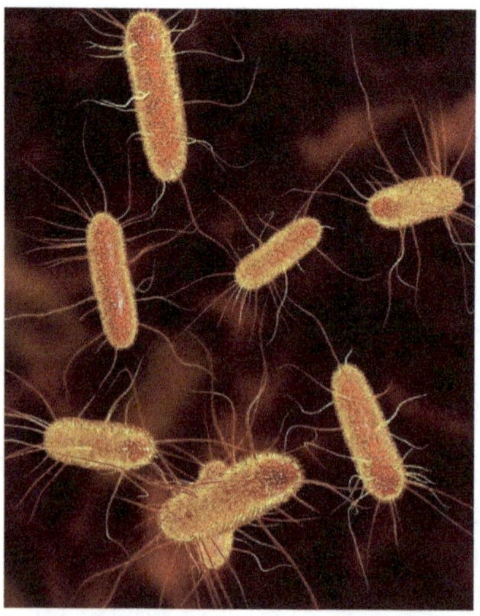

∧ Figure 3.2

Escherichia coli (commonly referred to as E. coli) is a prokaryote that lives in the human digestive tract.

∧ Figure 3.3

Saccharomyces cerevisiae, commonly known as baker's yeast or brewer's yeast, is a single-celled eukaryote.

Eukaryotes include all animals, plants, fungi, and protists. Eukaryotic cells have their DNA inside a nucleus. In addition, the DNA of eukaryotic cells is found in linear, rather than circular, **chromosomes**. Eukaryotic cells also have numerous **organelles**, structures that perform specific functions for the cell. Finally, eukaryotic cells are larger than prokaryotic cells—whereas prokaryotic cells measure 0.1 to 10 micrometers, eukaryotic cells usually measure 10 to 100 micrometers.

CONCEPT CHECK

Which of the following organisms are prokaryotes and which are eukaryotes: the bacterium that causes tuberculosis, a humpback whale, a honey mushroom?

CHECK YOUR ANSWER

The tuberculosis bacterium, like all bacteria, is a prokaryote. A humpback whale, like all animals, is a eukaryote. A honey mushroom, like all fungi, is a eukaryote.

More to Explore

To learn more about prokaryotic and eukaryotic cells:

https://www.livescience.com/65922-prokaryotic-vs-eukaryotic-cells.html

3.2 Cell Theory

Cell theory is the idea that the cell is the basic unit of life. This theory was several centuries in the making. It started in 1665, when Robert Hooke, an English scientist, coined the term *cell* and published the first description of cells in his book *Micrographia*. Hooke examined a piece of cork under a microscope and saw a series of small boxlike chambers (**Figure 3.4**). He called these chambers "cells" because they reminded him of monks' cells, which were small chambers where each monk could worship alone. We now know that the chambers Hooke saw were the dead cell walls that remain in dried plant matter.

READING CHECK

Who coined the scientific term cell and when?

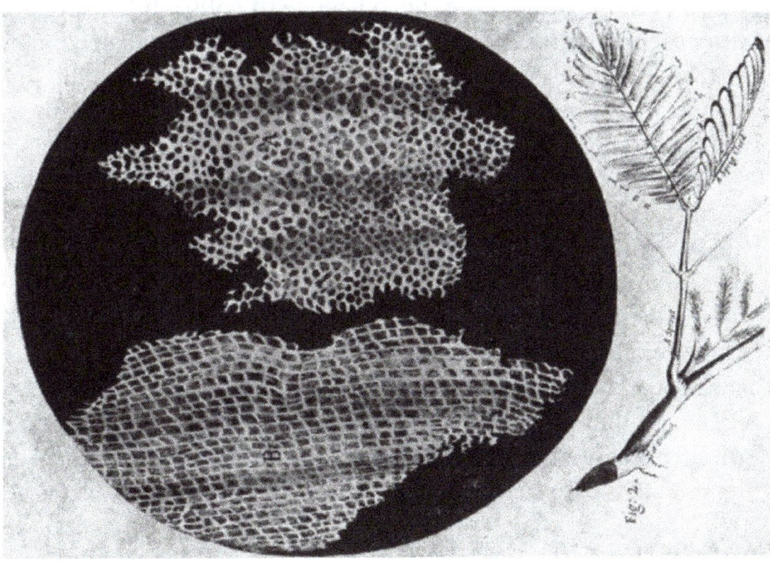

< Figure 3.4

Robert Hooke examined cork under a microscope and called the small chambers he saw "cells." This is Hooke's original drawing of what he saw.

It was not until the 1800s that the central importance of the cell was established. In 1838, studies of plants led German scientist Matthias Schleiden to conclude that all plants are made of cells. The following year, another German scientist, Theodor Schwann, came to the same conclusion about animals. The cell theory was finally completed in 1855 when German scientist Rudolph Virchow observed that all living cells come from other living cells.

In summary, the cell theory says:

1. All living things are made up of one or more cells.
2. All cells come from other cells.

CONCEPT CHECK

Rather than "cell theory," why not use the phrase "cell hypothesis"?

CHECK YOUR ANSWER

As discussed earlier in Chapter 1, a hypothesis is an educated guess at what might be true. Also, a hypothesis can be likened to the phrase "let us suppose." A theory, by contrast, is an overarching idea that explains much about the natural world. That all living organisms are made of cells is quite the generalization, which elevates this idea to a "theory".

More to Explore

To learn more about the history of cell theory, check out this TED-Ed video:
https://www.youtube.com/watch?v=4OpBylwH9DU

3.3 Looking at Cells

How can we look at cells? Microscopes are high-tech magnifying glasses that allow us to see very small objects with a fine level of detail. One type of microscope, the light microscope, has been around for centuries. In fact, Robert Hooke discovered the existence of cells while using one. A **light microscope** works by passing visible light through a specimen and then through a series of lenses. The lenses refract, or bend, the light in order to produce a magnified image of the specimen (**Figure 3.5a**).

Light microscopes are able to resolve objects on the order of a micrometer (10^{-6} meter) in size. This means that two lines closer together than 10^{-6} meter appear as a single line. (The resolving power of the human eye is about 1 to 10 millimeters, or 10^{-4} meter.) With a resolving power of 10^{-6} meter, light microscopes allow us to view cells and to make out the larger features inside them, such as the nucleus. However, they do not really allow us to see smaller cellular structures in detail.

READING CHECK

Is it possible to see cells with a light microscope?

To get around this problem, scientists illuminate very tiny objects with electron beams rather than with light. In an **electron microscope**, electric and magnetic fields, rather than optical lenses, are used to focus electron beams (**Figure 3.5b**). Electron microscopes are able to resolve objects about

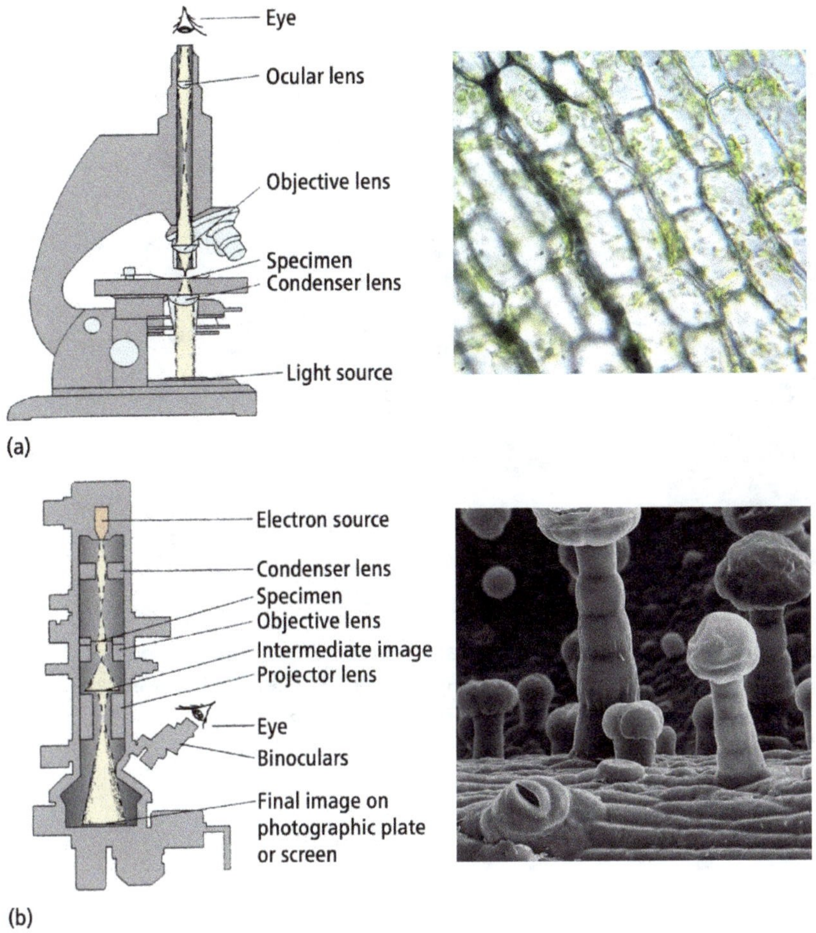

< Figure 3.5

Microscopes enable us to examine objects that are too small for the human eye to see. a) Light microscopes use glass lenses to bend light and magnify specimens. The photo shows plant cells seen through a light microscope. The small green bits inside the cells are plant structures called chloroplasts. (b) Electron microscopes use electric and magnetic fields to magnify specimens. This is an electron microscope photo of the surface of a walnut leaf. The opening of a stoma can be seen on the lower left.

a nanometer (10^{-9} meter) in size, which covers just about everything of biological interest (**Figure 3.6**).

There are two types of electron microscopes. Scanning electron microscopes (SEM) create a three-dimensional image of the surface of a specimen (see the photo in Figure 3.5b). Transmission electron microscopes (TEM) make images of thin sections through a specimen (**Figure 3.7**). To produce a TEM image, the specimen is treated with a preservative, fully dehydrated, and then set within a hard plastic. Ultra-thin slices of the specimen embedded within the plastic are then stained with a lead solution. The lead atoms bind to the structures of the specimen, which allows these structures to interact better with the electron waves. This increases the contrast of the final image providing stunning detail.

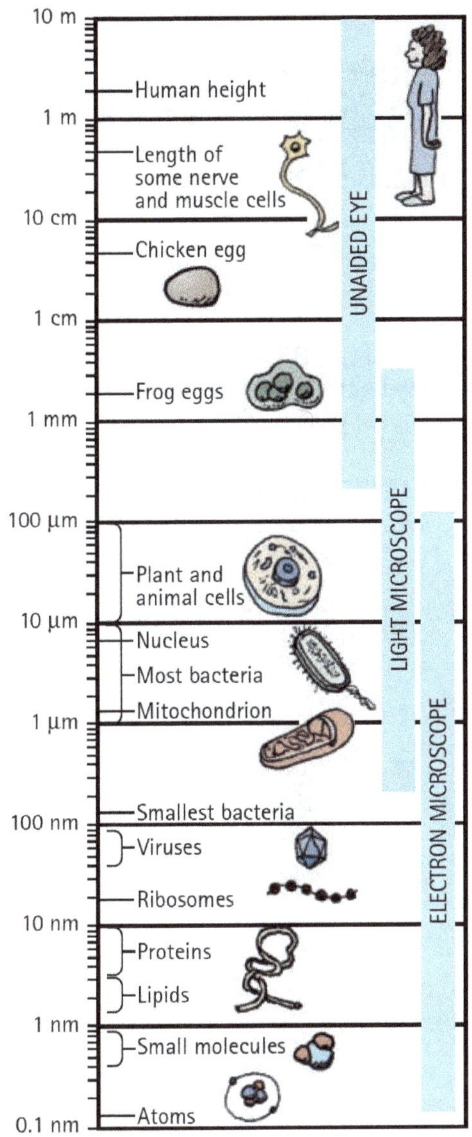

Measurement equivalents

1 meter (m) = 100 cm = 1,000 mm = about 39.4 inches
1 centimeter (cm) = 10^{-2} (1/100) meter (m) = about 0.4 inch
1 millimeter (mm) = 10^{-3} (1/1,000) m = 1/10 cm
1 micrometer (µm) = 10^{-6} m
1 nanometer (nm) = 10^{-9} m

Figure 3.6 >

Microscopes allow us to look into the world of the cell. Depending on the size of a feature, the naked eye, a light microscope, or an electron microscope may be used.

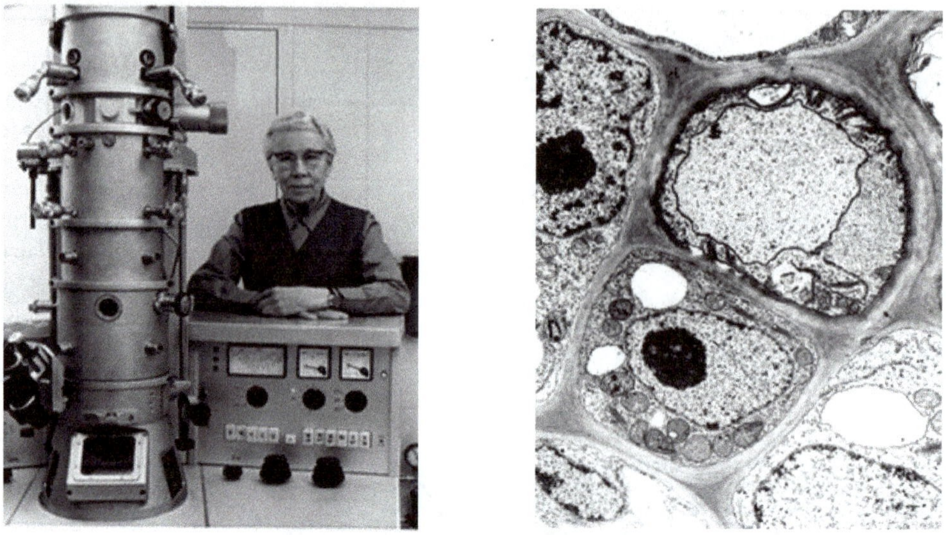

^ Figure 3.7

Katherine Esau of the University of California Santa Barbara was an early pioneer in the use of TEM for biological research. For example, she studied a special type of cell, called a *sieve element*, that conducts food within plants. She also used a TEM to reveal the interactions between plant cells and viruses. Source: University of California, Santa Barbara.

CONCEPT CHECK

Tiny bacteria called mycoplasmas have diameters as small as 0.1 micrometer. Can they be studied with light microscopes? With electron microscopes?

CHECK YOUR ANSWER

Mycoplasmas are too small to be examined with light microscopes, but they can be studied with electron microscopes.

More to Explore

To learn more about the history of electron microscopy, go to:

http://www.physics.emory.edu/faculty/weeks/lab/papers/bogner-micron07.pdf

You can also read more about how electron microscopes work here:

https://www.explainthatstuff.com/electronmicroscopes.html

3.4 Eukaryotic Cells

You, the tree outside, and your dog are all eukaryotes, organisms composed of eukaryotic cells. In fact, most of the living things we encounter on a daily basis—all the plants, animals, and fungi—are eukaryotes.

Two typical eukaryotic cells are shown in **Figure 3.8**, one animal cell and one plant cell. The first thing you might notice is that all eukaryotic cells are surrounded by a cell membrane. The **cell membrane** separates the inside of the cell from the outside and is responsible for controlling what goes into and out of the cell. Plant cells also have a rigid cell wall outside the cell membrane that is made of cellulose and other materials. The cell wall helps to protect and support the cell. All eukaryotic cells also have a *nucleus*, a structure within the cell that contains the cell's DNA. The nucleus is surrounded by a double membrane.

The portion of the cell that is inside the cell membrane but outside the nucleus is called the cell's **cytoplasm**. The cytoplasm is crisscrossed by fibers of the *cytoskeleton*. The cytoskeleton helps the cell hold its shape. The cytoplasm of eukaryotic cells also contains many structures called *organelles*. These structures are called organelles because, like the organs of the body, each performs a specific function in the cell. Each organelle is surrounded by a membrane and is attached to the cytoskeleton. Some of the different organelles are shown in Figure 3.8. All these structures working in unison are what bring the cell to life.

READING CHECK

Name a feature that all eukaryotic cells have in common?

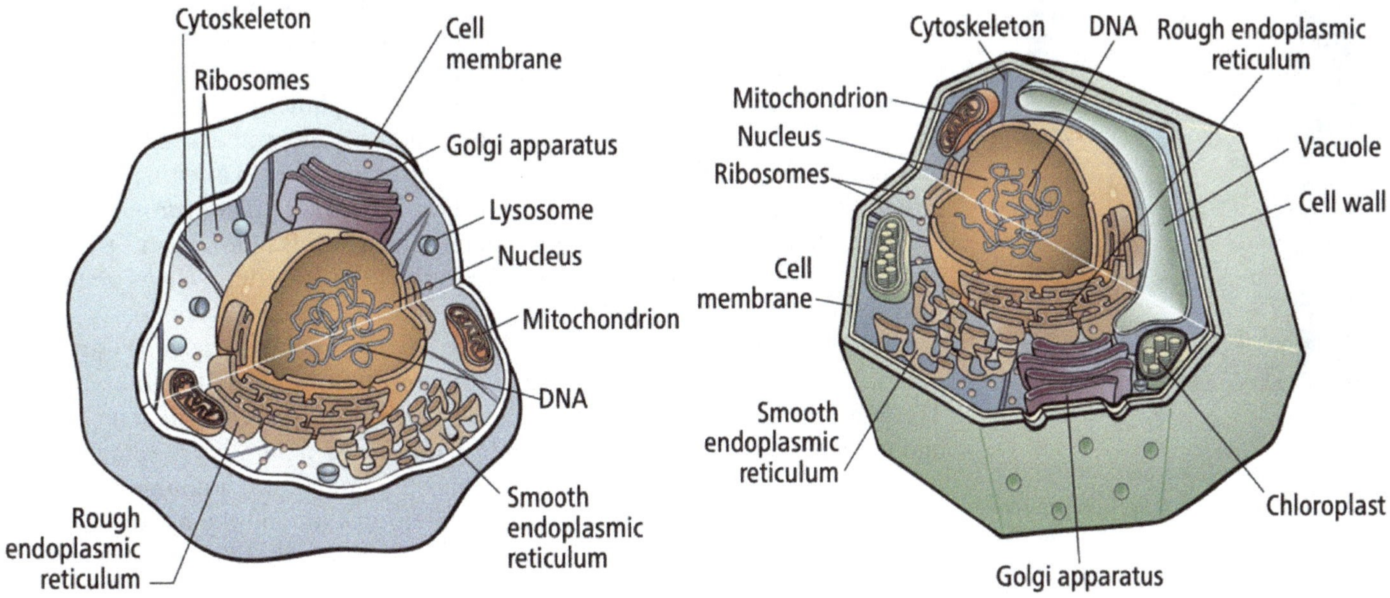

(a) An animal cell

(b) A plant cell

Figure 3.8

Eukaryotic cells have a cell membrane, a nucleus, and many different organelles. (a) This is a typical animal cell. (b) This is a typical plant cell. Plants cells have a cell wall outside the cell membrane.

CONCEPT CHECK

What is the function of the plant cell wall?

CHECK YOUR ANSWER

The cell wall of plant cells provides a rigidity that helps maintain the cell's shape and contributes to the building of the plant's structure.

More to Explore

For a virtual reality tour of a human cell:

http://naturedocumentaries.org/17217/virtual-tour-cell-xvivo-scientific-animation-2018/

3.5 The Cell Membrane

Figure 3.9

When oil and water are combined, they do not mix. Instead, there is a distinct layer of oil and a separate layer of water.

The cell membrane, which is found in both animal and plant cells, defines a cell's boundary, separating the inside of the cell from the outside.

One of the cell membrane's primary functions is to serve as a gatekeeper, controlling what goes into and out of the cell. Let's look at its structure to see how the cell membrane performs this function. The three primary components of the cell membrane are phospholipids, proteins, and short carbohydrates.

Phospholipids are molecules that are part hydrophilic and part hydrophobic. What does this mean? Have you noticed how oil and water separate after they have been combined? The oil floats on top of the water in a distinct layer, rather than mixing with it (**Figure 3.9**). This is because oil is *hydrophobic*, or insoluble in water. (*Hydrophobic* literally means "afraid of water.") The opposite of hydrophobic is *hydrophilic*, or soluble in water. (*Hydrophilic* literally means "loves water.")

Phospholipids have hydrophilic "heads" and hydrophobic "tails" (**Figure 3.10**). This determines how they orient themselves in the cell membrane. The hydrophilic heads are naturally drawn to the watery environment inside and outside the cell, whereas the hydrophobic tails naturally try to avoid this environment. The result is that the phospholipids form a double layer, or *bilayer*, with the hydrophobic tails pointing inward and the hydrophilic heads pointing outward (**Figure 3.11**). You can think of the phospholipids as making a sandwich, with the heads as two slices of bread and the tails as the peanut butter inside.

The cell membrane also includes a large number of *membrane proteins* embedded like straws through the phospholipid sandwich. Membrane proteins serve a variety of functions: They help cells communicate with other cells, control transport into and out of cells, and join cells to one another. Because different cells have different functions, membrane proteins also differ from one cell type to another.

READING CHECK

What is the molecular structure of a phospholipid?

Short carbohydrates are attached to the membrane proteins and phospholipids on the outside surface of the cell. These carbohydrates play an important role in cell recognition, the ability to distinguish one type of cell from another. For example, certain immune system cells use these short carbohydrates to identify foreign cells, such as disease-causing bacteria.

Because the cell membrane includes a mosaic of phospholipids and proteins, and because the phospholipids and many membrane proteins slide freely around the cell surface, the cell membrane is often described as a *fluid mosaic.* The fluidity of the cell membrane is essential to its ability to control the movement of materials into and out of the cell.

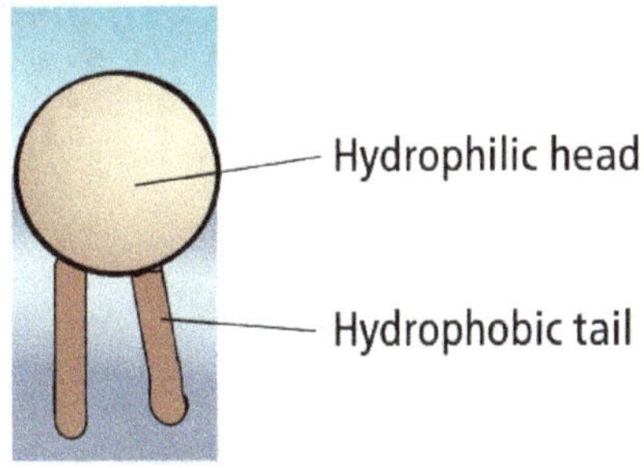

^ Figure 3.10

A phospholipid has a hydrophilic head and hydrophobic tails.

CONCEPT **CHECK**

What role do phospholipids play in the structure of a cell?

CHECK YOUR ANSWER

The phospholipids are molecules that are hydrophilic on one end and hydrophobic on the other. Within a solution of water, they'll naturally organize themselves into a bilayer where the hydrophilic regions face outward into the water while the hydrophobic regions are sandwiched inside. This bilayer creates the basic structure of the cell's membrane.

You can read more about the cell membrane at this website:
https://www.genome.gov/genetics-glossary/Cell-Membrane

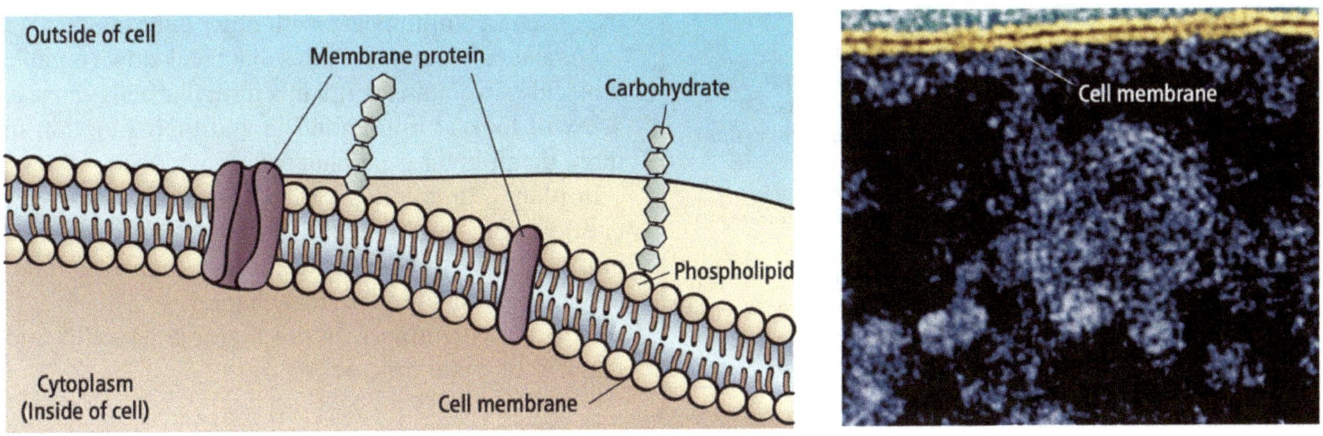

(a) (b)

^ Figure 3.11

(a) The cell membrane includes a phospholipid bilayer, with hydrophobic tails pointed inward and hydrophilic heads pointing toward the inside and outside of the cell. Proteins are embedded in the membrane, and short carbohydrates are attached to the outside of the membrane. (b) This photograph of a cell membrane shows the double layer of phospholipids. The hydrophilic heads are shown in yellow and the hydrophobic tails are shown in red.

3.6 Cell Organelles

More to Explore

More to explore about organelles:

https://microbenotes.com/cell-organelles/

READING CHECK

What organelle in human liver cells detoxifies drugs and other poisons?

The cytoplasm of eukaryotic cells contains many *organelles* that are attached to the cytoskeleton. You can see cell organelles in the cell diagrams in Figure 3.8. These are called "organelles" because, like the organs of the body, each performs a specific function in the cell. Each organelle, except for the ribosomes, is also surrounded by a membrane. Let's consider the organelles one at a time.

Ribosomes are organelles that assemble proteins. Some ribosomes are suspended in the fluid of the cytoplasm. These make proteins that will remain inside the cell.

Other ribosomes are attached to an organelle called the *rough endoplasmic reticulum*. These ribosomes assemble proteins that will go to the cell membrane or be exported from the cell. The rough endoplasmic reticulum appears "rough" because of the ribosomes embedded within it.

The *smooth endoplasmic reticulum* assembles membranes and may have additional functions depending on the cell. For example, the smooth endoplasmic reticulum of liver cells detoxifies drugs and other poisons.

The *Golgi apparatus* is sometimes described as the "post office" of a cell. It receives products from the endoplasmic reticulum, modifies them, and packages them for transport within or out of the cell.

Lysosomes are the garbage disposals of a cell. These organelles break down organic materials, such as damaged or worn-out organelles. Certain white blood cells of the immune system use lysosomes to destroy the bacteria they have engulfed.

Vacuoles are sacs surrounded by a membrane. Plant cells usually have a single large vacuole that can be used to store nutrients or other materials. In flowers, vacuoles store the pigments that provide color. Animal cells typically have smaller vacuoles, sometimes called *vesicles*, that are used to hold or transport a wide array of products. For example, neurons have many vesicles that contain the chemicals they use to communicate with other neurons.

Mitochondria are organelles that break down organic molecules to obtain energy in a form that cells can use. We will look at mitochondria and their function in more detail in the next chapter.

In plants, organelles called *chloroplasts* capture energy from sunlight and use it to build organic molecules. We will study this process, known as "photosynthesis," also in the next chapter.

Table 3.1 summarizes the major organelles and features of eukaryotic cells.

Table 3.1 Major Features of Eukaryotic Cells.

Nucleus	Contains the cell's DNA
Ribosome	Assembles proteins for the cell
Rough endoplasmic reticulum	Assembles proteins destined either to go to the cell membrane or to leave the cell
Smooth endoplasmic reticulum	Assembles membranes and performs other functions in specific cells
Golgi apparatus	Receives products from the endoplasmic reticulum and packages them for transport
Lysosome	Breaks down organic material
Mitochondrion	Obtains energy for the cell to use
Chloroplast	In plant cells, captures energy from sunlight to build organic molecules
Cytoskeleton	Helps cell hold its shape

CONCEPT **CHECK**

What are two places in which ribosomes are found in a cell?

CHECK YOUR ANSWER

Some ribosomes are found in the cytoplasm—these ribosomes assemble proteins that will remain in the cell. Other ribosomes are embedded in the rough endoplasmic reticulum—these ribosomes assemble proteins that will be sent to the cell membrane or exported by the cell.

Concept Review

Chapter 3

Detailed Chapter Summary

A cell is the basic unit of life. Life on Earth includes two types of organisms, prokaryotes and eukaryotes. Prokaryotes such as bacteria and archaea are made up of prokaryotic cells. Eukaryotes, which include plants, animals, fungi, and a diverse group called protists, are made up of eukaryotic cells. Prokaryotic cells have existed far longer than eukaryotic cells. Prokaryotic cells are also smaller and have their DNA in a single circular chromosome. Eukaryotic cells are larger and have a special structure called a nucleus that contains DNA. The DNA of eukaryotes is found in linear chromosomes. Eukaryotic cells also have a number of organelles that are not found in prokaryotes.

Cell theory emphasizes that cells are the basic units of life. Cell theory tells us two things: First, all living things are made up of one or more cells. And second, that all cells come from other cells.

Microscopes allow us to see into the world of the cell. Light microscopes allow us to see cells and the larger features within them. Electron microscopes are needed to obtain detailed views of the structures inside cells. Scanning electron microscopes create a three-dimensional image of the surface of a specimen, whereas transmission electron microscopes make images of thin sections through a specimen.

Eukaryotic cells have a number of key features. They have a cell membrane that separates the inside of the cell from the outside of the cell. Plant cells also have a cell wall outside the cell membrane. Eukaryotic cells have a nucleus that holds their DNA. The portion of the cell between the nucleus and cell membrane is called the cytoplasm, and it is supported by the criss-crossing fibers of the cytoskeleton. Organelles of eukaryotic cells include ribosomes, the rough endoplasmic reticulum, the smooth endoplasmic reticulum, the Golgi apparatus, lysosomes (in animal cells only), mitochondria, and chloroplasts (in plant cells only).

The cell membrane is composed of phospholipids, membrane proteins, and short carbohydrates. The phospholipids are arranged in a bilayer with their hydrophilic heads pointing to the outside and inside of the cell and the hydrophobic tails sandwiched between. Membrane proteins serve a variety of functions, including cell communication and the transport of molecules into and out of the cell. Short carbohydrates play a role in cell recognition. The cell membrane is often described as a fluid mosaic.

Key Terms

Cell membrane The structure that separates the inside of the cell from the outside of the cell.

Cells The basic units of life that make up all living organisms.

Chromosomes The DNA-containing structures within cells.

Cytoplasm The part of a cell inside the cell membrane but outside the nucleus.

Electron Microscope A device able to resolve structures down to a nanometer in size by using electron waves as a source of illumination rather than light.

Eukaryotes Organisms whose cells have a true nucleus, such as protists, animals, plants, and fungi.

Light microscope A device that produces a magnified image of a specimen by passing visible light through a specimen and then through a series of lenses.

Nucleus A structure within eukaryotic cells that is surrounded by a double membrane and that contains the cell's DNA.

Organelles Structures within the cytoplasm of eukaryotic cells that perform specific functions in the cell.

Prokaryotes Single-celled organisms, such as bacteria and archaea, whose cells lack a nucleus.

Review Questions

Simple Review Questions

3.1 What Is a Cell?

1. What are some examples of prokaryotes? What are some examples of eukaryotes?
2. Describe two or more differences between prokaryotic cells and eukaryotic cells.
3. How is the DNA of prokaryotes packaged differently from the DNA of eukaryotes?

3.2 Cell Theory

4. Why are cells called "cells"?
5. What are the two main ideas in cell theory?

3.3 Looking at Cells

6. Why are light microscopes of limited use to cell biologists?
7. Why are electron microscopes particularly useful to cell biologists?

3.4 A Closer Look at Eukaryotic Cells

8. What is the nucleus of a cell?
9. Which structure separates the inside of the cell from the outside?
10. Describe the cytoskeleton and the organelles, which are both found in the cytoplasm of cells.

3.5 The Cell Membrane

11. What are three components of the cell membrane?
12. How are phospholipids arranged in the cell membrane?
13. What are some of the functions of membrane proteins?

3.6 Cell Organelles

14. What is the function of mitochondria?
15. What is the function of lysosomes?
16. What is the function of chloroplasts?

Challenging Review Questions

3.1 What Is a Cell?

17. Are your cells more like those of yeast or those of bacteria? Defend your answer.
18. You look at a cell under a microscope and discover that the cell has mitochondria and chloroplasts. Is the cell a eukaryotic cell or a prokaryotic cell? Is the cell from a plant or from an animal? Defend your answers.

3.2 Cell Theory

19. Viruses are pieces of genetic material wrapped in a protein coat. Are viruses made up of cells? Based on what you know about cell theory, are viruses living things?

3.3 Looking at Cells

20. You hike near a pond, where you find strands of developing frog eggs. Each of the eggs is about 1 to 2 mm in diameter. You decide you want to study the eggs as they develop into tadpoles. What kind of microscope could you use?
21. What kind of microscope could you use to look at bacteria?
22. What kind of microscope could you use to look at viruses?

3.4 A Closer Look at Eukaryotic Cells

23. How are a cell's cytoskeleton and organelles like a person's skeleton and organs?
24. How is a cell wall's function different from a cell membrane's function? Do you have cell walls, cell membranes, both, or neither?

3.5 The Cell Membrane

25. Explain why phospholipids form a double layer in the cell membrane. Be sure to use the words hydrophilic and hydrophobic in your answer.
26. Why is the cell membrane called a fluid mosaic?

3.6 Cell Organelles

27. What organelle is found only in plants? What does this organelle do? Does this explain why animals have to eat, but plants don't?
28. Some cells in the body have more mitochondria than others. For example, nerve, muscle, and liver cells have lots of mitochondria. On the other hand, bone cells and fat cells have very few mitochondria. What do you think explains this difference?

Apply & Discuss Questions

29. A friend in your class is reading about biological cells that store energy in the form of tiny lipid droplets. He says, "It says here that the membranes around the droplets are just a single phospholipid layer instead of a double layer like everywhere else. That doesn't make sense, does it?" What is your reply?

Solutions to Odd-Numbered Questions

Simple Review Solutions

1. Prokaryotes now include two major lineages, the bacteria and the archaea. Eukaryotes include all animals, plants, fungi, and protists.

3. Eukaryotic cells have their DNA in a distinct nucleus, a feature that distinguishes them from prokaryotes. In addition, the DNA of eukaryotic cells is found in linear, rather than circular, chromosomes.

5. Cell theory says that all living things are made up of one or more cells and that all cells come from other cells.

7. Electron microscopes are able to resolve objects about a nanometer (10^{-9} meter) in size. This covers just about everything of biological interest.

9. The cell membrane.

11. The three primary components of the cell membrane are phospholipids, proteins, and short carbohydrates.

13. Membrane proteins serve a variety of functions—they help cells communicate with other cells, control transport into and out of cells, and join cells to one another.

15. Lysosomes are the garbage disposals of a cell. These organelles break down organic materials, such as damaged or worn-out organelles.

Challenging Review Solutions

17. Yeast is a eukaryote, and bacteria are prokaryotes. You are also a eukaryote, and for that reason, your cells are definitely more like those of yeast. For example, like yeast cells, your cells have a nucleus and other organelles missing from prokaryotes.

19. Viruses are not made up of cells. Their structure is very different from, and much simpler than, that of a cell. Based on cell theory, viruses are not living things because they are not made up of one or more cells.

21. Most bacteria are large enough to be seen using a light microscope. However, an electron microscope could allow you to see smaller details within bacteria that could be of great interest. So, it would be great to use both!

23. The cytoskeleton provides structure to a cell and helps it hold its shape. In other words, its function is similar to what your skeleton does for your body. And, each of a cell's organelles performs a specific function for the cell, similar to the way each of your organs performs a specific function in your body.

25. Phospholipids have hydrophilic "heads" and hydrophobic "tails." The hydrophilic heads are drawn to the watery environment inside and outside the cell. The hydrophobic tails, on the other hand, want to "get away from" that environment. The setup in which the hydrophilic areas and the hydrophobic areas of phospholipids are all in the environments they prefer is one in which the phospholipids form a double layer with the heads pointing outwards and the tails pointing inwards.

27. Chloroplasts are found only in plant cells. Chloroplasts are responsible for photosynthesis. This does explain why plants don't need to eat, but animals do—through photosynthesis, plants are able to make their own organic molecules that they can then break down for energy.

Apply & Discuss Solutions

29. Phospholipids around a tiny lipid droplet have their hydrophilic heads pointing out and their hydrophobic tails pointing in. The tails are, in this way, shielded from the watery environment.

Living Earth Essay

Life on Earth Connections

Geologic Time

According to the **principle of uniformitarianism**, the natural laws of physics, chemistry, and biology have been the same over the geologic past. This means that the processes we observe in nature today can be used to understand the past. The principle of uniformitarianism is a fundamental principle of geology. It can be summarized by the following statement:

The present is the key to the past.

Was your backyard once under the sea? Was your school campus once a dinosaur hiding place? Clues about Earth's past are recorded in the rocks. We call this the rock record. Learning how to interpret the rock record allows us to understand much about Earth's past.

To read the rock record accurately, it's important to know the age of the rock. Two ways of telling time in geology are absolute time and relative time. **Absolute time** is the time stated in years. **Relative time** is the order of events. For example, we could say that North America was mostly flooded 500 million years ago. That's absolute time. Or, we could say that North America was mostly flooded after the sea level rose worldwide. That's relative time.

Today, geologists use both absolute time and relative time. We have laboratory techniques involving radioactive isotopes for finding the ages of rocks in years. This is called *radioactive dating*. For relative dating, geologists follow a number of principles to determine the age of rock. For example, within a body of water, sediments sink to form a layer on top of older existing rock. Over time, those sediments compact to form sedimentary rock, which is necessarily younger than the rock beneath it. For this reason, the top layer in a cross-section of sedimentary rock is younger than the bottom layer. This demonstrates a principle of relative dating called *superposition*.

Figure A
The youngest layers are toward the top, while the oldest layers are toward the bottom.

To chart what they learn from absolute and relative dating, geologists use the geologic clock (**Figure B**). The geologic clock divides Earth's 4.6-billion-year history into various time units. We abbreviate millions of years ago as Ma (where "M" stands for "mega," and "a" stands for "ago") and billions of years ago as Ga (where "G" stands for "giga," and "a" stands for "ago"). Most divisions of the geologic clock mark major turning points in the history of geology or the history of life, such as mass extinctions.

The most important thing to note about the geologic clock is how much time this actually represents. Have you ever wondered how long it would take to count to a million without sleeping, counting by one every second? The

READING CHECK

What are two ways of telling geologic time?

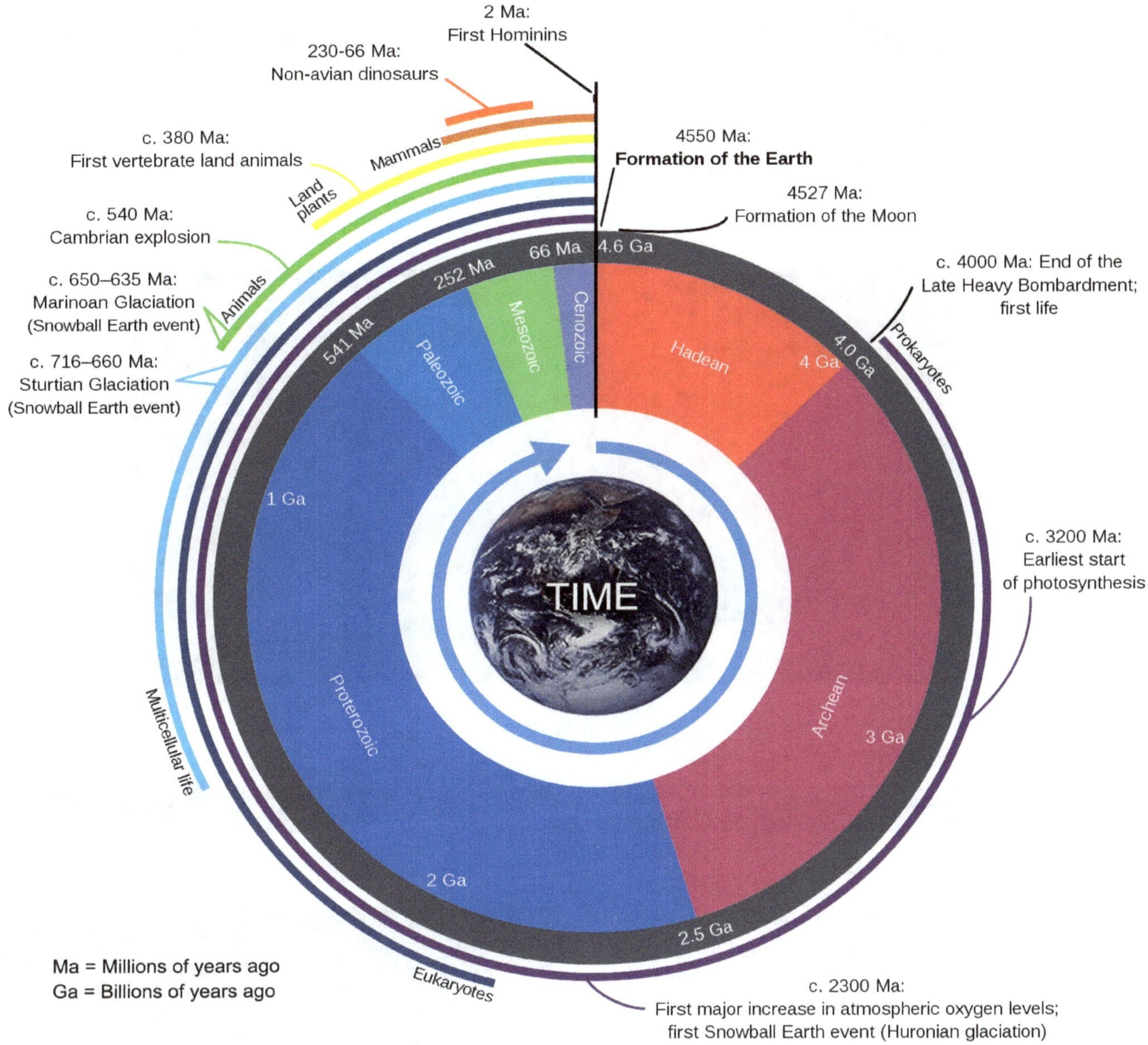

Figure B
The geologic clock divides Earth's history into units of time. For example, the Paleozoic era began about 542 Ma, or 542 million years ago.

answer is 11.6 days. But how long might it take to count to a billion? Multiply by 1000, and you get an astounding 32 years. Is a billion more than a million? Very much so. The geologic clock shows Earth to have been around for about 4.6 billion years! These time scales are next to impossible for us to grasp intuitively.

Consider that the Colorado River started carving out the Grand Canyon of Arizona "only" about 6 million years ago. Even the slowest process, such as a river moving silt downstream, can have profound effects when given vast amounts of time. As another example, consider that South America and Africa are moving away from each other. They do so about as fast as your fingernail grows, which tells us these two continents were once together about 250 million years ago.

The Grand Canyon of Northern Arizona

The Hadean (4.6 Ga - 3.8 Ga)

The name Hadean (hey-DEE-un) comes from the Greek word Hades, meaning "Hell." During the Hadean, Earth was a violent place, hot, and molten. However, the oceans began to form when water vapor escaping from volcanoes condensed to make rain. Scientists also speculate that water also came to Earth from comets.

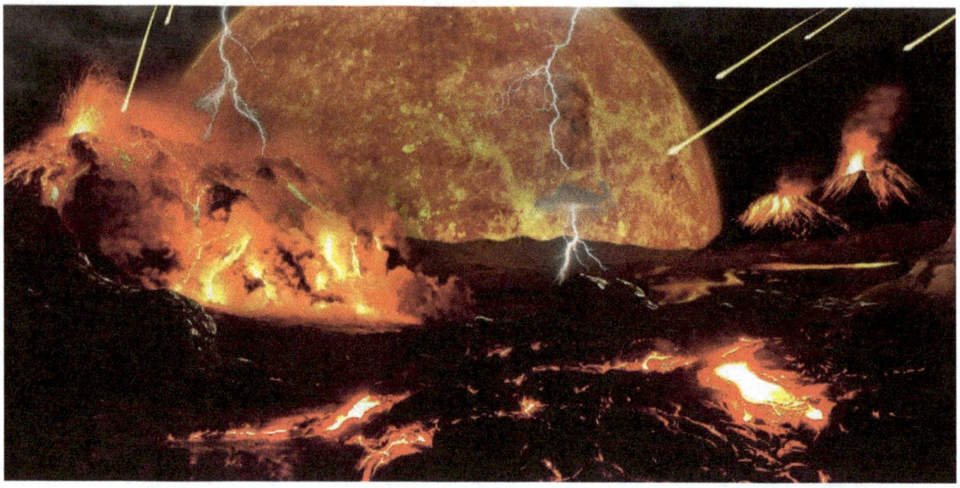

The Hadean, by Tim Bertelink

The Archean (3.8 Ga - 2.4 Ga)

The Archean (ar-KEE-un) Earth featured small patches of rocky crust called *microcontinents*. The earliest fossils we can confidently date are the stromatolites, large rocky structures built by cyanobacteria starting about 3.5 Ga when Earth's atmosphere had little to no free oxygen. The cyanobacteria changed that. Like green plants, cyanobacteria use photosynthesis to covert carbon dioxide and water to a carbohydrate plus free oxygen:

$$6\ CO_2 + 6\ H_2O + \text{light} \rightarrow C_6H_{12}O_6 + 6\ O_2$$

The conversion of carbon dioxide and water to oxygen was enormously important in the evolution of life. Without free oxygen in the atmosphere, life as we know it would have been confined to the oceans. The ozone shield that protects us from lethal doses of ultraviolet radiation from space would not have developed either.

Evidence that Earth's atmosphere began to accumulate oxygen comes from Archean rocks known as banded iron formations (BIFs), as shown in **Figure C**. These BIFs contain iron oxide that could not have been formed without free oxygen in the atmosphere.

Figure C
The rust in these banded iron formations, which date back to Archean time, indicates that Earth's atmosphere was beginning to accumulate oxygen.

The Proterozoic (2.4 Ga - 542 Ma)

The first evidence of plate tectonics appears in the Proterozoic. During this time, tectonic plates collided, merged, and broke apart as they do today. The planet began to cool and stabilize. The concentration of oxygen in the atmosphere increased.

Life continued to evolve in the Proterozoic. A diversity of microscopic life thrived in shallow seas that covered most of the planet. The first eukaryotes—cells containing a nucleus— developed. Then at the end of the Proterozoic, almost all this life was extinguished, potentially as the planet entered a state called *Snowball Earth.* Ice covered almost all of the planet, potentially due to runaway climate change instigated by a particular positioning of the continents.

When the Earth warmed up again, many new ecological niches became available, which set a favorable stage for multicellular organisms. This was a huge step forward in the evolution of life. Jellyfish, other soft-bodied animals, and even some small organisms with shells and other hard body parts appeared by the end of the Proterozoic.

Snowball Earth, by Oleg Kuznetsov

The Paleozoic (542 Ma - 248 Ma)

Supercontinents appeared at the start of the Paleozoic, beginning with the vast supercontinent of Gondwana. During the Paleozoic, this supercontinent began to break up and then come back together to form yet another supercontinent called Pangaea (pan-GEE-uh), which means "universal land."

The start of the Paleozoic was a time known as the Cambrian, during which conditions were optimal for a rapid increase in new life forms—a time known as the Cambrian explosion. Over the next 250 million years, life diversified and prospered. But this was interrupted by the largest extinction in Earth's history, known as the Permian extinction. This extinction coincided with the largest known volcanic eruption in Earth's "recent" history. This eruption lasted for about a million years and likely produced significant climate changes resulting in the extinction of about 90 percent of all life forms at the time.

Extinction is the fate of every species. A species can go extinct for any number of reasons. Its habitat may be destroyed, or environmental conditions may change in such a way that populations cannot adapt. Throughout most of Earth's history, species have gone extinct at a "background" rate. Sometimes, however, many species go extinct over a very short time. Such an event is called a **mass extinction**.

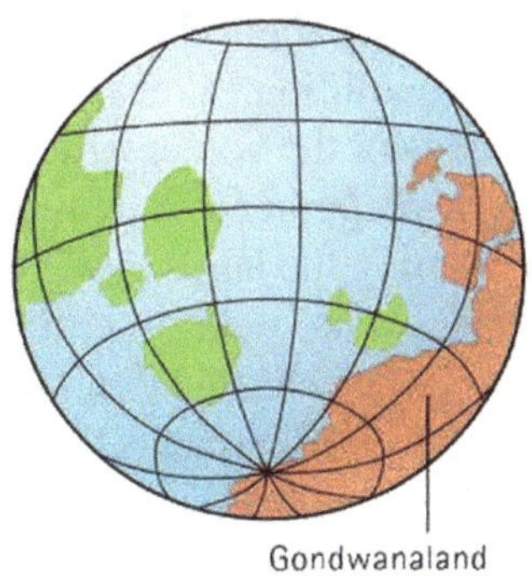

Figure D
Most of the world's land masses were joined together as a huge southern continent Gondwana during the Paleozoic.

The Mesozoic (248 Ma - 66 Ma)

During the Mesozoic, reptiles that survived the Permian extinction at the end of the Paleozoic evolved and diversified to become the dominant animal species of the world. The advent of dinosaurs occurred during the Mesozoic. Mammals also evolved early in the Mesozoic but were relatively small and insignificant compared with the dinosaurs.

During the Mesozoic, the Pangaea supercontinent began to break up to create the smaller continents we know today. Land plants greatly diversified during the Mesozoic. True pines and redwoods appeared. Flowering plants also evolved and soon became the dominant plant on land. Insects that pollinated these plants also proliferated during the Mesozoic.

The end of the Mesozoic is marked by yet another mass extinction—this one likely brought on by the collision of a massive comet or asteroid. The dinosaurs were wiped out, except for those that could fly, which we see today as birds.

The Mesozoic

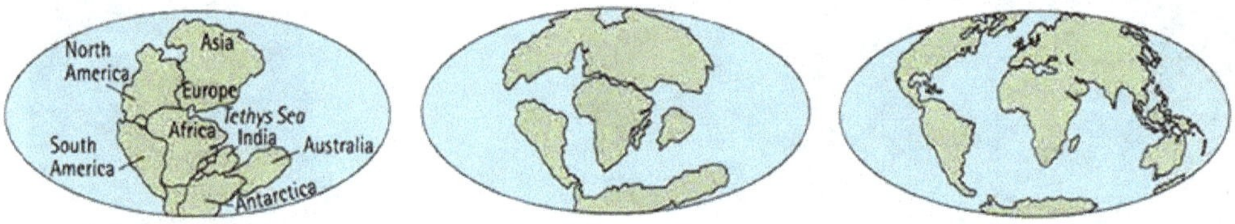

Figure E
Stages in the breakup of Pangaea.

The Cenozoic (66 Ma - Present)

After the mass extinction at the end of the Mesozoic, many environmental niches were once again left vacant. New species, including mammals, insects, and birds, filled them in the Cenozoic.

By about 6 million years ago, a lineage of mammals evolved to become the hominids. Fossil records show that our own species, *homo sapiens*, arose from this hominid line about 300,000 years ago. Then about 70,000 years ago, there was an intellectual revolution, likely brought on by an enhanced ability to control the voice, which led to stronger language skills and a strengthened community. We were primarily foragers roaming the lands until about 12,000 years ago when our ancestors opted to cultivate plants and animals. A foraging life means you don't carry much stuff with you. An agricultural life means you live in one area and can begin to collect many possessions. The sudden increase in human possessions (artifacts) starting about 12,000 years ago led archeologists to figure out the beginning of the agricultural revolution. Further, when people started living and working together in towns and cities, this marked the beginning of modern history.

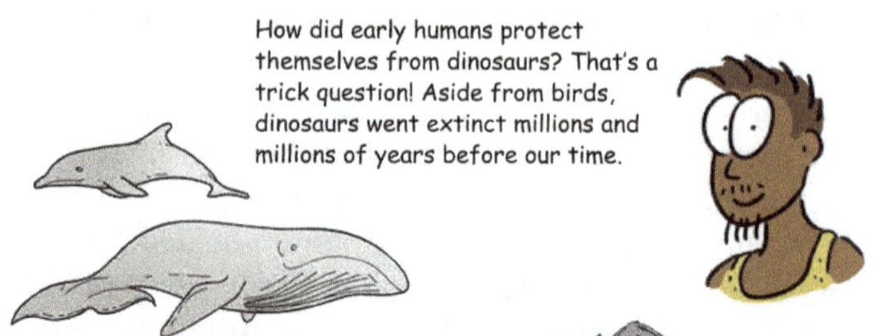

Figure F
Some of the creatures that evolved during the Cenozoic.

Chapter 4

How Cells Work

THE MAIN IDEA

Cells perform a wide variety of tasks in order to live and carry out their functions.

4.1 Cellular Transport
4.2 Cell Communication
4.3 ATP and Chemical Reactions
4.4 Enzymes
4.5 Photosynthesis
4.6 Cellular Respiration

Previously, we learned what cells are and what cells look like. In this chapter, we will look at how cells do their work. How do things move in and out of cells? How do cells get energy—and what does this have to do with baking bread and brewing beer? Why would life as we know it be impossible without photosynthesis, a process that happens in plant cells? Study onward to find out what cells do all day.

4.1 Cellular Transport

Cells need to take in a variety of resources, including water, oxygen, and organic molecules. Cells also generate wastes they must dispose of. Now that we know the structure of the cell membrane, we can discuss how it performs the important task of controlling how materials move into and out of cells. Transport across the cell membrane occurs in a number of ways—through simple diffusion, facilitated diffusion, active transport, endocytosis, and exocytosis.

Simple Diffusion

Some molecules are able to cross the phospholipid bilayer of the cell membrane directly. Hydrophobic molecules, such as the gases oxygen and carbon dioxide, can pass directly through the double layer of hydrophobic tails. Certain small hydrophilic molecules—such as water—can also cross the cell membrane this way. What governs the way these substances move into and out of cells? Answer: the physical process of diffusion. Diffusion is

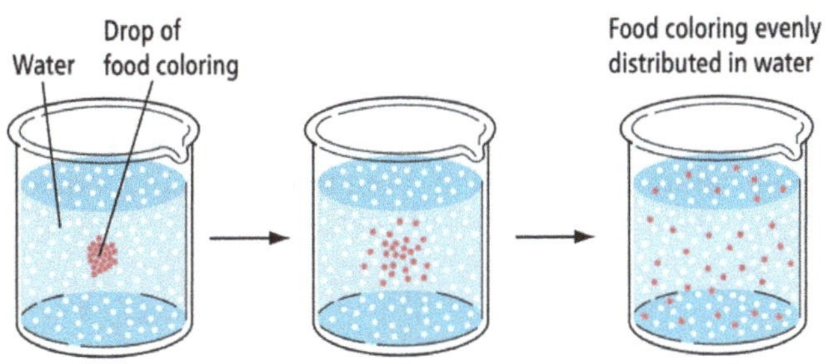

< Figure 4.1

A drop of food coloring diffuses in a beaker of still water. Over time, and without any stirring, molecules diffuse from where they are more concentrated to where they are less concentrated. Eventually, the water becomes uniformly colored.

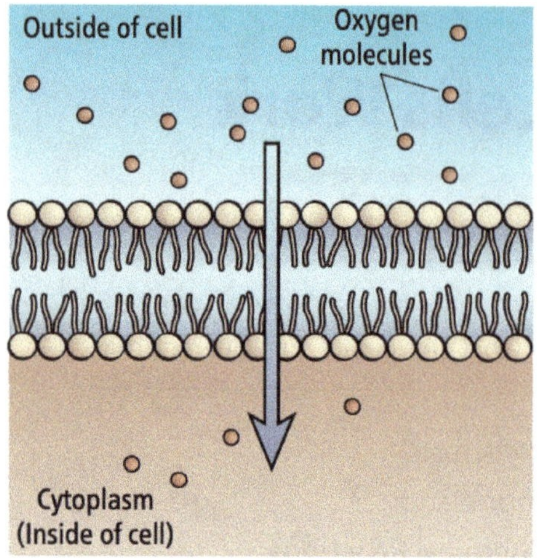

Figure 4.2

Molecules that diffuse freely across the cell membrane show a net overall movement from an area of higher concentration to an area of lower concentration.

the movement of molecules from an area of high concentration to an area of low concentration. This is movement down a concentration gradient. A familiar example of diffusion is the way a drop of food coloring spreads in a beaker of water (**Figure 4.1**).

The biological process of diffusion of molecules directly across the phospholipid bilayer of the cell membrane is called *simple diffusion*. For example, consider how oxygen diffuses from the fluid surrounding our cells into the cells themselves (**Figure 4.2**). Oxygen molecules are found both inside and outside the cell, and they move around randomly. Sometimes an oxygen molecule drifts from outside the cell to inside, and sometimes an oxygen molecule drifts from inside the cell to outside. However, because there is a higher concentration of oxygen molecules outside the cell than inside, the net effect of this simple diffusion is to move oxygen molecules into the cell. (Note that there are fewer oxygen molecules inside cells because cells use oxygen in cellular respiration, a process we will discuss later in this chapter.)

Diffusion works best over small distances. Because of this, bodily processes that depend on diffusion require very thin structures. The walls of our capillaries are very thin, so oxygen can diffuse efficiently from our blood to our tissues. Processes that depend on diffusion also require large surface areas. The intricate branching of the vessels in our lungs creates a greater surface area so that more oxygen can diffuse into the blood. Suppose you know that amphibians receive much of their oxygen from simple diffusion across the skin. Can you develop a hypothesis for why the giant salamander in **Figure 4.3** has such wrinkly skin?

The diffusion of water has a special name—*osmosis*. Like other substances, water diffuses from an area with a high concentration of water molecules to an area with a low concentration of water molecules. Because a higher concentration of water molecules means a lower concentration of solutes, and vice versa, another way to say this is that diffusion moves water from an area of lower solute concentration to an

Figure 4.3

The Chinese giant salamander is the largest amphibian in the world. These salamanders live in streams and obtain oxygen from diffusion across their wrinkly skin.

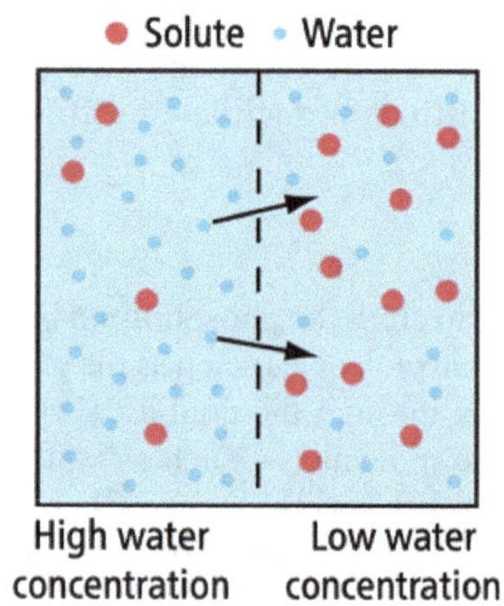

Figure 4.4

In osmosis, water molecules move from an area of lower solute concentration to an area of higher solute concentration. Note that the solute molecules cannot cross the barrier, but the water molecules can.

area of higher solute concentration (**Figure 4.4**). Osmosis is important in biology because controlling water flow is essential to all cells. With too much water, cells can burst; with too little water, cells can shrivel.

Facilitated Diffusion

Many of the molecules that cells need, including ions and large hydrophilic molecules such as proteins and carbohydrates, cannot pass freely across the phospholipid bilayer of the cell membrane. So how do these molecules get into and out of cells? They use special "gates" in the cell membrane. These gates are membrane proteins called *transport proteins*.

Transport proteins are very specific about the molecules they let across the cell membrane. A molecule fits into its transport protein the same way that a key fits into a lock—only the correct key will work in a given lock (**Figure 4.5**).

∧ Figure 4.5

Transport proteins and the molecules they transport fit together like a lock and key.

In *facilitated diffusion*, a transport protein moves molecules down a concentration gradient from an area of high concentration to an area of low concentration (**Figure 4.6**). One example of facilitated diffusion is the movement of the sugar glucose (the basic fuel that cells burn for energy) into red blood cells. In addition to diffusing directly across the phospholipid bilayer, water can also use facilitated diffusion to cross the cell membrane. The transport proteins used by water are called *aquaporins*. Aquaporins allow water to move more quickly across the cell membrane than it can through diffusion alone. Like diffusion, facilitated diffusion requires no energy from the cell. For this reason, simple diffusion and facilitated diffusion are both examples of **passive transport**.

READING CHECK

What do aquaporins do?

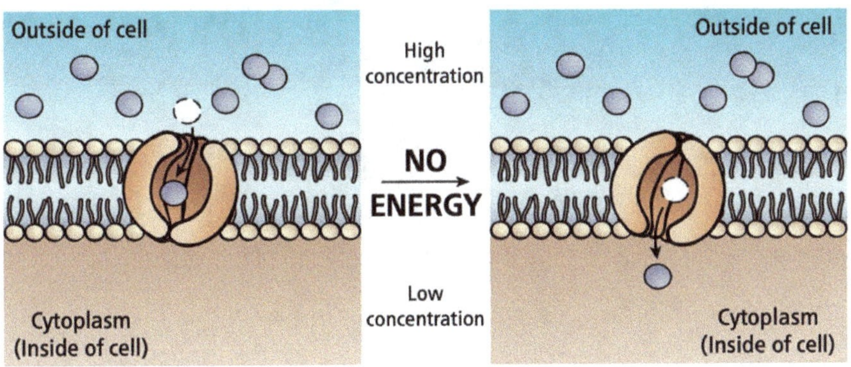

< Figure 4.6

In facilitated diffusion, molecules move from an area of high concentration to an area of low concentration using a "gate" provided by a transport protein.

Active Transport

In **active transport**, a transport protein moves molecules *against* a concentration gradient. This means the molecules move from an area of a low concentration to an area with a high concentration (**Figure 4.7**). In this case, energy from the cell is required. Active transport is used to move many organic molecules, including most proteins, into cells. Active transport also controls the concentration of many ions inside and outside cells. An example of active transport, the movement of sodium and potassium ions by the sodium-potassium pump, is described later in this chapter.

Figure 4.7 >

In active transport, molecules move from an area of low concentration to an area of high concentration—that is, against a concentration gradient. During active transport, a transport protein embedded within the membrane uses energy supplied by the cell.

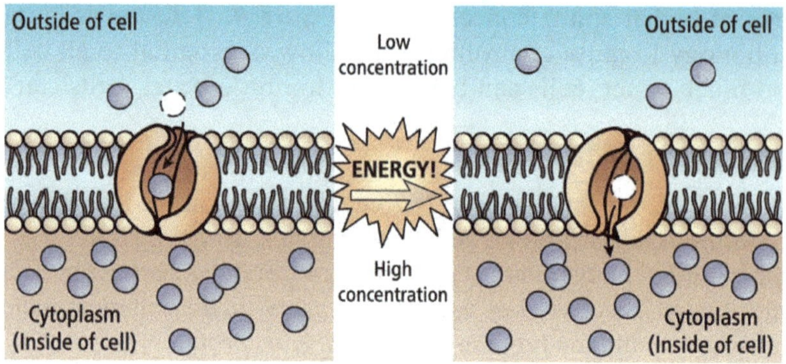

Endocytosis and Exocytosis

Larger amounts of material can be moved into and out of cells through *endocytosis* and *exocytosis* (**Figure 4.8**). Both these processes require the input of energy from the cell. In endocytosis, a portion of the cell membrane folds inward and pinches off, enclosing material within a vesicle inside the cell. Endocytosis is used by certain white blood cells of the human immune system to engulf invading bacteria. In exocytosis, the opposite process occurs—a vesicle fuses its membrane with the cell membrane and dumps its contents outside the cell. Many endocrine cells use exocytosis to release hormones into the bloodstream. Neurotransmitters—the chemicals that neurons use to signal one another—are also released through exocytosis.

Figure 4.8 >

(a) In endocytosis, a portion of the cell membrane pinches off to form a vesicle that brings materials into the cell. (b) In exocytosis, a vesicle inside the cell fuses with the cell membrane, dumping its contents outside the cell. (c) This photo shows vesicles dumping their contents through exocytosis.

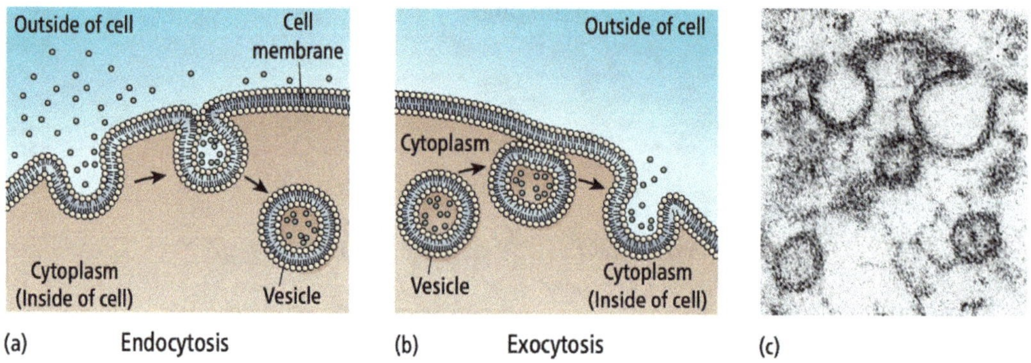

CONCEPT **CHECK**

Insects don't have lungs. Instead, they get oxygen from a series of small branching tubules in their bodies that are connected to the outside air. Oxygen diffuses through the tubules to reach their tissues. Use this information to explain why a mosquito could never grow to be 12 feet tall.

CHECK YOUR ANSWER

Because diffusion works well only at small distances, the insect respiratory system constrains insects to small body sizes. A 12-foot mosquito would not be able to get enough oxygen to the tissues deep inside its body.

More to Explore

You can read more about active and passive transport here:
https://microbenotes.com/difference-between-active-transport-and-passive-transport/

4.2 Cell Communication

The cells of multicellular organisms communicate with one another in order to coordinate their activities. The "messages" they send take the form of molecules. For example, nerve cells send special molecules to muscle cells, telling the muscle cells to contract. Likewise, the pituitary gland sends a different molecule, called growth hormone, to many different cells in the body, telling them to grow and divide.

In animals and plants, special structures allow local messages to pass directly from one cell to an adjacent cell. In animal cells, these structures are *gap junctions*, which are tiny channels between cells (**Figure 4.9a**). Gap junctions are found between many cells in the body. For example, how is it that the heart muscle is able to contract simultaneously to produce each powerful heartbeat? Answer: The unified contraction is orchestrated by messenger molecules moving through the gap junctions between adjacent heart muscle cells. In plant cells, these "in-between cell structures" are called *plasmodesmata* (**Figure 4.9b**). Plasmodesmata are slender threads of cytoplasm that link adjacent plant cells. It is through the plasmodesmata that one plant cell is able to send a chemical signal to an adjacent plant cell.

Message molecules may also travel through the bloodstream to faraway target cells. As mentioned earlier, an example of this is growth hormone, sent from the brain's pituitary gland to cells all over the body. When a message molecule reaches a target cell, it binds to a protein called a *receptor*. Some receptors are membrane proteins, whereas others are inside the cell. Receptors are extremely specific about the molecules they bind to. This is because a message molecule and its receptor fit together like a key in a lock—only the right combination will work. This lock-and-key fit also means that only cells with the appropriate receptors will "receive" and respond to specific message molecules. The binding of a message molecule to its receptor sets off a series of chemical reactions that results ultimately in the target cell's response to the message. As just one example, a cell may receive a message that tells it to grow and divide (**Figure 4.10**).

Problems with cell communication can be dangerous. In some cancer cells, a problem in the communication process causes cells to receive the "grow and divide" message continuously and divide out of control.

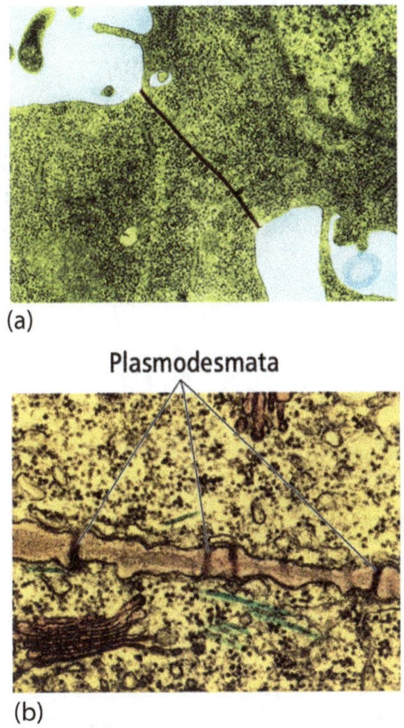

^ Figure 4.9

(a) Gap junctions are tiny channels between adjacent animal cells. These channels allow small molecules to pass. This photo shows a line of gap junctions between two rat cells. (b) This photo shows plasmodesmata, narrow passages that link the cytoplasm of adjacent plant cells.

READING CHECK

How is it that the heart muscle is able to contract simultaneously to produce each powerful heart beat?

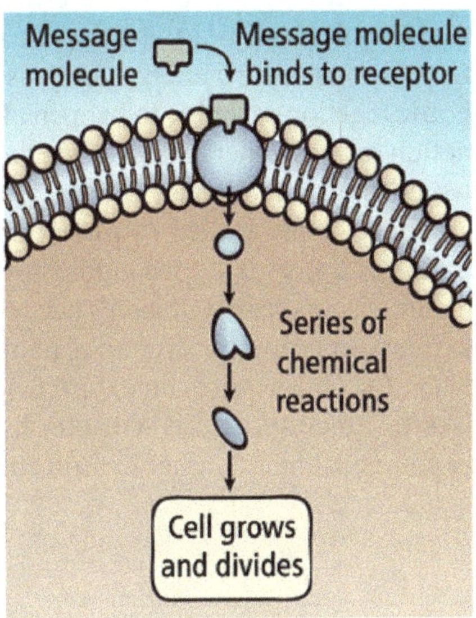

< Figure 4.10

A cell receives a message when a message molecule binds to a receptor in its cell membrane. This begins a chain of events that ends with the cell's response to the message. In this case, the cell responds by growing and dividing.

CONCEPT CHECK

Explain how a cell uses a chemical message molecule to communicate with another cell elsewhere in the body.

CHECK YOUR ANSWER

A cell releases a message molecule, which travels through the bloodstream to a target cell. There, the message molecule binds to a receptor, which might be a membrane protein or a protein inside the cell. The binding of a message molecule to its receptor sets off a series of chemical reactions that results ultimately in the target cell responding to the message in some manner. An example of a target cell response could include a muscle cell contracting in response to a message molecule from a neuron.

More to Explore

You can read more about cell communication here:

https://projects.ncsu.edu/project/bio183de/Black/communication/communication.html

4.3 ATP and Chemical Reactions

At any moment, countless chemical reactions are occurring in cells. These reactions sustain life by allowing cells to perform essential functions such as building proteins or transporting molecules across membranes. But what determines precisely which chemical reactions occur?

In order for any chemical reaction to occur, it must obey the laws of thermodynamics. According to these laws, energy is always conserved, which means that energy cannot be created nor destroyed. Energy, however, spontaneously transfers from where it is concentrated to where it is more dilute. Consider a campfire. Within that campfire, energy is initially concentrated near the burning wood. However, this energy travels from the campfire to its colder surroundings. So, although the campfire's energy is conserved, it also tends to disperse, demonstrating the laws of thermodynamics.

READING CHECK

What do endothermic reactions require?

There are two kinds of chemical reactions: *exothermic* reactions release energy, and *endothermic* reactions absorb energy. Because of the laws of thermodynamics, exothermic reactions are able to happen all by themselves. How so? Exothermic reactions *release* energy, and as the energy is released, it can spread outward, which is a spontaneous process. By contrast, endothermic reactions require energy to be pulled inward from some external source. This is a concentration of energy against a gradient, which, according to the laws of thermodynamics can happen, but won't happen spontaneously (on its own). For a cell to function properly, it requires numerous endothermic reactions. This is to say; the cell requires an external source of energy. Otherwise, the cell ceases to function, and it will die. In cells, the most common external energy source is the remarkable molecule adenosine triphosphate, or **ATP**.

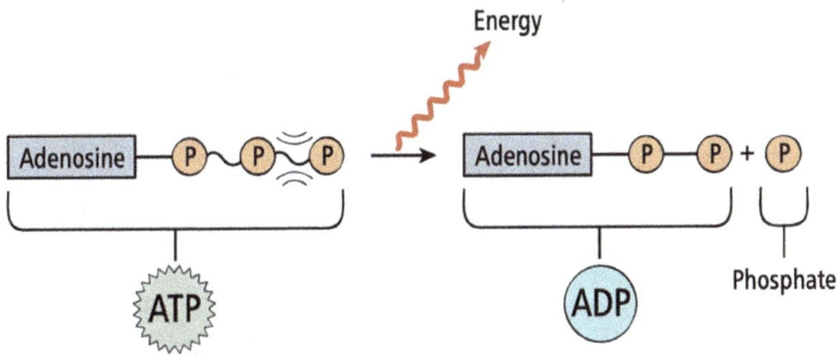

< **Figure 4.11**

Energy is obtained from ATP when one of its phosphate groups is removed, leaving ADP.

ATP provides energy for cellular processes. It consists of an adenosine molecule and three phosphate groups. Cells obtain energy from ATP when one of its phosphate groups is removed, leaving adenosine diphosphate, or ADP (**Figure 4.11**).

This process is fairly easy to explain. Each phosphate group carries negative charges. Negative charges repel other negative charges. Thus, when that third phosphate group is removed, it doesn't just float away. Rather, it accelerates away as if it had been shot from a loaded spring gun. As a result, its potential energy (energy of position) is transformed into kinetic energy (energy of motion). Cells are very good at making use of this released kinetic energy. This means that a cell goes through a significant amount of ATP just to survive. Each second, your body runs through about 125 moles of ATP to stay alive. This adds up to over 50 kilograms (110 pounds) of ATP per day, which is one of the most amazing facts of biology.

The cell eventually turns ADP back into ATP by adding a phosphate group during cellular respiration, a process discussed later in this chapter. As you can see in **Figure 4.12**, ATP is like a rechargeable battery that can be repeatedly used up and recharged.

The sodium-potassium pump provides an example of a biological process that requires precisely one molecule of ATP for each cycle. The sodium-potassium pump uses active transport to maintain a low concentration of sodium ions (Na^+) and a high concentration of potassium ions (K^+) inside animal cells. It does this by using ATP to pump sodium ions out and to pump potassium ions in. Maintaining appropriate ion concentrations helps to regulate water flowing into and out of the cell and is also critical for such processes as neuron firing and muscle contraction.

How does the pump work? The sodium-potassium pump is a transport protein in the cell membrane (**Figure 4.13**). It behaves like a swinging door, shuffling ions into and out of the cell. In its default state, the protein is open to the inside of the cell. There, it binds three sodium ions. An ATP

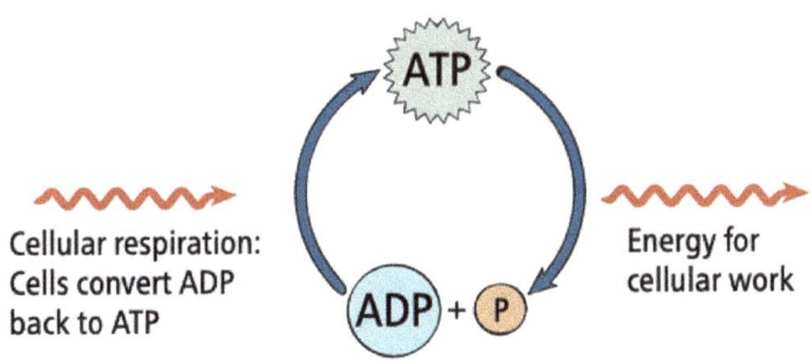

< **Figure 4.12**

Cells obtain energy from ATP when one of its phosphate groups is removed, leaving ADP. ADP is eventually turned back into ATP through the addition of a phosphate group during cellular respiration.

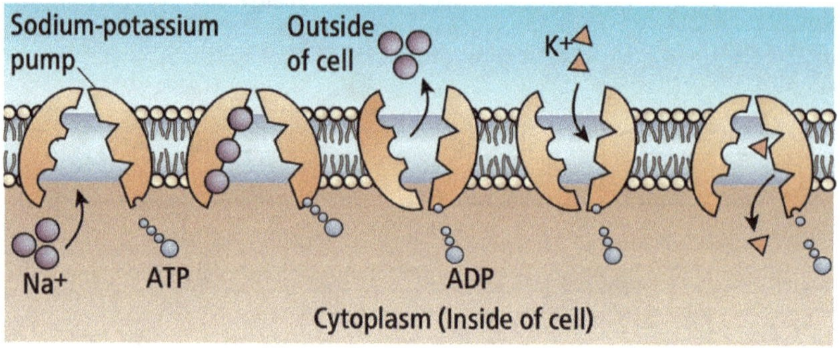

Figure 4.13 >

The sodium-potassium pump moves sodium ions out of cells and potassium ions into cells. One molecule of ATP allows the movement of three sodium ions out and two potassium ions in.

reaction, in which a molecule of ATP transfers a phosphate group to the protein, causes the protein to shift and open to the outside of the cell. The sodium ions are released, and two potassium ions are bound, causing the phosphate group to be released from the transport protein. The loss of the phosphate group causes the protein to shift back to its original position and to release the potassium ions inside the cell. Amazingly, more than a third of all the ATP consumed by animals (including humans) is used to run this pump.

CONCEPT CHECK

1. Each "pump" of the sodium-potassium pump moves three sodium ions out and two potassium ions in. If you have five molecules of ATP, how many sodium and potassium ions can be transferred from one side of the cell membrane to the other?

2. If ATP were temporarily unavailable, what would happen to a sodium potassium pump that was just about to release sodium ions outside the cell?

CHECK YOUR ANSWERS

1. Fifteen sodium ions can be moved out of the cell, and ten potassium ions can be moved into the cell.

2. The sodium-potassium pump would release the sodium ions, bind potassium ions, shift, release the potassium ions inside the cell, and bind more sodium ions. (None of these activities requires ATP.) At this point, the pump would be stuck, since ATP is required for it to proceed to the next step: shifting to open to the cell's exterior.

More to Explore

To learn more about ATP, check out this fun video:

https://www.youtube.com/watch?v=23Zzl6WZS28&vl=en

4.4 Enzymes

We now see how cells use energy from ATP to power chemical reactions. However, a second condition must be met before a chemical reaction can occur: The reacting molecules must collide with enough energy to get the reaction going. This energy, called *activation energy*, is needed for the initial breaking of bonds. In fact, the activation energy for many essential chemical reactions in living organisms is very high. As a result, some important reactions would take 100 years to happen on their own. Because of this, cells rely on *catalysts*, which lower the activation energy of reactions and allow them to happen more quickly. The catalysts in living cells are large, complex proteins called **enzymes**. With the help of enzymes, essential chemical reactions can happen in milliseconds instead of taking 100 years.

How does an enzyme work? An enzyme binds the reactants of a reaction—the enzyme's *substrate*—at its active site and then releases the products (**Figure 4.14**). Enzymes, like other catalysts, are not altered or destroyed in the reactions they catalyze. This means they are available to catalyze the same reaction over and over again. Enzymes are involved in nearly all of the chemical reactions in cells; in fact, several thousand unique enzymes have been identified in living organisms. Each of these enzymes is highly specific for a certain reaction—that is, each enzyme is very picky about the substrate to which it binds and, therefore, about the reaction it catalyzes.

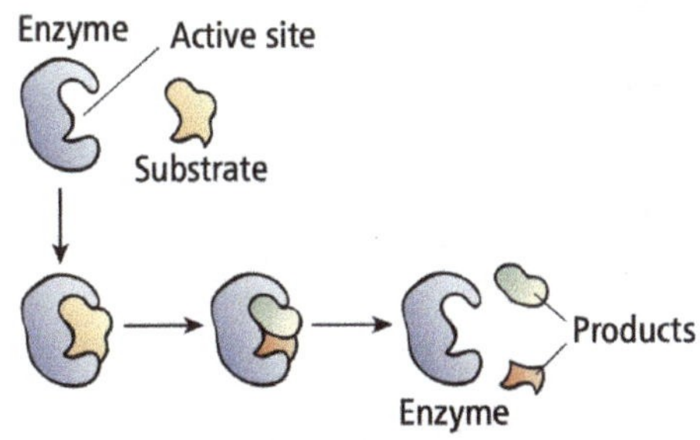

^ Figure 4.14

Enzymes are large proteins that function as catalysts. They are very specific, binding to only specific molecules and catalyzing only specific reactions. Note that the enzyme is not used up in the reaction it catalyzes, and so it is available to catalyze more of the same reaction again.

Cells regulate enzymes carefully in order to control chemical reactions. Regulation takes place in different ways. First, cells control the synthesis and degradation of enzymes. (Enzymes are not used up when they catalyze a reaction, so they must be actively degraded when no longer needed.) Second, how well an enzyme works depends on the chemical features of its environment—the temperature, pH, and other factors. For example, if the temperature is too high or too low, an enzyme's shape may change, and the reactants may not fit. Finally, enzymes can be blocked by inhibitors.

There are two types of inhibitors: Competitive and Non-competitive. In *competitive inhibition* (**Figure 4.15a**), the inhibitor binds to the active site of an enzyme, preventing it from binding to its substrate. (If a substrate fits into an enzyme like a key in a lock, the inhibitor is like bubblegum stuck in the keyhole.) This type of inhibition is called competitive inhibition because the inhibitor and the substrate compete for the active site. For example, cyanides are toxic because they competitively inhibit enzymes that are critical for cellular respiration. In non-competitive inhibition (**Figure 4.15b**), an inhibitor binds to an enzyme somewhere other than at the active site, changing the enzyme so that it can no longer bind to its substrate. Lead is toxic because it binds to a number of enzymes in the body and disrupts their structures.

READING CHECK

What happens to an enzyme soon after it initiates a chemical reaction?

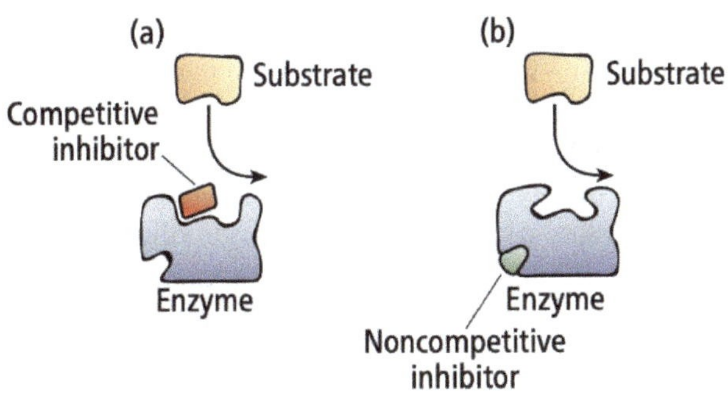

< Figure 4.15

(a) In competitive inhibition, the inhibitor binds to the enzyme's active site and blocks the substrate from binding. (b) In noncompetitive inhibition, the inhibitor binds to the enzyme somewhere other than at the active site. This changes the active site so that the enzyme is no longer able to bind to the substrate.

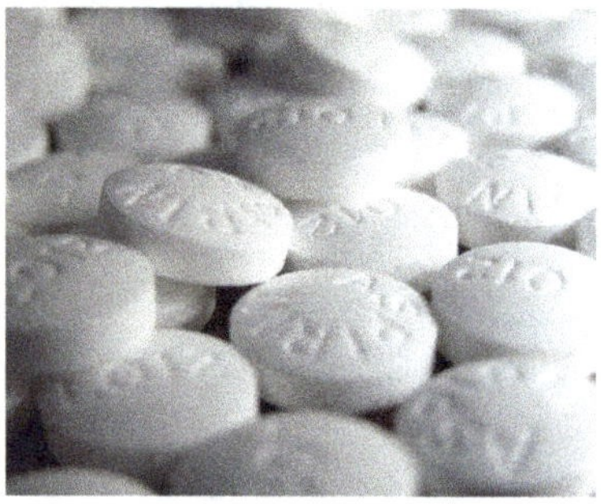

^ **Figure 4.16**

Aspirin is an effective painkiller because it inhibits an enzyme that is needed to make prostaglandins. Prostaglandins make us more sensitive to pain.

Many medications work by inhibiting enzymes. The antibiotic penicillin kills bacteria by inhibiting an enzyme that bacteria need to build cell walls. Aspirin (**Figure 4.16**) works as a painkiller because it inhibits an enzyme needed to make prostaglandins, molecules that make us more sensitive to pain.

CONCEPT CHECK

What does an enzyme bind to when it catalyzes a chemical reaction?

CHECK YOUR ANSWER

An enzyme binds the reactants of a reaction, also known as the enzyme's substrate when it catalyzes a reaction.

More to Explore

See some examples of human enzymes, at the following site:

https://www.medicalnewstoday.com/articles/319704#the-basics

4.5 Photosynthesis

Along the highways of California's Central Valley, billboards celebrate the work of farmers. Some of them say, *"Farmers: turning water and sunlight into food."* This is what crops do and, in fact, what all plants do. They turn sunlight and water into organic molecules, some of which are turned into living plant matter. (As we will see, plants also need carbon dioxide from the air.)

Plants and certain other organisms use **photosynthesis** to convert light energy from the Sun into chemical energy in organic molecules. Almost all life on Earth depends ultimately on photosynthesis for organic molecules and energy. This is because plants and other photosynthesizers (such as algae and certain bacteria) are food for herbivores, and herbivores are food for carnivores.

The chemical equation for photosynthesis is:

$$6\ CO_2 + 6\ H_2O + \text{sunlight} \rightarrow C_6H_{12}O_6 + 6\ O_2$$

Carbon dioxide, water, and sunlight go in; glucose and oxygen come out. Keep in mind that this is a summary of the overall process. In fact, many chemical reactions are needed before the reactants in photosynthesis are converted into the products. Ultimately, the oxygen released is the source of the oxygen we breathe—yet another way photosynthesis is fundamental to life as we know it.

Photosynthesis takes place in the plant organelles called **chloroplasts** (**Figure 4.17**). It occurs in two steps: *light-dependent reactions* and *light-independent reactions*. During light-dependent reactions, energy is captured

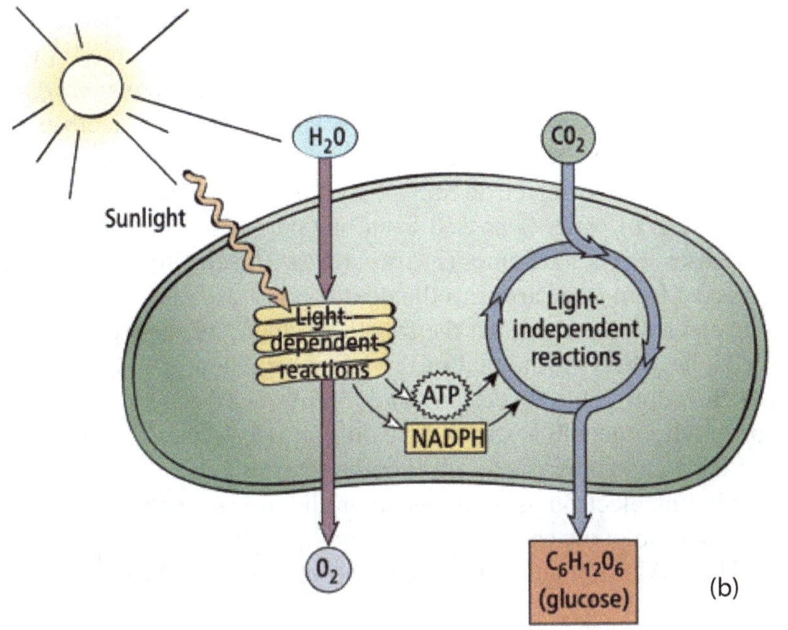

(b)

(a)

< Figure 4.17

Photosynthesis takes place in plant organelles called chloroplasts, found primarily in stems and leaves. (a) A model of a chloroplast with an inset photo taken by an electron microscope. (b) Photosynthesis occurs in two steps: the light-dependent and light-independent reactions.

from sunlight. During the light-independent reactions, carbon is fixed—that is, carbon atoms are moved from atmospheric carbon dioxide to the organic molecule glucose.

The light-dependent reactions begin when sunlight strikes a chlorophyll molecule inside a chloroplast. *Chlorophyll* is a pigment, meaning it is a molecule that absorbs light. Chlorophyll absorbs blue-violet light and red-orange light best. The light that chlorophyll does not absorb—green light—is reflected by plants. This is why plants are green.

When sunlight strikes a chlorophyll molecule, the energy in sunlight excites electrons in the chlorophyll molecule and knocks them out (**Figure 4.18**). The chlorophyll molecule, which is now missing electrons,

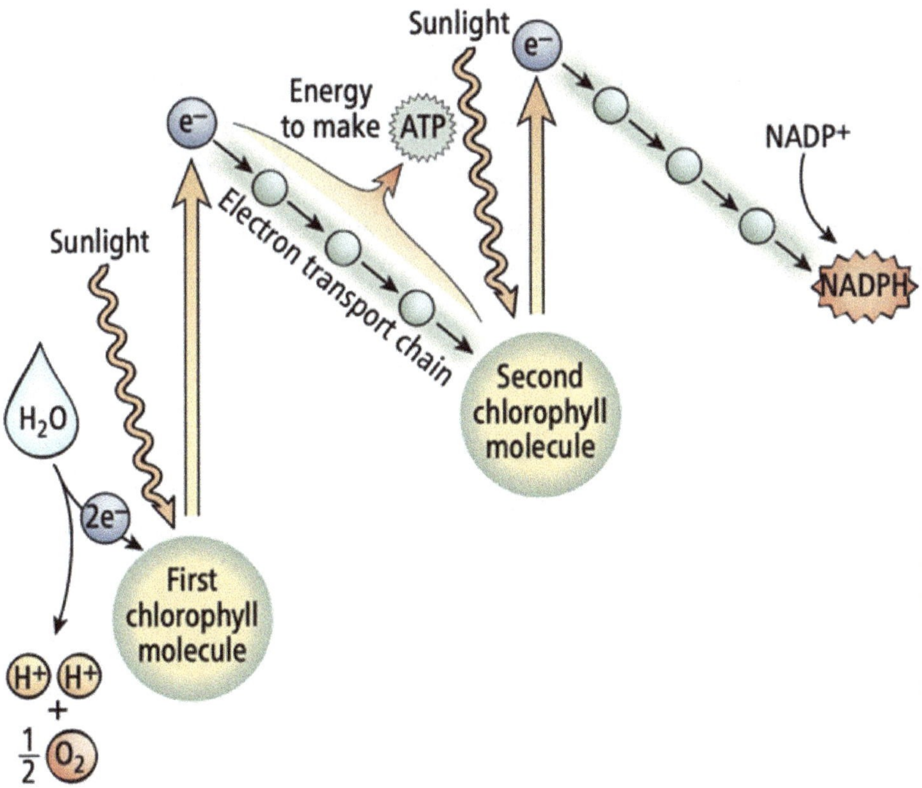

< Figure 4.18

In the light-dependent reactions of photosynthesis, two electron transport chains capture energy from sunlight and store it in ATP and NADPH. Oxygen is released.

^ **Figure 4.19**

In an electron transport chain, the electron is passed from one carrier to the next the way a bucket of water is passed from one person to the next in a cartoon fire brigade. The electron loses energy as it is passed down the chain.

READING CHECK

What happens to the energy of an electron as it is passed down the transport chain?

extracts them from water (H_2O), which causes oxygen (O_2) to be produced. Meanwhile, a high-energy electron knocked out of the chlorophyll molecule passes down an electron transport chain. The *electron transport chain* consists of a series of carrier molecules, each receiving the electron and then passing it on to the next carrier.

You can imagine the electron transport chain as a cartoon fire brigade, in which a bucket of water is passed from one person to the next until the last person tosses it over the flames (**Figure 4.19**). The electron is the bucket of water passed from one carrier to the next.

As the electron passes down the transport chain, its energy is used to convert ADP into ATP. At the end of the transport chain, it is received by a different chlorophyll molecule sensitive to different frequencies of light. This chlorophyll molecule is struck by sunlight, which energizes the electron before it's passed down a second electron transport chain. This time, the energy lost by the electron as it passes down the transport chain is used to convert another molecule called $NADP^+$ into its higher energy state called NADPH. The NADPH is used to drive the subsequent light-independent reactions.

To summarize, two important things happen during the light-dependent reactions. First, energy from sunlight is converted into chemical energy in the form of ATP and NADPH. Second, oxygen is released.

In the light-independent reactions, also known as the *Calvin cycle*, the cell uses the energy stored in ATP and NADPH to fix carbon (see Figure 4.17b). During the Calvin cycle, six molecules of carbon dioxide (CO_2) are taken in and used to make a molecule of glucose ($C_6H_{12}O_6$). Cells use glucose as a starting point for making other carbohydrates, lipids, and, with the addition of nitrogen, amino acids, and nucleic acids—in short, all the macromolecules of life.

CONCEPT CHECK

1. What three things does a plant need to photosynthesize? What two products are made during photosynthesis?
2. Is oxygen released during the light-dependent or the light-independent reactions of photosynthesis?

CHECK YOUR ANSWERS

1. Plants need sunlight, water, and carbon dioxide for photosynthesis. The two products are glucose and oxygen.
2. Oxygen is released during the light-dependent reactions.

More to Explore

Read more about photosynthesis here:

https://ssec.si.edu/stemvisions-blog/what-photosynthesis

4.6 Cellular Respiration

What do the following things have in common: heavy breathing during exercise, bread dough rising, and alcohol in wine and beer? They are all related to how cells obtain energy.

All cells need ATP, which provides the energy required for many essential cellular processes. In order to obtain ATP, cells break down glucose and other organic molecules. The *aerobic*, or oxygen-using, breakdown of glucose is known as **cellular respiration** (Figure 4.20). Cellular respiration takes place in the cell cytoplasm and in organelles called **mitochondria**.

The equation for cellular respiration is:

$$C_6H_{12}O_6 + 6\ O_2 + \text{about 38 molecules of ADP} \rightarrow 6\ CO_2 + 6\ H_2O + \text{about 38 molecules of ATP}$$

Glucose, oxygen, and ADP go in; carbon dioxide, water, and ATP come out. This equation summarizes the process of cellular respiration. The many chemical reactions that occur from the beginning to the end of cellular respiration can be divided into three steps: glycolysis, the Krebs cycle, and electron transport.

Glycolysis

The first step in breaking down glucose is glycolysis (which literally means "sugar splitting"). Glycolysis takes place in the cell cytoplasm. During glycolysis, the six-carbon glucose molecule is split into two pyruvic acid molecules, each containing three carbon atoms. In addition, two molecules of ATP are produced in the process.

The Krebs Cycle and Electron Transport

The Krebs cycle and electron transport occur in the mitochondria. Before entering the Krebs cycle, the pyruvic acid produced during glycolysis is converted into acetic acid and bound to a molecule of coenzyme A. This entire complex is called acetyl-CoA. During the Krebs cycle, acetyl-CoA is broken down into carbon dioxide. Two molecules of ATP are produced from those two pyruvic acid molecules as well molecules called NADH and $FADH_2$. The NADH and $FADH_2$ molecules are significant in that they are able to hold high-energy electrons used for the next step of *electron transport*.

During electron transport, high-energy electrons carried by NADH and $FADH_2$ are sent down *electron transport chains* (Figure 4.21). As electrons are passed from one carrier in the transport chain to the next, their

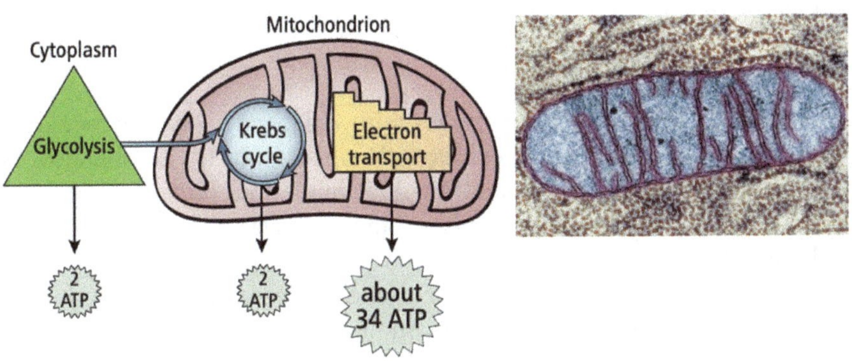

(a) Cellular respiration

(b) Mitochondrion

< Figure 4.20

(a) Cellular respiration takes place in three steps: glycolysis, the Krebs cycle, and electron transport. Glycolysis occurs in the cytoplasm. The Krebs cycle and electron transport occur in the mitochondria. (b) This is an electron microscope photo of a mitochondrion.

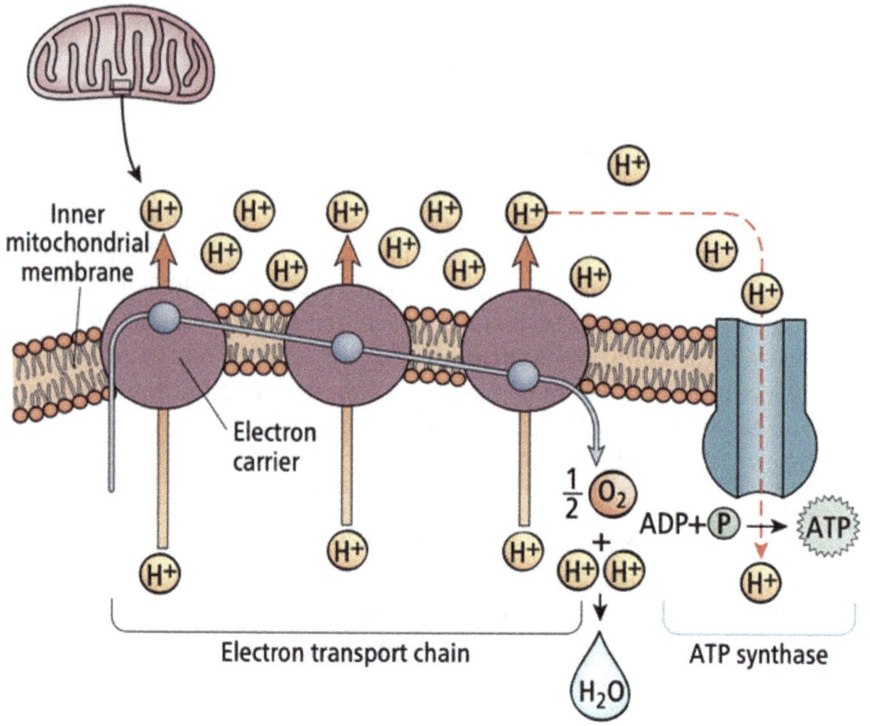

Figure 4.21 >

As electrons move down electron transport chains, hydrogen ions are pumped across the membrane inside mitochondria. When the hydrogen ions move back across the membrane, they generate ATP.

energy is used to pump hydrogen ions (H^+) across a membrane inside the mitochondrion. At the end of the electron transport chain, the low-energy electron needs to be moved away from the system. Recall that oxygen has a strong affinity for electrons. The oxygen, O_2, found in the mitochondria is thus there to capture those low-energy electrons. In doing so, the oxygen also combines with hydrogen ions to form water.

READING CHECK

What tends to happens to the concentration of oxygen within mitochondria?

What then happens to the concentration of oxygen, O_2, within the mitochondria? That oxygen is constantly being consumed as it transforms into water. This means that oxygen is always being depleted within the mitochondria. This results in an oxygen, O_2, concentration gradient that causes more oxygen to diffuse into the mitochondria. In turn, this is why oxygen is always diffusing into the cell from the bloodstream, which gets that oxygen from the lungs. This is what all the oxygen we breathe is for—it is needed to catch those electrons at the end of electron transport chains. Keep breathing, and you'll keep producing ATP from the electron transport chains and have the energy to stay alive.

The concentration gradient of hydrogen ions across the inner mitochondrial membrane is then used to make ATP. As hydrogen ions diffuse back across the inner mitochondrial membrane (down their concentration gradient), they pass through a protein complex called ATP synthase and turn ADP into ATP. This process generates most of the ATP harvested during cellular respiration.

Fermentation

In certain cells, glycolysis is sometimes followed by an *anaerobic* (non–oxygen-using) process known as **fermentation** instead of the aerobic process just described. Fermentation yields no ATP, but it does regenerate the molecules necessary to keep glycolysis going. In this way, cells can continue to obtain ATP through glycolysis.

Alcoholic fermentation is an anaerobic process used by yeast. It takes place in the cytoplasm of yeast cells. In alcoholic fermentation, the pyruvic acid from glycolysis is broken down into ethanol and carbon dioxide.

TABLE 4.1 comparing cellular respiration and fermentation.

Cellular Respiration	Fermentation
Aerobic (uses oxygen)	Anaerobic (does not use oxygen)
Steps: glycolysis, Krebs cycle, electron transport	Steps: glycolysis, fermentation
Takes place in cytoplasm (glycolysis) and mitochondria (Krebs cycle and electron transport)	Takes place in cytoplasm only
Makes 38 molecules of ATP	Makes 2 molecules of ATP (during glycolysis)

Alcoholic fermentation is essential both to making wine and baking bread. Yeast cells ferment the sugar in grape juice to turn it into wine. The same process makes bread rise—yeast cells ferment the sugars in bread dough, releasing tiny bubbles of carbon dioxide gas. (And, yes, fermentation also produces ethanol in bread dough, but it evaporates during baking. This is why you don't have to be 21 to purchase dinner rolls.)

Lactic acid fermentation is an anaerobic process that occurs in some animal cells as well as in certain species of bacteria and fungi. It takes place in the cytoplasm of these cells. In lactic acid fermentation, the pyruvic acid from glycolysis is broken down into lactic acid. In animal muscle cells, lactic acid fermentation occurs during strenuous exercise when the oxygen supply—despite heavy breathing—can't quite meet the demand. By regenerating the molecules required for glycolysis, lactic acid fermentation allows muscle cells to continue to make ATP without oxygen. The lactic acid produced during strenuous exercise causes a burning sensation in the muscles. Red blood cells, which lack mitochondria, rely on lactic acid fermentation for all their ATP. Finally, lactic acid fermentation by certain species of bacteria and fungi is used to make cheese and yogurt.

CONCEPT CHECK

1. About 38 molecules of ATP are obtained from 1 glucose molecule. Of these 38, how many come from glycolysis, the Krebs cycle, and electron transport?

2. How are the chemical reactions for photosynthesis and cellular respiration similar? How are they different?

CHECK YOUR ANSWERS

1. Of the approximately 38 ATP molecules, 2 come from glycolysis, 2 from the Krebs cycle, and about 34 from electron transport.

2. The reactions are similar in that products of one are reactants of the other, and vice versa. The reactions are different in that photosynthesis takes in energy and builds glucose, whereas cellular respiration breaks glucose down and releases energy.

More to Explore

You can review how cellular respiration happens here:
https://www.youtube.com/watch?v=00jbG_cfGuQ

Concept Review

Chapter 4

Detailed Chapter Summary

In this chapter, we looked at how cells work. We began by looking at how molecules are able to get into and out of cells. Small molecules are able to pass right through the phospholipid bilayer of the cell membrane. In *diffusion*, molecules move from a region of high concentration to one of lower concentration. This is a natural kind of motion, and no energy from the cell is required to make it happen. Molecules can also cross the cell membrane via *transport proteins* embedded within the cell membrane. When molecules move from an area of high concentration to one of lower concentration, this process is called *facilitated diffusion*. With or without the transport protein, these are examples of *passive transport*. Specialized transport proteins, however, are able to allow the passage of molecules from a region of low concentration to one of higher concentration. This process requires the use of energy, which can be supplied by an energy molecule such as *ATP*. This is called *active transport*. *Endocytosis* and *exocytosis* are examples of active transport on a grander scale, where larger groups of molecules are passed into or out of the cell all at once.

Cells can communicate with each other by the exchange of molecules. One cell may release a specific molecule that binds to a receptor on the surface of another cell. Upon binding, there is a cascade of chemical events that results in some change within that cell. This can happen over long distances within the body. Between adjacent cells, local messages pass directly through a *gap junction* in animals or through *plasmodesmata* in plants.

All chemical reactions follow the laws of thermodynamics. This includes chemical reactions occurring within the body. These laws specify that energy cannot be created nor destroyed and that energy tends to disperse from where it is concentrated to where it is dilute. Many chemical reactions within the cell are endothermic, which means they require the input of energy to occur. A common source of this energy is the molecule ATP. An example was provided to show how ATP promotes the movement of sodium and potassium ions against concentration gradients.

The amount of energy needed to drive endothermic reactions can be lowered using a catalyst. Large proteins behaving as catalysts work by binding a substrate to a portion of its surface called the active site. For example, it might take a large amount of activation energy to split a molecule into two smaller halves. When bound to the active site of the appropriate enzyme, however, this splitting of the molecule is much easier to do. As a catalyst, the enzyme itself is not destroyed by the process, which means it can facilitate many chemical reactions, one after the other. There are a number of ways the body controls these catalytic reactions. It can block the active site with an alternate molecule, which competes for the active site. This is called *competitive inhibition*. There may also be adjacent regions on the protein that, when bound, upsets the shape of the active site, which makes the active site no longer effective. This is called *noncompetitive inhibition*.

Photosynthesis is a process in plant cells in which the energy of sunlight is used to convert carbon dioxide and water into glucose and oxygen. Almost all life on Earth depends ultimately on photosynthesis for organic molecules and energy. Photosynthesis involves a pigment called chlorophyll. When hit with sunlight, the chlorophyll molecule is able to extract high-energy electrons from water, which produces oxygen. The electrons are then passed through a series of steps to produce the energy molecule ATP followed by NADPH. These energy molecules are then used to produce glucose from carbon dioxide. Cells use glucose as a starting point for making other carbohydrates, lipids, and, with the addition of nitrogen, amino acids, and nucleic acids—in short, all the macromolecules of life.

While photosynthesis allows for the production of organic molecules from energy (sunlight), there is an opposite process called *cellular respiration*, in which organic molecules are broken apart to create energy molecules, such as ATP. Cellular respiration takes place partly in the cell cytoplasm and partly in organelles called *mitochondria*. There are three distinct phases to cellular respiration. It begins with *glycolysis*, where the glucose molecule is split, yielding two molecules of ATP. This is followed by the Krebs cycle, where the chemical products of glycolysis (pyruvic acid) are broken down further to produce more energy molecules, including ATP, NADH, and $FADH_2$. During the third phase, called electron transport, hydrogen ions are pumped outside of the mitochondrial membrane. The energy for this to happen comes from electrons carried by the NADH and $FADH_2$ molecules produced through the Krebs cycle. The electrons are transported through a number of energetically downhill steps. With each step, more hydrogen ions are pumped outward. But all of these steps are made possible only in the presence of oxygen, O_2, which is needed to accept these electrons as they flow "downhill." That's why it's so important for us to keep breathing oxygen.

Then, much like flowing water allows for the moving of a water wheel, as the hydrogen ions naturally flow back across the mitochondrial membrane, they allow for the production of even more ATP molecules. In short, through cellular respiration, the cell breaks down glucose to make ATP.

Lastly, we looked at a process known as fermentation. Although fermentation produces no ATP, it does produce molecules that support continued glycolysis. This allows for the further generation of energy molecules but in the absence of oxygen, O_2, and is thus called anaerobic.

Key Terms

Active transport Transport across the cell membrane that requires energy from the cell.

ATP Adenosine triphosphate, the molecule that provides the energy for most cellular processes.

Cellular respiration The aerobic breakdown of glucose to produce ATP.

Chloroplasts The organelles within plant cells in which photosynthesis occurs.

Diffusion The passive movement of molecules from an area of high concentration to an area of low concentration.

Enzymes Complex proteins that catalyze chemical reactions in living organisms.

Fermentation The anaerobic breakdown of glucose that results in the production of ethanol and carbon dioxide (alcoholic fermentation) or lactic acid (lactic acid fermentation).

Mitochondria Eukaryotic organelles that break down organic molecules to produce ATP.

Passive transport Transport across the cell membrane that does not require energy from the cell.

Photosynthesis The process in plants and some other organisms in which light energy from the Sun is converted into chemical energy in organic molecules.

Review Questions

Simple Review Questions

4.1 How Things Get Into and Out of Cells

1. What is diffusion?

2. What is the difference between simple diffusion and facilitated diffusion?

3. What distinguishes passive transport from active transport?

4. How do endocytosis and exocytosis move materials into and out of cells?

4.2 Cell Communication

5. What are gap junctions? What function do they serve?

6. Describe what happens when a message molecule binds to a receptor on the cell membrane.

7. Why do only certain cells receive and respond to a specific message molecule?

4.3 ATP and Chemical Reactions in Cells

8. How does ATP provide energy for cells?

9. Describe how the sodium-potassium pump works. Which step in the active transport of sodium and potassium ions requires ATP?

4.4 Enzymes and Chemical Reactions

10. Why do cells need catalysts? What are the catalysts in cells called?

11. Describe two ways that cells can regulate enzymes.

12. How does penicillin kill bacteria?

4.5 Photosynthesis

13. Why is almost all life on Earth directly or indirectly dependent on photosynthesis?

14. What happens during the light-dependent reactions of photosynthesis? What happens during the light-independent reactions?

15. In the chemical reaction for photosynthesis, what are the reactants, and what are the products?

4.6 Cellular Respiration

16. Describe the process of glycolysis. How much ATP is produced during glycolysis?

17. About how many ATP molecules does a cell obtain from one molecule of glucose through cellular respiration? What other products result from cellular respiration?

18. What are the products of alcoholic fermentation?

19. Give two examples of cells in the human body that use lactic acid fermentation. Why does each of these cells use lactic acid fermentation?

Challenging Review Questions

4.1 How Things Get Into and Out of Cells

20. Transport proteins and the molecules they transport are described as fitting together like a lock and key. Why is it important that they have such a specific fit?

21. Glucose moves into many cells through facilitated diffusion. This works because there is usually a higher concentration of glucose molecules outside the cell than inside the cell. Why is this the case? (Hint: Do cells use up the glucose inside them?)

22. Why does oxygen diffuse into cells rather than out of them? Why does carbon dioxide diffuse out of cells rather than into them?

23. What is the difference between endocytosis and using a transport protein to cross the cell membrane?

4.2 Cell Communication

24. How are gap junctions and plasmodesmata similar? How do they differ?

25. Message molecules and their receptors are described as fitting together like a lock and key. Why is it important that they have such a specific fit?

26. If growth hormone is found in the bloodstream and travels all over the body, why do only certain parts of the body respond and grow?

4.3 ATP and Chemical Reactions in Cells

27. If ATP is like a rechargeable battery, how is its energy used up, and how is it recharged?

4.4 Enzymes and Chemical Reactions

28. Enzymes and their substrates are described as having a lock-and-key fit. Why is it important that they have such a specific fit?

29. The deadly nerve gas sarin binds to an enzyme called acetylcholinesterase, which breaks down acetylcholine in the body. If acetylcholine is not broken down, muscles are unable to relax after contracting. Without prompt

treatment, respiratory collapse and death follow. Sarin works by binding to acetylcholinesterase at the site where acetylcholine normally binds. What form of enzyme regulation does this represent?

30. Describe at least two different ways cells can regulate the function of enzymes.

4.5 Photosynthesis

31. What happens to a plant that is kept in the dark?

32. You collect some pond weed from a local pond to keep in your aquarium. One day, you notice that bubbles are coming out of your plant. Why is the pond weed releasing bubbles? What kind of gas is in the bubbles?

33. Global warming has resulted from the large amounts of carbon dioxide released when fossil fuels such as oil and coal are burned for energy. Carbon dioxide traps heat. Why might the loss of forests also contribute to global warming?

4.6 Cellular Respiration and Fermentation

34. Why can't you live without oxygen?

35. What are some differences between fermentation and cellular respiration? Which process produces more ATP? Why do some cells in the human body use fermentation?

36. You visit a friend who is a winemaker. Some of his wine is fermenting in a vat. When you take a look, you notice that the wine is covered with a layer of bubbles. Why are there bubbles? What kind of gas is inside the bubbles?

37. Why is champagne bubbly and wine is not? (Hint: In non-bubbly wines, fermentation happens with the grape juice exposed to air. In champagne, there is an extra round of fermentation during which the bottles are capped tight.)

38. Some animals that live in desert environments, such as this kangaroo rat, never drink water. Kangaroo rats live entirely on the starches and lipids in the dry seeds they eat. Yet we know that all living organisms need water, and, in fact, the bodies of kangaroo rats have about the same water content as those of other animals. How do kangaroo rats get their water?

Apply & Discuss Questions

39. You and a friend are eating lunch in the cafeteria. Suddenly, you both smell very strong perfume. "Wow," your friend says, "I wish they would turn the fan off. It's blowing that perfume right at us." You say, "Even if they turn the fan off, there would still be diffusion. You would still smell the perfume." Which of you is correct?

40. You and a friend go to a garden shop to buy fertilizer for your new plants. Your friend says, "Didn't we just learn that plants make their own organic molecules through photosynthesis? Plants shouldn't need fertilizer." Another customer overhears your discussion and looks at the bag of fertilizer you are holding. "Well," she says, "it looks like the main ingredients in fertilizer are nitrogen and phosphorus. I don't know if plants really need that stuff or not." What do you say? Do you buy the fertilizer?

41. You are talking with a winemaker about yeast. "I know that yeast can survive under both aerobic and anaerobic conditions," he says, "but what I don't understand is why they seem to need so much more sugar under anaerobic conditions. I think they gobble a hundred times as much sugar." Can you explain to him why yeast might need more sugar under anaerobic conditions than under aerobic conditions? Is his estimation of a hundred times as much sugar reasonable?

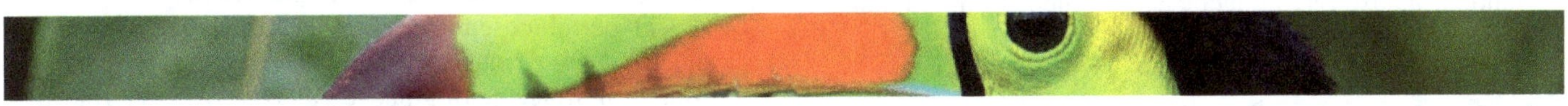

Solutions to Odd-Numbered Questions

Simple Review Solutions

1. Diffusion is the tendency for molecules to move from an area of high concentration to an area of low concentration—that is, down a concentration gradient.

3. Passive transport requires no energy from the cell. In active transport, a transport protein moves molecules against a concentration gradient. In this case, energy from the cell is required.

5. Gap junctions are tiny channels that connect adjacent animal cells. These allow very local messages to pass directly from one cell to an adjacent cell.

7. Cells receive and respond to a specific message molecule only if they have the receptor for that molecule. Receptors are extremely specific about the message molecules they bind to. This is because a message molecule and its receptor fit together like a key in a lock—only the right combination will work.

9. The sodium–potassium pump is a transport protein in the cell membrane. It behaves like a swinging door, shuffling ions in and out of the cell. In its default state, the protein is open to the inside of the cell. There, it binds three sodium ions. An ATP reaction, in which a molecule of ATP transfers a phosphate group to the protein, causes the protein to shift and open to the outside of the cell. (This is the step

where ATP is required.) The sodium ions are released, and two potassium ions are bound, causing the phosphate group to be released from the transport protein. The loss of the phosphate group causes the protein to shift back to its original position and to release the potassium ions inside the cell.

11. Cells regulate enzymes carefully in order to control chemical reactions. Regulation takes place in a number of different ways. First, cells control the synthesis and degradation of enzymes. (Enzymes are not used up when they catalyze a reaction, so they must be actively degraded when they are no longer needed.) Second, how well an enzyme works depends on the chemical features of its environment – the temperature, pH, and other factors. For example, if the temperature is too high or too low, an enzyme's shape may change, and the reactants may not fit. Finally, enzymes can be blocked by inhibitors.

13. Almost all life on Earth depends ultimately on photosynthesis for organic molecules and energy. This is because plants and other photosynthesizers (such as algae and certain bacteria) are food for herbivores, and herbivores are food for carnivores. So photosynthesizers are the ultimate source of all food.

15. Carbon dioxide, water, and sunlight go in (these are the reactants); glucose and oxygen come out (these are the products).

17. Cells obtain about 38 molecules of ATP from one molecule of glucose during cellular respiration. The other products of cellular respiration are water and carbon dioxide.

19. In animal muscle cells, lactic acid fermentation occurs during strenuous exercise when the oxygen supply—despite hard breathing—can't quite meet the demand. By regenerating the molecules required for glycolysis, lactic acid fermentation allows muscle cells to continue to make ATP without oxygen. Red blood cells, which lack mitochondria, also rely on lactic acid fermentation to obtain ATP.

Challenging Review Solutions

21. There is more glucose outside cells than inside cells because glucose molecules inside cells are quickly broken down to make ATP (through cellular respiration).

23. In endocytosis, a portion of the cell membrane folds inward and pinches off, enclosing the transported material within a vesicle inside the cell. With a transport protein, the molecule entering the cell binds to the transport protein and is released inside the cell.

25. The specific fit between a message molecule and its receptor allows for messages to be targeted to specific cells—only cells with the appropriate receptors respond to the message molecule.

27. The energy stored in ATP is "used" when one of its phosphate groups is removed, leaving adenosine diphosphate, or ADP. "Recharging" happens when a phosphate group is added to ADP to build ATP.

29. This is competitive inhibition since it binds at the enzyme's active site, where the substrate normally binds.

31. A plant that is kept in the dark will not have access to the sunlight energy it needs to perform photosynthesis. Eventually, the plant will die from its inability to make organic molecules.

33. When forests are destroyed, there are fewer trees to perform photosynthesis. Since carbon dioxide is removed from the atmosphere and converted to organic material during photosynthesis, the loss of forests contributes to increased carbon dioxide levels in Earth's atmosphere, which in turn worsens global warming.

35. Cellular respiration uses oxygen, whereas fermentation does not. Cellular respiration takes place in the cytoplasm (the first step—glycolysis) and mitochondria (the second and third steps—Krebs cycle and electron transport), whereas fermentation occurs entirely within the cytoplasm. Cellular respiration makes a lot more ATP than fermentation—38 ATP versus 2 ATP. However, muscle cells use fermentation when oxygen is unavailable to them, and blood cells use fermentation because they do not have mitochondria.

37. The bubbles come from carbon dioxide released during alcoholic fermentation.

Apply & Discuss Solutions

39. By stirring the air, the fan brings the smell of perfume to you more quickly. However, even without the fan, the perfume would diffuse all around the room eventually.

41. Yeast obtain a lot more ATP from glucose when they are in aerobic conditions and use cellular respiration—about 38 ATP per glucose molecule. When they are kept in anaerobic conditions and obtain ATP through fermentation, each molecule of glucose yields only 2 ATP (during glycolysis). This is why yeast need more sugar under anaerobic conditions—though only about 19 times as much.

Living Earth Essay

Life on Earth Connections

Plate Tectonics

Have you ever noticed on a world map that Africa and South America fit together like pieces of a giant jigsaw puzzle (**Figure A**)? The first person to provide a detailed explanation of this was the German scientist Alfred Wegener (1880—1930). Wegener's hypothesis, called *continental drift*, states that the world's continents move slowly over Earth's surface. Wegener hypothesized that the individual continents of today's world were once joined as a single, giant landmass. Wegener named this supercontinent Pangaea (pan-GEE-uh), which is Greek for "universal land."

Wegener supported his hypothesis of continental drift with impressive biological evidence. For example, fossils of the same species of *Glossopteris* plant, shown in **Figure B**, were found in places that are separated today by large oceans—India, Australia, South America, Africa, and Antarctica. Seeds from these plants were too heavy to have been blown from continent to continent. Also, fossils of *Mesosaurus*, a reptile, appeared in both Africa and South America. That creature couldn't swim and could not have evolved in exactly the same way in these widely separated regions. So why do we find fossil evidence of the plant and this reptile in such separate places?

Alfred Wegener

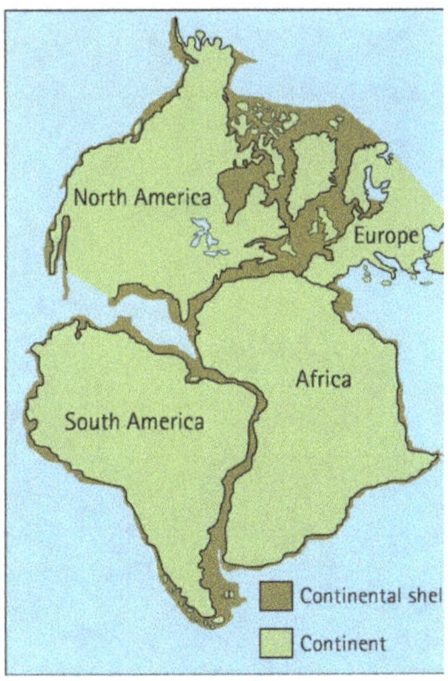

Figure A
The jigsaw-puzzle fit of the continents suggests that they were once joined. Earth was a puzzle to Wegener—literally!

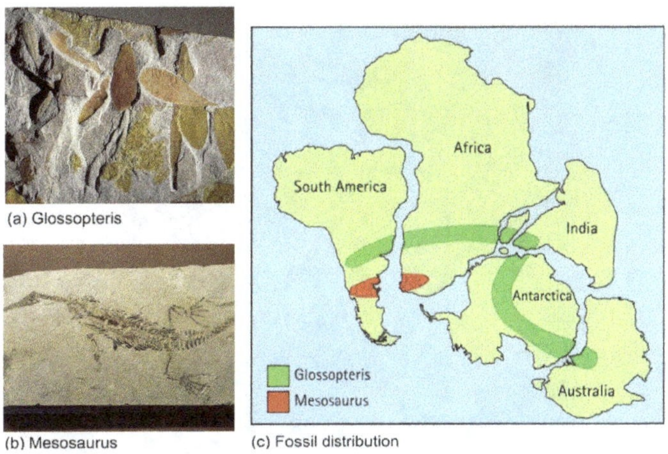

Figure B
(a) This rock contains a fossil of a *Glossopteris* fern, an ancient seed plant. Can you see its long leaves? (b) This is a fossilized reptile called *Mesosaurus*. (c) These species could not cross oceans. Why, then, were *Mesosaurus* and *Glossopteris* fossils distributed as shown here?

To explain the matching of *Glossopteris* and other plant and animal fossils, geologists of Wegener's time proposed that prehistoric land bridges once connected the continents. Yet there was no physical evidence that such land bridges ever existed. Continental drift, on the other hand, was supported by evidence, and it provided a neat explanation of matching fossils. The continents had once been a single landmass with a single community of animals and plants. When the continents later split along the present-day Mid-Atlantic Ocean, representatives of these species were divided among the new continents.

Wegener also used rock evidence to support his hypothesis of continental drift. For example, diamond-bearing rock formations occur where Africa and South America would meet if the Atlantic Ocean did not separate them. And in some places, mountain chains separated by oceans have patterns of folds that would be continuous if the mountains were brought together (**Figure C**).

Figure C
Wegener likened the fossil and rock matches to finding two pieces of torn newspaper with matching contours and lines of type. If the edges and the lines of type fit together, he said, the two pieces of the newspaper must have originally been one.

Besides the matching fossils and rock, ancient climate evidence supported Wegener's hypothesis. For example, continental drift explained puzzling evidence of ancient ice sheets on regions now located near the equator. How did these ice sheets form on land in such a warm location today? Continental drift provides an explanation: when the ice sheets formed on these landmasses, the landmass was located near the South Pole.

Yet Wegener's hypothesis was dismissed in scientific circles. No one, including Wegener, could explain what forced the continents apart. How could massive continents of solid rock "drift" across the Earth's surface? These questions put acceptance of continental drift on hold for decades.

CONCEPT CHECK
What evidence might lead someone with no scientific knowledge to suspect that the continents were once connected?

CHECK YOUR ANSWER
The most obvious evidence is found in world maps that show continents fitting together much like pieces in a jigsaw puzzle.

Seafloor Spreading

In the 1950s, many decades after Wegener proposed his hypothesis, scientists used sonar technology to map the seafloor for the first time. They discovered that the ocean bottom is not flat, as had been presumed, but has deep canyons, trenches, and underwater volcanoes.

An amazing discovery was the Mid-Atlantic Ridge, the longest and tallest mountain range in the world (**Figure D**). In the center of the ridge and all along its length, there

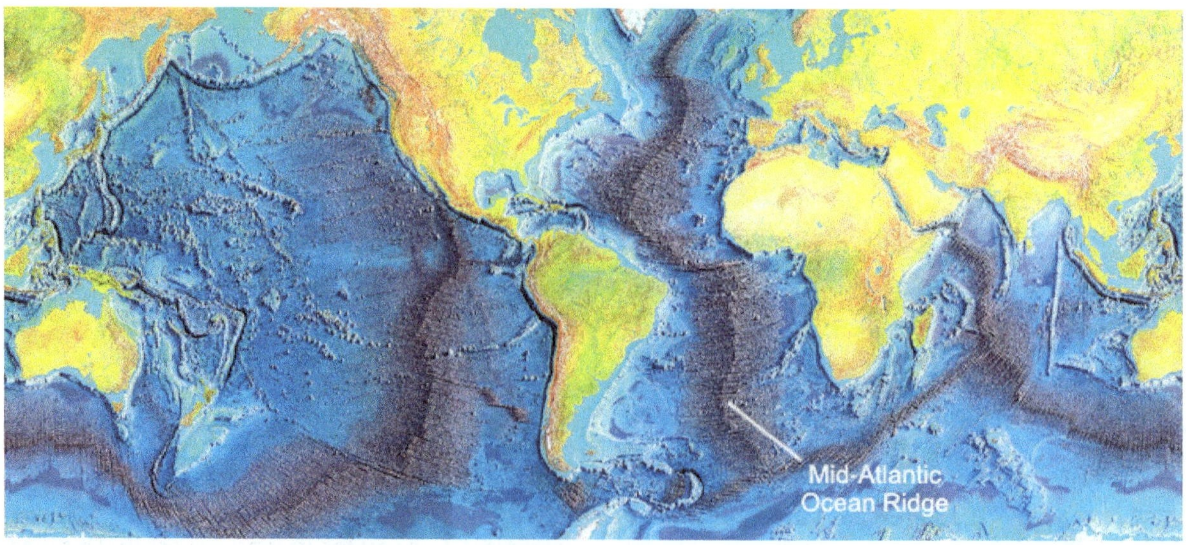

Figure D
The seafloor map shows that the ocean floor is not flat and calm. It is a dynamic place, crisscrossed with canyons, trenches, and mountain chains.

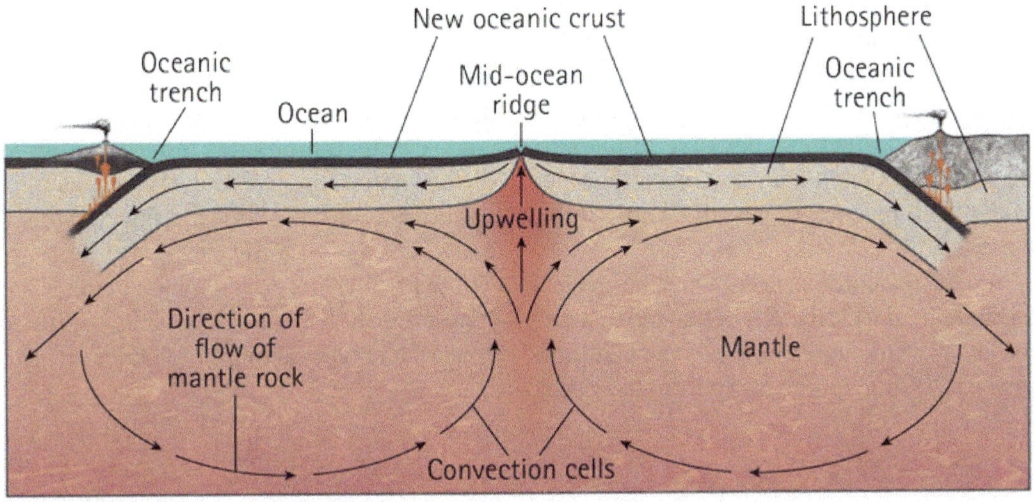

Figure E
In a conveyor-belt fashion, new lithosphere is formed at the rift zone of a mid-ocean ridge while old lithosphere is recycled back into the mantle. As a result, the ocean floor is renewed about every 200 million years. But why is Earth's interior warm enough to support convection currents? The main reason is the presence of naturally-occurring radioactivity.

is a valley, or rift, containing the youngest rocks of the seafloor. We know now that the entire global system of underwater mountains called the mid-ocean ridge, winds all around Earth like the seam on a baseball.

To explain the features of mid-ocean ridges, the American geologist Harry Hess proposed that convection currents within Earth's molten interior cause the top layers of Earth, known as the lithosphere, to spread away from each other in the area of a mid-ocean ridge (**Figure E**). The same action would create deep trenches where the slowly moving lithosphere could subduct beneath another slab of lithosphere.

The idea of seafloor spreading was strongly supported by the mirror image magnetic stripes found on each side of a ridge. As shown in **Figure F**, molten iron particles arising from a ridge align to Earth's magnetic field, which, itself, flips orientation about once every 700,000 years. The magnetic orientations of iron running parallel to the ridge serve as a definitive proof for the idea of seafloor spreading.

READING CHECK

Geologists of Wegener's day found *Glossopteris* and *Mesosaurus* fossils on widely separated landmasses. What was Wegener's explanation for this puzzling fossil find?

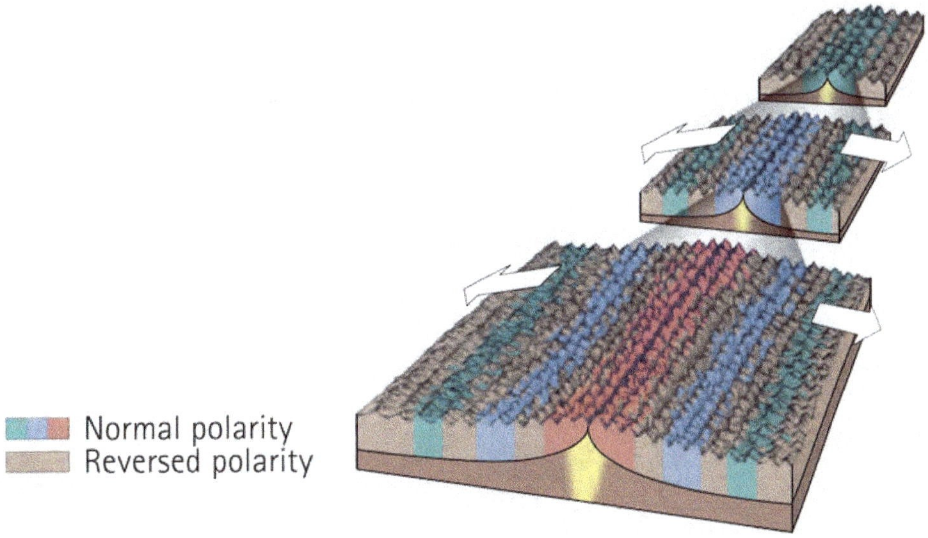

Figure F
The symmetrical pattern of magnetic strips on both sides of the spreading center shows that seafloor forms on both sides of a mid-ocean ridge at about the same rate.

Plate Tectonics

Plate tectonics is the grand, unifying theory of Earth Science. It explains Earth's dramatic and ever-changing features. Plate tectonics is to Earth science as the theory of evolution is to biology, Newton's laws are to physics, and the periodic table is to chemistry. As more than one geologist has said, "With plate tectonics, the whole world suddenly makes sense."

The theory of plate tectonics states that the lithosphere is divided into pieces called tectonic plates that move about on top of deeper layers. The plates move as the seafloors widen, which pushes the lithosphere away from the ridges that crisscross the planet. Plate tectonics provides what the hypothesis of continental drift lacked—a way for continents to move.

The tectonic plates move at different speeds and in different directions. The average speed of plate movement is about 5 cm/yr, about the rate at which your fingernail grows. Some tectonic plates are large; others are small. There are about a dozen major plates, as **Figure G** shows. Planet Earth—a small blue island of life in the vastness of space—is a very dynamic place.

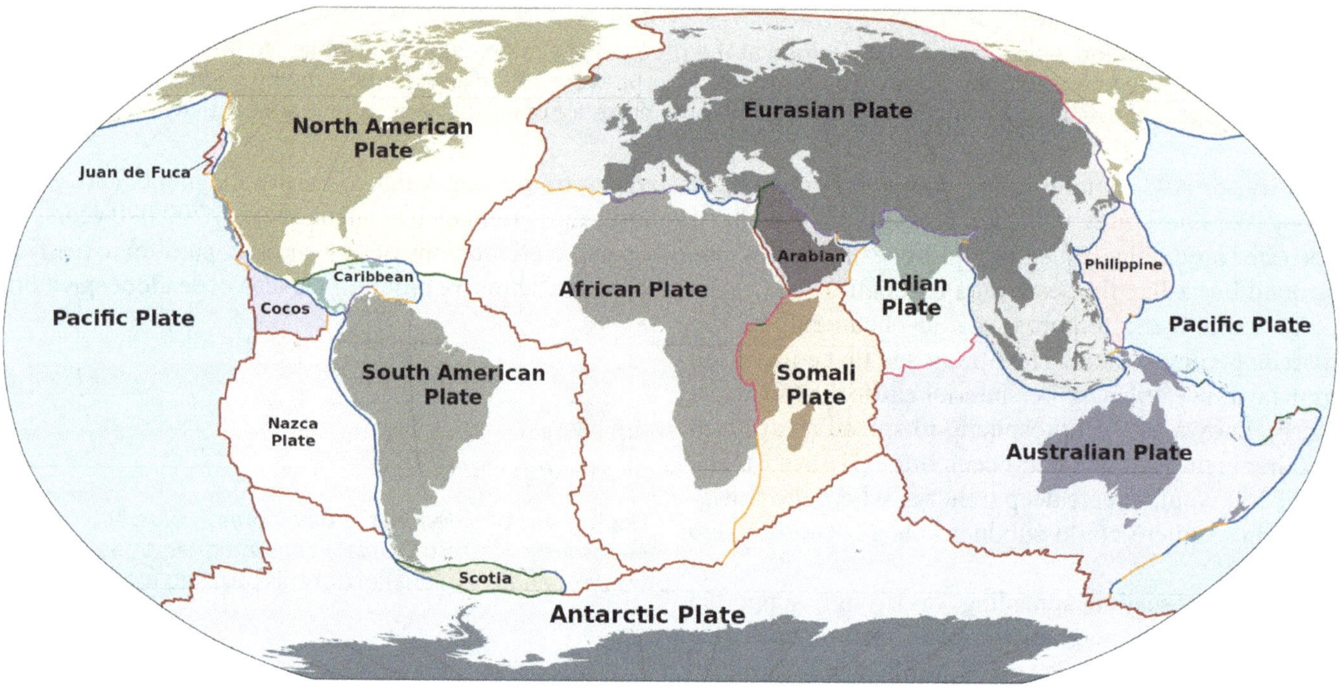

Figure G
Like a cracked, hard-boiled egg, Earth's rigid outer shell, the lithosphere, is a patchwork of 12 major tectonic plates. These move at different rates and in different directions.

Chapter 5

DNA and Genes

THE MAIN IDEA

The genes in DNA tell cells how to build proteins.

5.1 What is a Gene?
5.2 Chromosomes
5.3 The Structure of DNA
5.4 How DNA Is Copied
5.5 How Proteins Are Built
5.6 Genetic Mutations

Only in horror movies do humans give birth to aliens or giant flies. In real life, humans always give birth to human babies. But why is that? Why does a chicken egg always hatch a chick? Why not an eagle, or a bullfrog, or—just once!—a Tyrannosaurus rex? The offspring of living organisms are like their parents because they inherit their parents' genes. What are genes? What are genes made of? And how do genes determine what organisms are like? We'll learn the answers to these questions in this chapter.

5.1 What is a Gene?

Genes determine many traits in living organisms—the colors of a plant's flowers, the length of a cat's tail, and the substances that make up a crab's shell or a bacterium's cell wall. In humans, genes affect our eye color, whether we are tall or short, and whether our hair is straight or curly. Genes even influence our personalities. But what is a gene, and how does a gene determine a trait?

A **gene** is a section of DNA that contains the instructions for building a protein. An organism's collection of genes makes up the organism's **genotype**. The resulting traits of an organism make up its **phenotype**. Genotype and phenotype are illustrated in **Figure 5.1**.

READING CHECK

What is the "genotype" of an organism?

< Figure 5.1

An organism's genes, found in its DNA, make up its genotype. An organism's traits, such as this flamingo's pink feathers, curved beak, and yellow eyes, make up its phenotype.

Let's look at an example of genotype and phenotype in humans. Consider two genes for eye color, one gene calling for brown eyes and the other for blue eyes. If you have both these genes, then your genotype is one brown-eye gene and one blue-eye gene. But with this genotype, the color of your eyes will be brown. In this case, we say your phenotype is brown eyes. One subject of this chapter is how a genotype generates a particular phenotype—how genes become traits.

If genes contain instructions for building proteins, you might wonder: Why are proteins so important? Why do so many of our traits depend on proteins? The answer is that proteins do many important jobs in living organisms. Proteins provide structure, act as hormones, transport molecules, function in cell communication, and protect organisms from disease. In addition, enzymes are proteins, and enzymes are needed for practically every chemical reaction that occurs in living things.

CONCEPT **CHECK**

A boy has straight hair. Is that part of his genotype or phenotype?

CHECK YOUR ANSWER

The traits of an organism—such as the straight hair of a boy—are part of the boy's phenotype.

More to Explore

For more on genes, check out this website at the National Human Genome Research Institute:

https://www.genome.gov/genetics-glossary/Gene

5.2 Chromosomes

In eukaryotes, DNA is found in the cell nucleus, where it is packaged in structures called *chromosomes* (**Figure 5.2**). Each chromosome consists of a long DNA molecule wrapped around small proteins called *histones*. The histones keep DNA from getting tangled when it is not being used. This is needed because cells have a lot of DNA—if all the DNA in a single human cell were unwound, it would stretch to a length of seven feet! DNA is unwound from the histones when the cell needs to use it. Chromosomes are usually loosely packed, but they become condensed when a cell is dividing, something we will learn more about later.

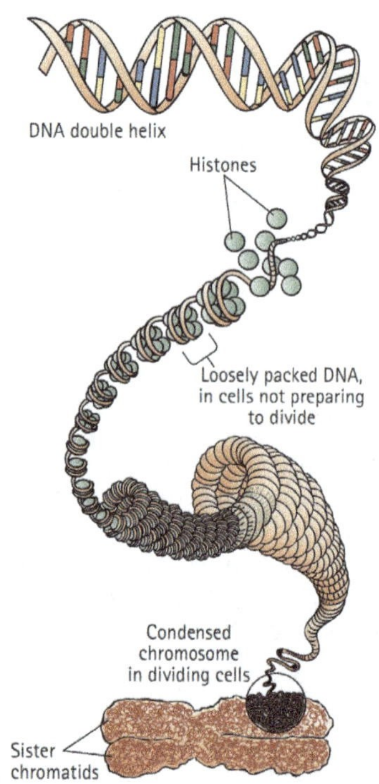

(a)

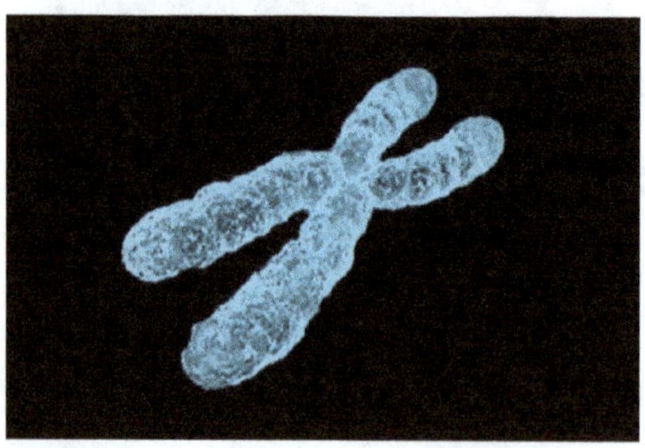
(b)

< Figure 5.2

(a) Chromosomes consist of DNA wrapped around histone proteins. Chromosomes are usually loosely packed but become condensed when a cell is dividing.
(b) This photo shows a condensed chromosome having duplicated itself into two sister chromatids in preparation for cell division.

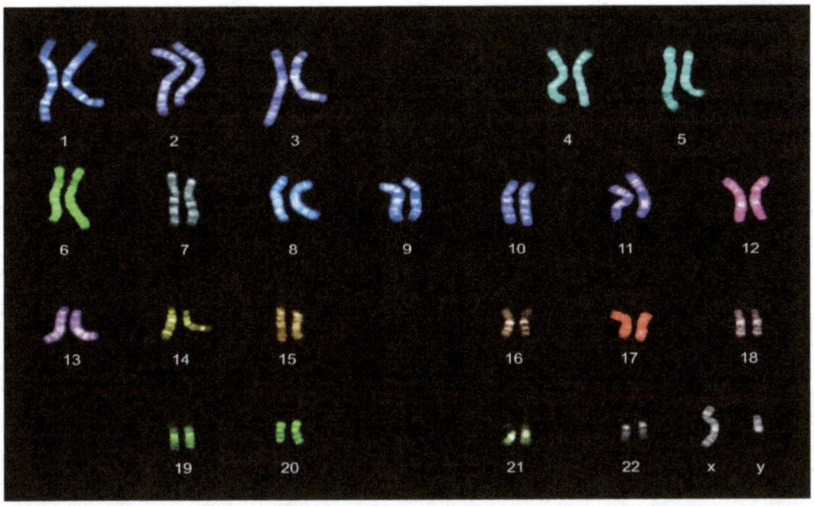

< Figure 5.3

Humans have 23 pairs of chromosomes. The last pair, the sex chromosomes, determine whether a person is male or female. Color was added to this image to help differentiate the pairs. Notice each chromosome is a single twisted strand.

Most cells have two of each kind of chromosome, like a pair of matched shoes. Cells with two of each kind of chromosome are **diploid**, and their matched chromosomes are called homologous chromosomes. Some cells—such as sperm and eggs—have only one of each kind of chromosome. These cells are **haploid**. Different organisms have different numbers of chromosomes. Chickens have 78 (39 pairs), mosquitoes have 6 (3 pairs), and yeast have 32 (16 pairs). In humans, there are 46 chromosomes, or 23 pairs (**Figure 5.3**). The last pair are called *sex chromosomes* because they determine sex. Females have two X chromosomes, whereas males have one X and one Y chromosome. The rest of the chromosomes are known as autosomes.

Most of our cells (except our haploid sperm and eggs) have the same 46 chromosomes, with the same DNA and the same genes. This might make you wonder: If all our cells have the same genes, what makes different kinds of cells so different? For example, what makes a brain cell so different from a muscle cell? The answer is that different genes are expressed in each type of cell. The proteins a brain cell makes from its DNA are very different from the proteins a muscle cell makes. To see how proteins are made from DNA, let's start at the beginning. What is the structure of DNA?

READING CHECK

How many pairs of chromosomes are there in humans?

CONCEPT CHECK

1. If a cell contains the chromosomes shown in Figure 5.3, is the cell haploid or diploid?
2. Do the chromosomes in Figure 5.3 belong to a male or female?

CHECK YOUR ANSWER

1. The cell is diploid because it contains two of each kind of chromosome.
2. They belong to a male because there is an X chromosome and a Y chromosome.

More to Explore

To read more about chromosomes, you can look at this website:

https://www.genome.gov/about-genomics/fact-sheets/Chromosomes-Fact-Sheet

5.3 The Structure of DNA

Now that we know genes are found in DNA let's look more closely at DNA itself. A molecule of **deoxyribonucleic acid**, or **DNA**, consists of two strands. Together, the two strands make a spiral ladder with two "sides" and a series of "rungs" (**Figure 5.4**). Because DNA is made up of two strands twisted into a spiral or helix, it is called a double helix.

READING CHECK

What is a nucleotide?

Let's look at a single strand of DNA, then at how DNA's two strands fit together. A DNA strand is a chain of nucleotides or structural building blocks, including a nitrogenous base, a sugar molecule, and a phosphate group. The sugar in DNA nucleotides is deoxyribose. So, a DNA strand has a backbone (or "side" of the ladder) made up of alternating molecules of deoxyribose sugar and phosphate. Sticking out from the backbone is a series of nitrogenous bases. Each nitrogenous base is half of a "rung" of the DNA ladder. The four nitrogenous bases in DNA are adenine (A), cytosine (C), guanine (G), and thymine (T).

Now let's put two strands of DNA together. Each nitrogenous base binds with a base on the other strand using chemical bonds known as *hydrogen bonds*. Because each nitrogenous base can best form hydrogen bonds with another specific nitrogenous base, the binding of bases always happens in a specific way. Adenine always pairs with thymine (A–T), and guanine always pairs with cytosine (G–C).

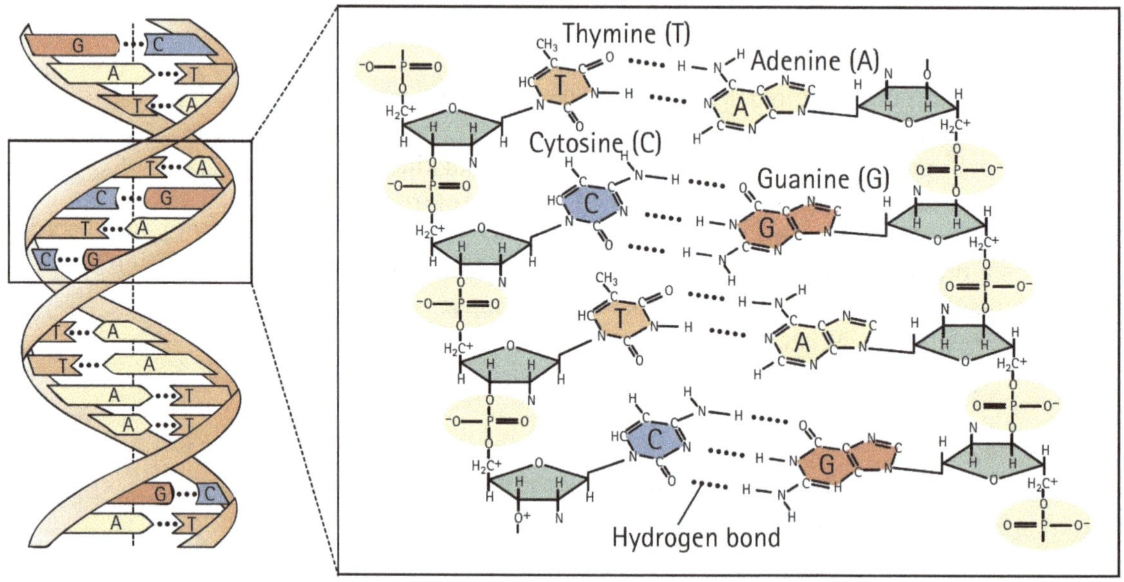

(a)

∧ Figure 5.4a

DNA is shaped like a spiral ladder with two sugar–phosphate strands as the "sides" of the ladder and paired bases as the "rungs." Hydrogen bonds hold the paired nitrogenous bases together.

Figure 5.4b >

This photograph of DNA shows that it has the shape of a double helix.

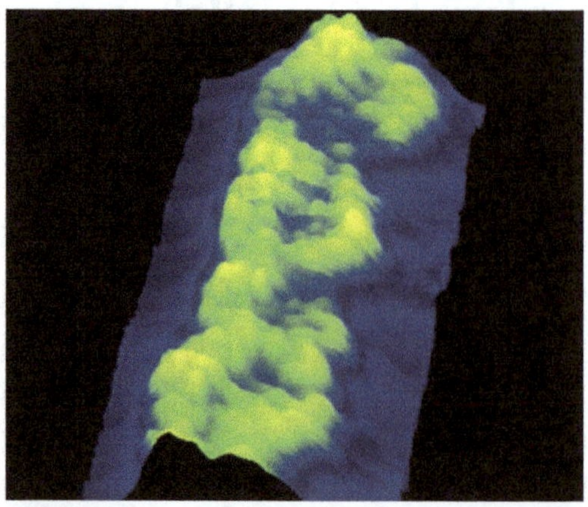

(b)

CONCEPT**CHECK**

If one strand of DNA contains the nucleotides ACCTGA, what are the nucleotides on the opposite strand?

CHECK YOUR ANSWER

Because of the way the nitrogenous bases pair, the opposite strand must have TGGACT.

More to Explore

You can read more about DNA structure here:

https://www.visiblebody.com/learn/biology/dna-chromosomes/dna-structure

5.4 How DNA Is Copied

DNA must be copied in order for cells to divide and reproduce. This process is called **DNA replication**. How does DNA replication happen? During DNA replication, DNA's two strands are separated as if the spiral ladder were unzipped down the middle. Each strand then serves as a template for building a new partner. This is possible because of the way the nitrogenous bases pair—because A always pairs with T, and G always pairs with C. As free nucleotides pair up with nucleotides on the template strand, they are attached to the new DNA strand. Each new DNA molecule includes one old strand and one new strand, and the new DNA molecules are identical to the original (**Figure 5.5**).

READING CHECK

How do nitrogenous base pairs pair up with one another?

DNA replication always begins at fixed locations within chromosomes. In eukaryotes, replication begins simultaneously at many different points in a chromosome, allowing the job to be completed efficiently. It takes a human cell a few hours to copy all of its DNA.

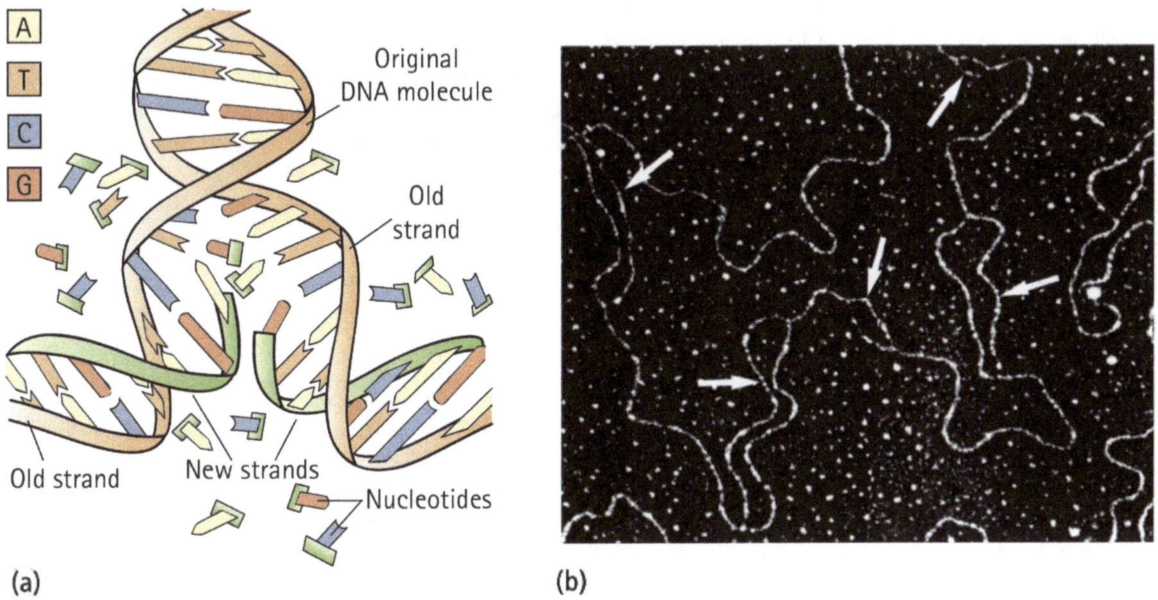

Figure 5.5

(a) When DNA is copied, the two strands are separated. Each old strand (orange backbone) serves as a template for building a new strand (green backbone). (b) This photo shows DNA replication in the cell of a fruit fly. Arrows show where the DNA strands have been separated and replicated.

CONCEPT CHECK

During DNA replication, the two strands of the DNA molecule are separated. If one strand has the nucleotides C and T. What nucleotides will bind to those?

CHECK YOUR ANSWER

Following the usual base-pairing rules, a G nucleotide will bind to the C, and an A nucleotide will bind to the T.

More to Explore

This video will help you review DNA structure and DNA replication:

https://www.youtube.com/watch?v=8kK2zwjRV0M&vl=en

5.5 How Proteins Are Built

We now know the structure of DNA and how DNA is copied. So we are almost ready to examine how DNA provides cells with the instructions for building proteins. Before we do that, though, we have to introduce one more molecule—RNA.

RNA

Like DNA, **ribonucleic acid**, or **RNA**, consists of a sugar–phosphate backbone attached to nitrogenous bases (**Figure 5.6**). But RNA differs from DNA in three ways: (1) RNA is single-stranded rather than double-stranded like DNA; (2) RNA uses the sugar ribose instead of deoxyribose; and (3) RNA uses the nitrogenous base uracil (U) instead of thymine (T).

How is the information in DNA used to build a protein? During **transcription**, DNA is used as a template for building a molecule of RNA. Then, d uring **translation**, this RNA molecule is used to assemble a protein (**Figure 5.7**).

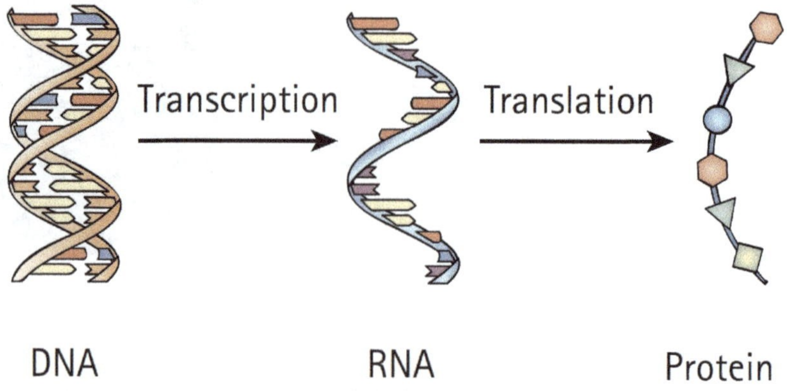

Figure 5.6 >

DNA is used to build a protein through transcription and then translation.

Transcription

In eukaryotes, transcription occurs in the cell nucleus. The two strands of DNA are separated, and one strand serves as a template for building an RNA transcript. Special nucleotide sequences in the DNA indicate where transcription should begin and end, marking the beginning and end of a gene. The construction of the RNA molecule follows the same base-pairing rules that exist for DNA, except that RNA uses uracil (U) instead of thymine (T). So, where DNA has the nucleotides A, C, G, and T, the RNA molecule will have the nucleotides U, G, C, and A, respectively (**Figure 5.8**).

As free RNA nucleotides pair up with complementary nucleotides on the DNA strand, they are added to the growing RNA molecule. Once the transcription is complete, the DNA zips back up, and the newly built RNA is ready for the next step, the processing phase.

The cell processes the RNA molecule into a mature **messenger RNA** molecule, or **mRNA** molecule. First, stretches of nucleotides that are not used to build the protein, called introns, are removed (**Figure 5.9**). The stretches of nucleotides that are used to build the protein, called *exons*, remain. It is as if a starting transcript of Shakespeare's Hamlet contained the line "aggfr uidosa to be dfjklsdf or rewerwe not to be," and the irrelevant parts had to be removed in order to leave the coherent "to be or not to be." (If you have trouble remembering which parts are removed and which parts are kept in the transcript, you can think to yourself, "EXons EXpressed, INtrons IN the trash.") Scientists are not sure why introns exist. However, a single RNA transcript can sometimes be processed in multiple ways, resulting in different mRNA molecules. Second, a "cap" and "tail" are added to the beginning and end of the RNA molecule. The cap and tail allow the cell to recognize the molecule as mRNA. Once processing is complete, the mRNA molecule moves from the nucleus of the cell to the cytoplasm, where it will be translated to build a protein.

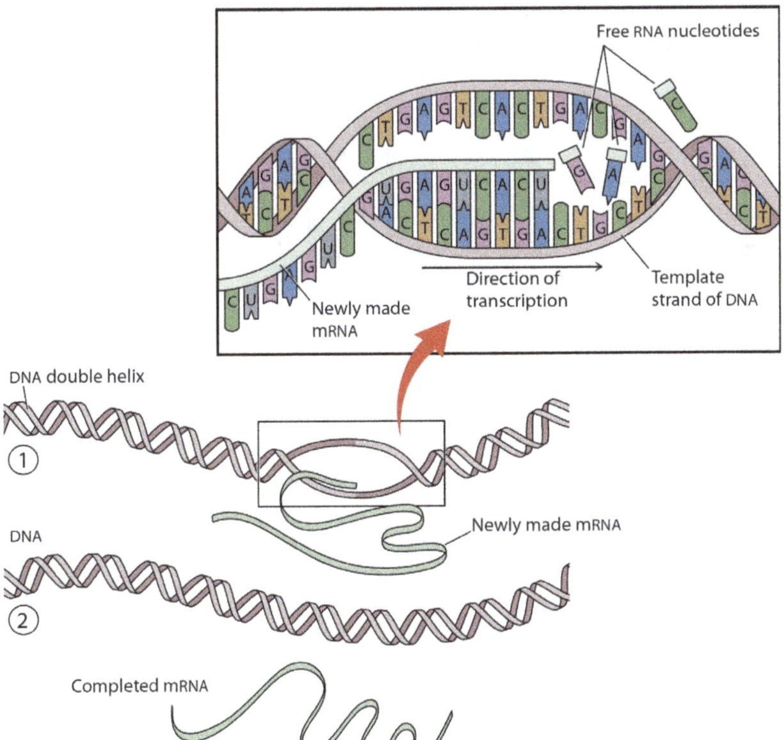

Figure 5.8

During transcription, information in DNA is used to build a molecule of RNA.

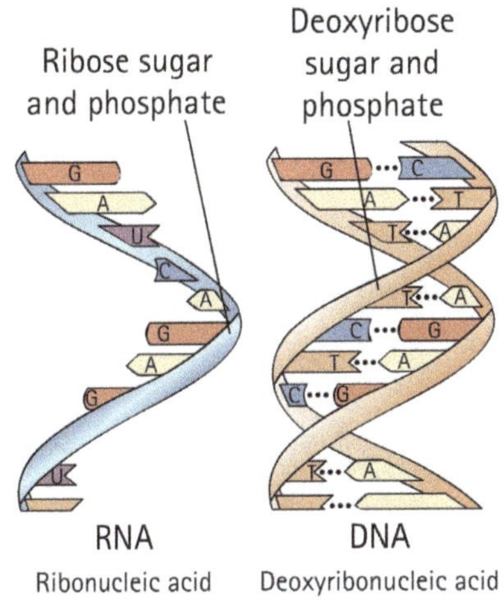

Figure 5.7

RNA is a single-stranded nucleic acid.RNA uses the sugar ribose and the nitrogenous base uracil instead of thymine.

READING CHECK

What are stretches of nucleotides that are not used to build mRNA called?

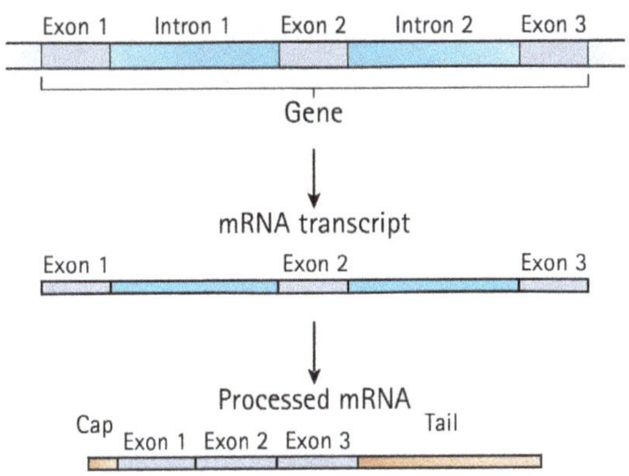

Figure 5.9

The RNA transcript undergoes a processing phase in which introns are removed, and a cap and tail are added. The result is an mRNA molecule that is ready for translation.

Translation

During translation, mRNA is used to build a protein. Translation takes place in the cell cytoplasm at the organelles called *ribosomes*. Ribosomes are composed of proteins and a type of RNA called *ribosomal RNA*, or *rRNA*. How does translation work? Remember that a protein is a folded chain of amino acids. In translation, three nucleotides along the mRNA strand form a three-letter "word" called a **codon**. During translation, codons are "read" from the mRNA molecule one at a time. Most of these codons tell the ribosome to add a specific amino acid to the protein being built. A few codons tell the ribosome that there are no more amino acids in the protein and that translation is complete.

The *genetic code* tells us what each codon stands for (**Table 5.1**). For example, the codon AGU translates to the amino acid serine, and the codon GUG translates to the amino acid valine. Certain codons, such as UAA, are called *stop codons* because they tell the ribosome that there are no more amino acids in the protein. The first codon to be translated from mRNA is always AUG—methionine. Interestingly, the genetic code is the same for practically all living organisms. So, you read the codons in mRNA, string together amino acids in the correct order, and presto! You have a protein. Now, how is all this accomplished? Translation requires another type of RNA, **transfer RNA**, or **tRNA**. A tRNA molecule includes a sequence of three nucleotides called an *anticodon* and carries a single, specific amino acid.

During translation, the mRNA molecule binds to a ribosome. The codon being translated is positioned at a specific site. A tRNA molecule with the appropriate anticodon binds to the codon (**Figure 5.10**). The binding of anticodon to codon follows the usual rules—A binds with U, and G binds with C. The amino acid carried by the tRNA is then added to the growing protein. After this, the ribosome shifts the mRNA so that a new codon can be translated. This process is repeated until a stop codon is reached. As translation proceeds, the growing protein folds up into its appropriate three-dimensional shape. It takes a ribosome less than a minute to build a typical protein molecule.

Table 5.1 >
The genetic code. For example, the codon CUA codes for the amino acid leucine, Leu.

	U		C		A		G		
U	UUU	Phe	UCU	Ser	UAU	Tyr	UGU	Cys	U
	UUC		UCC		UAC		UGC		C
	UUA	Leu	UCA		UAA	Stop	UGA	Stop	A
	UUG		UCG		UAG	Stop	UGG	Trp	G
C	CUU	Leu	CCU	Pro	CAU	His	CGU	Arg	U
	CUC		CCC		CAC		CGC		C
	CUA		CCA		CAA	Gln	CGA		A
	CUG		CCG		CAG		CGG		G
A	AUU	Ile	ACU	Thr	AAU	Asn	AGU	Ser	U
	AUC		ACC		AAC		AGC		C
	AUA		ACA		AAA	Lys	AGA	Arg	A
	AUG	Met Start	ACG		AAG		AGG		G
G	GUU	Val	GCU	Ala	GAU	Asp	GGU	Gly	U
	GUC		GCC		GAC		GGC		C
	GUA		GCA		GAA	Glu	GGA		A
	GUG		GCG		GAG		GGG		G

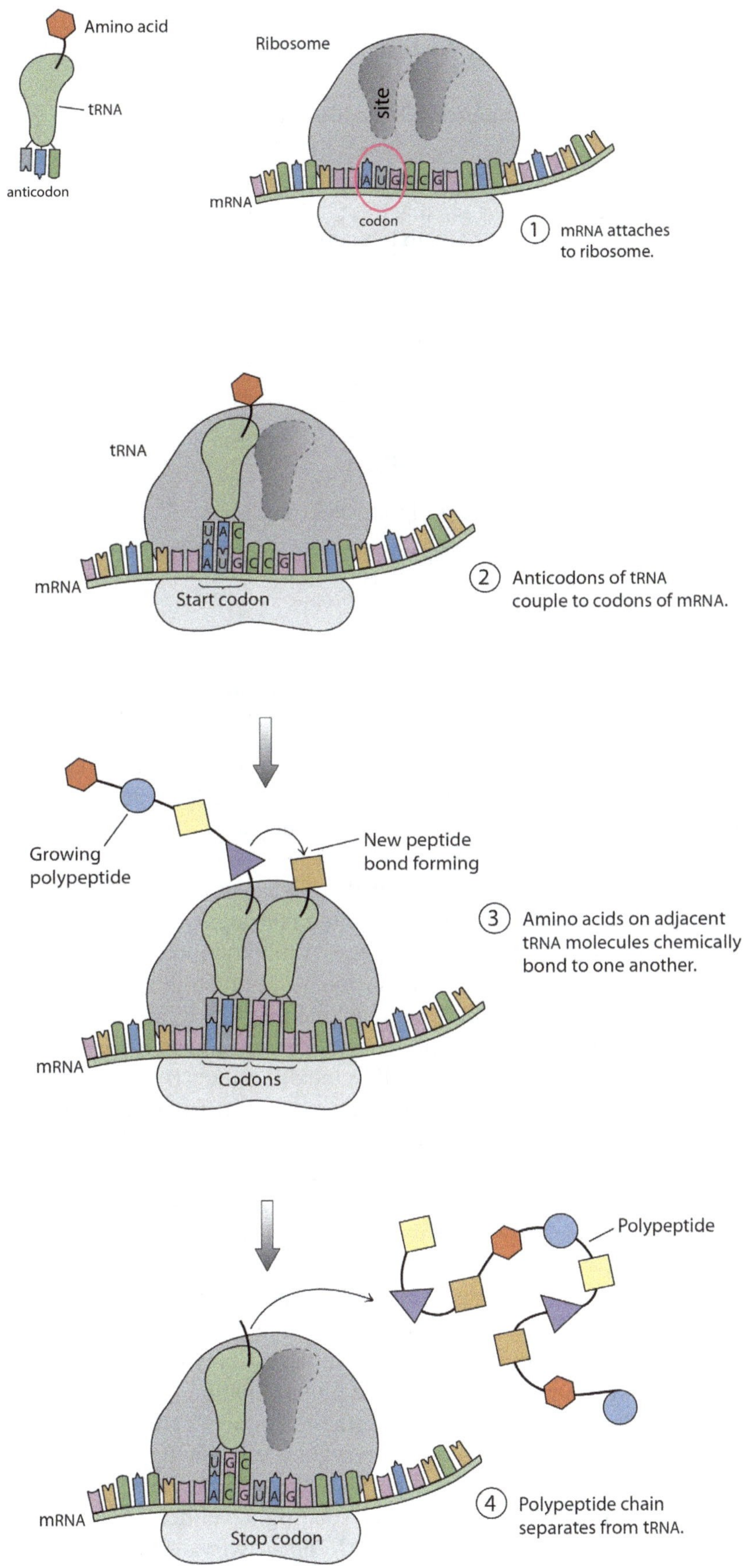

< Figure 5.10

Ribosomes are where messenger RNA is translated into a polypeptide chain through the action of transfer RNA.

CONCEPT **CHECK**

1. A molecule of mRNA has the codon GUC. What is the anticodon on the tRNA that binds to it during translation? What amino acid does the tRNA carry?
2. If a stretch of DNA has the nucleotide sequence ACCTGAT, what sequence will the RNA transcript have?

CHECK YOUR ANSWERS

1. The tRNA has the anticodon CAG. Looking at the genetic code table, we see that this tRNA molecule carries the amino acid valine. (When you use the genetic code table, remember that you need to look up the codon: GUC translates to valine.)
2. The RNA transcript will have the nucleotide sequence UGGACUA.

More to Explore

This video reviews transcription and translation:
https://www.youtube.com/watch?v=6YqPLgNjR4Q

5.6 Genetic Mutations

READING CHECK

Why do mutations in egg or sperm cells generally have a more significant effect than a mutation striking a skin cell on a toe?

A **genetic mutation** occurs when the sequence of nucleotides in an organism's DNA is changed. Genetic mutations have two leading causes: 1) errors during DNA replication and 2) exposure to things that damage DNA, such as ultraviolet light, X-rays, and chemicals.

A genetic mutation can have no effect at all, or it can have a significant effect. A mutation in a gene is likely to have a more consequential effect than one that affects DNA that is not part of a gene. A mutation in an exon is likely to have a bigger effect than one in an intron. A mutation in an egg or sperm may have a more considerable effect than one that strikes a skin cell on the big toe—this is because a mutation in an egg or sperm can be passed on to offspring and then appear in every cell in the offspring's body.

The majority of genetic mutations have no effect on proteins or organisms. However, a small number of mutations interfere with how proteins function, and so are disadvantageous for organisms. In rare cases, a mutation will produce something new and advantageous. Mutations are the source of all genetic diversity and provide the raw materials for evolution. For this reason, mutations are ultimately responsible for the diversity of life on Earth. Let's look at three kinds of mutations: point, frameshift, and nonsense.

A *point mutation* occurs when one nucleotide is substituted for another, such as when a C becomes a G. A point mutation may change a protein's sequence of amino acids. For example, a point mutation that changes AAC to AAG changes the amino acid asparagine to lysine. (Refer to the genetic code in Table 5.1.) Not all point mutations change amino acids, however. A point mutation that changes GCA to GCC has no effect because both codons stand for the same amino acid—alanine. You might wonder: Does changing one amino acid in a protein really matter? The answer is sometimes. Some amino acid changes do not affect protein function. Others have a significant effect—a single amino acid change is responsible for several serious human diseases, including sickle cell anemia, which we will discuss later. Nucleotides can also be inserted into or removed from DNA. If this happens in a gene, the codons that are "read" during translation can be shifted, producing a *frameshift mutation*. Frameshift mutations change a protein's amino acid sequence entirely, usually resulting in nonfunctional proteins. A *nonsense mutation* is a mutation that creates a stop codon in the middle of a gene. Nonsense mutations cause translation to stop before all the amino acids in a protein have been added. Nonsense mutations result in short, often nonfunctional, proteins. Both point mutations and frameshift mutations can result in a nonsense mutation. **Figure 5.11** shows examples of different mutations. **Figure 5.12** compares mutations to changes in English sentences—this may help you understand how different mutations can affect proteins.

(a) Original sequence:	AGC serine	CUG leucine	UAC tyrosine	UGG tryptophan	ACA threonine	UUG leucine	CCA proline
(b) Point mutation 1:	AGC serine	CUG leucine	UAC tyrosine	UGG tryptophan	ACU threonine	UUG leucine	CCA proline
(c) Point mutation 2:	AGG arginine	CUG leucine	UAC tyrosine	UGG tryptophan	ACA threonine	UUG leucine	CCA proline
(d) Nonsense mutation:	AGC serine	CUG leucine	UAG STOP	UGG	ACA	UUG	CCA
(e) Frameshift mutation:	AGC serine	UCU serine	GUA valine	CUG leucine	GAC aspartic acid	AUU isoleucine	GCC A alanine

^ Figure 5.11

Mutations can affect proteins in many different ways. (a) This is a gene's original nucleotide sequence. (b) This point mutation does not change the amino acid sequence of the protein. (c) This point mutation changes one amino acid into another. (d) The insertion of a nucleotide causes a frameshift mutation that completely changes a protein's sequence of amino acids. (e) This nonsense mutation produces a stop codon, which results in the production of a shorter protein.

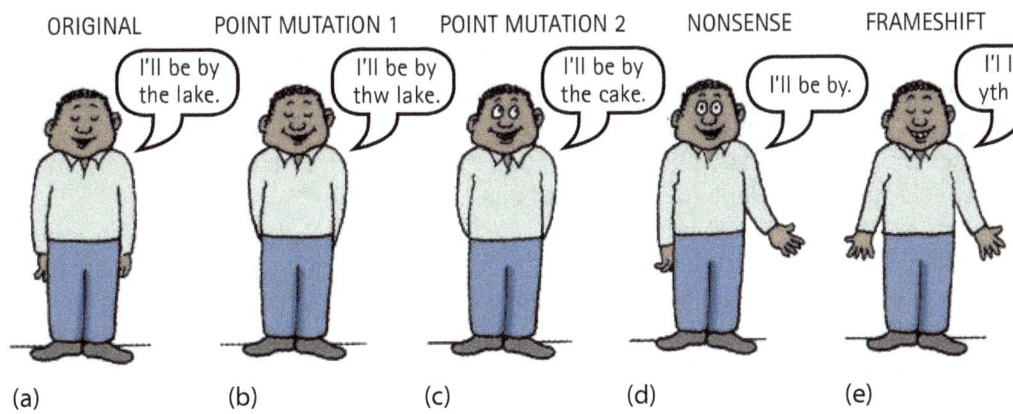

^ Figure 5.12

If a DNA nucleotide sequence were a sentence, different mutations would have different effects on the meaning of the sentence. (a) This is the original sentence (DNA nucleotide sequence). (b) This point mutation has only a minimal effect on the meaning (protein). (c) This point mutation affects the meaning (protein) significantly. (d) This frameshift mutation causes the meaning to be lost (the protein is nonfunctional). (e) This nonsense mutation causes the meaning to be lost (the protein is nonfunctional).

CONCEPT CHECK

Two different point mutations are shown below. What effect does each mutation have on the protein that is produced? (Use the genetic code in Table 5.1.) Original sequence: AGC CUG UAC UGG ACA UUG CCA.

1. AGC CUG UAC UGG ACC UUG CCA
2. AGC CUG UAA UGG ACA UUG CCA

CHECK YOUR ANSWERS

1. ACC and ACA code for the same amino acid (threonine), so this mutation has no effect on the protein.
2. This is a nonsense mutation—it produces a stop codon, shortening the protein to just serine–leucine. This is a serious mutation because it has a severe effect on the protein. A mutation like this in an important gene could cause illness or death.

More to Explore

You can read more about genetic mutations here:

https://www.nature.com/scitable/topicpage/genetic-mutation-1127/

Concept Review

Chapter 5

Detailed Chapter Summary

A *gene* is a section of DNA that contains the instructions for building a protein. Remember that proteins are essential in living organisms because they provide structure and functionality. For example, enzymes are proteins that help determine the chemical reactions in cells. An organism's genes make up its *genotype*, and an organism's traits make up its *phenotype*.

Genes are contained in structures called chromosomes. A *chromosome* consists of a single molecule of DNA wrapped around proteins called histones. Diploid cells have two of each type of chromosome. Haploid cells, such as sperm and eggs, have one of each type of chromosome. Humans have 23 pairs of chromosomes, including 22 pairs of autosomes and 1 pair of sex chromosomes. The sex chromosomes determine biological sex.

DNA has the structure of a double helix, that is, a spiraling ladder. DNA consists of two strands. Each strand has a backbone made of alternating molecules of deoxyribose sugar and phosphate (the "side" of the ladder) with a series of protruding nitrogenous bases (each representing one-half of a "rung" of the ladder). Four nitrogenous bases are used in DNA -- adenine, cytosine, guanine, and thymine -- or A, C, G, and T for short. Within the DNA molecule, the nitrogenous bases always bind with specific partners -- A always pairs with T, and G always pairs with C.

DNA replication describes the process through which a molecule of DNA is copied in preparation for cell division. During DNA replication, the two strands of DNA are separated, and each strand is used as a template to build a new strand. This is possible because of the specific base-pairing rules in DNA (A always pairs with T, and G always pairs with C). Because of how DNA replication occurs, each new DNA molecule consists of one old strand and one new strand.

Aside from DNA, the other nucleic acid essential to life is RNA. *RNA* is a single-stranded nucleic acid with a backbone consisting of alternating ribose sugar and phosphate molecules. Like DNA, RNA uses four different nitrogenous bases. These nitrogenous bases are similar to those found in DNA, except that RNA uses uracil (U) instead of thymine. Thus, the nitrogenous bases found in RNA are A, C, G, and U.

The process of building a protein from a section of DNA takes place in two steps, *transcription*, and *translation*. Transcription takes place in the cell nucleus. During transcription, DNA is used to build a molecule of RNA.

The two strands of DNA are separated, and one strand is used to build the RNA molecule. This process relies on the usual base pairing rules. Thus, where DNA has A, C, G, and T, respectively, the RNA transcript will have U, G, C, and A. A processing phase converts the RNA transcript to a mature messenger RNA, or *mRNA*, molecule. During the processing phase, a cap and tail are added to the molecule, introns are removed from the sequence, and exons are left in the sequence.

Transcription is followed by the process of translation, which occurs at ribosomes in the cell cytoplasm. During translation, a *codon* of three nucleotides in the mRNA molecule provides instructions on the assembly of the protein. These instructions include signaling the start of translation, specifying the addition of a single specific amino acid, and signaling the end of translation. The *genetic code* describes the specific instruction provided by each three-nucleotide codon. Transfer RNA, or *tRNA* molecules, assist in translation. Each tRNA molecule carries a three-nucleotide sequence called an *anticodon* as well as a specific amino acid. During translation, the appropriate anticodon binds to the codon in the mRNA molecule via base-pairing, and the associated amino acid is added to the growing protein. Translation ends when a stop codon is reached.

A *genetic mutation* occurs when the sequence of nucleotides in an organism's DNA is changed. The effects of genetic mutations on organisms vary widely—a mutation can have no effect at all, or it can be fatal. Genetic mutations are also the source of all genetic diversity. Some specific types of genetic mutations include point, nonsense, and frameshift mutations. A point mutation occurs when one nucleotide is substituted for another. A nonsense mutation creates a stop codon in the middle of a gene. A frameshift mutation occurs when the insertion or deletion of nucleotides results in a shifted reading frame during translation.

Key Terms

Codon A sequence of three nucleotides in an mRNA molecule that either stands for an amino acid or ends translation.

Deoxyribonucleic acid (DNA) The cell's genetic material: a double-stranded molecule consisting of sugar–phosphate backbones attached by pairs of matched nitrogenous bases in a double helix.

Diploid Describes a cell that has two of each kind of chromosome.

DNA replication The process through which a DNA molecule is copied in order for cells to divide and reproduce.

Gene A section of DNA that contains the instructions for building a protein.

Genetic mutation A change in the nucleotide sequence of an organism's DNA.

Genotype The genetic makeup of an organism.

Haploid Describes a cell that has one of each kind of chromosome.

Messenger RNA (mRNA) An RNA molecule made during transcription that carries genetic information from DNA to the ribosomes.

Phenotype The traits of an organism.

Ribonucleic acid (RNA) A single-stranded molecule consisting of a sugar–phosphate backbone attached to a series of nitrogenous bases.

Transcription The first step in building a protein, in which a molecule of RNA is assembled from information contained in DNA.

Transfer RNA (tRNA) An RNA molecule that transfers an amino acid to a growing protein during translation.

Translation The assembly of a protein based on information contained in an mRNA molecule.

Review Questions

Simple Review Questions

5.1 What Is a Gene?

1. What is a gene?

2. Why do proteins determine many of an organism's traits?

5.2 Chromosomes: Packages of Genetic Information

3. How is DNA packaged into chromosomes?

4. What is the difference between diploid and haploid cells? What types of cells are haploid?

5.3 The Structure of DNA

5. Why is DNA described as a double helix?

6. How is DNA like a ladder? What are the "sides" of the ladder, and what are the "rungs"?

7. What are the four nucleotides found in DNA? How do they pair?

5.4 How DNA Is Copied

8. How is DNA copied?

9. Is a new molecule of DNA put together using two newly made strands?

5.5 How Proteins Are Built

10. How does RNA differ from DNA?

11. What base-pairing rules are followed in making an RNA transcript from a DNA template?

12. What is a codon?

13. Describe the role of tRNA in translation.

5.6 Genetic Mutations

14. What are the two leading causes of genetic mutations?

15. What is a point mutation?

16. What is a frameshift mutation and its effect on a protein?

Challenging Review Questions

5.1 What Is a Gene?

17. What is the difference between genotype and phenotype? Is your height part of your genotype or phenotype?

18. How do your genes determine what kinds of chemical reactions occur in your cells?

5.2 Chromosomes: Packages of Genetic Information

19. Is your finger made of diploid cells or haploid cells?

20. Why do some of your cells have only 23 chromosomes rather than 46?

21. What kind of sex chromosomes do you have? Where in your body are sex chromosomes found?

22. Do different types of cells in your body have different genes? If not, what makes them so different from one another?

5.3 The Structure of DNA

23. If you know the nucleotides on one strand of DNA, explain how you can tell what nucleotides are on the other strand of DNA.

24. A DNA nucleotide is made up of a nitrogenous base, a sugar molecule, and a phosphate group. Explain how these three types of molecules form the DNA "ladder."

25. If DNA is made up of nitrogenous bases, sugar molecules, and phosphate groups, why is all the genetic information in a DNA molecule contained in the nitrogenous bases rather than in the sugar molecules or phosphate groups?

5.4 How DNA Is Copied

26. When DNA is copied, why isn't there an "old" molecule and a "new" molecule?

27. Can RNA replicate the way DNA replicates? Why or why not?

5.5 How Proteins Are Built

28. How is transcription similar to DNA replication? How is it different?

29. We compared mRNA processing to editing "aggfr uidosa to be dfjklsdf or rewerwe not to be" to obtain "to be or not to be." In this comparison, is "aggfr" an exon or an intron? Is "not" an exon or an intron?

30. What is the genetic code?

31. Do all codons code for amino acids? If not, what else can a codon code for?

5.6 Genetic Mutations

32. Are point mutations in the first, second, and third positions of a codon equally likely to cause a change in the amino acid sequence of a protein? What type of point mutation is least likely to change the amino acid

sequence? Use the genetic code in Table 5.1 to answer these questions.

33. Why is a frameshift mutation more likely to disrupt a protein's function than a point mutation?

34. You are studying two different mutations in a gene that codes for a protein. In the first, a nonsense mutation occurs near the beginning of the gene. In the second, a nonsense mutation occurs near the end of the gene. Which mutation is more likely to disrupt protein function?

35. You are studying two different mutations in a gene that codes for a protein. In one mutation, a single nucleotide is inserted near the beginning of the gene. In the other mutation, three nucleotides are inserted near the beginning of the gene. Which mutation is more likely to disrupt protein function? Why?

Apply & Discuss Questions

36. What are the three types of RNA, and what is the function of each type?

37. Your friend Stacie says, "I understand how a point mutation can result in a nonsense mutation, but I don't get how a frameshift mutation can result in a nonsense mutation. Can it?" What do you say? Can you come up with an example of a frameshift mutation that results in a nonsense mutation?

38. "Listen to this," your friend says, reading from a newspaper article. "'Scientists believe that most genetic mutations have no effect on organisms, a small number have a disadvantageous effect, and an even smaller number have an advantageous effect.'" She looks at you. "Do you agree, and if so, why?"

Solutions to Odd-Numbered Questions

Simple Review Solutions

1. A gene is a section of DNA that contains the instructions for building a protein.

3. Each chromosome consists of a long DNA molecule wrapped around small proteins called histones.

5. Because DNA is made up of two strands twisted into a spiral or helix, it is called a double helix.

7. The four nucleotides in DNA are adenine (A), guanine (G), cytosine (C), and thymine (T). Adenine always pairs with thymine (A–T), and guanine always pairs with cytosine (G–C).

9. Each new DNA molecule includes one old strand and one new strand.

11. Where DNA has the nucleotides A, C, G, and T, the RNA transcript will have the nucleotides U, G, C, and A, respectively.

13. A tRNA molecule includes a sequence of three nucleotides called an anticodon and carries a single, specific amino acid. During translation, the mRNA molecule binds to a ribosome. The codon being translated is positioned at a specific site. A tRNA molecule with the appropriate anticodon binds to the codon. The binding of anticodon to codon follows the usual rules—A binds with U, and G binds with C. The amino acid carried by the tRNA is then added to the growing protein.

15. A point mutation occurs when one nucleotide is substituted for another, such as when a C becomes a G.

Challenging Review Solutions

17. An organism's genes, found in its DNA, make up its genotype. The traits of an organism make up its phenotype. Your height is one of your traits, so it is part of your phenotype.

19. Your finger is made of diploid cells, like most of your body except for your sex cells.

21. Females are XX, males are XY. Sex chromosomes are found in every cell in your body (with the exception of cells that lack nuclei, such as red blood cells).

23. Each nucleotide can best form hydrogen bonds with another specific nucleotide. Because of this, the binding of nucleotides between two strands of DNA occurs in a specific way. Adenine always pairs with thymine (A–T), and guanine always pairs with cytosine (G–C). These base-pairing rules allow the nucleotides of the second strand to be deduced from the nucleotides on the first strand.

25. The sugar molecules and phosphate groups are found in a fixed, constant pattern throughout a single organism's DNA and even from one organism to another. The nitrogenous bases, however, differ and so are able to hold information.

26. Every new DNA molecule has one old strand and one new strand because during DNA replication, the original DNA molecule is unzipped, and each strand is used as a template for putting together a new strand.

Apply & Discuss Solutions

37. A frameshift mutation can result in a nonsense mutation. Suppose mRNA has the sequence CCUAAC. This would code for two amino acids: Proline-Asparagine. Then an insertion between the first two C's produces the new sequence CCGUAAC. The new sequence codes for Proline-STOP. In other words, it results in a nonsense mutation.

Living Earth Essay

Life on Earth Connections

The Great Oxygenation

When a tree is cut down, you can see rings in its stump, showing how it grew in response to seasonal changes over the tree's life. The number of rings gives you a reasonable estimate of how many years the tree lived. Tree rings also reveal information about past climate. Trees generally grow faster when there is a lot of moisture, which is reflected in widely spaced tree rings. Drought produces rings that are closer together.

In a similar manner, Earth's rocks reveal much about our planet's history. An excellent example of such rocks is the **banded iron formations** (Figure A). While tree rings provide information about the recent past, banded iron formations provide information about the deep past from almost 4 billion years ago until almost 2 billion years ago.

When you look at a banded iron formation, the first thing you may notice is its beautiful layers. The red layers are composed of iron oxide, a form of rust. Figure B shows how the iron-containing layers formed in the sea during Earth's early history. The rock also has other layers in shades of white and gray. Why are there so many layers in a banded iron formation? They reflect alternating time periods when there was more and less oxygen, O_2, in the ocean. The key is the following: There was plentiful iron in the early oceans from sources such as undersea volcanoes, but there was no free oxygen. As oxygen built up due to biological processes, it reacted with the free iron to form iron oxide, which sank to the ocean floor. Therefore, each band of iron within a BIF is a signature of the presence of oxygen, O_2.

While tree rings can tell us about recent weather patterns, banded iron formations reveal the history of oxygen, O_2. For insight into this history, it's essential to recognize

Figure A
Like tree rings (left), banded iron formations (right) provide important clues about the past.

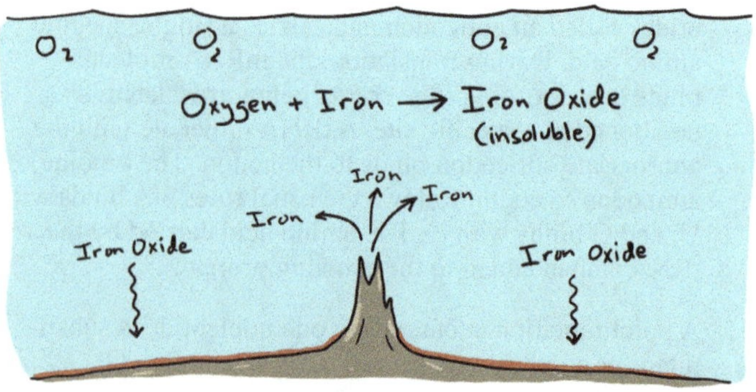

Figure B
Iron introduced to a body of water will react with any available oxygen to form insoluble iron oxide, which sinks to the floor.

that oxygen, O_2, is a very reactive molecule. There must be a continual source for large amounts of oxygen to accumulate. By analogy, consider a bubble machine creating bubbles at a birthday party. Bubbles easily pop. You'll only get lots of bubbles as long as the machine generates them. Likewise, oxygen molecules quickly react. To have lots of oxygen molecules, they need to be generated perpetually.

READING CHECK

What portion of Earth's history do banded iron formations tell us about?

TThe first BIFs did not start forming until almost 4 billion years ago when free oxygen started accumulating on Earth. Oxygen became plentiful due to the actions of Earth's early photosynthesizers, the oxygen-producing cyanobacteria. However, due to its great reactivity, oxygen can be quite toxic. So the early photosynthetic bacteria could have poisoned themselves with the oxygen they were producing. Fortunately for them, though, the iron in the seawater could quickly react with and remove the poisonous oxygen. Thus the cyanobacteria became highly dependent upon iron.

A cycle was set up that lasted for millions of years. The cyanobacteria produced lots of oxygen that was removed by the iron. Periodically, however, the iron in seawater became depleted. When this happened, toxic oxygen built up, which would kill the cyanobacteria. But when the concentration of iron increased again due to various geological forces, the cyanobacteria would be able to prosper once again. From these cycles, iron oxide bands accumulated on the ocean floor (**Figure C**). The banded iron formations in the rock record tell the story.

Eventually, the cyanobacteria evolved ways to detoxify the oxygen they produced. This was a crucial moment in the history of life on Earth. No longer so dependent upon the iron, the cyanobacteria were able to proliferate. As a result, the oxygen levels in the water increased dramatically. Any iron that was introduced to the water would immediately react to form iron oxide and sink to the ocean floor. Although the ocean sediments still contained iron oxide, the degree of banding greatly decreased as the cyanobacteria thrived.

The absence of banded iron formations in the rock record at about 2 billion years ago suggests that this was a time of great oxygenation. It is sometimes referred to as the **Great Oxygenation Event**. The oxygen build-up in ocean water resulted in the diffusion of oxygen into the atmosphere, which until then had been mostly empty of oxygen (**Figure D**).

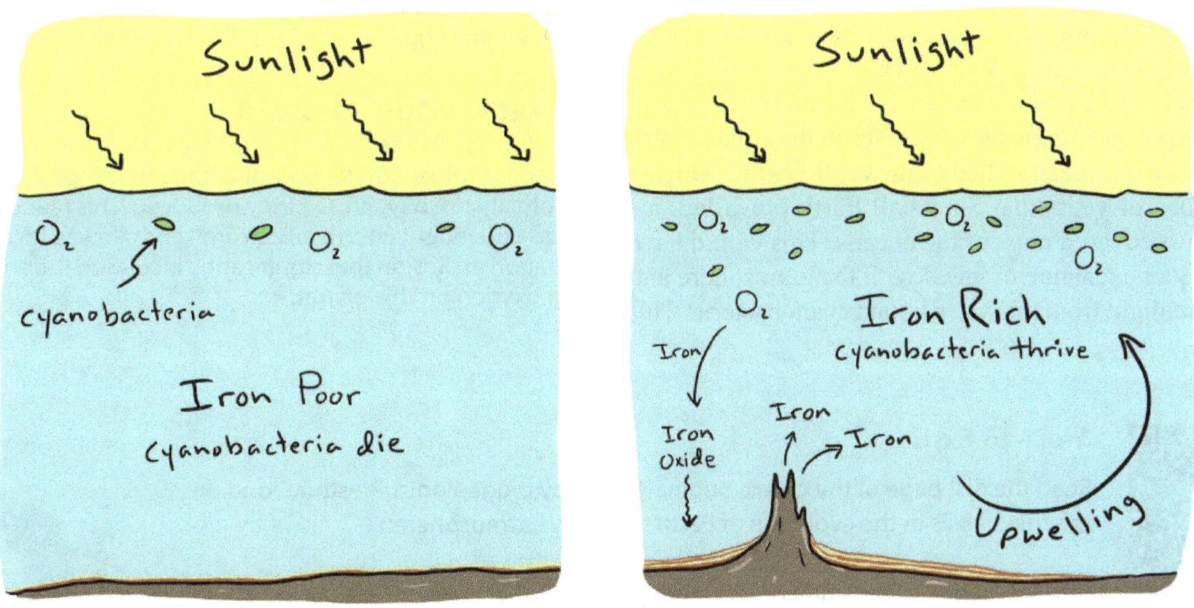

Figure C
(a) Early cyanobacteria within iron-poor waters would die from the toxicity of the oxygen they produced. (b) Within an iron-rich environment, the cyanobacteria could thrive as the oxygen would be removed by the iron.

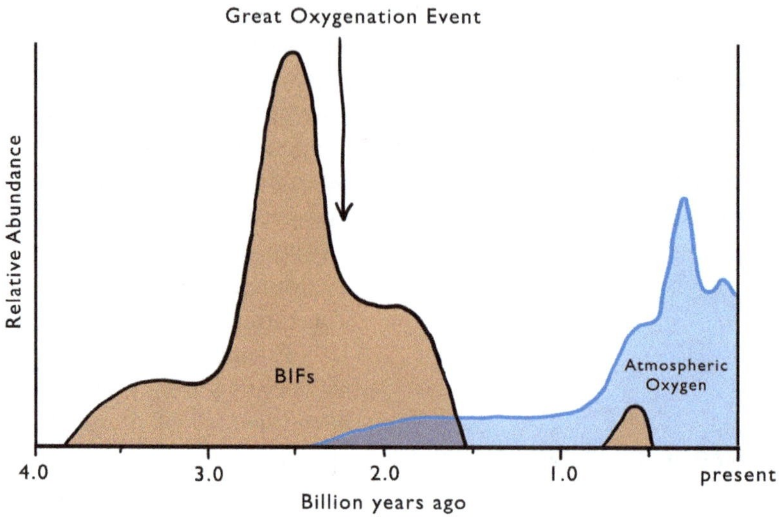

Figure D
The brown graph shows the relative abundance of Banded Iron Formations (BIFs) dating back about 4 billion years ago. The Great Oxygenation Event likely began when the formation of BIFs started to decline rapidly. In blue are the relative amounts of oxygen found in the atmosphere.

As oxygen moved into the atmosphere, it started to react with minerals on land. Atmospheric oxygen levels were still relatively low for many years, on the order of 5%, compared to today's 21%. Only after all the minerals in the oceans, lakes, and land were fully oxidized did oxygen levels start to rise above the 5% mark. The rise in atmospheric oxygen set the stage for the further evolution of oxygen-requiring multicellular living organisms.

Extensive banded iron formations have not formed since the Great Oxygenation Event. But there is an intriguing exception ranging from about 800 to 600 million years ago. During this time there was a small resurgence of BIFs. This raises the question: With oxygen freely available in the atmosphere, how could these have formed? One theory is that the polar ice caps moved closer to the equator producing what has been called "Snowball Earth." During the millions of years that Snowball Earth prevailed, ice would have covered most of the oceans. This would have prevented an exchange of gases with the atmosphere and blocked sunlight from the water-bound cyanobacteria. This would favor anoxic waters, a condition needed for banded iron formations.

There is still much to learn about our deep past. As the example of banded iron formations shows, figuring out what happened often requires understanding how biology and chemistry interact with Earth's materials.

To Life!

CONCEPT CHECK

According to our current understanding, what evolutionary event triggered the Great Oxygenation Event?

CHECK YOUR ANSWER

The Great Oxygenation Event may have been triggered when photosynthetic cyanobacteria developed ways to detoxify the oxygen, O_2, they produced. This made them less dependent upon available iron and allowed for a population explosion that significantly increased the amount of oxygen on this planet.

More to Explore

Read the first page of the classic original article by biogeologist Preston Cloud on the role of BIFs in the evolution of Earth's primitive atmosphere:

https://science.sciencemag.org/content/160/3829/729

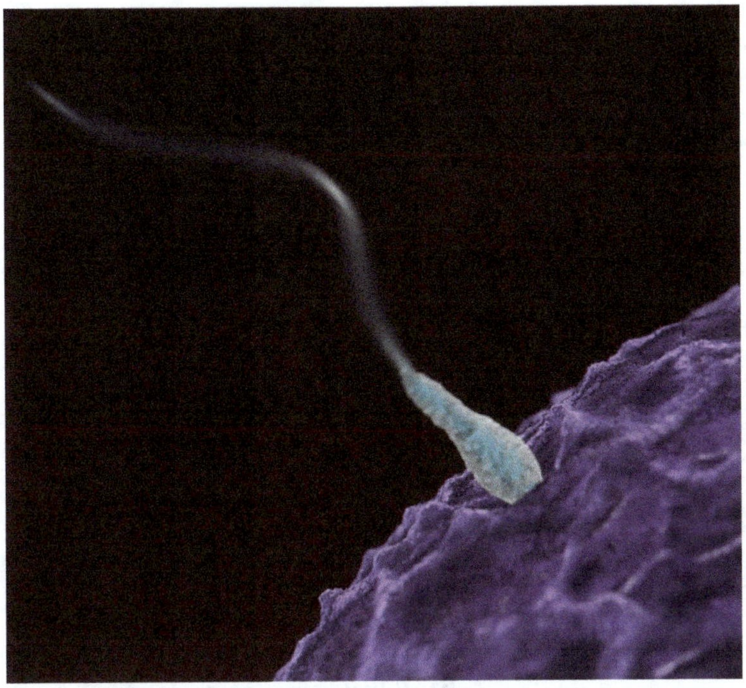

Chapter 6

Inheritance

THE MAIN IDEA

Genes determine a lot about what we are like.

6.1 How Cells Reproduce
6.2 Cell Division and Genetic Diversity
6.3 Traits and Inheritance
6.4 First Law of Inheritance
6.5 Second Law of Inheritance
6.6 Beyond Mendel

He has his mother's eyes and his father's nose... but why those, and not his *father's* eyes and his *mother's* nose? In this chapter, we'll focus on how traits are inherited. We'll also learn how changes in our genes are ultimately responsible for all the different kinds of life on Earth—as well as for challenging diseases such as cancer.

6.1 How Cells Reproduce

Cells reproduce by dividing. Cell division allows single-celled organisms to reproduce themselves entirely. It allows multicellular organisms to develop, grow, and maintain their tissues (**Figure 6.1**).

Mitosis is a form of cell division in which one parent cell divides into two daughter cells, each containing the same genetic information as the parent cell. How does this cell division occur? Cells that are preparing to divide enter the *cell cycle*, shown in **Figure 6.2**.

The cell cycle is divided into four stages—gap 1, synthesis, gap 2, and mitosis and cytokinesis. (If you need help remembering these in the correct order, you can try using the mnemonic phrase "**G**o, **S**ally, **G**o! **M**ake **C**ake!") Gap 1, synthesis, and gap 2 are collectively known as *interphase*.

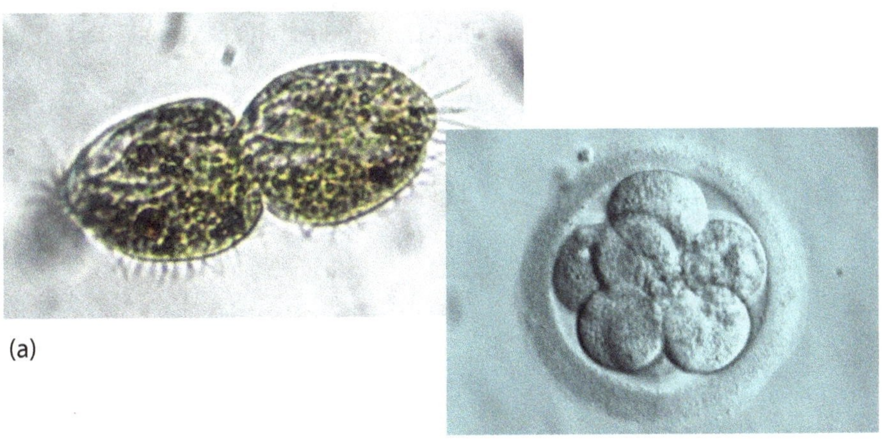

(a)

(b)

< Figure 6.1

Cell division is essential for reproduction, growth and development, and maintenance. (a) A paramecium, a single-celled organism, reproduces by dividing into two. (b) The early development of a human embryo involves multiple divisions of the fertilized egg. After day 3, the fertilized egg has divided several times to produce this early embryo, which has eight cells.

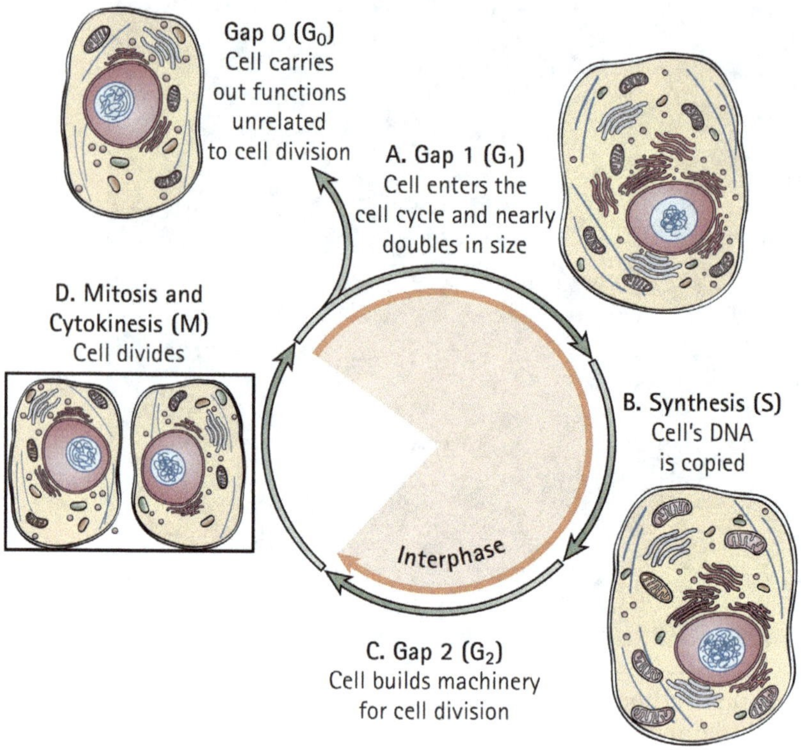

Figure 6.2 >

The cell cycle has four stages.
(A) During gap 1, the cell grows to about double its original size.
(B) During synthesis, an exact copy of the cell's DNA is made.
(C) During gap 2, the cell builds the machinery required for mitosis.
(D) During mitosis and cytokinesis, the cell divides.

During interphase, the cell makes the necessary preparations for division. During mitosis and cytokinesis, the cell divides. Let's look at each step in more detail.

During gap 1 (G_1), a cell prepares to divide by growing to approximately double its original size. All the important components of the cytoplasm, including the mitochondria and other organelles, also double in number. You might notice that calling this stage a "gap" is a little misleading in suggesting that nothing is going on. In fact, important events occur during both "gap" stages.

They are gaps only from the point of view of someone focused exclusively on whether the cell's DNA is doing anything obvious, such as replicating or being divided up.

During synthesis (S), the cell creates an exact copy of its genetic material—its DNA. This occurs through the process of DNA replication, which was described in Chapter 5.

During gap 2 (G_2), the cell builds the machinery necessary for cell division. This includes the structures that will separate the two copies of the genetic material and divide the cell into two daughter cells.

During mitosis and cytokinesis (M), the cell divides. Mitosis describes the division of the nucleus, which takes place in four steps, or "phases" (**Figure 6.3**). During *prophase*, the normally loosely packed chromosomes

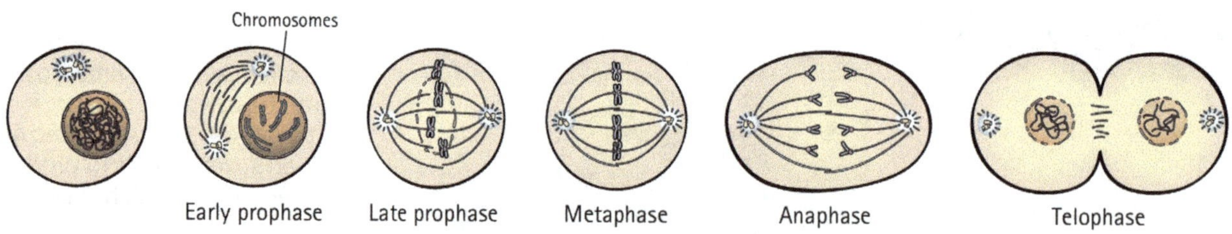

∧ Figure 6.3

There are four phases of mitosis: prophase, metaphase, anaphase, and telophase. During cell division, mitosis is followed by cytokinesis.

condense, and the membranes surrounding the nucleus break down. The condensed chromosomes each consist of two identical sister chromatids attached at a point called the centromere (**Figure 6.4**).

The *mitotic spindle* forms during prophase. The mitotic spindle consists of a series of fibers that attach to the chromatids and split the identical genetic material between the two daughter cells. This happens during *metaphase*, which begins when the chromosomes line up at the equatorial plane—the plane that passes through the imaginary "equator" of the cell. Then, during *anaphase*, the two sister chromatids are pulled apart by the shortening of the mitotic-spindle fibers and move to opposite poles of the cell. New nuclear membranes form around each set of chromosomes during *telophase*, and the chromosomes return to their loosely packed state. The division of the nucleus is followed by cytokinesis, the division of the cytoplasm to yield two separate daughter cells. There is a mnemonic for the phases of mitosis, too: "**P**eople **M**eet **A**nd **T**alk **C**ars."

Note that cells are not always in the cell cycle. Many cells are neither dividing nor preparing to divide; they are simply carrying out their regular functions. These cells are said to be in gap 0 (G_0). Some cells are in G_0 temporarily and will eventually re-enter the cell cycle. Other cells, such as many neurons, are in permanent G_0 and will never divide again.

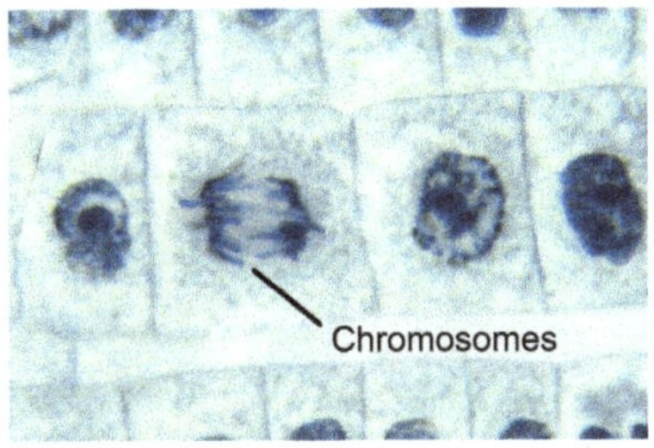

^ Figure 6.4

This closeup of chromosomes as seen during mitosis, shows that the DNA replicated during the synthesis phase becomes condensed.

READING CHECK

What part of the cell divides during cytokinesis?

CONCEPT CHECK

During which stages of the cell cycle does the cell have twice its normal amount of genetic material?

CHECK YOUR ANSWER

The cell has twice the normal amount of genetic material during G_2 and mitosis, M, stages—that is, after synthesis, S, stage and continuing up until cytokinesis, C, is complete. The cell has double the amount of genetic material after synthesis, S, because it copies its genetic material during synthesis, S. During mitosis, M, all that genetic material is still in one big cell. Only after cytokinesis, when two new cells are formed, does each new cell have the normal amount of genetic material.

More to Explore

Your genome is produced by the public engagement team at the Wellcome Genome Campus in Cambridge, United Kingdom

https://www.yourgenome.org/facts/what-is-mitosis/

6.2 Division and Genetic Diversity

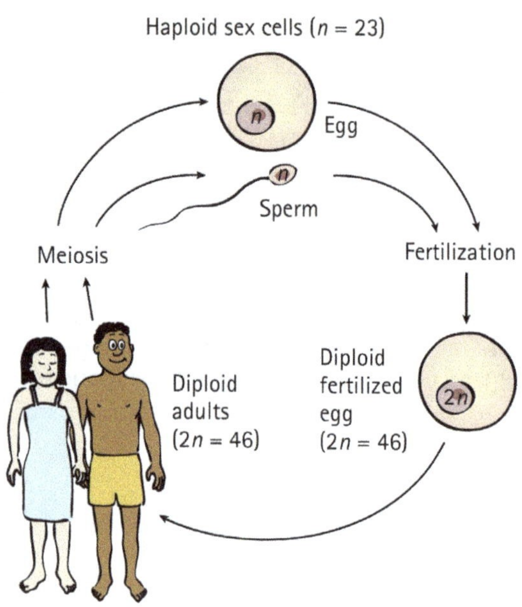

Figure 6.5

Sperm and eggs are haploid cells produced during meiosis. At fertilization, sperm and egg fuse to make a diploid cell. This diploid cell develops into a diploid individual (right).

You have learned that some cells reproduce through a type of cell division called *mitosis*. In mitosis, one cell divides into two daughter cells, each of which contains the same genetic information as the original cell. **Meiosis** is another type of cell division. Meiosis is used to make haploid cells, such as eggs and sperm. In meiosis, one diploid cell, with two of each kind of chromosome, divides into four haploid cells, each with only one of each kind of chromosome. The normal diploid chromosome number is restored during sexual reproduction when sperm and egg fuse at fertilization (**Figure 6.5**).

When meiosis begins, the diploid cell has already copied its DNA. Meiosis takes place in two steps: *meiosis I* and *meiosis II* (**Figure 6.6**). During meiosis I, the original cell divides into two cells. During meiosis II, these two cells divide again to produce four haploid cells.

Meiosis I begins with prophase I. During *prophase I*, the chromosomes condense, and the membranes of the nucleus break down. As in mitosis, each chromosome includes two identical sister chromatids. Homologous chromosomes then line up with each other, and a process known as *crossing over* occurs. In crossing over, a chromosome exchanges parts with its homologous chromosome (**Figure 6.7**). As a result, the chromosomes in the dividing cell are no longer identical to the ones in the original cell. Instead, many chromosomes now include parts of *each* of the two original homologous chromosomes. On a genetic level, crossing over results in **recombination**, the production of new combinations of genes different from those found in the original chromosomes. We'll see why this is important when we consider how meiosis produces genetic diversity.

In *metaphase I*, homologous chromosomes line up in the middle of the cell. In *anaphase I*, the homologous chromosome pairs separate. In *telophase I*, the chromosomes move to opposite ends of the cell. Cytokinesis occurs, producing two cells.

Meiosis II is similar to meiosis I, except for a few key differences. During *metaphase II*, the single, unpaired chromosomes (as opposed to homologous chromosome pairs) move to the center of the cell. In *anaphase II*, the sister chromatids separate. In *telophase II*, the sister chromatids move to opposite ends. Cytokinesis occurs, producing four haploid cells.

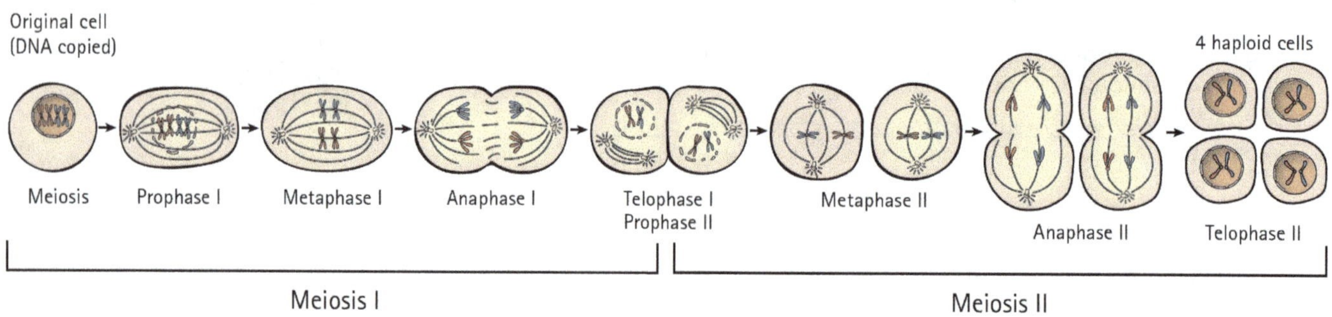

Figure 6.6

Meiosis produces four haploid cells.

A mistake during meiosis can cause a sperm or egg to end up with the wrong number of chromosomes. In humans, embryos produced by these sex cells usually do not survive, but there are exceptions. One of the most common chromosomal abnormalities is having three copies of chromosome 21, which is called *trisomy 21*. Trisomy 21 causes Down syndrome, a condition characterized by mental retardation and defects of the heart and respiratory system.

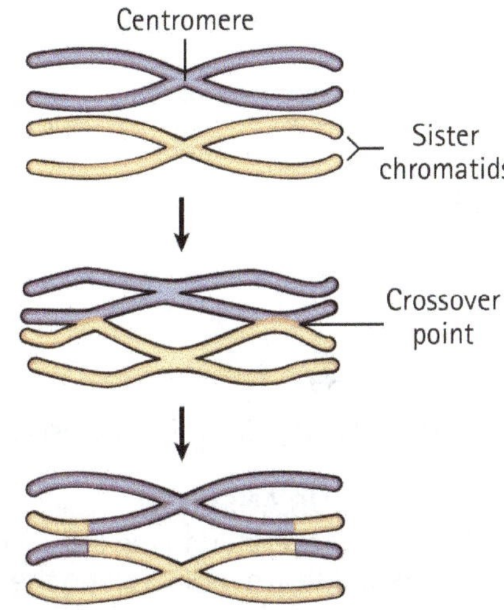

∧ Figure 6.7

During meiosis, homologous chromosomes exchange genetic material in a process called crossing over. The result of crossing over is that chromosomes end up with parts of each of the two original homologous chromosomes, as indicated in this illustration by the exchanged colors.

Meiosis and Sexual Reproduction Produce Genetic Diversity

Siblings often resemble each other, but they're not exactly alike. This is because the eggs and sperm of their parents are all different, so the siblings receive different genes. Many different factors contribute to this genetic diversity.

First, crossing over during meiosis increases the genetic diversity of an individual's sperm or eggs. How? At the start of meiosis, our cells have two of each kind of chromosome—a maternal chromosome from our mother and a paternal chromosome from our father.

Every time meiosis occurs, crossing over takes place. But—here's the key—crossing over occurs at a different point along the chromosome every time. This means that each chromosome in our eggs and sperm is a unique mix of our maternal and paternal chromosomes.

Even without crossing over, though, all our sex cells would be different. This is because homologous chromosomes separate independently during meiosis I. One egg might get maternal chromosomes 1, 3, 4, 5, and so on, and paternal chromosomes 2, 6, 7, 8, and so on. A second egg is almost certain to get different chromosomes, perhaps maternal chromosomes 2, 3, 5, 6, and so on, and paternal chromosomes 1, 4, 7, 8, and so on (**Figure 6.8**). There are a huge number of possibilities!

Finally, during fertilization, when an egg and sperm join, genetic material is brought together in different ways. Each of the many different possible eggs can join with each of the many different possible sperm.

READING CHECK

How can crossing over during meiosis affect an individual's sperm or eggs?

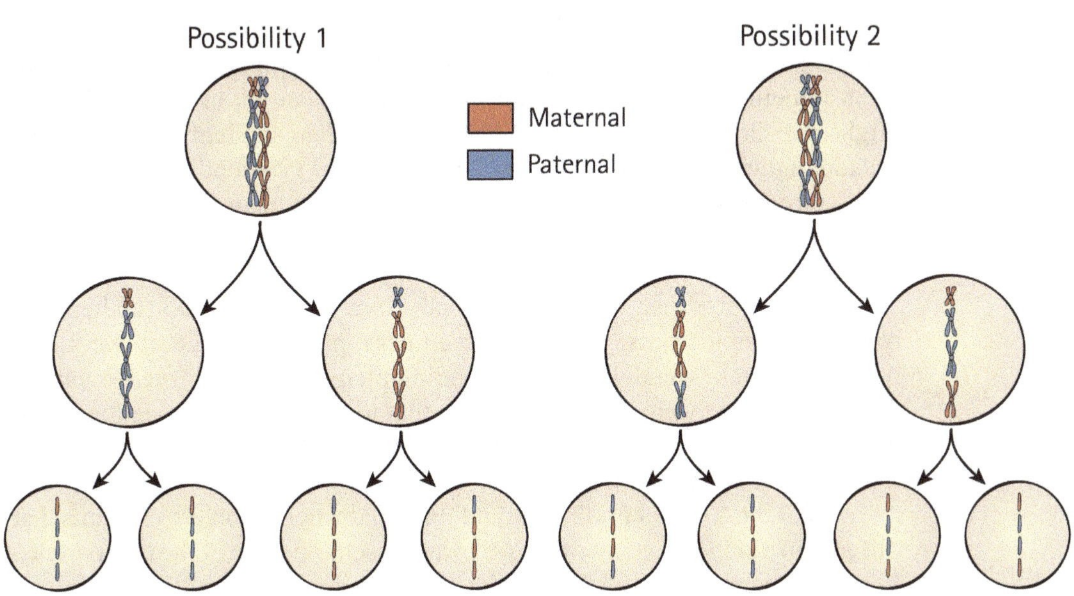

< Figure 6.8

Each time meiosis occurs, the resulting cells receive different combinations of maternal and paternal chromosomes. This figure shows two possibilities for meiosis in an organism with 4 pairs of chromosomes. The number of possibilities is exceedingly large, with the 23 pairs found in humans.

In meiosis I, crossed-over homologous chromosomes separate. In meiosis II, individual chromatids separate.

Comparing Mitosis and Meiosis

Mitosis and meiosis have three key differences: (1) Mitosis produces two cells; meiosis produces four cells. (2) Mitosis produces diploid cells; meiosis produces haploid cells. (3) The cells produced by mitosis are identical to one another and to the original cell; the cells produced by meiosis are all different.

CONCEPT**CHECK**

Name two ways in which meiosis contributes to genetic diversity in the eggs or sperm of an individual.

CHECK YOUR ANSWER

1. Crossing over during meiosis increases the genetic diversity of an individual's sperm or eggs. Because crossing over occurs at different points along the chromosome every time, chromosomes in eggs and sperm end up as unique mixes of maternal and paternal chromosomes.

2. Homologous chromosomes separate independently during meiosis I so that every cell receives a different combination of maternal and paternal chromosomes.

More to Explore

Information about meiosis from the National Human Genome Research Institute

https://www.genome.gov/genetics-glossary/Meiosis

6.3 Traits and Inheritance

Λ Figure 6.9

Gregor Mendel (1822 – 1884) was the founder of modern genetics. Here, he is examining a plant.

We look like our parents because we inherit our parents' genes. But why do we resemble our parents in certain ways but not others? For example, why do we have our mom's dimples but not her curly hair? Why do brown-eyed parents sometimes have a blue-eyed child? And, strangest of all, why do traits sometimes skip generations? For example, a crook in Grandpa's nose might not show up in any of his children and then suddenly appear in his granddaughter's. How can we explain that? The 19th-century Augustinian monk Gregor Mendel (**Figure 6.9**) discovered the answers to these questions. Mendel did not make his discoveries by studying people, though. He studied peas.

Mendel bred pea plants, which vary in a number of traits. There are round or wrinkled peas, yellow or green peas, purple or white flowers, and other variations. Mendel started by finding plants that "bred true," meaning these plants always produced offspring that looked like them. He then bred two plants that differed in a single trait— for example, a round-pea plant and a wrinkled-pea plant.

What did Mendel find? Mendel found that when he bred two plants that differed in a single trait, all of the offspring resembled one of the two parents. For example, when a round-pea plant was bred with a wrinkled-pea plant, all the offspring had round peas. Traits that were expressed in the offspring

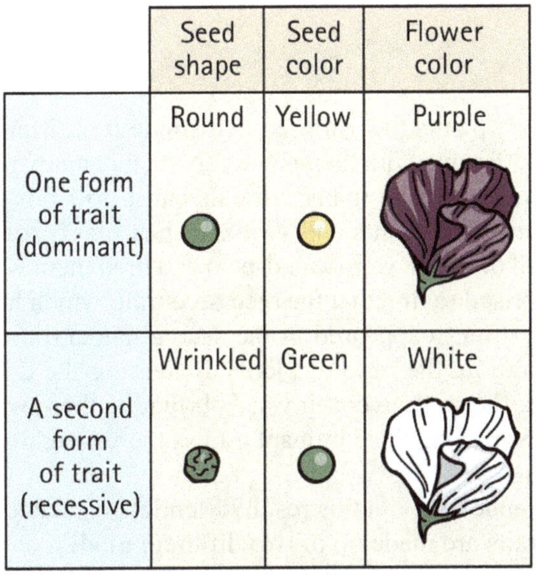

< **Figure 6.10**

Mendel bred pea plants that varied in a number of traits. For every pair of traits he looked at, one trait was dominant and the other was recessive.

(such as round peas) Mendel called *dominant.* Traits that were not expressed in the offspring (such as wrinkled peas) Mendel called *recessive.* In every pair of the traits Mendel looked at, one was dominant, and the other was recessive (**Figure 6.10**). Humans have dominant and recessive traits too (**Figure 6.11**).

READING CHECK

When Mendel bred two plants that differed in a single trait, did the offspring resemble just one or both of the parent plants?

< **Figure 6.11**

Dimples are a dominant human trait. So is a hairline that comes to a peak in the middle—a widow's peak hairline.

CONCEPT CHECK

When Mendel bred green-pea plants and yellow-pea plants, all of the offspring had yellow peas. Which trait is dominant? Which trait is recessive?

CHECK YOUR ANSWER

Yellow peas are dominant. Green peas are recessive.

More to Explore

Learn more about Gregor Mendel's life from the publisher of the scientific journal "Nature", a Collaborative Learning Space for Science

https://www.nature.com/scitable/topicpage/gregor-mendel-a-private-scientist-6618227/

6.4 First law of Inheritance

After breeding two plants that differed in a single trait, Mendel's next step was to breed the offspring with themselves. (Note that many plants can breed with themselves, something many animals can't do.) For example, after Mendel bred round-pea plants and wrinkled-pea plants together, he took their offspring (all of which were round-pea) and bred them with themselves. Mendel was surprised to find that the recessive trait, which had disappeared in the first generation, reappeared in the second generation. Moreover, in the second generation, the ratio of plants expressing the dominant trait to plants expressing the recessive trait was 3:1–that is, there were three times as many plants expressing the dominant trait as there were plants expressing the recessive trait.

How did Mendel explain this result? Mendel postulated that the genes that determine traits are made up of two different **alleles**, or versions of the gene. One allele is inherited from each parent. Then Mendel came up with his first law: the principle of segregation. The *principle of segregation* says that when an individual makes sex cells (sperm or eggs), half the sex cells carry one allele, and the other half carry the other allele. To summarize:

1. Mendel postulated that each inheritable trait is determined by entities he called "alleles."
2. Mendel further postulated there are two alleles for each inheritable trait, one of which is inherited from each parent.
3. Today, we know that alleles correspond to genes located within DNA.

Let's see how the principle of segregation explains Mendel's results by reviewing his breeding experiments. Remember that Mendel started with plants that differed in a single trait—say, round peas or wrinkled peas. The round-pea plant had two round alleles, a genotype we write as *RR*. The wrinkled-pea plant had two wrinkled alleles, or genotype *rr*. (A capital letter is used for a dominant trait, and a lowercase letter for a recessive trait.) The *RR* and *rr* plants are *homozygotes* for this trait—a **homozygote** has two identical alleles for a given trait.

When Mendel bred the *RR* (round-pea) and *rr* (wrinkled-pea) plants together, the first-generation offspring inherited an *R* allele from the round parent and an *r* allele from the wrinkled parent. These first-generation offspring had the genotype *Rr*. An *Rr* plant is a **heterozygote** for this trait because it has two different alleles. Which trait does an *Rr* plant have? An *Rr* plant has round peas because the *R* allele is dominant. The **dominant** allele is expressed in the heterozygote. The **recessive** allele is not expressed in the heterozygote; it is hidden.

Mendel then bred the *Rr* plants with themselves. According to the principle of segregation, *Rr* plants make equal numbers of *R* and *r* sperm and eggs. What alleles will the offspring inherit? It is equally likely that an offspring will get (1) *R* egg and *R* sperm, (2) *R* egg and *r* sperm, (3) *r* egg and *R* sperm, or (4) *r* egg and *r* sperm. This means that a quarter of the offspring are RR, half are Rr, and a quarter are rr (**Figure 6.12**). *RR* and *Rr* plants have round peas; *rr* plants have wrinkled peas. So, three-quarters of the offspring have round peas, and one-quarter have wrinkled peas—the ratio of round-pea plants to wrinkled-pea plants is 3:1.

READING CHECK

Mendel bred heterozygous Rr plants with each other. How many of the offspring ended up with the genotype RR, how many with the genotype Rr, and how many with rr?

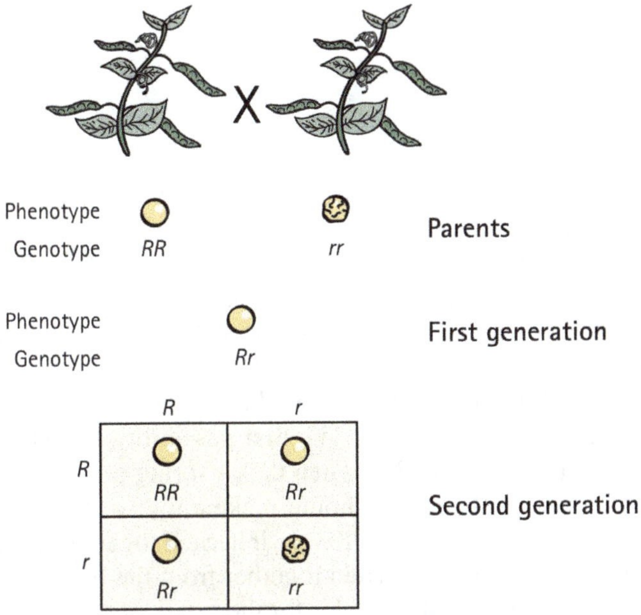

< Figure 6.12

Mendel bred a round-pea plant with a wrinkled-pea plant. In the first generation, all the offspring had round peas. Then Mendel bred the offspring with themselves. The result was round-pea plants and wrinkled-pea plants in a ratio of 3:1.

CONCEPT CHECK

1. In humans, "dimples" is dominant and "no dimples" is recessive. You can use D and d to represent the two alleles. Which allele represents "dimples"? Which allele represents "no dimples"?

2. What is the genotype of a person who is a heterozygote for dimples? What is this person's phenotype—that is, does she have dimples?

3. If a person has dimples, what are the two possibilities for his or her genotype? If a person has no dimples, what is his or her genotype?

CHECK YOUR ANSWERS

1. We use capital letters for dominant traits and lowercase letters for recessive traits. So D represents "dimples," and d represents "no dimples."

2. A heterozygote has two different alleles, so a heterozygote has the genotype Dd. She has dimples because D is dominant.

3. A person with dimples must have at least one dominant allele, so the genotype is either DD or Dd. A person with no dimples must have two recessive alleles, so the genotype is dd.

CONCEPT CHECK

1. You breed a tall pea plant (TT) with a short pea plant (tt). What allele is found in the sex cells of each plant? What alleles do the offspring inherit? Are the offspring tall or short?

2. When you breed Tt plants with themselves, what alleles do the offspring have? Are they tall or short? In what ratio?

CHECK YOUR ANSWERS

1. The tall plant produces sex cells that carry the T allele. The short plant produces sex cells that carry the t allele. The offspring are Tt. They are all tall.

2. A quarter of the offspring are TT, half are Tt, and a quarter are tt. Three-quarters are tall, and one-quarter are short.

More to Explore

Learn more about alleles from the Jackson Laboratory, a non-profit biomedical research institution.

https://www.jax.org/news-and-insights/minute-to-understanding/what-is-an-allele

6.5 Second Law of Inheritance

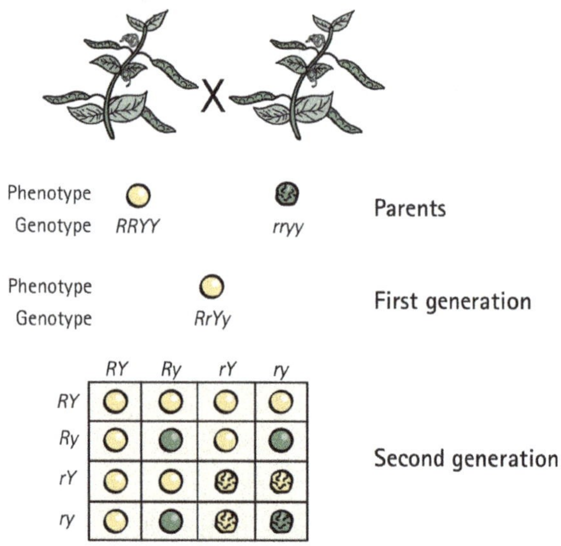

^ Figure 6.13

When Mendel bred plants that differed in two traits, the traits were inherited independently. Breeding plants with round yellow peas and plants with wrinkled green peas produced plants with round yellow peas in the first generation. In the second generation, Mendel saw all the different combinations of traits in a ratio of 9:3:3:1.

Mendel's next experiment was to breed plants that differed in two traits. He bred plants with round yellow peas (*RRYY*) to plants with wrinkled green peas (*rryy*). In the first generation, all the offspring inherited *R* and *Y* alleles from the first parent and *r* and *y* alleles from the second parent. The offspring were *RrYy*. They had round yellow peas.

Next, Mendel bred the *RrYy* plants with themselves. He wanted to test two possibilities. The first possibility is that alleles for the two genes stay together when the offspring make sex cells. If this is the case, *RrYy* plants should make only two kinds of sex cells: *RY* and *ry* (because *R* and *Y* were inherited together from one parent, and *r* and *y* were inherited together from the other parent). The second possibility is that alleles for the two genes behave independently when sex cells are made. In this case, *RrYy* plants should make sex cells with all combinations of alleles—*RY, Ry, rY*, and *ry*.

When the *RrYy* plants were bred with themselves, the offspring had every type of pea: round yellow, round green, wrinkled yellow, and wrinkled green. This couldn't happen if all the sex cells had been either *RY* or *ry*, because then all the offspring would have either round yellow peas or wrinkled green peas. You could never get wrinkled yellow peas or round green peas. Also, the ratio of round yellow:round green:wrinkled yellow:wrinkled green peas was 9:3:3:1, exactly what you expect if the sex cells carry all combinations of alleles (**Figure 6.13**). Focusing on each trait, we see the same 3:1 ratio; that is, a 3:1 ratio of yellow to green and a 3:1 ratio of round to wrinkled. The inheritance of one trait is independent of the inheritance of a second trait. This is Mendel's second law: the principle of independent assortment.

READING CHECK

What is another name for Mendel's second law of inheritance?

CONCEPT **CHECK**

Your father gave you an allele for no dimples (d) and an allele for a widow's peak (W). Your mother gave you a dimples allele (D) and a straight hairline allele (w).

a. What is your genotype? Are you a homozygote or a heterozygote for each of these traits?

b. What is your phenotype? That is, do you have dimples? Do you have a widow's peak?

c. What will the genotypes of your sex cells be?

CHECK YOUR ANSWERS

a. Your genotype is DdWw. You are a heterozygote for both traits.

b. You have dimples, and you have a widow's peak. Both those traits are dominant.

c. You make sex cells with all possible combinations of alleles: DW, Dw, dW, and dw.

More to Explore

From the School of Biomedical Sciences Wiki:

https://teaching.ncl.ac.uk/bms/wiki/index.php/Principle_of_Independent_Assortment

6.6 Beyond Mendel

Mendel's work provided the vital first steps in our understanding of inheritance. But inheritance can also be more complicated than what Mendel described. Let's consider some additional wrinkles.

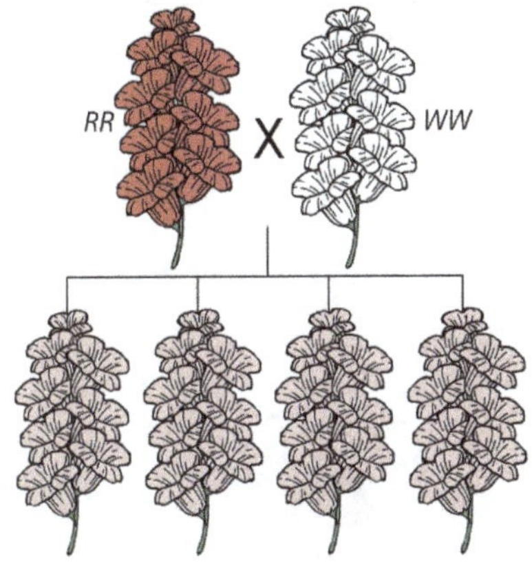

^ Figure 6.14
The flower color in snapdragons shows incomplete dominance. When a snapdragon with red flowers (RR) is bred with a snapdragon with white flowers (WW), the offspring (RW) has pink flowers.

Incomplete Dominance

In **incomplete dominance**, there are two alleles for a trait, and neither is dominant. The heterozygote has an intermediate trait. For example, when you breed a red snapdragon with a white snapdragon, you get pink snapdragons (**Figure 6.14**).

Codominance

In **codominance**, a heterozygote with two different alleles expresses the traits of *both* alleles. An example can be found in the human blood type. Your blood type describes molecules on the surface of your red blood cells. Knowing your blood type can be a matter of life and death. If you ever receive blood cells with a surface molecule you don't normally have, your body attacks them. This makes the blood cells clump together, which can be fatal.

Depending on your blood type, you can have the A molecule (blood type A), the B molecule (blood type B), neither (blood type O), or both (blood type AB). There are three blood type alleles: *A, B,* and *O.* A person with genotype *AA* or *AO* has A molecules (blood type A). A person with genotype *BB* or *BO* has B molecules (blood type B). A person with genotype *OO* has neither molecule (blood type O). A person with genotype *AB* has both A and B molecules (blood type AB)—both the *A* trait and the *B* trait are expressed. The *A* and *B* alleles are codominant (**Figure 6.15**).

Polygenic Traits

Polygenic traits are determined by more than one gene. (*Poly* means "many," so *polygenic* refers to "many genes.") Human height and skin color are both polygenic. Polygenic traits show more of a continuum than traits determined by a single gene. To see why, consider human height.

Suppose the three genes, A, B, and C, determine height. (More than three are actually involved.) Each gene has a tall allele and a short allele, which we will indicate with *T* and *S*. The shortest people have all short alleles—genotype $A_SA_SB_SB_SC_SC_S$. People with genotype $A_SA_SB_TB_SC_SC_S$ (five short alleles and one tall allele) are somewhat taller, those with four short alleles and two tall alleles are even taller, and so on. In this way, multiple genes produce a gradation of heights (**Figure 6.16**).

Blood type (Phenotype)	O	A	B	AB
	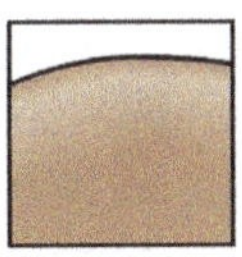	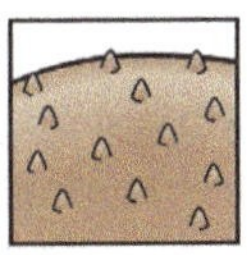	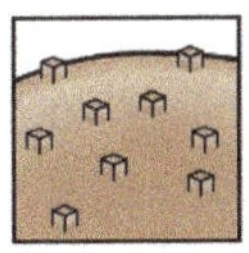	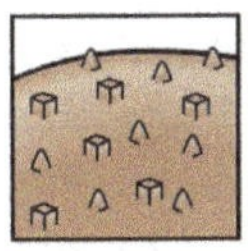
Genotype	*OO*	*AA* or *AO*	*BB* or *BO*	*AB*

< Figure 6.15
Human blood type offers an example of codominance. Combinations of the A, B, and O alleles determine blood type. A and B are both dominant to O, but A and B are codominant.

Figure 6.16 >

Human height is a polygenic trait. Taller people have more tall alleles. (Height is also affected by nongenetic factors such as diet and nutrition.)

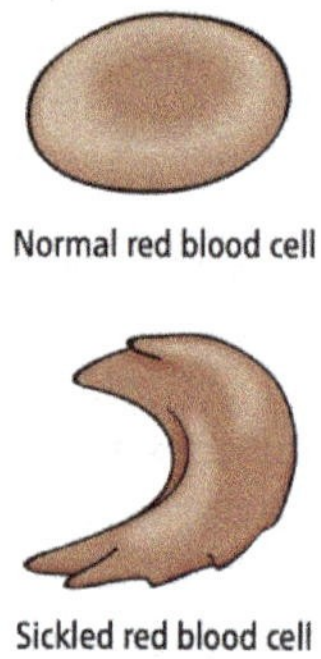

∧ Figure 6.17

The sickle-cell allele offers an example of pleiotropy in humans. The allele can cause normally round red blood cells to assume a sickle shape, resulting in pain and tissue damage. However, the allele also offers protection from malaria.

Pleiotropy

In **pleiotropy**, a single gene affects more than one trait. The disease sickle cell anemia provides an example of pleiotropy in humans. The sickle-cell allele for hemoglobin makes red blood cells turn sickle-shaped under certain conditions, causing tissue damage and pain (**Figure 6.17**). Homozygotes (people who have two sickle-cell alleles) are severely affected. Heterozygotes (people who have one sickle-cell allele and one normal allele) have only mild symptoms. Given the sickle-cell allele's harmful effects, scientists were puzzled by how common it is, especially among people of African descent. Why hadn't evolution through natural selection caused the sickle-cell allele to disappear from human populations?

Pleiotropy provides the answer. It turns out that the sickle-cell allele also protects people from malaria. This is an example of pleiotropy because the sickle-cell allele affects more than one trait—it affects the shape of red blood cells *and* resistance to malaria. The fact that the sickle-cell allele protects against malaria—along with the allele's mild symptoms in heterozygotes—explains why the allele is common in populations where malaria has been a danger in the past.

Linked Genes

Linked genes are frequently inherited together, in apparent contradiction to Mendel's principle of independent assortment. How can genes be linked? Independent assortment occurs when alleles found on *different* chromosomes separate independently at meiosis. But, if the alleles for two genes are on the *same* chromosome, they are often inherited together—though not quite all the time, because sometimes crossing over shuffles them up. In fact, the closer two alleles are to each other on a chromosome, the more likely they are to be inherited together. (The genes for Mendel's pea traits happened to be on different chromosomes—that's why he concluded that genes sort independently.) Body color and wing size in fruit flies are linked genes.

READING CHECK

Give an example of linked genes in fruit flies.

Sex-Linked Traits

Sex-linked traits are determined by genes found on the X chromosome. Because males have only one X chromosome, they have only one allele for these traits. This means that males need only one recessive allele to express a recessive sex-linked trait, whereas females need two recessive alleles. Because of this, recessive sex-linked traits are seen more often in males than in females. In humans, two recessive sex-linked traits are red-green color blindness (**Figure 6.18**) and the blood disease hemophilia. Both conditions affect far more males than females.

(a) (b)

∧ Figure 6.18

Red-green color-blindness is a sex-linked trait that affects many more males than females. (a) This image is used to test for red-green color blindness. Do you see the letter? (b) This photo shows how these flowers (lower right) would appear to someone with red-green color blindness.

Horizontal Gene Transfer

We inherit traits from our parents. This much is clear. But because nature is rather creative, it's also quite possible to receive genes from elsewhere, including even another species. For example, microorganisms such as bacteria sometimes receive genes through the action of viruses that infect them. This type of transmission is called *horizontal gene transfer*. (The term "horizontal" distinguishes this from the "vertical" transmission from parents to offspring we have been studying up to now.) In many cases, horizontal gene transfer occurs between species, such as when a virus transmits a gene to a bacterium. Horizontal gene transfer in bacteria explains, in part, how bacteria can quickly evolve to become resistant to our antibiotics. As we will explore in the next chapter, viral DNA found in bacteria has also led to a revolutionary gene editing tool called *CRISPR-Cas9*.

Other living things are likewise impacted by horizontal gene transfer. In fact, the mitochondria found in all eukaryotic cells (remember that mitochondria are organelles that break down organic molecules and generate energy in the form of ATP) are likely the remnants of an ancient organism that infected our cells only to become permanently incorporated. Furthermore, it's estimated that about 8 percent of the human genome arrived not through traditional inheritance but through viral infection, that is, horizontal gene transfer. The main point is that nature is creative, and there is still much we have to learn when it comes to genetic inheritance.

CONCEPT CHECK

Your blood type is O, your mother's blood type is A, and your father's blood type is B. What is your genotype? What about your mother's and your father's? Explain your reasoning.

CHECK YOUR ANSWER

Because your blood type is *O*, your genotype must be *OO*. Your mother's blood type is *A*, so she must have the genotype *AO*. (She must have an *O* allele because she passed one to you.) Similarly, your father must be *BO*.

More to Explore

An early scientific research review article on the nature of Horizontal Gene Transfer. Note the 2008 publication date.

https://www.nature.com/articles/nrg2386

Concept Review

Chapter 6

Detailed Chapter Summary

Cells reproduce when they make genetically identical copies of themselves. Dividing cells move through the cell cycle stages: gap 1, synthesis, gap 2, and mitosis and cytokinesis. During gap 1, a cell grows to double its original size. During synthesis, the cell creates an exact copy of its DNA. During gap 2 (G_2), the cell builds the machinery necessary for cell division. During mitosis and cytokinesis (M), the cell divides.

Mitosis describes the division of the cell nucleus and takes place in a series of phases: prophase, metaphase, anaphase, and telophase. During *prophase*, the chromosomes condense, and the membranes surrounding the nucleus break down. The mitotic spindle also forms. During *metaphase*, the chromosomes line up at the equatorial plane. During *anaphase*, the two identical chromatids are pulled apart by the shortening of the mitotic-spindle fibers and move to opposite poles of the cell. During *telophase*, new nuclear membranes form around each set of chromosomes, and the chromosomes return to their loosely packed state. Mitosis is followed by cytokinesis, the division of the cytoplasm. This results in the production of two daughter cells.

Meiosis is another type of cell division. Meiosis is used to produce haploid sperm and eggs. During meiosis, a single diploid cell divides to produce four haploid cells. Meiosis takes place in two steps, meiosis I and meiosis II, each of which is divided into further steps. The cells produced by meiosis are all genetically different for several reasons. First, there is crossing over, which results in the production of new combinations of genes. Second, there is the independent separation of chromosomes during meiosis.

Gregor Mendel's work helped establish how genes and traits are inherited. A gene is made up of two separate alleles. A homozygote has two identical alleles for a trait, and a heterozygote has two different alleles. The alleles for many genes are either dominant or recessive. The dominant allele is expressed in a heterozygote, and the recessive allele is not expressed in a heterozygote.

Mendel performed a series of breeding experiments in peas that led him to formulate his two "laws" of inheritance. Mendel's principle of segregation says that when an individual makes sex cells (such as sperm and eggs), half the sex cells carry one allele, and the others carry the other allele. This explains the 3:1 ratio seen in breeding heterozygotes for a single trait.

Mendel's principle of independent assortment says that the inheritance of one trait is independent of the inheritance of a second trait. This explains the 9:3:3:1 ratio seen in breeding heterozygotes for two different traits.

More complex patterns of inheritance take us beyond Mendelian genetics. In *incomplete dominance*, there are two alleles where neither is dominant, and the heterozygote has an intermediate trait. In *codominance*, there are two alleles where the heterozygote shows the traits of both alleles. In *polygenic traits*, a trait is determined by more than one gene. In *pleiotropy*, a single gene affects more than one trait. *Linked genes* are frequently inherited together because they are found on the same chromosome. *Sex-linked traits* are determined by alleles found on the X chromosome so that inheritance patterns differ in males and females. Sex-linked traits are seen in more males than females since males have only one X chromosome. Finally, in horizontal gene transfer, genes are not inherited from parents but obtained from other means.

Key Terms

Alleles Versions of a gene.

Codominance Describes a form of inheritance in which the traits of both alleles are expressed in the heterozygote.

Dominant Describes an allele that is expressed in a heterozygote.

Heterozygote An organism that has two different alleles for a given gene.

Homozygote An organism that has two identical alleles for a given gene.

Incomplete Dominance Describes a form of inheritance in which the heterozygote has a trait that is intermediate between the traits of the two homozygotes.

Linked Genes When two or more genes are frequently inherited together, typically because they are found on the same chromosome.

Meiosis A form of cell division in which one diploid cell divides to produce four haploid cells.

Mitosis A form of cell division in which one cell divides into two daughter cells, each of which contains the same genetic information as the original cell.

Pleiotropy A situation in which a single gene affects more than one trait.

Polygenic Traits A trait that is determined by more than one gene.

Recessive Describes an allele that is not expressed in a heterozygote.

Recombination The production of new combinations of genes that differ from combinations found in the parental chromosomes resulting from crossing over during meiosis.

Sex-Linked Traits A trait determined by a gene found on the X chromosome that is typically more commonly expressed in males due to the fact that males have only one X chromosome.

Review Questions

Simple Review Questions

6.1 How Cells Reproduce

1. What are the stages of the cell cycle? What happens during synthesis (S)?
2. Describe the phases of mitosis.
3. What are the end products of mitosis?

6.2 Cell Division and Genetic Diversity

4. What is crossing over? Why is crossing over important to genetic diversity?
5. What are the products of meiosis?

6.3 Traits and Inheritance

6. What is the difference between dominant and recessive traits?

6.4 Mendel's First Law of Inheritance

7. When Mendel bred pea plants that differed in a single trait, what did he see in the offspring? When Mendel allowed these offspring to self-fertilize, what did he see?

6.5 Mendel's Second Law of Inheritance

8. What does Mendel's Second Law of Inheritance say?

6.6 Beyond Mendelian Inheritance

9. What is codominance? Provide an example of codominance.
10. What is a polygenic trait? Provide at least one example of a polygenic trait.
11. What is pleiotropy?

Challenging Review Questions

6.1 How Cells Reproduce

12. Imagine that a cell goes through all the stages in the cell cycle, except that cytokinesis doesn't happen. How will that cell be different from normal cells?
13. This photo shows a cell in the process of cell division. Which stage of the cell cycle does it show?

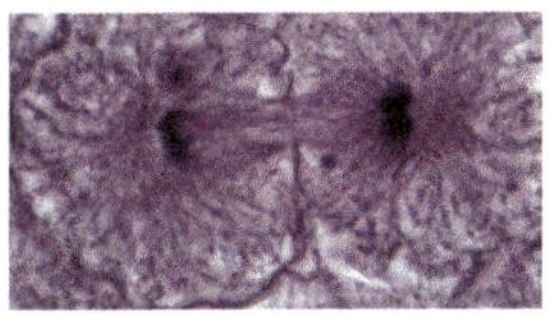

6.2 Cell Division and Genetic Diversity

14. How can the same two parents produce children that are all different?

15. Describe three differences between mitosis and meiosis.

16. The figure below shows a set of human chromosomes. What is unusual about this person's genetic makeup? How does a problem during meiosis relate to this condition? What health issues might this person suffer from?

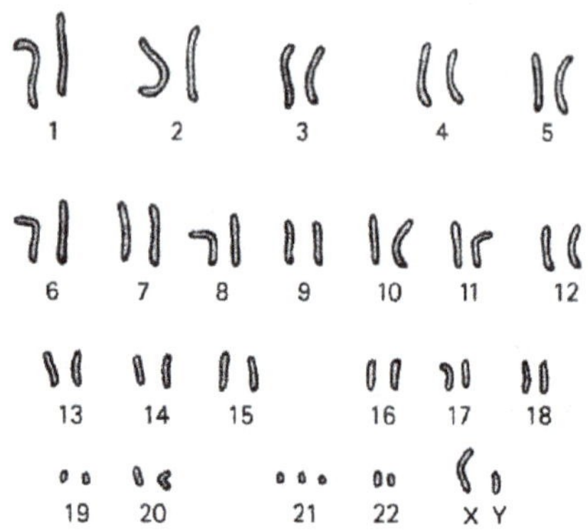

6.3 Traits and Inheritance

17. Some fruit flies have straight wings, whereas others have curly wings. A fruit fly biologist takes a line of curly-winged flies that breeds true and a line of straight-winged flies that breed true. She then performs a cross, mating a curly-winged fly with a straight-winged fly. She finds that all the offspring have curly wings. Is the curly-wing trait dominant or recessive?

6.4 Mendel's First Law of Inheritance

18. Explain how a trait can skip generations.

19. If you have dimples, will all your children have dimples? (Remember that "dimples" is a dominant trait.)

20. Is it possible for two parents with widow's peaks to have a child who has a straight hairline? Is it possible for two parents with straight hairlines to have a child with a widow's peak? (Remember that a widow's peak hairline is dominant, and a straight hairline is recessive.)

6.5 Mendel's Second Law of Inheritance

21. Suppose that you repeat Mendel's breeding experiments. First, you breed plants with round yellow peas (*RRYY*) to plants with wrinkled green peas (*rryy*) to obtain *RrYy* pea plants. Then, you cross the *RrYy* plants with themselves and collect the resulting peas. You find only 3 wrinkled green peas. About how many wrinkled yellow peas should you expect? What about round green peas? What about round yellow peas?

6.6 Beyond Mendelian Inheritance

22. Can you tell what alleles a pea plant with round seeds has? Can you tell what alleles a red snapdragon has? Why are your answers different?

23. You need a blood transfusion, but you don't know your blood type. Which type of blood should you be given? Does this explain why people with type O blood are called universal donors?

24. Universal receivers are people who can safely receive any blood type during a blood transfusion. Which blood type do universal receivers have? Explain your answer.

25. Why are there more color-blind men than color-blind women?

Apply & Discuss Questions

26. Does the process of meiosis explain Mendel's two laws: the principle of segregation and the principle of independent assortment?

27. In the case of linked genes, explain why two genes are more likely to be inherited together the closer together they are on a chromosome.

28. You learned that when red snapdragons are bred with white snapdragons, the offspring are pink snapdragons—an example of incomplete dominance. What happens when you breed a pink snapdragon with another pink snapdragon?

Solutions to Odd-Numbered Questions

Simple Review Solutions

1. The cell cycle is divided into four stages—gap 1, synthesis, gap 2, and mitosis and cytokinesis. During synthesis (S), the cell creates an exact copy of its genetic material—its DNA.

3. The end products of mitosis are two daughter cells, each of which contains the same genetic information as the parent cell.

5. In meiosis, one diploid cell, with two of each kind of chromosome, divides into four haploid cells, each with only one of each kind of chromosome.

7. Mendel found that when he bred two plants that differed in a single trait, all of the offspring resembled one of the two parents. Then, Mendel bred the offspring with themselves. Mendel was surprised to find that the recessive trait, which had disappeared in the first generation, reappeared in the second generation. Moreover, in the second generation, the ratio of plants expressing the dominant trait to plants expressing the recessive trait was 3:1–that is, there were three times as many plants expressing the dominant trait as there were plants expressing the recessive trait.

9. In codominance, there are two alleles, and the heterozygote expresses the traits of both alleles. An example can be found in the human blood type. Your blood type describes molecules on the surface of your red blood cells. You can have the A molecule (blood type A), the B molecule (blood type B), neither (blood type O), or both (blood type AB). There are three blood type alleles: *A*, *B*, and *O*. A person with genotype *AA* or *AO* has A molecules (blood type A). A person with genotype *BB* or *BO* has B molecules (blood type B). A person with genotype *OO* has neither molecule (blood type O). A person with genotype *AB* has both A and B molecules (blood type AB)—both the *A* trait and the *B* trait are expressed. The *A* and *B* alleles are codominant.

11. In pleiotropy, a single gene affects more than one trait.

Challenging Review Solutions

13. Since it is dividing, the cell is in mitosis.

15. The key differences between mitosis and meiosis include: (1) Mitosis produces two cells. Meiosis produces four cells. (2) Mitosis produces diploid cells. Meiosis produces haploid cells. (3) The cells produced by mitosis are identical to one another and to the original cell. The cells produced by meiosis are all different.

17. Since all the offspring of the cross have curly wings, the curly-wing trait is dominant.

19. Not necessarily. You may be a heterozygote with genotype *Dd*. You may pass a *d* allele to your child. If your child also receives a *d* allele from your spouse, he or she will have no dimples.

21. In this cross, you expect to obtain round yellow:round green:wrinkled yellow:wrinkled green peas in a ratio of 9:3:3:1. So, if you find 3 wrinkled green peas, you expect to find three times as many wrinkled yellow peas—so, about 9. You also expect to find three times as many round green peas—again, about 9. You expect to find nine times as many round yellow peas, so about 27.

23. Anyone can receive type O blood, so this is the type they should give you. The red blood cells in Type O blood have neither A nor B molecules that could cause the cells to be rejected and attacked. This is why people with type O blood are called universal donors—they can donate blood to anyone.

25. Colorblindness is a recessive sex-linked trait. These traits affect more males than females because males have only one X chromosome. Males are affected if they have only one recessive allele, whereas females are only affected if both alleles are recessive.

Apply & Discuss Solutions

27. Genes that are on the same chromosome will be inherited together unless crossing over occurs at a point between their two locations. The closer two alleles are on a chromosome, the less likely it will be for crossing over to occur between them.

Living Earth Essay

Life on Earth Connections

Life in the Trenches

According to plate tectonics, one plate subducting beneath another can create a deep depression in the ocean floor—an ocean trench. Ocean trenches are the deepest places on Earth. The Mariana Trench is the deepest trench of them all. It reaches about 11,000 m (7 mi) below sea level. If the world's tallest surface mountain, Mt. Everest, were sunk to the bottom of the Mariana Trench, there would still be a mile of ocean above it!

Figure A
The black anglerfish has a huge jaw and teeth, making it an effective predator. This fish needs to be an effective predator because food is scarce in the ocean trenches. This fish also has a glowing or bioluminescent "lure" attached to its head to attract prey. Microbes that glow in the dark produce the light for the lure.

Conditions in trenches are extreme. Water is nearly frozen. Pressure reaches more than a thousand atmospheres (equal to about 16,000 lb per square inch). Sunlight is totally absent in these pitch-black depths. Yet life thrives in the trenches. Trenches are inhabited by unusual organisms, sometimes called *extremophiles*. Trench extremophiles include species of worms, shrimp, fish, crabs, and microbes. **Figure A** shows one trench species, the black anglerfish, which has a number of special adaptations for living in its habitat. **Figure B** shows another trench species, the world's deepest-dwelling fish, Abyssobrotula galathaea, which was found in the Puerto Rican Trench at a depth of 8,372 m.

Figure B
This fish was found at a record-holding depth of over 8 kilometers deep.

The adaptations of deep trench extremophiles are not well understood in great part because of the technical difficulty of reaching such depth. Trenches are also vast regions, hundreds, if not thousands of miles long. The trench life, by contrast, appears to be limited to small and hard-to-find areas where the conditions are just right. This includes the presence of "black smokers," which are volcanic vents spewing out ultra-hot water with ample minerals (**Figure C**). Most trench life can be found along the boundaries of these black smokers. The communities come and go with the black smokers, which means one spot may be rich in life upon a first visit. But by the time of a second visit, it may be gone.

Figure C
The energy from the subduction of one plate under another can lead to "black smokers," which are vents spewing nutrients and ultra-hot water.

Most notable within these deep trench communities are the giant tube worms, which can grow over two meters in length (**Figure D**). Within their gut, these creatures harbor bacteria that produce chemical energy from the reduction of hydrogen sulfide into elemental sulfur. These chemical-consuming bacteria also produce oxygen as a waste product. The process is called *chemosynthesis*. The free oxygen and chemical energy can then be used to produce more complicated biomolecules that support the tubeworm and other organisms within the community. Thus, a food chain develops in a hydrothermal vent despite the lack of light for photosynthesis.

It's now clear that life can thrive in the extreme conditions of an ocean trench where the pressure is immense, the water super cold, there is no sunlight, and therefore no photosynthesis. This raises the possibility that similar extreme life forms might also exist on select moons of both Jupiter and Saturn. Moons such as Europa, Ganymede, and Enceladus are covered in ocean water—frozen at the surface but liquid underneath. Evidence of life on these worlds might come from satellites that swoop through geysers gushing from the surface. Stay tuned!

READING CHECK

What is an example of chemosynthesis?

CONCEPT CHECK

1. What are the physical conditions in the trenches that make life difficult there?
2. Where does the heat that escapes from hydrothermal vents come from?

CHECK YOUR ANSWER

1. Lack of light, extreme cold, and extremely high pressure.
2. The heat comes from Earth's interior.

More to Explore

Learn more about ocean trenches from the Woods Hole Oceanographic Institution:

https://www.whoi.edu/know-your-ocean/ocean-topics/how-the-ocean-works/seafloor-below/ocean-trenches/

Figure D
Tube worms were found in a trench close to the Galapagos Islands. The tubeworm can retreat into the hardened white tube at its base for safety. The blood within the tubeworm gives its head a red color. The mouth is not a traditional "mouth" that ingests food. Rather the mouth is a route by which the tubeworms pull in nutrients needed by the bacteria. Therefore, its "mouth" is more akin to a lung.

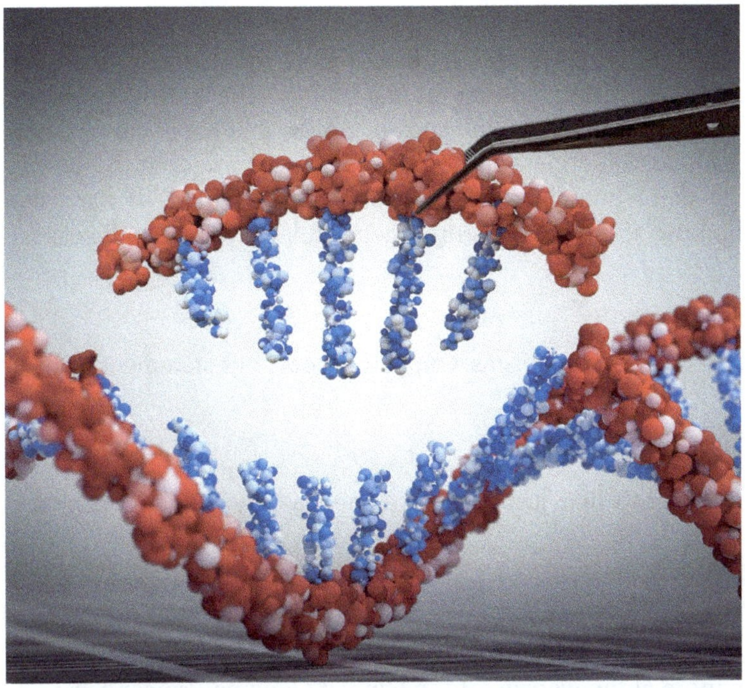

Chapter 7

Genetic Technologies

THE MAIN IDEA

Understanding how genetics works at the molecular level has provided some exceedingly powerful tools.

7.1 The Human Genome
7.2 Genetic Testing
7.3 Cancer
7.4 DNA Technology
7.5 Genetically Engineered Mosquitoes
7.6 Genome Editing with CRISPR-Cas9
7.7 Concerns about DNA Technology

The total genetic material of an organism is known as its genome. By the early 2000s, the DNA sequence of the entire human genome was mapped. How has this helped us manage genetic diseases like sickle cell anemia and cancer? How is a pedigree used to provide genetic counseling? How exactly are scientists able to manipulate the DNA of a living organism? What is CRISPR, and how has it led to a revolution in genetic engineering? And what ethical issues might be involved? These are topics we address within this chapter.

7.1 The Human Genome

As soon as the structure of DNA was known, scientists began to develop tools for studying—as well as manipulating—this genetic material. Advances in DNA technology now allow scientists to study the total genetic material of a living organism—the organism's **genome**. One of the first genomes scientists chose to study was the human genome.

The Human Genome Project, which was completed in 2003, determined the DNA sequence of the entire human genome. As a result of this project, we now know that there are about 3.2 billion nucleotide pairs in the human genome. Amazingly, more than 99.9% of these are identical in all humans. We also know that the human genome is not made up exclusively of genes. In fact, less than 2% of the genome carries genetic instructions for building proteins. Most of the genome—at least 50%—consists of repeat sequences that appear over and over again. Although repeat sequences do not have a direct function, they may occasionally rearrange the human genome, creating new genes or reshuffling existing ones.

We also know that humans have a total of about 19,000 protein-producing genes. Genes are often found in clusters at random spots along our chromosomes, with large sections of non–protein-coding DNA between them. Some genes give rise to RNA transcripts that are processed in different ways so that one gene can provide the instructions for building multiple proteins. The function of more than half of our genes is still completely unknown.

READING CHECK

What is an SNP?

Scientists have found numerous locations in the genome where the nucleotide sequence differs among human beings. These differences are called *single-nucleotide polymorphisms*, or *SNPs*. SNPs make every person's DNA unique, which is how forensic scientists are able to identify individuals based on samples containing any trace DNA. SNPs are also helpful in allowing scientists to identify genes related to human diseases.

CONCEPT CHECK

Scientists believe that most mutations have little effect on genes or organisms. Can you support this statement using what you learned about the human genome?

CHECK YOUR ANSWER

Genes make up less than 2% of the human genome. This suggests that, just by chance, the majority of mutations that strike human DNA will not affect any genes.

More to Explore

Learn more at the website of the Human Genome Project here:

https://www.genome.gov/human-genome-project

7.2 Genetic Testing

Many prospective parents have questions about different inherited diseases and the possibility of genetic testing. For example, what is their risk of having a child with a genetic disease? What tests are available that show whether a developing fetus has a genetic disease? What is it like to raise a child with a genetic disease? Trained genetic counselors can help people answer questions like these.

READING CHECK

What is often the first step that a genetic counselor will take to help couples who have a family history of gnetic disease?

Genetic counseling is especially important for couples with a family history of genetic disease. Genetic counselors often begin by creating a pedigree, a family tree that shows which relatives are affected by genetic disease. A sample family tree for the occurrence of Tay–Sachs disease within a family is shown in **Figure 7.1**. Information about the inheritance pattern of the disease in question—for example, whether the disease allele is dominant or recessive—can then be used to assess a couple's risk of having an affected child.

For some diseases, medical tests can provide additional information. Those with a family history of Tay–Sachs disease (a fatal disease of the central nervous system) or cystic fibrosis (a serious disease characterized by mucus buildup in the lungs, digestive system, and other organs) can undergo simple tests to determine whether they are *carriers* of the disease—that is, whether they possess a disease allele in addition to a normal allele. Because these diseases affect only people with two recessive alleles, a couple is at risk of having an affected child only if both parents are carriers.

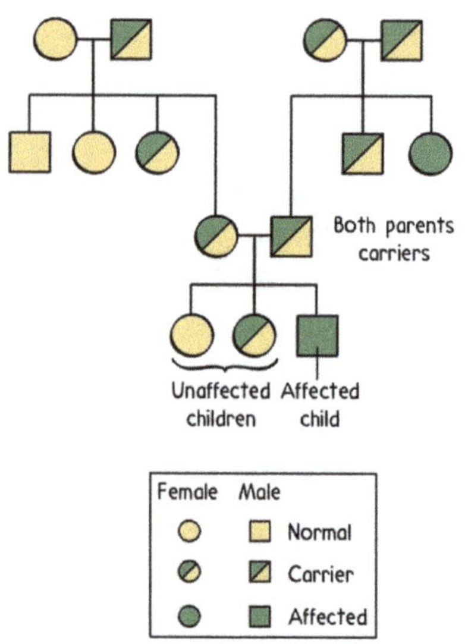

< Figure 7.1

This three-generation family tree (pedigree) shows the occurrence of Tay–Sachs disease within a family. The allele for Tay–Sachs disease behaves as a recessive—that is, only people who are homozygous for the allele are affected. This type of analysis allows a genetic counselor to assess the risk of genetic disease in a family.

If both parents are, in fact, carriers, procedures such as *chorionic villus sampling* or *amniocentesis*, in which fetal cells are collected and examined during pregnancy, can be used to determine whether a fetus is affected. Couples may also have the option of undergoing *in vitro* fertilization, testing the embryos, and implanting only unaffected ones.

Genetic testing is also useful for families without a history of genetic disease. For Down syndrome (**Figure 7.2**), a condition associated with mental retardation and other health issues, risk depends primarily on the mother's age. Older women have a greater risk of having children with Down syndrome. Down syndrome is caused by trisomy 21, the presence of three copies of chromosome 21 instead of two. Tests during pregnancy can examine the chromosomes in a fetus's cells to determine whether they show trisomy 21.

Figure 7.2 >

Down syndrome occurs in about one out of every 1000 births. It is caused by trisomy 21.

CONCEPT **CHECK**

What does it mean for someone to be a "carrier" of a genetic disease?

CHECK YOUR ANSWER

For genetic diseases that are caused by recessive alleles, a carrier is a person who possesses one disease allele and one normal allele. Carriers are not affected by the disease themselves, but they are able to pass a disease allele to their offspring, who will inherit the disease if they also inherit a disease allele from the other parent.

More to Explore

For more information on family pedigrees as well as the opportunity to create your own, check out the following website:

https://www.genome.gov/Pages/Education/Modules/YourFamilyHealthHistory.pdf

7.3 Cancer

Cell division is normally a carefully orchestrated process controlled by the proteins produced by a large number of genes. Cancer occurs when mutations in these genes lead cells to divide out of control. A mutation in a single gene is not usually enough to produce cancer—mutations in many important genes are typically required.

Cancer tends to strike older people more often than younger people because mutations have had more time to accumulate in their cells. People who have been exposed to mutation-causing agents accumulate mutations more rapidly, making cancer more likely to develop. Finally, some people are prone to developing certain types of cancers because they have inherited mutations in cancer-related genes. For example, mutations in the *BRCA* genes (named for *BR*east *CA*ncer) are linked to aggressive breast and ovarian cancers.

READING CHECK

What does is mean to say that a cancer has "metastisized"?

In a deadly cancer, tumor cells do more than divide out of control. Cancers are said to have metastasized if tumor cells have developed the ability to spread around the body and give rise to secondary tumors. This is a crucial point in disease progression because cancers typically become much harder to treat once they have metastasized.

Environmental Causes of Cancer

Although cancer is a genetic disease, its causes are usually environmental. For example, when people migrate from one place to another, their risk of developing cancer is determined mainly by where they move to rather than where they moved from. In fact, a person's environment is probably responsible for about 80%–90% of the mutations that result in cancer, with inherited genetic factors accounting for the other 10%–20%.

^ Figure 7.3

Reducing exposure to UV light, including wearing hats, sunglasses, and sunscreen, will help you avoid skin cancer.

Environmental factors that increase the risk of cancer include smoking, diet, radiation, ultraviolet light, chemicals, and infection by certain viruses and bacteria. The roles of factors such as smoking and radiation are well understood: Scientists know how and why these things increase genetic mutations. The roles of other factors are still being studied. For example, scientists know that diet is important in the development of many cancers. However, people's diets are such complicated mixes of nutrients that it can be hard to connect cancer with any particular foods. Yet, most of the research shows that a diet rich in fruits and vegetables can help protect against cancer.

A well-known environmental risk factor for skin cancer is ultraviolet (UV) light, which is abundant in sunlight. The ozone in Earth's upper atmosphere protects life by absorbing most of this UV light from our Sun. In the 1980s, large holes in the atmospheric ozone over the polar regions of our planet were discovered. Ozone loss was associated with the use of chemicals called *chlorofluorocarbons* (CFCs) in refrigerants and aerosol sprays. The chlorine in CFCs speeds up the breakdown of ozone, causing it to become depleted. An international agreement to phase out the use of CFCs has succeeded in shrinking the ozone hole. This is considered one of the world's major environmental successes. Staying out of sunlight and using sunscreen also help reduce exposure to UV light (**Figure 7.3**).

CONCEPT**CHECK**

If cancer is usually caused by environmental factors, why is it considered a genetic disease?

CHECK YOUR ANSWER

Cancer is a genetic disease because it is the result of genetic mutations. But, unlike other genetic diseases, cancer is not usually an inherited genetic disease. This is because the mutations that cause cancer are not usually inherited from one's parents. Instead, they are caused mainly by environmental factors.

More to Explore

You can learn more about cancer at the website for the National Cancer Institute:

https://www.cancer.gov/about-cancer/understanding/what-is-cancer

7.4 DNA Technology

Advances in DNA technology have led to many practical applications as well as some serious concerns. **Genetic engineering** allows us to change the traits of organisms by directly manipulating their DNA. In this section, we will look at two applications of genetic engineering: genetically modified organisms and cloning.

Genetically Modified Organisms

A **genetically modified organism**, or GMO, is an organism whose DNA has been changed through genetic engineering. GMOs are usually engineered for some practical purpose.

Scientists produced one of the first beneficial GMOs when they placed the gene for human insulin into the bacterium *Escherichia coli*, a common inhabitant of the human digestive tract. At the time, insulin, which is used to treat some forms of diabetes, had to be collected from the carcasses of pigs and cows. When genetically modified *E. coli* began to crank out human insulin, the advantages were clear—human insulin could now be produced easily and in large amounts within a laboratory setting. In 1982, insulin became the world's first genetically engineered pharmaceutical product (**Figure 7.4**). Genetically engineered bacteria now produce many other products, including many different human molecules and medicines, and scientists continue to work to expand this list.

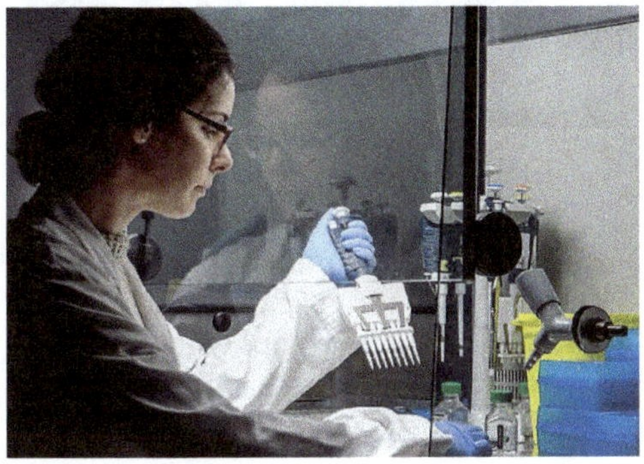

^ Figure 7.4

Genetically engineered bacteria (and other microorganisms) make a wide variety of important pharmaceutical products. This is an insulin production facility.

Figure 7.5 shows a typical process for developing genetically modified bacteria. In this example, a human gene is extracted from human DNA and combined with bacterial DNA. The bacterial DNA, which now contains the human gene, is then re-introduced into bacteria.

Many genetically modified (GM) plants and animals have also been developed by scientists. Genetically engineered plants have been developed for a variety of purposes. Some produce useful products—for example, GM tobacco plants are used to make medicines for certain autoimmune diseases. Other GM plants have genes that offer resistance to pests, diseases, or herbicides.

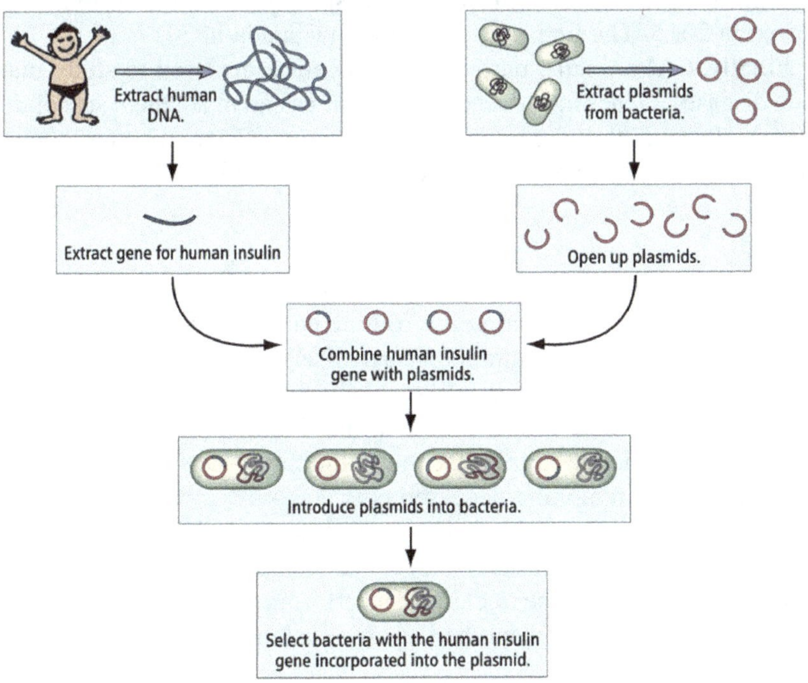

< Figure 7.5

This figure shows how bacteria can be genetically modified to produce human insulin. Genetically modified bacteria often carry their foreign genes in plasmids, small circles of DNA separate from the main bacterial chromosome.

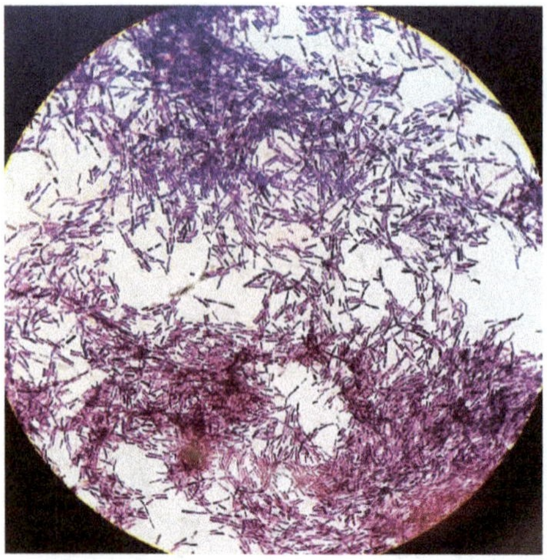

^ **Figure 7.6**

The soil bacterium Bacillus thuringiensis (orange) makes protein crystals, shown in purple, that are toxic to many insects. Genetically modified crops with this gene also make the toxic crystals, making them resistant to many pests.

Genetically engineered varieties of many crop plants, including corn, soybeans, and cotton, have been given a gene from the soil bacterium *Bacillus thuringiensis*, shown in **Figure 7.6**. This gene codes for a toxic protein that kills certain insects and makes the crops resistant to these pests. Many other GM plants carry foreign DNA that makes them resistant to the herbicide Roundup. Farmers who plant these "Roundup Ready" crops can use Roundup to kill weeds without damaging their crops.

Still, other GM plants have been engineered to grow in harsh environments. Some GM varieties of tomato, canola, and rice are able to grow in salty soils with the help of a special ion transport protein from a naturally salt-tolerant plant. (Salty soils are a common side effect of intensive agriculture.) Considerable effort is now being made to develop plants that are resistant to drought and high temperatures, both of which have worsened with global warming. Drought-tolerant species such as wild barley and the resurrection fern (**Figure 7.7**) have special genes that can be transferred to crops such as wheat. GM wheat carrying drought-related wild barley genes needs only one-eighth as much water as regular wheat. Other research focuses on developing genetically engineered plants for biofuels and for use in cleaning up pollutants.

^ **Figure 7.7**

Naturally drought-resistant plants such as the resurrection fern provide genes that can be transferred to crop plants, enabling them to survive the high temperatures and severe droughts. These ferns appear dead but will revive as soon as it rains.

Genetically engineered animals make proteins for human use, provide agricultural products, and aid in the study of human diseases.

Certain GM sheep and goats make foreign proteins in their milk, and researchers are working on a GM chicken that can produce foreign proteins in its eggs. In agriculture, researchers increased wool production in sheep by inserting special DNA sequences from mice. Other scientists are trying to use a roundworm gene to engineer healthier pork—GM pigs that carry the gene produce higher amounts of healthy omega-3 fatty acids.

The first GM animal to arrive at supermarkets was an Atlantic salmon which grows twice as fast as regular salmon and requires less food. The GM "AquAdvantage" salmon has DNA from two other species—a growth hormone gene from a different salmon species and some DNA from a fish called ocean pout. With this DNA, the AquAdvantage salmon can grow year-round rather than seasonally, and it can reach market weight in half the time of regular Atlantic salmon. It took 25 years for the U.S. Food and Drug Administration (FDA) to approve the salmon for consumption, which occurred in 2015. The first sales occurred in Canada in 2017.

Finally, GM animals, particularly GM mice, are used to study many human diseases. Genes introduced into GM mice can make them susceptible to human diseases that they don't otherwise get. These mice can then be used to study these human diseases as well as possible cures.

Cloning

Cloning is the creation of an organism that is genetically identical to one that already exists. Plant cloning is a routine part of agriculture. However, cloning is much harder in animals—particularly mammals. Cloning has a number of potential uses. In agriculture, cloning could create herds of genetically identical animals with desirable traits. Endangered or even extinct animals could be cloned to increase their numbers. And some people have expressed interest in cloning deceased pets.

In 1997, Dolly the sheep became the first cloned mammal (**Figure 7.8**). Cloning was done through a process called nuclear transplantation. In *nuclear transplantation*, the nucleus of a cell from the animal being cloned is placed into an egg cell that has had its own nucleus removed (**Figure 7.9**).

READING CHECK

How can weeds be selectively killed while crops are protected?

^ Figure 7.8

Dolly the sheep was the first cloned mammal.

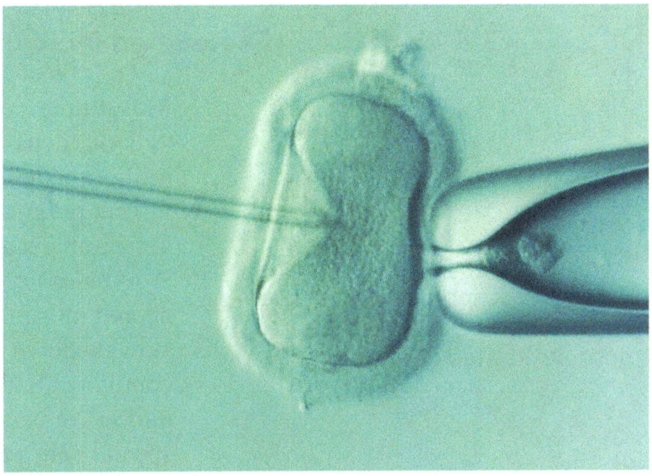

^ Figure 7.9

This photo shows a key step in the process of animal cloning. In nuclear transplantation, a cell nucleus from the animal being cloned is transplanted into an egg that has had its own nucleus removed. The egg is held in place by a vaccum from the pipet shown on the right.

The resulting embryo is then implanted into a surrogate mother. Since Dolly, other mammals have been cloned, including dogs, cats, mice, and horses, as well as rare species such as the European mouflon (a small wild sheep) and the banteng (a type of wild cattle). A group of Japanese, Russian, and American scientists is even hoping to use tissue from a frozen mammoth to clone a live woolly mammoth. The plan is to insert the nucleus of a mammoth cell into an elephant egg cell that has had its own nucleus removed. If an embryo results, it could then be implanted into a female elephant.

Unfortunately, cloned animals often suffer from developmental abnormalities or other health problems. Today, there is only limited use of animal cloning in agriculture to replicate a few particularly desirable animals.

In January 2018, a pair of monkeys—long-tailed macaques—became the first cloned primates. Human cloning, however, has never been performed and would certainly run into serious ethical obstacles. More than 30 countries have formally banned human cloning. Although there is no such ban in the United States, laws currently prevent federal funds from being used to work on human cloning.

CONCEPT CHECK

Is an animal produced through nuclear transplantation genetically identical to the animal that provides the nucleus, the animal that provides the egg, or the surrogate mother? Explain.

CHECK YOUR ANSWER

The animal is genetically identical to the animal that provides the nucleus. DNA is contained within the nucleus.

More to Explore

You can learn more about the history of genetically modified organisms here:

http://sitn.hms.harvard.edu/flash/2015/from-corgis-to-corn-a-brief-look-at-the-long-history-of-gmo-technology/

7.5 Genetically Engineered Mosquitoes

READING CHECK

How do most approaches that use genetic engineering to reduce mosquito-related diseases work?

Mosquitoes can be very harmful to humans. They spread diseases such as malaria, dengue fever, and the more recent Zika. Malaria kills nearly half a million people every year, and Zika can cause severe birth defects when it infects pregnant women. Could genetically engineered mosquitoes finally help humans avoid the impacts associated with these insects?

Scientists are working on a variety of approaches to genetically engineered mosquitoes. Most of these aim to reduce mosquito populations, thus slowing down the spread of mosquito-borne diseases. For example, the British company Oxitec has developed genetically engineered "friendly male" mosquitoes. When these males mate with females, their offspring are affected by special genes that cause them to die as larvae, preventing their development into biting adults (**Figure 7.10**). Over time, the release of enough "friendly males" causes mosquito populations to decline. In fact, field studies show that Oxitec's "friendly males" can reduce mosquito populations by as much as 95%.

^ Figure 7.10

Scientists are working on a variety of genetically engineered mosquitoes that may help prevent the spread of diseases such as malaria, dengue fever, and Zika. Show here are mosquito larvae.

Another group, based at Imperial College London, has developed a "gene drive" that spreads female infertility in mosquitoes. Gene drives cause certain alleles to be inherited more often than expected under Mendelian inheritance, resulting in the alleles spreading very quickly. The Imperial College team's gene drive includes three genes that prevent females from laying eggs. Again, the release of mosquitoes with the gene drive can cause mosquito populations to decline.

But what are the potential consequences of wiping out an entire insect population? How would this affect other species that interact with mosquitoes? Mosquitoes, for example, are an important source of nutrition for many bat species. And could mosquitoes end up being replaced by an even worse pest? These are important questions to be considered.

A third genetic engineering approach aims to develop mosquitoes that are themselves disease-resistant. This way, the mosquitoes will be unable to spread disease. The advantages of this approach are that wiping out mosquito populations is not necessary, and unintended consequences are less likely. Groups at the University of California, Irvine, and the University of California, San Diego, are developing genetically engineered mosquitoes with high activity in the part of the immune system that fights the malaria parasite. These mosquitoes are resistant to malaria and are less likely to transmit the disease to humans.

Whether any of these strategies will defeat mosquitoes is yet to be determined. However, genetic engineering could provide powerful tools in the battle against an age-old pest.

CONCEPT CHECK

How might genetically engineered "friendly male" mosquitoes help prevent the spread of mosquito-transmitted diseases?

CHECK YOUR ANSWER

When the "friendly males" mate with females, their offspring are affected by special genes that cause them to die as larvae, preventing their development into biting adults. As a result, the release of enough "friendly males" can, over time, reduce mosquito population sizes.

More to Explore

You can read more about one genetically engineered mosquito, and the danger of its genes spreading into the natural population, here:

https://www.sciencemag.org/news/2019/09/study-dna-spread-genetically-modified-mosquitoes-prompts-backlash

7.6 Genome Editing with CRISPR-Cas9

It's been called the biotechnology discovery of the century. But what exactly is CRISPR-Cas9? CRISPR-Cas9 is a powerful tool for genome editing. It allows scientists to add, remove, or change DNA at specific locations in the genome. The CRISPR-Cas9 system was developed in 2012 by Jennifer Doudna at the University of California, Berkeley, and Emmanuelle Charpentier at the Max Planck Institute for Infection Biology in Berlin (**Figure 11**), Germany. The term "CRISPR" is short for Clustered Regularly Interspaced Short Palindromic Repeats. A CRISPR is a sequence of nucleotides in the DNA commonly found within prokaryotes, such as bacteria. The "Cas" is a CRISPR-associated protein that works with the CRISPR sequence. In 2020, Doudna and Charpentier were awarded the Nobel Prize in Chemistry for their work.

^ Figure 7.11

Emmanuelle Charpentier and Jennifer A. Doudna discovered one of gene technology's sharpest tools: the CRISPR/Cas9 genetic scissors.

CRISPR-Cas9 is a system used by some bacteria to defend themselves from viruses. When viruses attack, these bacteria save some of the virus's DNA in special "CRISPR arrays." If the viruses ever return, DNA from the CRISPR arrays is used to make RNA. This RNA locates matching DNA sequences in the attacking viruses. Then, the Cas9 enzyme cuts the DNA apart, destroying the virus.

READING CHECK

How do bacteria use CRISPR-Cas9?

The CRISPR-Cas9 genome editing tool works in a similar way. Scientists create "guide RNA" with the desired target sequence and attach it to the Cas9 protein. The RNA locates the matching DNA, and the Cas9 enzyme cuts it (**Figure 7.12**). Specific genes can then be added or removed from that precise location in the genome.

The CRISPR-Cas9 system is exciting because of its many potential uses. For example, many severe human diseases, including hemophilia and cystic fibrosis, are the result of a small number of faulty genes. CRISPR-Cas9 may allow scientists to eliminate and replace faulty genes with normal, working versions. Genome editing could also lead to treatments for more genetically complex diseases, such as cancer.

Although CRISPR-Cas9 has yet to be used to treat human diseases, it has already been used to slow the growth of cancer in mice, to spread infertility in female mosquitoes, and to remove HIV, the virus that causes AIDS, from living organisms.

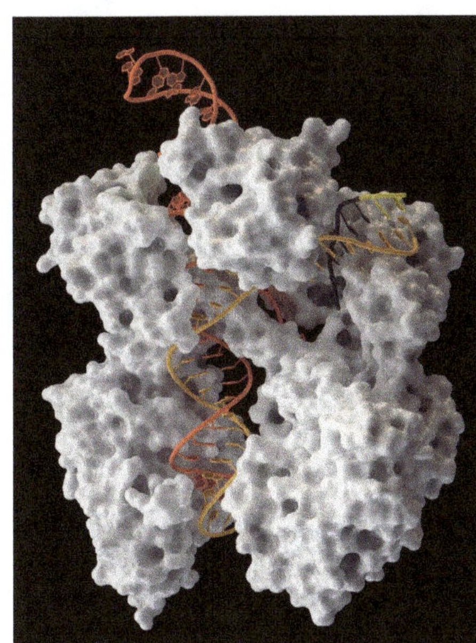

^ Figure 7.12

The CRISPR-Cas9 genome editing system includes a guide RNA (red) that helps to locate a target in the DNA genome (yellow). The Cas9 enzyme (white) linked to the RNA then cuts the DNA at the appropriate spot.

This powerful technology raises some ethical concerns, however. First, there are issues of safety. More research is needed to determine what could go wrong if CRISPR-Cas9 were to be used to treat human diseases. For example, in one experiment, the Cas9 enzyme cut the human genome at several unintended locations in addition to the target location. An important question then becomes, what limits, if any, should be placed on editing the human genome?

Although most people agree that using a technology such as CRISPR-Cas9 to treat disease is reasonable, people disagree about whether other uses are ethical. For example, is it ethical to use genes to improve or "enhance" human traits? Many people also favor a ban on the use of genetic technologies in human eggs and sperm, which would produce genetic changes that would then be passed from one generation to the next.

CONCEPT **CHECK**

Why is the CRISPR-Cas9 system a valuable tool for scientists?

CHECK YOUR ANSWER

The CRISPR-Cas9 system allows scientists to manipulate DNA effectively. It can be used to cut DNA at specific locations in the genome and then add or delete DNA sequences.

More to Explore

You can read more about CRISPR-Cas9 here:

https://medlineplus.gov/genetics/understanding/genomicresearch/genomeediting/

7.7 Concerns about DNA Technology

Although DNA technology has been the source of great excitement, it has also given rise to serious concerns. Could scientists accidentally release a deadly bacteria or virus? Are GM plants and animals safe to eat? Are GM crops damaging the environment and giving rise to dangerous "superweeds"? Also, which DNA technologies could one day be applied to humans, and what impact will this have on our society and culture? Finally, who will have access to DNA technologies?

Some genetically engineered bacteria and viruses are dangerous to human health or natural habitats. To help keep these organisms from leaving the lab, scientists give them genetic mutations that prevent them from performing crucial natural functions. This makes it harder for them to survive in the wild. (In the lab, scientists keep them alive by providing conditions that make up for the genetic deficiencies.) Despite these precautions, accidental releases remain a legitimate concern.

Genetically modified crops are probably the most controversial product of DNA technology today. In the United States, GM crops are very common. In 2014, more than 90% of the soybeans, cotton, and corn planted were of genetically modified varieties. In many European countries, however, GM crops are banned or strictly controlled. A basic question is, are GM plants safe to eat? In 2010, the European Commission reviewed 25 years of research to address the issue. That investigation indicated that GM plants are not riskier for human health than plants produced through conventional breeding methods.

However, scientists and consumer groups worry that safety is not always adequately tested. In 2009, a group of two dozen scientists from public research institutions in 17 states warned the U.S. Environmental Protection Agency that the companies that develop GM plants routinely act to "inhibit public scientists from pursuing their mandated role on behalf of the public good" and that these companies had made independent analyses of GM crops impossible.

A related issue is the labeling of GM foods (**Figure 7.13**). Labeling is not currently required in the United States, although a large majority of consumers favor it. Labeling is opposed by the industry, which believes that consumers may avoid labeled products.

< Figure 7.13

Are genetically modified foods safe? Should they be labeled? Some people, including these protestors, think that additional regulation is needed.

Environmental concern

Another concern about GM crops is their effect on the environment. For example, crops that contain *Bacillus thuringiensis* genes, which are toxic to pest species, can also kill non-target species. "Roundup Ready" crops encourage farmers to use more chemical herbicides, which damage the environment. Some weeds can also evolve resistance, becoming "superweeds." Roundup has also been linked to the decline of honeybee populations. A 2018 study found that the herbicide Roundup makes honeybees vulnerable to deadly infections by harming the beneficial bacteria in the honeybees' guts.

READING CHECK

How is the herbicide Roundup linked to the decline of honeybee populations? .

Contamination

A further worry is contamination by genetically engineered plants or their genes. In a 2007 test of "Roundup Ready" creeping bentgrass, developed for use on golf courses, some plants escaped from a field in Idaho and ended up in irrigation canals miles away in Oregon. Hundreds of escaped plants, including many that had already produced their own seeds, were found. GM crops could also breed with wild plants and transfer their herbicide-resistance genes to the wild varieties, creating instant "superweeds." One such plant, a resistant mustard weed, was discovered in 2005 in a field used for trials of GM oilseed rape (a plant used to produce canola oil). Before the discovery, such transfers were described as nearly impossible.

Genes from GM crops can also contaminate traditional crops. In 2004, the Union of Concerned Scientists studied corn, canola, and soybeans and found that the traditional seed supplies for all three were contaminated with DNA from genetically modified varieties. Contamination is a particular worry for organic farmers because organic produce cannot include GM varieties. In 2011, an organic farmer in Australia sued his neighbor after GM canola blew onto his fields, resulting in the loss of his organic license.

Cost

A final concern about GM crops is cost. GM seeds are too expensive for farmers in many parts of the world. In India, many farmers were encouraged to take out large loans to buy "insect-proof" GM seeds. When the crops failed due to drought (some GM crops require more water than traditional crops), many of the farmers went bankrupt.

More to Explore

For more on genetic engineering and potential controversies, you can check out this YouTube video:

https://www.youtube.com/watch?v=FY_ZUEKWhBc

What about DNA technology and human society? What will happen when we apply DNA technologies to ourselves? Curing a genetic disease is one thing, but what about enhancing our abilities? Should constraints be placed on DNA technology? In addition, as genetic technologies are developed, who will control access to them? Should important technologies be available only to the wealthy and privileged? For example, when a private company began to clone cats for people who missed their deceased pets, the price tag was $50,000.

On the other hand, could the ability to manipulate DNA become too readily available—in easy, do-it-yourself kits? Will we have enough time to think through a technology's social and ethical implications or to legislate adequate controls? And what if something unexpected occurs? We will be facing questions like these for years to come. So that our answers to these questions can be based upon understanding rather than ignorance, it's upon each of us to become familiar with the basic biology concepts underlying genetic engineering.

CONCEPT CHECK

Is there evidence that some genetically modified crops could be harmful to species in natural environments?

CHECK YOUR ANSWER

Yes, in some cases. Crops that contain Bacillus thuringiensis genes, which are toxic to pest species, can also kill non-target species. "Roundup Ready" crops encourage farmers to use more chemical herbicides, which damage the environment. And Roundup has also been linked to the decline of some honeybee populations.

Concept Review

Chapter 7

Detailed Chapter Summary

The human genome, consisting of 3.2 billion nucleotide pairs, was sequenced during the Human Genome Project. More than 99% of nucleotide sequences are identical in all humans. The human genome contains a total of about 19,000 protein-producing genes. However, these genes represent only 2% of the genome. The rest of the genome consists of other sequences, including many repeat sequences with no clear function. Single-nucleotide polymorphisms, or SNPs, describe locations in the genome that vary among humans. There are millions of SNPs in the genome, and some of these are expected to be important to human diseases.

There are many inherited genetic diseases that affect humans. Pedigrees, which are charts showing the members of a family affected by a genetic disease, can help determine a couple's risk of having a child with a genetic disease. Creating a pedigree is an important first step in assessing risk. If a disease requires two recessive alleles in order to affect a person, then it becomes essential to know whether one or both parents are carriers. Carriers of a disease are heterozygotes who possess one disease allele and one normal allele, meaning that they are not affected by the disease but can pass it on. Genetic testing can also determine whether a fetus is affected. Chorionic villus sampling and amniocentesis are two techniques used to collect fetal cells for testing during pregnancy. Down syndrome is a genetic disease correlated with the age of the mother. It is caused by trisomy 21.

Cancer occurs when mutations in the genes that control cell division allow cells to divide out of control and spread through the body. Mutations in many important genes are usually required for cancer to develop. In addition, many environmental factors increase the risk of developing cancer, including smoking, diet, radiation, ultraviolet light, chemicals, and infection by certain bacteria and viruses. Certain inherited mutations can also make cancer more likely to develop.

Genetic engineering describes the process of changing the traits of organisms by directly manipulating their DNA. A genetically modified organism is one whose

DNA has been changed through genetic engineering. Many genetically modified bacteria, plants, and animals have been developed. Some of these produce beneficial compounds, such as human insulin, whereas others have valuable traits, such as disease resistance or rapid growth.

Cloning describes the creation of an organism that is genetically identical to one that already exists. Many animals have been cloned, but human cloning has so far never been performed and would be ethically problematic.

A number of scientists are working on using genetic engineering to address human diseases carried by mosquitoes. One group is working on genetically engineered mosquitoes whose offspring die. Another group is developing mosquitoes that spread infertility. Still, others are working to make the mosquitoes themselves resistant to diseases such as malaria so that they will not pass these diseases to humans.

CRISPR-Cas9 is a powerful tool for genome editing. This tool allows scientists to add, remove, or change DNA at specific locations in the genome. Scientists hope to use this tool to treat certain genetic diseases, including cancer.

Despite the promise of many DNA technologies, there are also some concerns. One worry is that deadly bacteria or viruses could be released. Another is that genetically modified crops could be unsafe for human consumption or could harm natural environments. DNA from genetically engineered plants could also contaminate natural populations. The technology can also be seen as a distraction from the development of more sustainable agricultural practices. Because DNA technologies are generally costly, there is also worry that only the wealthy and privileged will have access to important technologies. Finally, there are numerous ethical questions regarding the responsible use of DNA technologies that still need to be answered.

Key Terms

Genome The total genetic content of an organism.

Genetic engineering The process of changing the traits of organisms by directly manipulating their DNA.

Genetically modified organism An organism whose DNA has been changed through genetic engineering.

Review Questions

Simple Review Questions

7.1 The Human Genome

1. How many genes does a human have?
2. What is an SNP?

7.2 Inherited Diseases and Genetic Testing

3. What is a pedigree?
4. Explain what a carrier of a disease is.

7.3 Cancer: A Genetic Disease

5. Is cancer usually the result of a single genetic mutation?
6. What does it mean for a cancer to have metastasized?
7. What are some of the most important cancer-related environmental risk factors?
8. How is ozone loss from Earth's atmosphere related to skin cancer?

7.4 DNA Technology and Genetic Engineering

9. What is a genetically modified organism?
10. Use examples to describe how genetically modified organisms are useful to humans.
11. Explain how a mammal is cloned.

7.5 Genetically Engineered Mosquitoes

12. Explain how genetically engineered "friendly male" mosquitoes can help address mosquito-borne diseases.
13. Why might it be better to develop mosquitoes resistant to malaria than drastically decrease mosquito populations?

7.6 Editing with CRISPR-Cas9

14. What does the CRISPR-Cas9 technology allow scientists to do?

7.7 Concerns About DNA Technology

15. Describe some of the safety, social, and ethical concerns that relate to genetically modified crops.

Challenging Review Questions

7.1 The Human Genome

16. Does all your DNA code for proteins?

17. Explain how studying SNPs could help scientists identify genes that are related to specific human diseases.

7.2 Inherited Diseases and Genetic Testing

18. The allele for Tay-Sachs disease behaves as a recessive. A couple would like to know if their children are at risk for the disease and go in for genetic testing. The test reveals that one of them is a carrier of Tay-Sachs, but the other is not. Is this couple at risk of having an affected child?

19. The allele for Tay-Sachs disease behaves as a recessive. A couple would like to know if their children are at risk for the disease. Both are carriers of Tay-Sachs. Are their children at risk of being affected by Tay-Sachs disease?

7.3 Cancer: A Genetic Disease

20. Cancer is caused by "genes gone awry,"—yet cancer is not usually an inherited genetic condition. Explain why.

21. Are older people or younger people more likely to develop cancer? Defend your answer.

22. What are some factors that could make certain people more likely to develop cancer than others?

23. Explain how the chlorofluorocarbons (CFCs) in refrigerants and aerosol sprays affect Earth's ozone layer. Have international agreements to phase out the use of CFCs helped restore the ozone layer?

24. Australia has one of the highest skin cancer rates in the world. Scientists believe two factors behind this are the high UV radiation levels and the popularity of tanning. Explain how these factors could lead to unusually high skin cancer rates.

7.4 DNA Technology and Genetic Engineering

25. How can genetic engineering help humans adapt to global warming and climate change?

26. How might genetic engineering affect the human diet in the future?

7.5 Genetically Engineered Mosquitoes

27. Describe three different genetically engineered mosquitoes and the advantages of each in helping to control mosquito-borne diseases such as malaria.

7.6 Genome Editing with CRISPR-Cas9

28. How does the CRISPR-Cas9 genome editing technology work?

29. Are there any ethical concerns regarding the use of CRISPR-Cas9 technology?

7.7 Concerns About DNA Technology

30. What is a "superweed"? Explain the benefits and dangers of genetically engineered crops that are resistant to herbicides.

31. Write an imaginary letter to your grandfather telling him about genetically engineered organisms. Tell him why the development of this technology could result in many practical applications. Also, tell him about the potential dangers of this technology. Are you more excited or more worried about the possible consequences of DNA technology?

Apply & Discuss Questions

32. Artemisinin is a powerful antimalaria drug. Unfortunately, it is expensive and in short supply because it must be extracted from the leaves of the sweet wormwood tree, a rare plant with a very limited distribution. As a result, many populations suffering most from malaria cannot afford artemisinin. Could DNA technology be used to address this problem?

33. Does this cartoon make you excited about DNA technologies or worried about it? What constraints would you place on DNA technology if you were a lawmaker? Should there be different rules regarding applying DNA technology to humans and applying it to other species?

Solutions to Odd-Numbered Questions

Simple Review Solutions

1. Humans have a total of about 19,000 genes.

3. A *pedigree* is a family tree that shows which relatives are affected by genetic disease.

5. A mutation in a single gene is not enough to produce cancer—mutations in many important genes are required.

7. Environmental factors that increase cancer risk include smoking, diet, radiation, ultraviolet light, chemicals, and infection by certain viruses and bacteria.

9. A genetically modified organism, or GMO, is an organism whose DNA has been changed through genetic engineering.

11. Cloning is done through a process called nuclear transplantation, in which the nucleus of a cell from the animal being cloned is placed in an egg cell that has had its own nucleus removed. The resulting embryo is then implanted in a surrogate mother.

13. The advantages of this approach are that wiping out mosquito populations is not necessary and unintended consequences are less likely.

15. DNA technology has been the source of great excitement but has also given rise to serious concerns. Could scientists accidentally release a deadly bacteria or virus? Are transgenic plants and animals safe to eat? Are transgenic crops damaging the environment and giving rise to dangerous "superweeds"? Also, which DNA technologies could one day be applied to humans, and how will this impact our society and culture? Finally, who will have access to DNA technologies?

Challenging Review Solutions

17. Scientists can look for SNPs that are unique to people who share a certain disease or condition. Such SNPs would likely be found within genes that are related to that disease or condition.

19. Yes. There is a ¼ chance their child would inherit two recessive disease alleles and suffer from Tay-Sachs disease

21. Cancer occurs as a result of mutations in many of the key genes that control cell division. Older people are more likely to develop cancer than younger people because mutations have had more time to build up in their cells.

23. The chlorine in CFCs speeds up the breakdown of ozone, causing it to become depleted. An international agreement to phase out the use of CFCs has succeeded in shrinking the ozone hole. This is considered one of the world's major environmental successes.

25. Much effort is currently going toward developing genetically engineered plants that are resistant to drought and high temperatures, both of which have worsened with global warming.

27. The British company Oxitec has developed genetically engineered "friendly male" mosquitoes. These males mate normally with females, but they have special genes that cause their offspring to die as larvae, so that they never grow to become biting adults. A group at Imperial College London has developed a "gene drive" that spreads female infertility in mosquitoes. Both of these have the potential to reduce mosquito populations. Groups at the University of California, Irvine, and the University of California, San Diego, are developing genetically engineered mosquitoes with high activity in the part of the immune system that fights the malaria parasite. These mosquitoes are resistant to malaria and will not transmit the disease to humans.

29. A big unanswered question is, what limits should be placed on editing the human genome? Although most people agree that using a technology such as CRISPR-Cas9 to treat disease is reasonable, people disagree about whether other uses are ethical. For example, is it ethical to use genes to improve or "enhance" human traits? Many people also favor a ban on the use of genetic technologies in human eggs and sperm, which would produce genetic changes that would then be passed from one generation to the next.

31. Dear Grandpa, A genetically engineered organism is one whose DNA has been changed through technology. Many genetically engineered organisms have practical applications. Examples include E. coli, which produces human insulin; plants that can grow under difficult conditions—for example, in unusually salty soils or during drought and high temperatures; and animals that can be used to study human diseases. Although this sounds good so far, there are potential dangers as well. Some people worry about the accidental release of deadly bacteria or viruses. Others worry that genetically engineered plants and animals are dangerous to human health or that they will damage the environment by spreading their genes to native populations. Finally, some scary things could happen if DNA technologies are used on humans.

Apply & Discuss Solutions

33. Answers will vary, but this is an important question!

Living Earth Essay

Life on Earth Connections

Supervolcanoes

In 1971, the young geologist Mike Voorhies and his wife, Jane, were strolling along a gully on the edge of a farm in northeast Nebraska not far from where they lived. Since he was a little kid, Mike had been fascinated by fossils, having found his first ancient camel tooth at the age of 8. On that day, by the gully, he happened to notice what looked like an animal skull protruding from the eroding edges. Within minutes, he and Jane unearthed not just the skull but the entire skeletal remains of a 12 million-year-old rhinoceros. They had discovered what has since become known as the Ashfall Fossil Beds of Nebraska.

Figure A
Mike Voorhies discovered the Ashfall Fossil Beds of Nebraska.

Further explorations revealed the remains of hundreds of large vertebrate animals who, around a water hole, had died together upon being buried by a meters-thick layer of ash. The source of this ash? A supervolcano 1000 miles to the west—an earlier version of the very same supervolcano that now resides beneath Yellowstone National Park in northwest Wyoming.

Recall that the outer layers of Earth are made up of a series of tectonic plates, looking much like a cracked, hard-boiled egg. About 95% of Earth's volcanoes arise along these cracks. These volcanoes can be destructive, but they pale in comparison to the fewer "supervolcanoes" found on Earth.

Figure B
The Grand Prismatic Spring of Yellowstone is one of many bodies of water heated by an underlying pool of magma.

Supervolcanoes can form over a "hot spot," which is where there is a direct line between Earth's molten mantle and the surface. These hot spots can occur in the middle of a tectonic plate rather than along the edge of the plate. Yellowstone resides over such a hot spot, which explains its many natural hot springs along with frequent earthquake activity (**Figure C**).

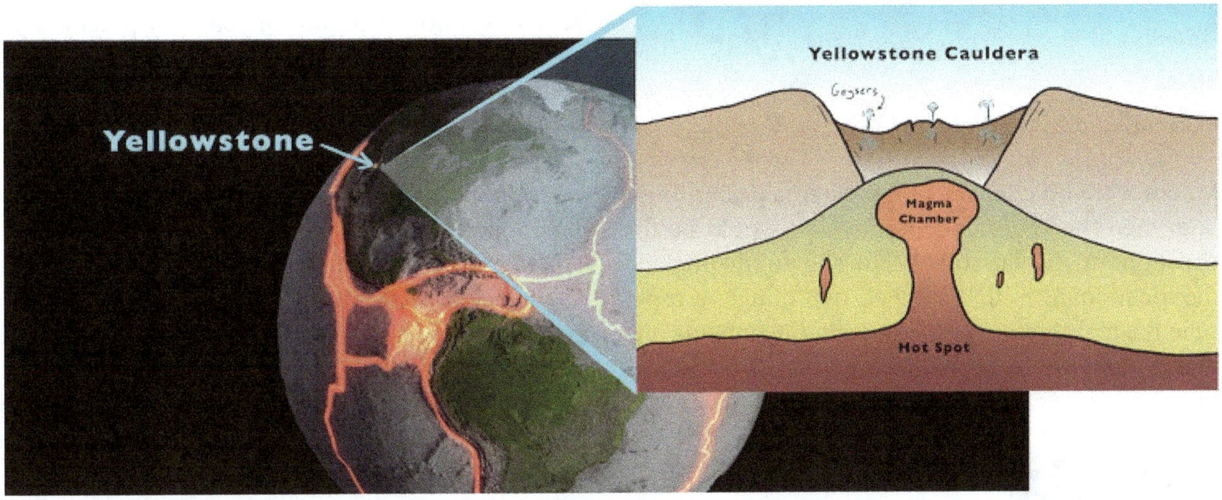

Figure C
Yellowstone is found over a hot spot within the North American tectonic plate.

Clues in the Phenocrysts

The Yellowstone supervolcano has a history of blowing up once every 600,000 years or so. Notably, the last mega-explosion occurred about 630,000 years ago, which might suggest it is due for another super-eruption any day now. However, scientists have learned that volcanoes are not so well-behaved or predictable. Yellowstone is one of the most closely studied and monitored volcanoes around the planet. As of this writing, the Yellowstone supervolcano is showing no unusual signs of impending doom.

But that doesn't stop scientists from wanting to learn as much as they can about this volcanic system. For example, geology graduate student Hannah Shamloo, and her advisor, Christy Till, from Arizona State University, published research showing that the build-up to the last major eruption of Yellowstone may have occurred over a matter of only years or even shorter as months, as opposed to thousands of years. But how exactly did they come to this conclusion? After all, we're talking about a supervolcano that erupted some 630,000 years ago.

Hannah and her team traveled to Yellowstone to collect samples of ash from the layer corresponding to the last mega-explosion (**Figure D**). Back in the laboratory, she used instruments to measure the chemical composition of micro-crystals known as phenocrysts—tiny crystals that form as magma cools as it slowly rises toward the surface. She had learned that as the crystal grows, trace elements, such as barium, Ba, get embedded within the crystal. The gradient from the center of the crystal to the outer edges thus provides a storyline of the changing conditions beneath the volcano prior to eruption.

Professor Till's Research Group on-site in Yellowstone

Professor Hannah Shamloo

Figure D
As a graduate student, Hannah Shamloo and her team looked at how crystals formed within Yellowstone's magma formed just prior to the last eruption. The chemical composition of these crystals held important clues to how this eruption unfolded. Hanna is now a professor of geology at Central Washington University.

READING CHECK

Where do supervolcanoes tend to form?

If there were no changes in the magma's conditions over time, then the chemical composition would be the same throughout the crystal. What she found instead were chemical changes that showed two things: a rapid increase in the temperature of the surrounding magma and an increasing amount of crystallized barium (**Figure E**).

Figure E
Hannah Shamloo by the area in Yellowstone where samples containing the phenocrysts were collected.

The problem with this is that barium content in a crystal is typically expected to increase while temperature decreases. Because Shamloo and her team observed a rapid temperature increase, they knew that the magma must have experienced some quick change right before the eruption Here was important evidence within these tiny phenocryst crystals. And like a thoughtful Sherlock Holmes, Shamloo and Till realized this pointed to a likely alternate mechanism of the last mega-explosion. Their observations of a sudden quick increase in temperature could be explained by the rapid influx of a large quantity of magma from deep below over not thousands of years, by potentially as short as months to years. If true, this means that present-day Yellowstone could go from its current conditions to a major explosion within a century.

As Hannah and her advisor are quick to point out, much more research is required to support or refute these conclusions. Further, as far as scientists can track, the subterranean magma chambers are currently not undergoing major movements. Thus, geologists estimate that the chances of a mega-explosion occurring within a year are about 1 in 760,000. On the other hand, the slow release of lava, which would devastate only the area around the national park, has a greater chance of occurring at about 1 in 10,000.

Figure F
The Pinatubo volcanic eruption of 1991 within the Philippines.

The main point of all of this is that we live on a planet that is very much alive. When it comes to volcanoes, earthquakes, tsunamis, wildfires, tornados, hurricanes, and other destructive forces, the more we learn about these systems, the better we can prepare ourselves. Learning more about these natural hazards also helps us to respect the wonders and powers of this living planet upon which we depend.

To Life!

CONCEPT CHECK

If the chemical composition of phenocrysts is uniform throughout the crystal, then what does this say about the condition in which the phenocrysts formed?

CHECK YOUR ANSWER

A uniform chemical composition indicates little changes in conditions within which the phenocryst formed.

More to Explore

Follow this link to learn more about the Yellowstone supervolcano:

https://www.usgs.gov/observatories/yellowstone-volcano-observatory

Chapter 8

Natural Selection

THE MAIN IDEA

Evolution is the genetic change within a population over time.

8.1 The Origin of Life

8.2 Is there life on Mars or Venus?

8.3 Charles Darwin

8.4 Natural Selection

8.5 Examples of Natural Selection

8.6 Adaptation

How might life have originated on Earth? And might we one day find signs of alien life elsewhere within our solar system? In the 1800s, Charles Darwin sailed to this planet's far reaches, collecting specimens of numerous species. Over the following decades, he came to recognize a mechanism by which different species could evolve under purely natural conditions. He called this Natural Selection. What exactly is natural selection, and how does it operate?

8.1 The Origin of Life

How did life originate? We know from fossils that life has existed on Earth for at least 3.5 billion years. This means that the Earth on which life evolved was very different from the Earth of today. It contained vast, lifeless oceans, violent volcanoes, and a turbulent atmosphere filled with lightning storms and incoming meteors (**Figure 8.1**). The atmosphere of the early Earth included no oxygen, which came later, as a result of the activity of living things. Although this environment seems a hostile place for life today, it may have been appropriate for producing the first life. Why do we think the early Earth environment could produce life from nonliving materials? A famous experiment suggests exactly this.

< Figure 8.1

Early Earth was very different from the Earth of today. Volcanoes spewed hot fumes into the atmosphere, and meteors regularly bombarded Earth's surface.

In 1953, Stanley Miller and Harold Urey built a model of the early Earth in a chemistry lab (**Figure 8.2**). A flask containing a mixture of simple compounds— including water vapor, ammonia, methane, and hydrogen gas—simulated Earth's early atmosphere. Liquid water was added to represent Earth's oceans. Electric sparks sent through the gases simulated lightning. When this model of early Earth was assembled, an amazing thing happened. Many complex organic molecules were formed, including amino acids, the building blocks of proteins. Not only had these molecules formed quickly, but they also formed in huge numbers. Further experiments showed that all the essential organic molecules that makeup life—not just amino acids but also sugars, lipids, and the nitrogenous bases found in RNA and DNA—can be generated in a similar way.

However, some scientists today question the importance of the Miller–Urey experiment. They think Earth's early atmosphere was actually quite different from the model atmosphere Miller and Urey used and that organic molecules may not have been so easy to generate. These scientists have proposed two other hypotheses for how Earth got its first organic building blocks. One hypothesis is that organic molecules were brought to Earth by meteorites. Meteorites steadily bombarded Earth during its early history, and some of the meteorites recovered here do, in fact, contain a wide variety of complex organic molecules, presumably formed in outer space. For example, a meteorite found in Australia in 1969 contained nearly one hundred different amino acids. A second hypothesis is that large numbers of organic molecules were formed in deep-sea environments on Earth, similar to the hydrothermal vent habitats of today.

The next question is how these many separate organic molecules advance to become living cells. Scientists do not know the entire story but have discovered some clues. For example, when certain lipids are added to water, for reasons easily explained by chemistry, they spontaneously form tiny hollow spheres called *liposomes*. Liposomes have double membranes similar to cell membranes. Although not alive, liposomes sometimes behave like living cells—they grow, shrink, and divide (**Figure 8.3**). Liposomes also run chemical reactions inside their membranes and control what molecules move into and out of them, two key features of living cells.

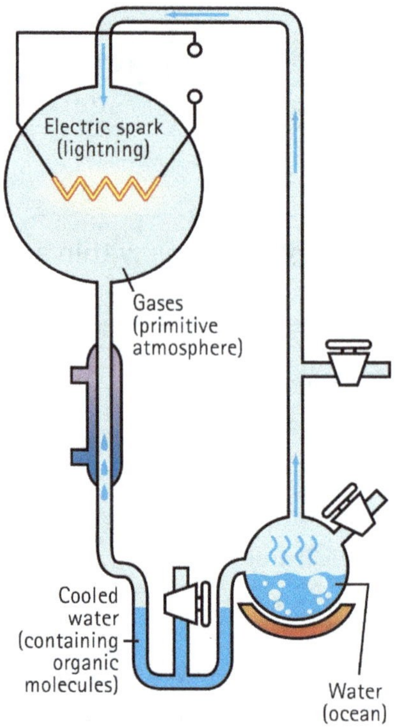

(a)

(b)

Figure 8.2

Stanley Miller and Harold Urey built a model of the early Earth and showed that complex organic molecules could be formed during lightning storms. (a) Miller and Urey built a model of Earth's early atmosphere and oceans. (b) Miller in front of a Miller-Urey apparatus.

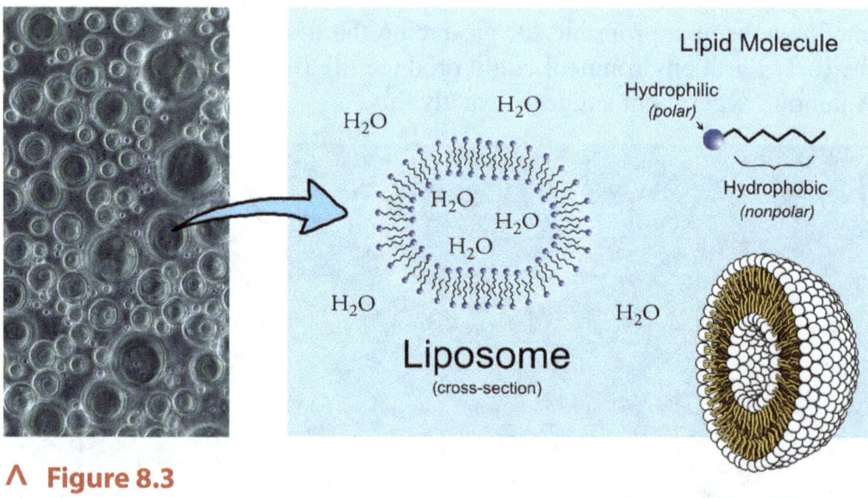

Figure 8.3

A liposome is a spherical organization of lipid molecules. Hydrophobic regions of each lipid molecule are sandwiched between the hydrophilic regions forming a bilayer membrane that encapsulates an interior of water. Liposomes show some cell-like behaviors. Some of these liposomes (shown left) are growing and dividing.

READING CHECK

Why do some scientists question the importance of the Miller-Urey experiment?

< **Figure 8.4**

Ancient fossils of bacteria show that life has existed on Earth for a long time. Fossilized stromatolite, left by mats of photosynthetic bacteria, are up to 2.72 billion years old.

Some liposomes may have eventually captured nucleic acids—that is, primitive genes. These early genes were probably made of RNA, not DNA. This is because, even in the absence of cells and enzymes, short strands of RNA can spontaneously assemble from individual nucleotides and even reproduce themselves. With a few more changes, RNA-containing liposomes may have become the very first cells—the first organisms on Earth. However it occurred, the transition to living cells was completed 3.5 billion years ago, the time of the earliest known fossil organisms. **Figure 8.4** shows some bacteria fossils from early in the history of life.

The origin of DNA is a complex topic, but several scientific theories provide potential explanations. One of the most widely accepted theories is the RNA world hypothesis. This theory suggests that RNA (ribonucleic acid) was the first molecule to carry genetic information and that DNA evolved from RNA at a later point. As described above, early Earth was rich in organic molecules, and these molecules combined to form the RNA polymer, which was able to store genetic information and catalyze chemical reactions. Over time, some of these RNA molecules evolved into DNA, which is more stable and better suited for the long-term storage of genetic information.

Another theory proposes that DNA may have originated from simple organic molecules, such as formaldehyde and hydrogen cyanide, which were present in the early Earth's atmosphere. These molecules could have combined to form deoxyribonucleotides, which are the building blocks of DNA. A third theory suggests that DNA may have originated from extraterrestrial sources, such as comets or meteorites. Some scientists believe these objects may have brought key organic molecules to Earth, including the building blocks of DNA. Indeed, spectral studies of galactic dust reveal the presence of organic molecules throughout our galaxy.

It's important to note that the origin of DNA is still a topic of active research, and scientists continue to explore and develop new theories to explain how this molecule may have emerged.

More to Explore

For meteorites, the Miller-Urey experiment, and more:

https://www.pbs.org/exploringspace/meteorites/murchison/index.html

CONCEPT CHECK

Miller and Urey found that organic molecules are easily formed in large quantities from nonorganic materials. But why do some scientists question the importance of their experiment?

CHECK YOUR ANSWER

Some scientists think the conditions on early Earth differed from what Miller and Urey modeled. Specifically, they think the composition of gases in Earth's atmosphere differed from what Miller and Urey assumed.

8.2 Is there life on Mars or Venus?

People have wondered for a very long time whether there is life elsewhere in the Universe. One obvious place to look is beyond Earth in our own solar system. Scientists quickly gravitated to Venus and Mars. Being closest to Earth, they seemed likely to have the most Earth-like conditions. But early explorations suggested that Venus would be hostile to life as we know it. Venus has almost no water, a crushingly dense carbon dioxide atmosphere, clouds of sulfuric acid, and a surface temperature of 462°C (864°F)! As a result, attention focused on Mars.

Mars appears inhospitable to life. The planet is arid, and its average temperature is only about −60°C (−80°F), though some parts can warm to a comfortable 20°C (68°F). The thin Martian atmosphere also lets through ultraviolet light from the Sun that would kill most living things on Earth. However, many geological features on Mars were clearly carved out by water. Consider, evidence suggests that billions of years ago, Mars once had oceans, warmer temperatures, and a thicker atmosphere. Perhaps remnants of ancient life forms on Mars can still be found.

READING CHECK

Why has the search for life on Mars involved looking for water?

Searching for life on Mars has meant searching for liquid water since life as we know it cannot exist without water. Using orbiters around Mars and rovers on the Martian surface, the National Aeronautics and Space Administration (NASA) has looked for places on the planet where pockets of water might still exist. Moreover, they have found evidence for it! Specifically, they have seen seasonal streaks of damp soil in certain craters. Further, a recent meteor collision along the equator revealed water ice beneath the surface. In 2021, NASA landed the new *Perseverance* rover on Mars (**Figure 8.5**). During the rover's early years of collecting samples on Mars, Perseverance has turned up carbon-bearing materials—the basis of life on Earth—in every sample it has examined.

Perseverance continues to look for water and uses ultraviolet light and X-rays to search for carbon molecules and other organic compounds. Even if there is no Martian life today, there could once have been life on Mars. This

Figure 8.5 >

Since it landed in 2021, the Mars rover Perseverance, which is about the size of a small car, has been exploring the Martian surface. Its mission is to search for evidence of current or past Martian life.

< **Figure 8.6**

Many features of Mars's surface provide evidence that there once was a large amount of water on the planet. This satellite image of a crater shows that it once held a large lake. The sediments that remain could be a good place to search for Martian fossils.

life may have thrived in the old Martian oceans and then gone extinct as the planet dried and the atmosphere thinned (**Figure 8.6**). If so, the living things may have left fossils behind. Scientists plan to search for fossils in Mars's sedimentary rocks, the same types of rocks that are often rich in fossils on Earth.

Of course, Martian life could be very different from life on Earth, in which case scientists will have to completely rethink what to look for. Interestingly, according to NASA scientists, there is one way in which we do know whether life can exist on Mars. There is life on Mars, and we know that because we sent it there. A small number of microbes from Earth have traveled to Mars along with NASA's Martian rovers. These microbes survived NASA's best efforts to decontaminate spacecraft before they took off, as well as the harsh conditions of the journey itself. This means the microbes have already survived high temperatures, radiation, and arid environments with little or no food. They also can hibernate for long periods until conditions are favorable again. Having hitched a ride to Mars, the microbes are now sitting on the Red Planet and could well survive there for millions of years.

Although Venus initially seemed like a less likely place to find life than Mars, some recent discoveries have suddenly brought Venus back into the spotlight. In September 2020, scientists announced they had found the chemical phosphine, PH_3, in the Venusian atmosphere. On Earth, phosphine is associated with the activity of certain bacteria, although there are also inorganic processes that could produce it. Yet phosphine is an unstable chemical—it doesn't last very long. Therefore, a stable amount of phosphine in the Vesuvian atmosphere means that it's somehow being continuously generated. But how so? With this discovery, scientists began to speculate about the possibility of microbial life in Venus's atmosphere, where temperatures might not be as high as on the surface. The research continues.

More to Explore

For more on NASA's Mars missions
https://mars.nasa.gov

CONCEPT**CHECK**

Why was Mars originally considered a more likely place to find life than Venus? What recent discovery has recently up-ended this?

CHECK YOUR ANSWER

Venus has almost no water and is extremely hot. Mars, on the other hand, once had plenty of liquid water. Although Mars is now much drier, there is evidence that some amount of water still exists on the planet. Renewed interest in the possibility of life on Venus came from the possible discovery of phosphine in the atmosphere, a chemical that is associated with bacterial activity on Earth.

8.3 Charles Darwin

How has life on Earth changed over time? For example, how did we get from tiny, primitive cells to humans, hippos, redwoods, and all the amazing diversity of life on Earth today?

For thousands of years, people believed that life on Earth did not change. They believed that Earth had always had the same species—and always would. Then fossils were discovered in Earth's rocks, and people began to wonder. Fossils suggested that the kinds of species living on Earth changed over time—old species disappeared, and new species appeared. Also interesting was that fossil organisms sometimes showed a distinct resemblance to modern species, as you can see in the fossil horse in **Figure 8.7**. Could some fossils actually be the ancestors of modern species?

French naturalist Jean-Baptiste Lamarck (1744–1829) was one of the first to argue that this was the case. Lamarck believed that modern species were descended from ancestors that had evolved—changed over time—to become better adapted to the environments they lived in.

According to Lamarck, organisms acquired new characteristics during their lifetimes and then passed these characteristics to their offspring. For example, ancestral giraffes stretched their necks to grab the high leaves on a tree, and their necks became longer. These giraffes then passed these longer necks to their offspring. The offspring reached for even higher leaves, stretching their necks even further, and so on (**Figure 8.8a**). Lamarck's theory for how change occurs, called the *inheritance of acquired characteristics*, proved to be incorrect: Organisms cannot pass characteristics acquired during their lifetimes to their offspring because these acquired characteristics are not genetic. However, Lamarck's fierce support for the idea that organisms evolve set the stage for Charles Darwin.

English naturalist Charles Darwin (1809–1882), shown in **Figure 8.9**, set forth his theory of evolution in his book *The Origin of Species by Means of Natural Selection*, published in 1859. Darwin proposed that **evolution**—inherited changes in populations of organisms over time—had produced all the living forms on Earth.

Figure 8.7 >

Could fossils be the ancestors of modern species? This fossil, found in Germany, is about 50 million years old. It has a clear resemblance to a horse, yet as a mature adult, it is only the size of a fox.

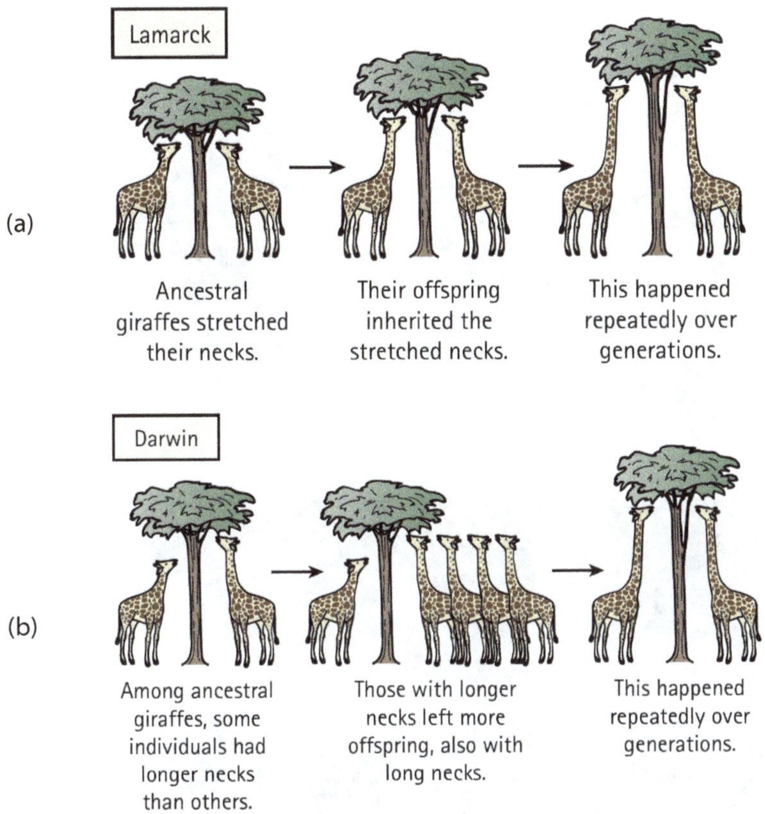

< Figure 8.8

a) Lamarck believed that organisms acquired new characteristics during their lifetimes and passed these characteristics to their offspring. (b) In his theory of evolution by natural selection, Darwin argued that organisms with advantageous traits left more offspring than organisms with other traits. As a result, advantageous traits became more common in a population.

Λ Figure 8.9

Charles Darwin developed the theory of evolution by natural selection.

Darwin's theory of evolution grew out of the observations he made aboard the H.M.S. *Beagle*, which sailed around South America from 1831 to 1836. As the ship's official naturalist, Darwin studied South American species, collecting large numbers of plants, animals, and fossils. While on the *Beagle*, Darwin collected 1,529 alcohol-preserved specimens and 3,907 skins, bones, and dried specimens. He also recorded 2,000 pages of notes on plants, animals, and geology.

Darwin became increasingly intrigued by the question of how species got to be the way they were. He was particularly struck by the living things he encountered on the Galápagos Islands, 950 kilometers from the South American continent. Darwin took particular note of the approximately 13 species of Galápagos finches—now known as Darwin's finches. Darwin's finches showed remarkable variation in the size and shape of their beaks, with each beak being suited to, and used for, a different diet (**Figure 8.10**). How had the beaks of these finches come to differ in this way? Darwin wrote, "Seeing this gradation and diversity of structure in one small, intimately related group of birds, one might really fancy that from an original paucity of birds in this archipelago, one species had been taken and modified for different ends." (From Charles Darwin, *The Voyage of the Beagle*, 1909.) Darwin was also inspired by the work of two of his contemporaries, Charles Lyell and Thomas Malthus. Lyell, a geologist, argued that Earth's geological features were created not by major catastrophic events—the favored theory of the time—but by gradual processes that produced their effects over long time periods. For example, the formation of a deep canyon did not require a cataclysmic flood but could result from a river's slow erosion of rock over millennia. Darwin realized this could be true for organisms as well: The accumulation of gradual changes over long periods could produce all the diversity of living organisms on Earth as well as all their remarkable features.

READING CHECK

Why was Darwin so interested in the finches he observed in the Galapagos Islands?

(a) (b) (c)

Figure 8.10 >

The finches Darwin saw on the Galápagos Islands—now called Darwin's finches—show remarkable variation in the size and shape of their beaks. Each is suited to a different diet. (a) The cactus finch has a pointy beak that it uses to eat cactus pulp and flowers. (b) The large ground finch has a blunt, powerful beak that it uses to crack seeds. (c) The woodpecker finch has a woodpecker-like beak that it uses to drill holes in wood. It then uses a cactus spine to pry out insects.

The economist Thomas Malthus was a second important influence on Darwin, and the one who led Darwin to his great idea of how evolutionary change happened. Malthus observed that human populations grow much faster than available food supplies, and he concluded, with despair, that famine was an inevitable feature of human existence. Darwin applied Malthus's idea to the natural world and argued that because there are not enough resources for all organisms to survive and reproduce as much as they can, living organisms are involved in an intense "struggle for existence." As a result, organisms with advantageous traits leave more offspring than organisms with other traits, causing populations to change over time. To return to the giraffe's long neck: Darwin argued that ancestral giraffes with longer necks were better at reaching the high leaves on trees. Because longer-necked giraffes got more food, they survived and left more offspring than ancestral giraffes with shorter necks. This happened repeatedly over generations. Over time, the giraffe population had more longer-necked giraffes **(Figure 8.8b)**. This process, which Darwin called *natural selection*, has been found to be the major driving force behind evolution.

CONCEPT **CHECK**

1. If Lamarck had been correct and evolutionary change occurred through the inheritance of acquired characteristics, what trait might a bodybuilder pass to his offspring?

2. Many animals that live in the Arctic, such as Arctic hares, have white fur. How could natural selection explain the evolution of their white fur color?

CHECK YOUR ANSWERS

1. If Lamarck were correct, the bodybuilder's children would inherit the increased muscle mass that the bodybuilder had acquired over a lifetime of weightlifting. Because Lamarck's theory turned out to be incorrect, however, the children will have to do their own bodybuilding.

2. Animals that were harder to see in their snowy environments had an advantageous trait—predators were less likely to spot them. Arctic hares with whiter fur were more likely to survive to adulthood, reproduce, and leave offspring. These offspring would also have inherited whiter fur. As a result, whiter fur became more common in the Arctic hare population. Over many generations, natural selection produced a white coat that matches the Arctic snow.

More to Explore

Read more about Darwin's voyage on the Beagle at:
https://www.amnh.org/exhibitions/darwin/a-trip-around-the-world

8.4 Natural Selection

Rabbits were introduced into Australia in 1859 when Thomas Austin released 24 individuals onto his property in the southeastern part of the continent. The rabbits quickly became pests, devastating farmlands, and natural habitats. Breeding "like rabbits," they spread across the continent in such large numbers that they were described as a "gray blanket" that covered the land. Many attempts were made to control the rabbit population, including the construction of an 1822-kilometer-long "rabbit-proof" fence—still the longest fence in the world. Unfortunately, by the time the fence was completed in 1907, the rabbits had already passed through. (The fence wouldn't have worked anyway—even after it was completed, rabbits would pile up so thickly behind it that some could eventually hop right over their companions' backs to the other side.)

In the early 1950s, the government decided to try to control the rabbit population by releasing the myxoma virus, a virus deadly to rabbits. Initially, the virus was a wonder, killing more than 99.9% of infected rabbits. Within a few years, however, fewer rabbits were dying. What had happened? Within the original rabbit population, a small number of individuals happened to be resistant to the myxoma virus. These resistant individuals survived the disease and reproduced, producing more disease-resistant offspring (**Figure 8.11**). Over time, the number of disease-resistant rabbits increased, and the virus became less and less effective. The rabbit population had evolved resistance to the myxoma virus through natural selection.

READING CHECK

Why did the virus kill fewer rabbits after a few years?

Figure 8.11 >

Deadly myxoma virus initially killed 99.9% of infected rabbits. However, a small number of naturally disease-resistant rabbits (blue) survived and reproduced, passing their trait of myxoma resistance to their offspring. As a result, the population became more resistant, and the virus became less effective.

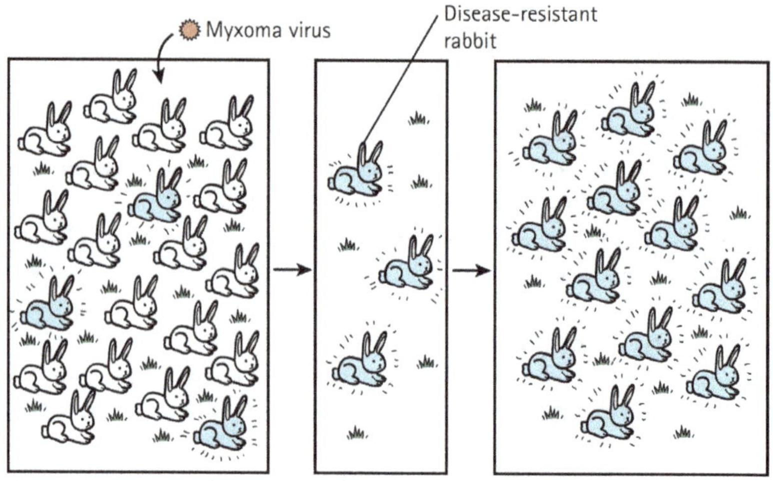

Natural selection occurs when organisms with advantageous traits leave more offspring than organisms with other traits, causing populations to change over time. Let's look more carefully at the process of natural selection.

1. *Genetic variation.* In any population of organisms, individuals have many traits that show variation—that is, they vary from individual to individual. Some of these traits are determined at least partly by genes and so are inherited traits—that is, they are passed from parents to offspring. Differences in the inherited traits of different individuals are known as **genetic variation**. In humans, for example, there is genetic variation in height, hair color, eye color, foot size, blood type, and many other traits.
2. *Natural selection.* Some inherited traits are advantageous. Organisms with inherited, advantageous traits are able to leave more offspring than organisms with other traits. This is known as **natural selection**. The **fitness** of an organism describes the number of offspring it leaves over its lifetime compared to other individuals in the population. An organism that leaves more offspring than other individuals in the population is said to have greater fitness.
3. *Evolution.* Because organisms with advantageous traits leave more offspring, advantageous traits are "selected for" and become more common in a population. What is the result? The population changes—evolves—to become better adapted to its environment.

Figure 8.12 summarizes the process of natural selection. Note that although natural selection acts on individuals within a population, allowing some individuals to leave more offspring than others, it is the population as a whole that evolves and becomes adapted to its environment.

V Figure 8.12

Genetic variation and natural selection lead to evolution in a population.

(1) GENETIC VARIATION | (2) NATURAL SELECTION | (3) EVOLUTION

In a population, there are differences in the inherited traits of different individuals.

Variation in heritable traits can result in some organisms leaving more offspring than others. This is called natural selection.

Advantageous traits become more common in a population. The population evolves to become better adapted to its environment.

CONCEPT CHECK

1. Which of these traits show genetic variation in cats: fur color, tail length, and number of eyes?

2. The cheetah is the fastest land animal on Earth. It can run 112 kilometers/hour (70 miles/hour)! Cheetahs prey on Thomson's gazelles that can run almost as fast, 80 kilometers/hour (50 miles/hour). How might natural selection have produced the cheetah's fast running speed?

CHECK YOUR ANSWERS

1. Fur color and tail length show genetic variation in cats. They are inherited traits that differ among cats. The number of eyes does not show genetic variation in cats—all cats have two eyes.

2. Faster cheetahs were better at catching Thomson's gazelles. Being better at catching food made faster cheetahs better at surviving and reproducing. As a result, faster cheetahs left more offspring, which were also fast. This resulted in a cheetah population with faster individuals. Over many generations, natural selection produced the remarkably fast cheetah we know today.

More to Explore

To read more about natural selection:

https://www.nhm.ac.uk/discover/what-is-natural-selection.html

https://evolution.berkeley.edu/evolibrary/article/evo_25

8.5 Examples of Natural Selection

The Peppered Moth

During the Industrial Revolution, coal was the primary fuel in England. Burning coal slathered dark soot on trees, rocks, and ground. And then a startling thing happened to the moths.

Peppered moths in England had always been light in color, with the scattering of dark peppery flecks that gave them their name. Their coloration made them hard to see in a habitat of lichen-covered trees and rocks. (Lichens are a combination of fungi and photosynthetic algae or bacteria growing together; they form crust-like growths on rocks, trees, and other surfaces.) This camouflage was believed to protect the moths from birds, their main predators. As the Industrial Revolution progressed, pollution killed the lichens, leaving the trees first bare and then darkened with soot. In 1848, the first dark peppered moth was found in the industrial center of Manchester, England. Dark moths had probably always existed in the population, but they had been extremely rare. Over the following decades, as more coal burned and the environment became increasingly sooty, more and more dark moths were seen. By 1895, 98% of peppered moths in industrialized areas were dark. Then, in the second half of the 20th century, antipollution laws were passed, and soot disappeared. Light moths increased in number, and today the dark moths have all but disappeared.

Did natural selection cause the coloration shifts in the peppered moth? Biologists hypothesized that natural selection favored light moths in lichen-covered habitats because they were better camouflaged. In sooty habitats, on the other hand, natural selection favored dark moths (**Figure 8.13**).

A series of experiments in the 1950s by the geneticist Bernard Kettlewell tested this hypothesis. First, Kettlewell released equal numbers of marked dark and light moths in polluted and unpolluted areas. Then, after a while,

Figure 8.13 >

Can you find the moths? Light peppered moths are well camouflaged on lichen-covered trees. Dark peppered moths are well-camouflaged on soot-covered trees.

he tried to recapture the moths. In polluted areas, Kettlewell recaptured more dark moths than light moths, which suggested that dark moths had survived better. The opposite was true in unpolluted habitats, where Kettlewell recaptured more light moths. Kettlewell also placed moths on tree trunks and filmed birds eating the moths. He found that birds ate what they could see: Birds ate more light moths in polluted habitats and more dark moths in unpolluted habitats.

Kettlewell's work became a classic example of natural selection. Eventually, however, certain aspects of his experiments were challenged. For example, moth experts pointed out that peppered moths don't usually sit on tree trunks, where Kettlewell had placed them. Instead, they usually rest on the undersides of branches. In addition, Kettlewell released the normally nocturnal moths during the daytime.

This may have affected the moths' ability to find resting spots. Finally, Kettlewell used a mix of lab-raised and wild-caught moths, which could differ in their behavior. These doubts led Michael Majerus of Cambridge University to conduct a new set of experiments between 2001 and 2007. Majerus's work confirmed that bird predation was the key factor affecting the relative numbers of light and dark peppered moths. Interestingly, a shift from light to dark forms in polluted areas (and back again, as pollution is cleaned up) has been reported in more than 70 other moth species in England and the United States alone.

Antibiotic-Resistant Bacteria

A patient is ill with pneumonia and gets a prescription for penicillin. After three days, he feels better and stops taking his pills. A few days later, his symptoms return. He quickly finds his pills and starts taking them again, but this time they have no effect. What happened? This frightening phenomenon is called antibiotic resistance. Antibiotic resistance is caused by natural selection: Penicillin killed most of the pneumonia bacteria, but a few penicillin-resistant bacteria survived. These bacteria multiplied, and the patient's infection returned— only this time, the bacteria were resistant to penicillin.

Antibiotics are wonder drugs. When penicillin, the first antibiotic, appeared, it dramatically cut the number of illnesses and deaths resulting from bacterial infections. After only a decade of use, however, the first penicillin-resistant bacterial strains appeared. Since then, antibiotic resistance has spread, with more and more bacterial populations becoming resistant to more and more different antibiotics. Diseases once easy to treat—tuberculosis, pneumonia, even common childhood ailments such as ear infections—are now often resistant to multiple antibiotics. In 2016, the World Health Organization reported that about 600,000 new cases of multi-drug-resistant tuberculosis appear each year, resulting in about 240,000 deaths.

READING CHECK

How many cases of multi-drug-resistant tuberculosis appear each year around the world?

Some of the most dangerous antibiotic-resistant bacteria are found in hospitals, where the use of many different types of antibiotics allows widely resistant strains to evolve. Methicillin-resistant Staphylococcus aureus (MRSA), a bacterial strain resistant to most antibiotics currently available, is responsible for more than 90,000 life-threatening infections and 20,000 deaths in the United States each year. And some MRSA strains are beginning to show resistance to the antibiotic vancomycin, often considered "the drug of last resort." Another worrisome development is the emergence of MRSA in the wider community.

The Centers for Disease Control (CDC) estimates that 1 in 3 people carry S. aureus in their nose and that 1 in 50 people carry MRSA. Sometimes, these can turn into infections. Community-based MRSA infections usually start as skin infections and spread through skin-to-skin contact. Some cases turn into "flesh-eating" disease, and others are halted only by drastic measures such as amputation. Environments with a higher risk for community-based MRSA infections include athletic facilities, dorms, prisons, and daycare centers. Compared to people whose infections respond to antibiotics, people who have antibiotic-resistant infections require longer hospital stays and are more likely to die from their infections.

All antibiotic use has the potential to contribute to resistance. However, resistance has been dramatically accelerated by the overuse of antibiotics. Under pressure from patients, physicians may prescribe antibiotics for illnesses not caused by bacteria. (Many common illnesses, such as colds, flu, and most sore throats, are caused by viruses.) These antibiotics select for resistance in our bodies' normal (non–disease-causing) bacterial populations, transferring resistance genes to disease-causing bacteria that later invade the body. In addition, the fact that patients sometimes stop taking their medications too soon contributes to the problem; this selects for antibiotic-resistant strains without providing the sustained dose that would actually kill all the bacteria.

Antibiotics have also seen heavy use in the livestock industry, where animals have long been given antibiotics regularly—even when they are healthy—to promote growth. Unfortunately, this practice greatly promotes the evolution of antibiotic resistance. The Centers for Disease Control report that each year in the U.S., about 6,200 cases of antibiotic-resistant salmonella infections are reported. For example, in August 2011, an antibiotic-resistant salmonella outbreak in ground turkey caused at least 79 illnesses and one death. In 2017, an antibiotic-resistant salmonella outbreak was linked to contact with infected dairy calves, some of which were used in 4-H youth agricultural programs nationwide. This outbreak resulted in 56 illnesses and 17 hospitalizations.

What can be done about antibiotic resistance? First, humans must learn to use antibiotics wisely, taking them only when needed—that is, for bacterial infections—and then taking the entire course of treatment. Second, physicians and veterinarians can promote a socially responsible approach to antibiotics by educating patients and agriculturalists on properly using these drugs. Third, the use of antibiotics in livestock should be controlled. In early 2017, the U.S. Food and Drug Administration finally banned antibiotics to help livestock gain weight. Farmers now have to obtain prescriptions for antibiotics from veterinarians rather than being able to buy them at feed stores or online. However, some scientists are concerned that legal loopholes will mean heavy use of antibiotics in livestock could continue. Finally, since many antibiotics are less effective because of resistance, scientists must search for new antibiotics to replace those that no longer do the job.

CONCEPT **CHECK**

How did Bernard Kettlewell test whether natural selection was acting on peppered moth populations?

CHECK YOUR ANSWER

Kettlewell released equal numbers of marked dark and light moths in different habitats, waited for a while, and then tried to recapture the moths. Kettlewell then compared the number of dark and light moths he recaptured to determine which moths had survived better.

More to Explore

You can read more about peppered moth evolution here:

https://www.nationalgeographic.com/science/phenomena/2013/10/09/evolution-in-color-from-peppered-moths-to-walking-sticks/

And you can learn more about antibiotic resistance here:

https://www.cdc.gov/drugresistance/about.html

8.6 Adaption

Natural selection leads to the evolution of **adaptations**—traits that make organisms well-suited to living and reproducing in their environments. You have already seen one example of an adaptation—the cheetah's speed. The cheetah's speed helps it catch the food to survive and reproduce.

Adaptations can relate to various aspects of an organism's life. Some adaptations help organisms survive. Survival is usually an essential factor toward relative success with reproduction. Survival requires that organisms be able to acquire food and other necessary resources. It also requires that organisms avoid becoming food for someone else. Anti-predator adaptations include camouflage, toxicity, or the ability to hide or run away (**Figure 8.14**).

(a) (b)

^ Figure 8.14

Almost every organism has adaptations that help prevent it from becoming food for someone else. (a) The spines of this cactus prevent most animals from eating it. (b) When threatened, this octopus releases a cloud of dark ink that may confuse a predator long enough for the octopus to escape.

(a) (b)

∧ Figure 8.15

Some adaptations for acquiring mates. (a) The beautiful feathers of this male bird of paradise help attract female mates. (b) These male deer are fighting for control of territory as well as mates.

Other adaptations have evolved to help organisms acquire mates. These include the beautiful feathers of male peacocks and birds of paradise (**Figure 8.15a**), the sexy "rib-bits" of male frogs, and the enchanting songs of many male birds. Males have evolved these "sexy" traits because females of the species find them attractive. In other species, females don't choose their mates based on attractive traits. Instead, males fight with other males to obtain mates. The adaptations of these males may include large size, great strength, or fighting structures such as antlers (**Figure 8.15b**). Natural selection that favors individuals best able to acquire mates is also called *sexual selection*. Finally, some adaptations relate to bearing and raising young. **Figure 8.16** shows one such adaptation—parental care. Parental care evolved because natural selection favored organisms that were able to help their offspring survive and thrive. Parental care is found in many animals, including humans.

READING CHECK

Why do parents care for their young in the case of many species?

Natural selection has produced remarkable adaptations over time. Nature does not plan ahead—it does not plan to make a falcon or a polar bear. Instead, adaptations are built step by step through the never-ending selection of the most successful forms.

< Figure 8.16

Parental care occurs in many species. This male poison dart frog is carrying his tadpoles on his back.

CONCEPT CHECK

Mating is very dangerous for a male praying mantis. Quite often, the female will eat him as he mates with her.

a. What advantage does the female gain from eating the male?

b. Would it be more advantageous ("adaptive") for the male not to mate at all?

CHECK YOUR ANSWERS

a. The female gets nutrients when she eats the male.

b. A male praying mantis that never mates is more likely to survive to old age. However, he won't leave any offspring if he doesn't mate. Remember that adaptations are traits that make organisms good at living and reproducing in their environments. It's not enough to survive—you also have to reproduce! This male praying mantis may not have long to live, but at least he has a good chance of leaving offspring.

Staying Warm and Keeping Cool

In this section, we will see how being the right size and shape can be an adaptation. Most mammals maintain a fairly constant body temperature. We humans, for example, have a body temperature that always stays around 37°C (98.6°F). Animals that live in extremely hot or extremely cold habitats need to be able to maintain appropriate body temperatures in those environments—to *thermoregulate*. In deserts, animals have to be able to lose heat to avoid overheating. In cold habitats, animals have to be able to retain heat. The heat an animal produces depends upon its size. In general, the larger an animal, the more heat it produces. Meanwhile, the heat an animal loses to its environment depends on its surface area because heat is lost through the body's surface. How effective an animal is at losing heat to the environment can be gauged by looking at the ratio of its surface area to volume. A large animal will have a low amount of surface area to volume. This means it will have trouble releasing heat to its environment. Think of it this way: All the heat generated within its body can't so easily escape.

Λ Figure 8.17
The amount of caramel on this caramel-covered apple is determined by its surface area. The amount of apple in this caramel-covered apple is determined by its volume. The surface-area-to-volume ratio is the amount of caramel divided by the amount of apple. A big caramel-covered apple has a low surface-area-to-volume ratio. A small caramel-covered apple has a high surface-area-to-volume ratio.

A smaller animal, by contrast, has more surface area compared to its volume. Heat is, therefore, more easily lost. **Figure 8.17** explains the surface-area-to-volume ratio using the example of a caramel-covered apple.

As a result, animals that live in cold habitats are often larger than related species that live in warm habitats. For example, the smallest bear in the world is the sun bear, found in the tropical forests of Southeast Asia (**Figure 8.18a**). Adult sun bears weigh between 27 and 65 kilograms (60 to 140 pounds). On the other hand, t he largest bear in the world is the polar bear (**Figure 8.18b**), which ranges throughout the Arctic. Adult polar bears weigh between 200 and 800 kilograms (440 to 1760 pounds).

Animals adapted to hot versus cold climates also vary in shape. Desert species often have long legs and large ears, increasing the surface area for heat dissipation. Their legs and ears are also covered with extensive blood vessels that carry heat from the body's core to the skin, where convection, the heat transfer by moving air, cools the animal. Arctic species often have short appendages and small ears that help conserve heat. An example is the rabbits shown in **Figure 8.19**.

(a) (b)

Figure 8.18

(a) The small sun bear is found in tropical forests in Southeast Asia. A small animal is better able to lose heat because of its high surface-area-to-volume ratio. (b) The polar bear, the largest terrestrial carnivore in the world, is found throughout the Arctic. A large animal is better able to retain heat because of its low surface-area-to-volume ratio.

(a) (b)

Figure 8.19

(a) This black-tailed jackrabbit lives in a California desert. Extensive blood vessels in its ears help it release heat. It also has long legs. (b) This arctic hare is a relative of the black-tailed jackrabbit. Its ears are much smaller.

CONCEPT CHECK

On cold days, people often bundle up babies and small children carefully. Are babies more likely to need the extra bundling than adults? Why or why not?

CHECK YOUR ANSWER

Babies have higher surface-area-to-volume ratios than adult humans because they are smaller. So, yes, they are likely to appreciate the extra bundling.

More to Explore

You can read more about adaptations, and see some fabulous examples, here:

https://www.nationalgeographic.org/encyclopedia/adaptation/

Concept Review

Chapter 8

Detailed Chapter Summary

How did life originate? At present, scientists do not know exactly how life originated, but they have some clues. The experiments of Miller and Urey suggest that organic molecules could form spontaneously under conditions present on early Earth. Organic molecules may also have been brought by meteorites or formed in deep-sea environments similar to the hydrothermal vent habitats still in existence today. Liposomes are structures that form spontaneously from certain lipids and may have served as early cell membranes. Liposomes show a variety of behaviors similar to those of cells. The first genetic material was probably RNA rather than DNA. RNA has the ability to spontaneously assemble into short strands that can reproduce themselves.

For thousands of years, people believed that life on Earth did not change. The discovery of fossils raised questions, however. Lamarck was one of the strongest early proponents of evolution. He argued that organisms evolve through the inheritance of acquired characteristics. This turned out to be incorrect.

Charles Darwin defined evolution as inherited changes in populations of organisms over time. Darwin's theory of evolution grew out of his experiences on board the H.M.S. Beagle, which sailed around South America. Darwin was also influenced by the work of two contemporaries, Lyell and Malthus. Lyell was a geologist who argued that the Earth's geological features arose through gradual processes that produced their effects over long time periods. Darwin argued that gradual processes acting over long periods could also account for changes in populations of organisms. Malthus was an economist who observed that human populations grew much faster than their food supplies and that famine was an inevitable feature of human existence. This led Darwin to his idea of natural selection.

Darwin argued that natural selection is the driving force behind evolution. In any population of organisms, individuals show genetic variation in many traits. Organisms with advantageous inherited traits survive better and reproduce more than organisms with other traits. As a result, these advantageous inherited traits become more common in the population. This is natural selection. As a result of natural selection, organisms evolve to better adapt to their environments.

We next looked in detail at two examples of natural selection. The first example was the evolution of peppered moth coloration during the Industrial Revolution. As the environment became polluted, the number of dark peppered moths increased. Experiments performed by Bernard Kettlewell showed that birds tended to eat more moths that were *not* camouflaged in their habitat – that is, birds ate more dark moths in unpolluted habitats and more light moths in polluted habitats. The second example was the evolution of antibiotic resistance in bacteria. Heavy use of antibiotics has resulted in the evolution of resistance in many different bacterial populations, with serious consequences for our ability to control disease-causing bacteria.

Adaptations can relate to different aspects of an organism's existence. For example, they can help organisms survive, acquire more mates (sexual selection), or more successfully raise offspring. One specific set of adaptations relates to maintaining an appropriate body temperature in hot and cold environments. Mammals in cold habitats tend to be larger, with smaller appendages than their close relatives. Mammals in hot habitats tend to be smaller, with larger appendages than their close relatives. This is due to the observation that these features result in favorable surface-area-to-volume ratios.

Key Terms

Adaptations Evolved traits that make organisms well-suited to living and reproducing in their environments.

Evolution Inherited changes in populations of organisms over time.

Fitness The number of offspring an organism produces in its lifetime compared to other organisms in the population.

Genetic variation Differences in the inherited traits of different individuals.

Natural selection The process in which organisms with inherited, advantageous traits leave more offspring than organisms with other traits, causing these advantageous traits to become more common in a population over time.

Review Questions

Simple Review Questions

8.1 The Origin of Life

1. What experiment did Miller and Urey perform? What were their results?
2. What are three hypotheses for how Earth's organic molecules originated?
3. What are liposomes? What characteristics of liposomes make scientists think that liposomes may have been the precursors of real cells?
4. Why is RNA, rather than DNA, believed to be the first genetic material?

8.2 Is There Life on Mars? Or Venus?

5. Why do scientists think life on Mars is more likely than life on Venus?
6. Why are scientists interested in searching for water on Mars?
7. Is there evidence that Mars once had water? Is there evidence that Mars currently has water?
8. What information regarding Venus caused scientists to take the possibility of life on Venus more seriously?

8.3 Charles Darwin and The Origin of Species

9. What was Lamarck's theory about how evolutionary change occurred?
10. What impressed Darwin about the finches on the Galápagos Islands?
11. How did the work of Thomas Malthus influence Darwin?
12. How did Charles Lyell's work influence Darwin?

8.4 Natural Selection

13. What is an inherited trait?
14. What is genetic variation?
15. Describe how natural selection occurs.

8.5 Examples of Natural Selection

16. What hypothesis did Bernard Kettlewell test using mark and recapture experiments on peppered moths?
17. What conclusion did Kettlewell reach after seeing the results of his mark and recapture experiments on peppered moths?
18. What practices have resulted in increased antibiotic resistance?

8.6 Adaptation

19. From the point of view of natural selection, why is it important for an organism to survive?
20. Define sexual selection and provide some examples of adaptations that evolved as a result of sexual selection.
21. Why is parental care adaptive in certain species?
22. Why is the surface-area-to-volume ratio important in thermoregulation?
23. You are studying a species of tropical goat and comparing it with a related Arctic species. Based on your knowledge of thermoregulation in mammals and its effect on the size and shape of organisms, predict some of the differences you might see between the tropical and Arctic species.

Challenging Review Questions

8.1 The Origin of Life

24. Why do scientists consider Miller and Urey's experiment important? Why do some scientists question the assumptions Miller and Urey made when they designed their experiment?

25. What role might environments similar to hydrothermal vents have played in the origin of life?

26. How are liposomes similar to cells? How are they different from real cells?

27. Explain why the earliest life on Earth may have resembled liposomes containing RNA.

8.2 Is There Life on Mars? Or Venus?

28. Of the planets in the Solar System other than Earth, why did scientists initially focus on Mars and Venus as the most likely to harbor life?

29. What features of Venus would make it difficult for life as we know it to live there?

30. What features of Mars would make it difficult for Earth's living things to survive there?

31. How much evidence is there that Mars currently supports life or that it once supported life? What kind of evidence is NASA searching for to determine whether there is life on Mars?

8.3 Charles Darwin and The Origin of Species

32. What role did fossils play in the development of Lamarck's ideas—and later Darwin's ideas—about evolution?

33. How did Darwin's experience on the Beagle influence his thinking and later work?

34. Horned lizards defend themselves with spiky horns on their heads and bodies. Some individuals will even jerk their heads to stab at predators with their horns. How might Darwin explain the evolution of these horns?

8.4 Natural Selection

35. What are some human traits that do not show genetic variation? What are some that do show genetic variation?

36. How would you determine whether a trait you are interested in studying is an inherited trait?

37. How is the story of the myxoma virus and Australian rabbits similar to the story of antibiotic resistance in bacteria?

38. Nancy Burley of the University of California, Irvine, ran the following experiment: She placed red color bands on the feet of some male birds and green color bands on the feet of other male birds. Females preferred to mate with males that had red color bands. Is this an example of natural selection? Why or why not?

39. What does it mean to say that one rabbit has greater fitness than another?

40. In recent decades, average human height has increased in many parts of the world. Do you think this is an example of evolution?

41. On islands, many large animals—such as elephants—evolve to become miniaturized. On the other hand, many small animals—including some rodents—evolve to be exceptionally large. Why might natural selection produce these results?

42. A single sunflower can have hundreds or even thousands of seeds. Will each of these seeds grow into a sunflower plant? Why or why not? What would happen if every one of these seeds grew into a sunflower plant?

8.5 Examples of Natural Selection

43. A biologist releases 200 marked light moths and 200 marked dark moths in a habitat. A few days later, she recaptured 13 light moths and 36 dark moths. What does this suggest about natural selection in the population? What might you guess about the trees in this habitat?

44. In 2017, the U.S. Food and Drug Administration banned the use of antibiotics to help livestock gain weight. How might this help prevent the evolution of antibiotic resistance?

8.6 Adaptation

45. Male birds of many species have brighter feathers than females. For example, male cardinals are bright red, whereas females are brownish. Bright colors on males are often adaptations for winning mates. Is being less colorful adaptive for female birds? Defend your answer.

46. John Endler studied guppies that lived in ponds in Trinidad. He noticed that the males in different populations of guppies had different coloration patterns. Males that lived in pools with no predators tended to have bright colors and large spots. Males that lived in pools with predators tended to have drab colors. Endler knew that female guppies preferred to mate with brightly colored males. Explain the role of natural selection and sexual selection in these guppy populations.

47. Not all brightly colored animals are trying to attract mates. Some animals use bright colors as a warning. For example, consider bees and wasps (yellow and black), coral snakes (red, yellow, and black), and poison dart frogs (many different bright colors). Explain how bright colors are adaptations in these animals.

48. Explain why mammals that live near the poles are likely to be larger than related mammals that live near the equator.

49. Two species of foxes are shown here. One is a kit fox in Arizona. The other is an Arctic fox. Which is which? How can you tell? Describe at least two traits that make each animal well-adapted to its environment.

Apply & Discuss Questions

50. If there is life on Mars (and this life had evolved on Mars), what adaptations might the organisms have? How might Martian life differ from life on Earth?

51. If life evolved on Venus, what adaptations might the organisms have?

52. During a drought, the supply of seeds available to a finch population decreases. The smaller, softer seeds, which are easier to crack, are quickly eaten up. Finches with larger, stronger beaks are better able to crack the larger seeds that remain. What evolutionary changes do you expect to see in this finch population?

53. Caterpillars of the Monarch butterfly eat plants that are toxic to other animals. As a result, their tissues become toxic. Birds that try to eat Monarchs vomit and then avoid the striking orange-and-black pattern in the future. The viceroy is another species of butterfly. Viceroys resemble monarchs, but they are not toxic. Is the appearance of the viceroy adaptive? How could you test this hypothesis?

Can you tell which is the Monarch and which is the Viceroy? Viceroys have a black stripe in the hind wing that goes across the other stripes. Monarchs do not.

54. You are eating a salad when you almost bite down on a green caterpillar hidden among the lettuce leaves. The friend who is eating with you says, "That would have been gross, but I don't think it's one of those super-poisonous caterpillars." Do you agree?

55. Bird eggs vary tremendously in color. Do you think the color of a bird's eggs is adaptive? What factors may have shaped the evolution of egg color in different species?

56. Broad-spectrum antibiotics are effective against a wide variety of bacteria. Narrow-spectrum antibiotics are effective against only certain types of bacteria. Public health officials have suggested that one way to combat antibiotic resistance is to use narrow-spectrum antibiotics whenever possible. Do you agree? Explain how the use of narrow-spectrum antibiotics instead of broad-spectrum antibiotics could slow the evolution of resistance in bacteria.

Solutions to Odd-Numbered Questions

Simple Review Solutions

1. In 1953, Stanley Miller and Harold Urey built a model of the early Earth in a chemistry lab. A flask containing a mixture of simple compounds—including water vapor, ammonia, methane, and hydrogen gas—simulated Earth's early atmosphere. Liquid water was added to represent Earth's oceans. Electric sparks sent through the mixture simulated lightning. When this model of early Earth was assembled, an amazing thing happened. Many complex organic molecules were formed, including amino acids, the building blocks of proteins. Not only had these molecules formed quickly, but there were also lots of them.

3. Liposomes have double membranes similar to cell membranes. Although they are not alive, liposomes sometimes behave like living cells—they grow, shrink, and divide. Liposomes also run chemical reactions inside their membranes and control what molecules enter and exit them, two key features of living cells.

5. Early explorations showed that Venus would be hostile to life as we know it. Venus has almost no water, a crushingly dense carbon dioxide atmosphere, clouds of sulfuric acid, and a surface temperature of 462 °C (864 degrees F)!

7. Many of Mars's geological features were clearly carved out by water, and the evidence suggests that Mars once had oceans. NASA has looked for places on the planet where pockets of water might still exist. And they have found evidence for it! Specifically, they have seen seasonal streaks of damp soil in certain craters. This has encouraged scientists to look for larger water reservoirs that might still exist under the Martian surface.

9. According to Lamarck, organisms acquire new characteristics during their lifetimes and then passed these characteristics to their offspring. For example, ancestral giraffes stretched their necks to grab the high leaves on a tree, and their necks became longer. They then passed these longer necks to their offspring. The offspring reached for even higher leaves, stretching their necks even further, and so on.

11. Malthus observed that human populations grow much faster than available food supplies and concluded, with despair, that famine was an inevitable feature of human existence. Darwin applied Malthus's idea to the natural world and argued that because there are not enough resources for all organisms to survive and reproduce as much as they can, living organisms are involved in an intense "struggle for existence." As a result, organisms with advantageous traits leave more offspring than organisms with other traits, causing populations to change over time.

13. Some traits are determined at least partly by genes, and so are inherited traits—that is, they are passed from parents to offspring.

15. Some inherited traits are advantageous. Organisms with inherited, advantageous traits are able to leave more offspring than organisms with other traits. This is known as natural selection.

17. Kettlewell found that birds ate what they could see: Birds ate more light moths in polluted habitats and more dark moths in unpolluted habitats.
19. Survival is usually an important first step in successful reproduction.

21. Parental care evolved in certain species because natural selection favored organisms that were able to help their offspring survive and thrive. Parental care is found in many animals, including humans.

23. The tropical species is likely to be smaller, with longer limbs and larger ears. The Arctic goat is likely to be larger, with shorter limbs and smaller ears.

Challenging Review Solutions

25. Some scientists think that, before there was life large numbers of organic molecules may have formed in deep-sea environments similar to today's hydrothermal vent habitats. So, the origin of life may have occurred in a hydrothermal vent-like habitat.

27. Liposomes could have provided a distinct, controlled environment for the earliest cells. Cells would also have needed genetic material, and scientists think it is likely the first genes were made of RNA rather than DNA. This is because RNA can spontaneously assemble from individual nucleotides and even reproduce without cells or enzymes.

29. Venus has almost no water, a crushingly dense carbon dioxide atmosphere, clouds of sulfuric acid, and a surface temperature of 462 °C (864 degrees F).

31. There isn't much evidence—yet—that life evolved on Mars or currently exists there, aside from any microorganisms that were accidentally transported from Earth to Mars that are still surviving there. NASA is focused on searching for water on Mars since water is required for life as we know it. NASA also aims to search for fossils and actual living things in locations where there is, or once was, water on Mars.

33. Darwin's voyage on the Beagle started him thinking about the questions that he would work on for the rest of his life. These were largely questions about how life on Earth had gotten to be the way it was. Darwin's experiences studying life in South America and the Galápagos also began to suggest answers to him about how living things evolve or change over time.

35. Answers will vary. Traits that do not show genetic variation include having a four-chambered heart, five fingers on each hand, two arms, two legs, one nose, etc. Traits that do show genetic variation include height, weight, arm length, foot length, eye color, hair color, etc.

37. In both cases, the administration of a lethal agent (antibiotics or myxoma virus) led to natural selection for resistant individuals in the population. These resistant individuals survived and reproduced. As a result, over time, populations evolved resistance to the lethal agent.

39. The fitness of an organism describes the number of offspring it leaves over its lifetime compared to other individuals in the population. An organism that leaves more offspring than other individuals in the population is said to have greater fitness. So, if one rabbit has greater fitness than another, then it has left more offspring than the other rabbit.

41. One leading hypothesis is that large animals become miniaturized because of limited food availability on relatively small islands and that small animals evolve to larger sizes because of the absence of predators.

43. Since she recaptured more dark moths, dark moths survived better in this population than light moths. This means that in this population, there is natural selection for dark peppered moths. This suggests that, if camouflage explains the difference in survival, the trees are probably dark rather than light or lichen-covered.

45. Bright colors on certain male birds help them attract female mates. However, bright colors also make these males more visible to predators. It is adaptive for females to be less colorful because they don't need the bright colors to attract mates, and their duller feathers make it harder for predators to see them.

47. Bright colors serve to warn other species that these organisms are dangerous. Other species either learn through experience or evolve an instinctual response to the bright appearance of these poisonous or toxic species. The bright colors make the toxic or dangerous animal easy to see and identify so that other species can avoid disturbing them or trying to eat them.

49. The kit fox is the one with pale brownish fur, large ears, and long limbs. The fur color helps it stay camouflaged in its environment. The large ears and long limbs help it increase surface area available for heat dissipation. The Arctic fox has white fur, which helps it stay camouflaged, and small ears and short limbs that help it decrease the surface area from which heat is lost.

Apply & Discuss Solutions

51. Many answers are possible! Surface life would need some way to withstand the very high temperatures and huge atmospheric pressure on Venus's surface, as well as the acidic clouds and rain. Life in the Venusian atmosphere, perhaps more likely, would need a way to stay within the area of favorable conditions within the atmosphere. There is also little to no water on Venus, so the chemistry of Venusian life would have to be pretty different from that of life on Earth.

53. The appearance of the viceroy is adaptive—birds may mistake it for a foul-tasting monarch and not try to eat it. One way to test this hypothesis would be to offer viceroys and other edible butterflies to birds that were known to be familiar with monarchs. Do the birds eat the viceroys less often than they eat the butterflies that do not look like toxic monarchs?

55. This is a very open question! Scientists have proposed many different ideas, and different factors may have shaped egg color evolution in different bird species. Egg color may be related to camouflage or heat absorption (darker eggs will absorb more heat). Or, they may allow some birds to recognize their own eggs in cases where there are nest parasites (birds that lay their eggs in the nests of other birds, such as the cuckoo). Other answers are possible!

Living Earth Essay

Life on Earth Connections

Magnetoreception

The term *magnetism* comes from the region of Magnesia, a province of Greece. Certain magnetic stones were found by the Greeks who lived there more than 2000 years ago. These stones, called lodestones, could attract pieces of iron. When lodestones were rubbed on pieces of iron, magnets were made. Magnets were first made into compasses and used for navigation by the Chinese in the twelfth century. The metal needle of these compasses would align with the Earth's magnetic field, which arises from the circulation of molten metals within its core (**Figure A**). Today much of modern industry—from motors to computers—relies on magnetism.

Figure A
Earth's magnetic fields migrate over time. Along with this, the location of the north and south magnetic poles typically wander within the vicinity of Earth's geographic poles. Earth's magnetic field is particularly important for shielding us from energetic ionized particles coming at us from the Sun. Interestingly, Earth's magnetic field regularly flips orientation about once every 700,000 years.

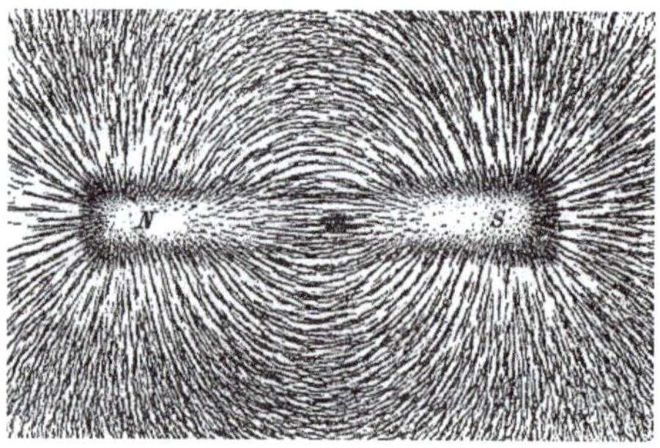

Figure B
Top view of iron filings sprinkled around a magnet. The filings trace out a pattern of magnetic field lines in the space surrounding the magnet. Interestingly, the magnetic field lines continue inside the magnet (not revealed by the filings), and form closed loops.

When you sprinkle some iron filings on a sheet of paper placed on a magnet, you'll see that the filings trace out an orderly pattern of lines that surround the magnet. Just as a gravitational field surrounds a massive body, a **magnetic field** occupies the space around the magnet. The shape of the field is shown by the filings. As shown in **Figure B**, the filings line up with the magnetic field lines that spread out from one pole and return to the other. As shown in **Figure C**, compass needles also line up with this field.

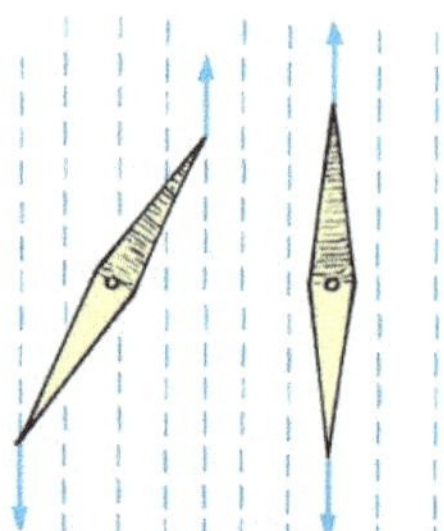

Figure C
(Left) When the compass needle is not aligned with the magnetic field, the oppositely directed forces on the needle produce a pair of torques (twisting forces) that (right) twist the needle into alignment.

A magnetic field is produced by the motion of electric charge. Where, then, is this moving electric charge in a common bar magnet? The answer is, in the electrons of the atoms that make up the magnet. These electrons, like the electrons in all matter, are in constant motion. Investigation reveals that electrons spin about their own axes like tops. The spinning of an electron's charge creates north and south poles about the electron. Thus, every spinning electron behaves as a tiny magnet (**Figure D**).

Figure D
Because every electron is charged, the spinning motion of every electron produces a magnetic field. Note: the electron is represented here as a yellow sphere, which is inaccurate because 1) electrons are not yellow, and 2) electrons are points in space, not spheres in space. That said, it has been found that all electrons have this property called spin that generates a tiny magnetic field.

Typically, the electrons within a material spin in all directions, which means their magnetic poles work against each other. With atoms such as iron, nickel, and cobalt, however, the fields do not cancel each other entirely. Most common magnets are therefore made from alloys containing iron, nickel, and cobalt in various proportions. Rather strong magnetic alloys can be made by including the rare earth metal neodymium, Nd. These neodymium magnets have many useful applications. Wind turbines fitted with neodymium magnets, for example, are able to produce electricity much more efficiently.

Life Sensing Magnetic Fields

Certain bacteria biologically produce grains of magnetite (a compound equivalent to iron ore) that they string together to form internal compasses. They then use these compasses to detect the Earth's magnetic field. Equipped with a sense of direction, the organisms are able to locate food supplies.

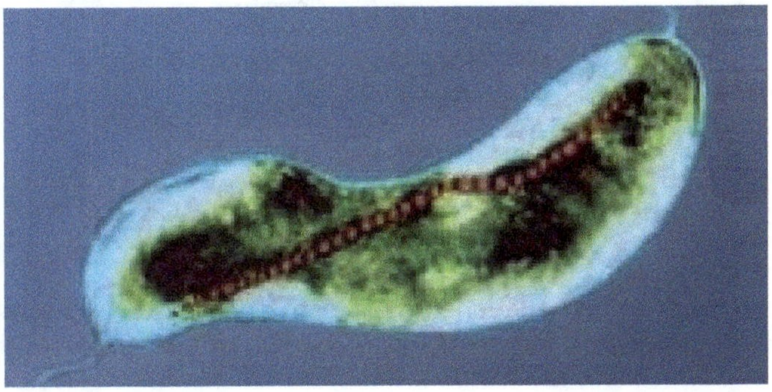

Figure E
First discovered in the 1960s, magnetosensitive bacteria develop chains of magnetic minerals within their cells. They have even been sent into space to study their magnetosensitivity while in orbital free-fall.

Bacteria are not the only living organisms with built-in magnetic compasses. Magnetic material is also in the abdomens of bees, whose behavior is affected by magnetic fields. Some wasps, Monarch butterflies, sea turtles, and fish join the ranks of creatures that have a magnetic sense. We say they are *magnetoreceptive*.

Most fascinating, however, is the magnetoreception demonstrated by birds who use Earth's magnetic field to help them to navigate, such as during migrations. Evidence suggests there are at least two mechanisms. The first relates to magnetite crystals found within their beak, which may modify nerve signals sent to the bird's brain. Earth's magnetic field, however, is relatively weak. Thus, they evolved a far more sensitive mechanism.

Birds may sense magnetic fields through their eyes.

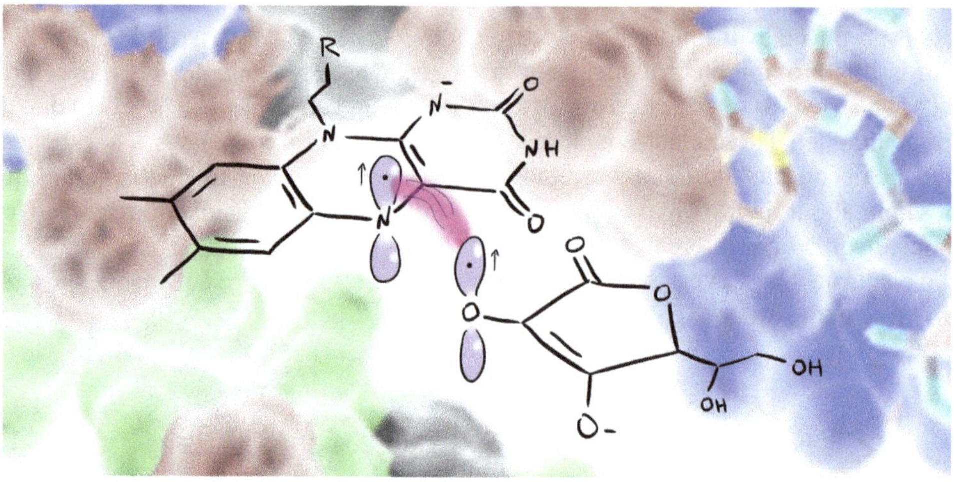

Figure F
The interaction between singlet electrons within a bird's eyes is most sensitive to the orientation of an external magnetic field, including Earth's magnetic field. Changes in orientation as the bird turns result in a signal sent to the bird's brain telling the bird which way is which.

READING CHECK

What produces a magnetic field?

As shown in **Figure F**, specialized biomolecules within the bird's eye are able to hold a single electron all by itself. This is an unstable situation because, as we learn in chemistry, electrons generally like to remain paired. Within this system, single electrons on neighboring molecules become quantum entangled, which means they somehow "know" what the other electron is doing despite the distance between them. As such, they rapidly oscillate between various energy states in response to each other's actions. These responses are super-sensitive to changes in external magnetic fields, including Earth's magnetic field. As the bird changes orientation, a signal is generated from this system setting off a cascade of neurotransmitters, which, in turn, initiate a signal to the bird's brain. In this way, the bird becomes strongly sensitive to Earth's magnetic field.

What the bird actually experiences is unknown. But it might be akin to polarized sunglasses. Look to a clear blue sky with polarized sunglasses, a must-do experiment. Tilt your head, and you'll see a change in the shades and colors. This is due to the sky itself emanating polarized light, which is a subject we discuss in physics. It's all connected.

To Life!

CONCEPT CHECK

What are two potential mechanisms for magnetoreception within living organisms?

CHECK YOUR ANSWER

One potential mechanism for magnetoreception may involve magnetite minerals, somehow affecting nerve signals. A second potential mechanism may involve nerve signals resulting from disruptions in the communication between quantum entangled singlet electrons within biomolecules, such as those within a bird's eye.

More to Explore

High-level scientific article describing potential mechanisms for a "chemical compass".

https://royalsocietypublishing.org/doi/full/10.1098/rsif.2013.1063

Dogs have been shown to have a geomagnetic sense as well!

Chapter 9

Evidence of Evolution

THE MAIN IDEA 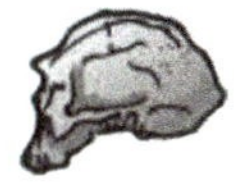

Evolution—inherited changes over time—has produced all the different kinds of living things on Earth.

9.1 Mechanisms of Evolution
9.2 How New Species Form
9.3 Natural Selection in Action
9.4 Fossils
9.5 Body Structures and Genetics
9.6 Biogeography
9.7 The Evolution of Humans

Flies taste their food with the hairs on their feet. Bats catch insects in midair. Cacti grow sharp spines to keep animals from eating them. These adaptations, and countless other ways in which organisms are structured to survive and reproduce, make up the incredible story of evolution. How do organisms change to fit their environment? After all, a giraffe can't grow a long neck just because it wants to. So how do adaptations (such as long necks) actually come about? Another question is: If organisms today all evolved from previous ones, how did life get started in the first place? Study onward to discover these secrets of life.

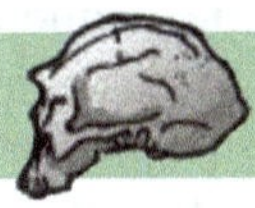

9.1 Mechanisms of Evolution

So far, we've seen how natural selection acts on organisms' traits—giraffe neck length, cheetah speed, peppered moth color, and so on. Traits are only part of the story, though, because what gets passed from parents to offspring are not traits—they are genes. A unification of modern genetics and Darwin's theory of evolution took place in the middle of the twentieth century and produced many new insights into how populations evolve.

Evolution and Genetics

Consider the peppered moths we discussed in Chapter 8. There, you learned that some peppered moths have light wings and others have dark wings. You also learned that, depending on the colors and conditions of the environment, one color pattern in peppered moths might be advantageous over the other. But how do genes determine the colors of peppered moths?

Peppered moths have a light color allele (a) and a dark color allele (A). The dark A allele is dominant over the light a allele. This means that aa moths have light wings, whereas Aa and AA moths have dark wings.

We already know that natural selection, through the action of predators such as birds, can make a particular phenotype, or trait, advantageous in a peppered moth population. For example, you learned that in a non-polluted area, light wings are advantageous because moths with light wings are better camouflaged from birds. What does this mean in terms of genes and alleles? If light wings are advantageous, then the light color allele a is advantageous in the population. If dark wings are advantageous, then the dark color A is advantageous.

Now let's think about all the wing color alleles in an entire population of peppered moths. In a population with many light moths and fewer dark moths, we might see that overall, the population has 92% *a* alleles and 8% *A* alleles. We use the term *allele frequency* to describe how common different alleles are in a population. So this population has allele frequencies of 92% *a* alleles and 8% *A* alleles.

If this population's habitat became more polluted, dark moths would become more common, and the dark allele would increase in frequency. The allele frequencies might change to 5% *a* and 95% *A*. The evolution of that population over time is seen in the changing of allele frequencies.

We can look again at natural selection and evolution and describe these processes in terms of genetics. Remember from Chapter 8 that the key features are genetic variation, natural selection, and evolution:

1. *Genetic variation.* Genetic variation exists when there are multiple alleles for a gene within a population. For example, in peppered moths, there are two alleles for color, *a* and *A*.
2. *Natural selection.* A specific allele may give an organism an advantage, allowing it to reproduce more than other organisms in the population. In a polluted habitat, for example, the A allele is advantageous. In an unpolluted habitat, on the other hand, the a allele is advantageous.
3. *Evolution.* As a result of natural selection, more copies of the advantageous allele are passed to the next generation and the frequency of the advantageous allele increases in the population. In a polluted habitat, the frequency of the *A* allele increases. In an unpolluted habitat, the frequency of the *a* allele increases.

Notice that although natural selection affects genes and allele frequencies, natural selection does not act directly on genes. Here's another way to say this: Natural selection acts on an organism's phenotype (its traits), not its genotype (its genes). To see why, go back to the peppered moth. In peppered moths, the dark allele (A) is dominant, and the light allele (a) is recessive. This means that both AA moths and Aa moths have dark wings (**Figure 9.1**). Whether a bird is likely to eat the moth depends on the moth's phenotype (whether it is dark or light), not its genotype. A bird is equally likely to eat a dark moth whether it has genotype AA or Aa.

READING CHECK

Besides natural selection, what other mechanisms cause populations to evolve?

Natural selection is the driving force behind evolution, and it causes populations to become adapted to their environments. However, natural selection is not the only mechanism that causes populations to evolve. Populations also change over time because of mutation pressure, genetic drift, and gene flow. In order to understand these other mechanisms of evolution, let's consider each using the example of peppered moths again.

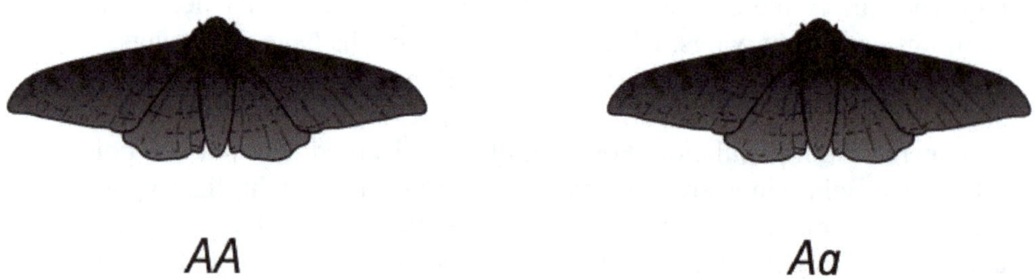

Figure 9.1 >

Natural selection acts on the phenotype, not the genotype. In the case of these two dark moths, the phenotype (dark wing color) matters, not the genotype (AA versus Aa).

Mutation Pressure

Remember that an allele is a version of a gene—in other words, it is a specific sequence of nucleotides within an organism's DNA. Any gene can mutate. This happens when the sequence of nucleotides is altered, sometimes due to environmental factors and sometimes spontaneously due to errors in the copying of DNA. Thus, it is possible for a dominant allele to mutate into a recessive allele and vice versa.

Mutation pressure exists if an allele is more likely to mutate in one direction than the other. For example, a genetic mutation may be more likely to turn a dark allele into a light allele than vice versa. That is, an *A* allele is more likely to mutate into an *a* allele than the other way around. If this is the case, then over time, the frequency of the dark allele A will decrease, and the frequency of the light allele a will increase.

Genetic Drift

Genetic drift is the evolution of populations due to chance. Imagine a half-polluted town where light moths and dark moths are equally successful—neither allele is advantageous in this town. Now suppose a sudden storm wiped out part of the town's peppered moth population. It *just might happen*—just by chance—that more dark moths survive. If so, then the frequency of the dark allele will increase. Notice that genetic drift produces evolution (inherited changes in a population over time) but that this evolution is not the result of natural selection—the dark allele was not advantageous.

Genetic drift can also occur when, just by chance, more alleles of one type are transmitted to the next generation than alleles of the other type. For example, even if light and dark moths have equal fitness, light moths might just happen to leave more offspring (and therefore more light alleles) one year. Genetic drift works a lot like a coin flip (**Figure 9.2**). In our imaginary town, light and dark moths are equally likely to survive and reproduce the same way you are equally likely to get heads or tails when you flip a coin. But if you flip a coin 100 times, you won't always get *exactly* 50 heads and 50 tails. Similarly, of 100 moths born in the next generation, there won't necessarily be *exactly* 50 light ones and 50 dark ones.

Genetic drift is particularly important in small populations because chance is more likely to change allele frequencies significantly in a small population. To see why, consider flipping a coin 10 times (a small population) versus 1000 times (a large population). With 10 flips, it is not at all unlikely that you'll get heads 60% of the time—that is, 6 heads and 4 tails. On the other hand, a similar result with 1000 flips—600 heads and 400 tails—is most unlikely.

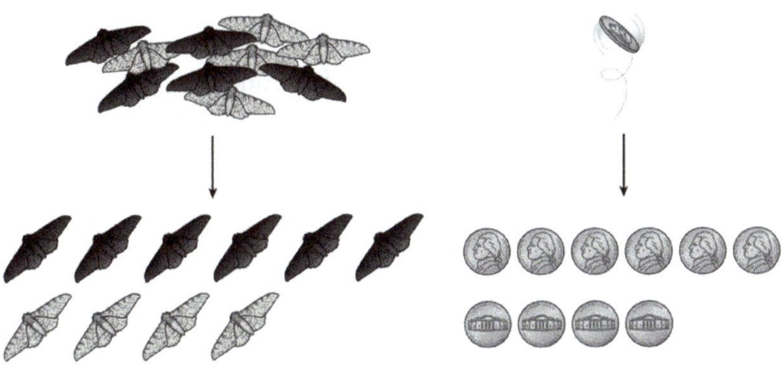

< Figure 9.2

If light and dark moths have equal fitness, we would expect them to leave the same number of offspring. However, just by chance, one group could leave more offspring than the other, resulting in genetic drift. This is true for the same reason that even though you expect the same number of heads and tails when you flip a coin 10 times, you might get more heads or more tails.

Gene Flow

Gene flow describes changes in allele frequencies that result from a net movement of alleles into or out of a population. For example, our half-polluted town may be next to a woodland that is home to a population of light moths. If a few of these light moths migrate from the woodland into town, the frequency of the light allele in the town will increase.

Where Genetic Variation Comes From

Natural selection cannot happen without genetic variation. Furthermore, populations with more variation have a better chance of adapting to a changing environment. This is because, in a population with more genetic variation, it is more likely that there are alleles that will allow some individuals to survive under new conditions. For instance, what would have happened to peppered moths during the Industrial Revolution if all the moths had been light and none were dark? In polluted areas, populations with only light moths might have died out.

But where does genetic variation come from? An understanding of genetics enabled biologists to answer this question. Genetic mutations constantly create new variations within populations. For example, when a genetic mutation changes the amino acids in a protein, it may produce a new allele for a given gene. Sexual reproduction also contributes to variation by bringing together alleles for different traits in new combinations. Then, any of the mechanisms of evolution—natural selection, mutation pressure, genetic drift, or gene flow—may affect whether that new allele becomes more or less common in the population.

CONCEPT**CHECK**

1. In a peppered moth population, genetic drift causes the frequency of the dark allele to increase one year. Will genetic drift also cause the frequency of the dark allele to increase the following year?

2. Twenty dark moths migrate into a peppered moth population, and 30 light moths migrate out. What effect does gene flow have on this population?

CHECK YOUR ANSWERS

1. Genetic drift is evolution due to chance, and there is no guarantee that chance will have the same effect the following year. The situation is the same as flipping coins—when you flip a coin 100 times and then do it again, you may get more heads the first time and more tails the second time.

2. Gene flow causes the frequency of the dark allele to increase in this population.

More to Explore

Read more about the mechanisms of evolution

https://evolution.berkeley.edu/evolibrary/article/evo_14

9.2 How New Species Form

We have seen how evolution through natural selection and other mechanisms causes populations to change over time. Can evolution also explain how different kinds of living things came to live on Earth? How does evolution produce new species?

A **species** is a group of organisms whose members can breed with one another but not with members of other species. Notice that this definition works well for organisms that reproduce sexually. For asexually reproducing organisms, species are recognized primarily by their similar characteristics and ways of life. The formation of a new species is called **speciation**. For sexually reproducing organisms, an important key to speciation is the evolution of *reproductive barriers* that prevent two groups of organisms from interbreeding.

There are two kinds of reproductive barriers: ones that work before fertilization and ones that work after fertilization. These are called prezygotic and postzygotic reproductive barriers. (A *zygote* is a fertilized egg, so *prezygotic* means "before fertilization," and *postzygotic* means "after fertilization.")

Prezygotic reproductive barriers prevent individuals of different species from mating in the first place or prevent fertilization from occurring if they do mate. There are many types of prezygotic barriers—organisms may differ in when they breed, where they breed, or in the details of their courtship rituals. In addition, their sex organs may not fit together properly, preventing successful sperm transfer, or other factors may prevent fertilization if sperm is transferred. **Figure 9.3** shows an example of a prezygotic reproductive barrier.

Postzygotic reproductive barriers act after fertilization has taken place. Postzygotic barriers occur when mating produces hybrids that either don't survive or are sterile—unable to breed themselves. The mule, the offspring of a horse and a donkey is sterile and cannot reproduce. Likewise, a liger (**Figure 9.4**), the product of the mating of a lion and a tiger, is sterile.

READING CHECK

How do prezygotic barriers affect the formation of new species?

< Figure 9.3

During courtship in red-crowned cranes, the birds dance around each other, bob their heads, stretch their necks, extend their wings, and leap straight into the air, singing in unison. Unless all these behaviors are performed just right, another bird species would have little hope of convincing a red-crowned crane to mate with it.

Figure 9.4 >
This is an example of a postzygotic reproductive barrier—the liger, a lion–tiger hybrid, is sterile.

Now let's consider how reproductive barriers evolve and, therefore, new species. In *allopatric speciation*, new species are formed after a geographic barrier divides a single population into two isolated populations (**Figure 9.5**). A geographic barrier could be a mountain range, a river, an ocean, a canyon, or—for aquatic organisms—a piece of land.

Once two populations are geographically isolated from each other, they evolve independently. Over time, natural selection and genetic drift may contribute to the evolution of key differences that prevent interbreeding. If a reproductive barrier evolves, the different populations become separate species.

Figure 9.5 >
Geographic barriers isolate populations and allow them to evolve independently. Sometimes, a reproductive barrier will evolve, resulting in allopatric speciation. In this example, the courtship song of birds divided by a mountain range diverges, resulting in a prezygotic reproductive barrier.

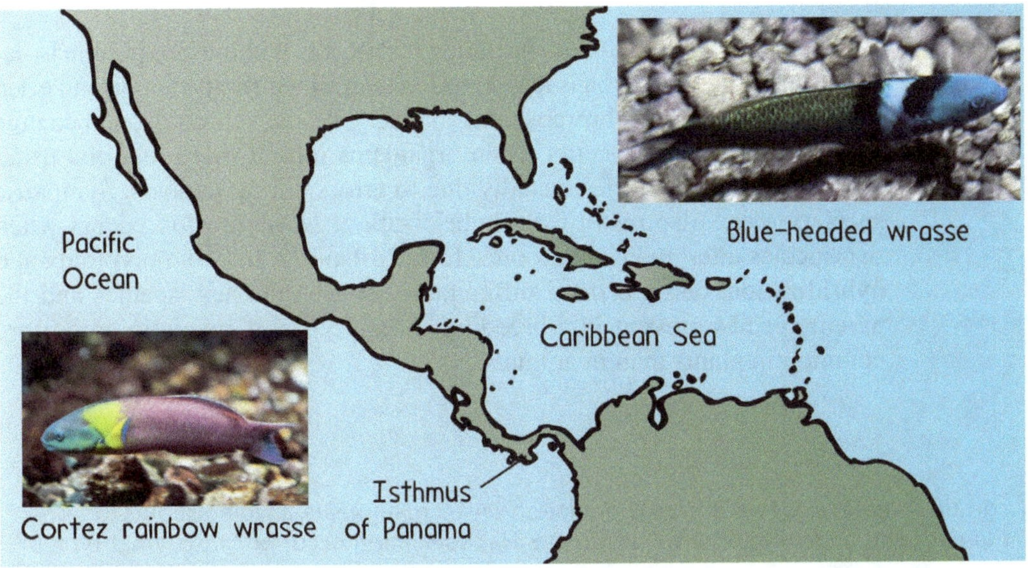

< Figure 9.6

The formation of the Isthmus of Panama 3 million years ago isolated Pacific and Caribbean marine populations, producing numerous instances of allopatric speciation. The blue-headed wrasse (Caribbean) and the Cortez rainbow wrasse (Pacific) are descended from a single ancestral species that formerly spanned Pacific and Caribbean waters.

Numerous instances of allopatric speciation have been recorded. For example, the rise of the Isthmus of Panama 3 million years ago divided the Caribbean Sea from the Pacific Ocean, splitting hundreds of marine organisms into separate Caribbean and Pacific populations. Most of these populations subsequently speciated by evolving reproductive barriers, including the wrasses in **Figure 9.6**.

Adaptive radiations are spectacular examples of allopatric speciation. They typically occur when a population is geographically isolated from its ancestral group, allowing it to adapt to a new ecological niche. The new environment presents unique selective pressures that favor individuals with certain traits, leading to the emergence of new species that are adapted to distinct ways of life. One example is the adaptive radiation of Darwin's finches on the Galapagos Islands. The finches diversified into different species, each with a unique beak shape that allowed them to exploit different food sources.

Another example is the Hawaiian honeycreepers, which underwent a remarkable adaptive radiation resulting in more than 30 distinct species (**Figure 9.7**). The honeycreepers exhibit a variety of beak shapes and sizes that are adapted to different feeding habits, such as nectar feeding, insect hunting, or seed cracking. These examples demonstrate how natural selection can drive speciation, leading to new species adapted to unique ecological niches.

V Figure 9.7

The Hawaiian honeycreepers represent an adaptive radiation of more than 30 bird species. The honeycreepers differ in plumage, beak shape, size, and diet. Unfortunately, many species are extinct or endangered because of habitat destruction and the introduction of non-native species such as rats, pigs, mongooses, cats, and mosquitoes.

More to Explore

More about speciation here:

https://www.nationalgeographic.org/encyclopedia/speciation/

Sympatric speciation is speciation that occurs without geographic isolation. Sympatric speciation is less common than allopatric speciation and often results from a sudden chromosomal change. One such chromosomal change is *polyploidy*, which occurs when organisms inherit more than the usual two sets of chromosomes, usually due to errors during meiosis. Sympatric speciation can also result from hybridization. *Hybridization* occurs when two species interbreed and produce fertile offspring. In both polyploidy and hybridization, chromosomal differences between the new species and the parent species prevent interbreeding. These types of speciation are more common in plants than in animals.

CONCEPT CHECK

1. A small river forms, dividing a group of moles into two isolated populations. After many years, a biologist puts together moles from opposite sides of the river and finds that they will not mate. Has speciation occurred? If so, what type of speciation was it?

2. Do you think the same river would cause a population of birds to become two separate species?

3. Two species of frogs do not interbreed because one species breeds in the spring, and the other breeds in the fall. Is this a prezygotic or postzygotic reproductive barrier?

CHECK YOUR ANSWERS

1. The moles on the two sides of the river now represent two different species because they don't interbreed. This was allopatric speciation because it occurred after a geographic barrier (the river) separated the populations.

2. Probably not, since a small river is not much of a geographic barrier for flying animals.

3. Prezygotic because it prevents mating.

9.3 Natural Selection in Action

All scientific theories make predictions about what we should observe in nature. If these predictions are confirmed, the theory is supported. The theory of evolution through natural selection has been tested repeatedly against observations of the natural world, and the evidence is overwhelming. In the following sections, we will look at some of this evidence.

Observations of Natural Selection

In many cases, scientists have seen natural selection produce evolutionary changes within relatively short time frames; they have observed and measured the actual changes in populations. These include some of the examples we looked at in the previous chapter:

1. Australian rabbits evolved resistance to the myxoma virus so that over time, a smaller fraction of individuals died from the disease.
2. Peppered moths evolved to become better camouflaged in their environments—dark moths became more and more common as habitats became polluted. Then they became less and less common as pollution was cleaned up.
3. Bacteria evolved resistance to certain antibiotics so that these antibiotics no longer controlled infections.

^ Figure 9.8

Artificial selection has produced great diversity in dogs.

Scientists have also studied how the beaks of Darwin's finches evolve after a drought, how insects evolve resistance to pesticides, and natural selection in a wide variety of other populations.

Artificial selection is the selective breeding of organisms with desirable traits in order to obtain organisms with similar traits. Humans artificially select for desirable traits in domesticated animals and crops all the time: We breed fast racehorses to try to get faster racehorses; different types of dogs to produce superior hunters, herders, or sled pullers (**Figure 9.8**); and varieties of strawberries to grow the largest and sweetest fruit. In artificial selection, humans control the reproductive success of different organisms and bring about distinct evolutionary changes in populations over time.

These changes can be dramatic—think how much a Chihuahua differs from the animal it is descended from, the wolf. Or look at **Figure 9.9** to see the difference between the corn we eat today and teosinte, the plant from which corn was bred. Artificial selection has produced countless forms of domestic animals and crops, all with traits valued by humans.

READING CHECK

What is the method by which human beings cause evolutionary changes in numerous species?

< Figure 9.9

Corn, one of the most important agricultural crops in the world, was laboriously bred through artificial selection from teosinte (above). Teosinte has tiny cobs, only a few rows of kernels, and inedible hard coverings on its seeds.

CONCEPT CHECK

How do observations of natural selection in action and artificial selection provide evidence for evolution?

CHECK YOUR ANSWER

Scientists have documented natural selection in many different wild populations, including peppered moths, Galápagos finches, Australian rabbits, and many other species. They have documented how natural selection leads to measurable evolutionary changes in these populations, offering direct support for Darwin's theory of evolution.

Artificial selection also provides evidence for evolution. In the case of artificial selection, humans, rather than nature, do the "selecting," but this also results in measurable evolutionary changes in populations over time. Most examples of artificial selection are found in domesticated animals and crops.

More to Explore

An article discussing how Darwin's understanding of artificial selection helped him to formulate his ideas

https://evolution-outreach.biomedcentral.com/articles/10.1007/s12052-008-0114-z

9.4 Fossils

Evolution has left a record in Earth's rocks in the form of fossils. Scientists can date fossils from the age of the rock formations they belong to and follow the evolution of groups of organisms over time. For example, did you know that whales are descended from mammals with hooves? Fossil whales show this. Fossil whales also tell us how many key whale traits evolved.

Fossil whales also show how whales lost their hind legs as they became better and better adapted to an aquatic existence. The oldest whale fossils, such as the 50-million-year-old *Ambulocetus*, have large hind legs that were used both on land and for swimming (**Figure 9.10a**). Also, in **Figure 9.10b**, we can see how, over time,

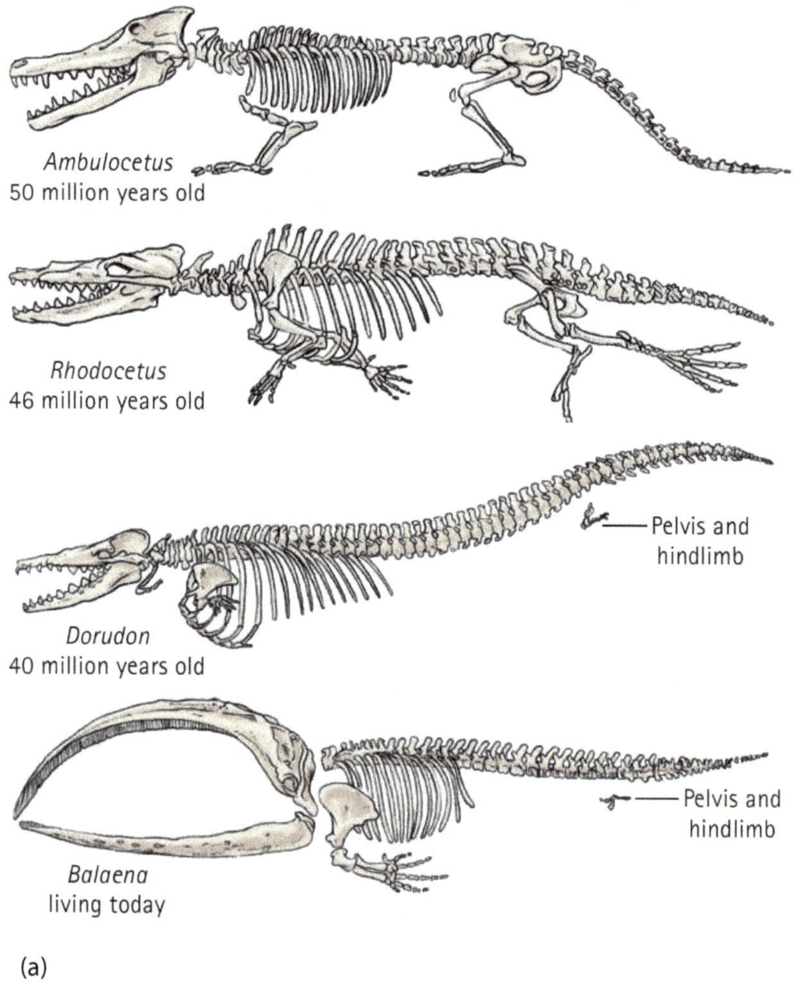

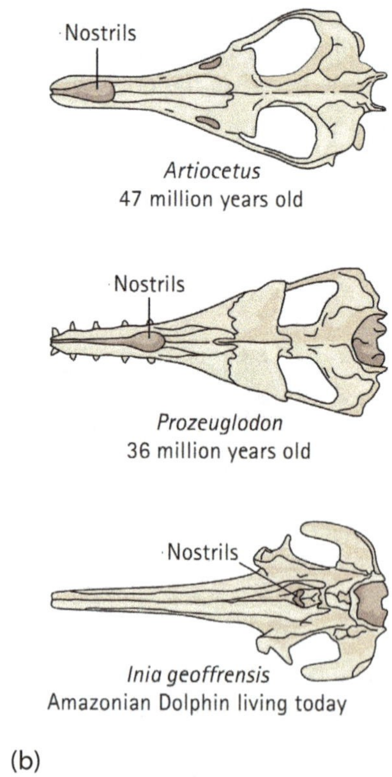

^ Figure 9.10

Fossil whales show how key features of these marine mammals evolved over time. (a) Fossil whales showing the reduction and loss of hind legs over time. (b) These fossil skulls show that the location of the nostrils shifted over time, from a position in front of the skull to a position on top of the skull—the "blowhole" seen in modern species.

whale nostrils moved from the front of the skull to the top of the skull, forming a blowhole.

Ambulocetus also has small hooves on its front legs, providing clear evidence that whales are descended from hoofed mammals. *Rhodocetus*, a 46-million-year-old fossil whale, shows reduced hind legs—these are not attached to the backbone and so could not have supported much weight. *Rhodocetus* also shows prominent tail muscles that would have been effective for swimming. In the 40-million-year-old *Dorudon*, hind limbs are present, but they are tiny: *Dorudon* was clearly a fully aquatic species. In modern whales, there is no evidence of hind limbs on the outside of the body, although tiny remnants of the pelvis and sometimes femurs remain inside the body.

READING CHECK

Do modern whales show any evidence that their ancestors had hind limbs?

Archaeopteryx, the famous 150-million-year-old fossil bird (**Figure 9.11**), also shows intermediate traits in the evolution of birds from their dinosaur ancestors. *Archaeopteryx* has many birdlike features, such as feathers, wings, and a wishbone. However, it also has dinosaur-like features absent in modern birds, including wing claws, bones in its tail, and teeth.

< Figure 9.11

Archaeopteryx, an early bird, has features of both the dinosaurs it evolved from and of modern birds.

CONCEPT CHECK

How do fossil whales provide evidence for evolution?

CHECK YOUR ANSWER

Fossils show how key traits evolved in whales. For example, the whale fossils that have been found show traits that are intermediate between the features of the ancestors (nostrils in front of the skull and large functional hind legs) and present-day whales (a blowhole on top of the skull and tiny vestigial hind limbs).

More to Explore

Read more about fossil whales here:

https://evolution.berkeley.edu/evolibrary/article/evograms_03

Learn more about Archaeopteryx here:

https://www.livescience.com/24745-archaeopteryx.html

9.5 Body Structures and Genetics

We can see evidence of species' evolutionary histories in their bodies' structures. Consider the limbs of different mammals. Different mammals use their front limbs for different purposes: Humans use theirs as arms and hands for manipulating tools, cats use theirs to walk, whales use theirs as flippers, and bats use theirs as wings. If each of these animals had originated independently, we would expect their limbs to look completely different. Each limb would have been designed from scratch to best perform its function. But, despite the different functions of human hands, cat legs, whale flippers, and bat wings, all these limbs show the same arrangement of bones, as you can see in **Figure 9.12**. This suggests that the limbs were inherited from a common ancestor and then modified through natural selection for different functions.

Another way to say this is that a mouse and a whale are about as different as two mammals can be. Yet just about every bone in a mouse corresponds to a specific bone in a whale. These similarities suggest that mice and whales had a common ancestor and that their skeletons were modified over time by natural selection to fit different environments and ways of life.

Vestigial Organs

An organism's evolutionary history often leaves traces in its body. In some organisms, these traces take the form of vestigial organs. *Vestigial organs* are not functional—they are just the remains of an organ found in the organism's ancestor. For example, we think of snakes as legless. But certain snakes actually have tiny, partial hind legs.

READING CHECK

Why do some organisms have vestigial organs?

The tiny stubs have no purpose—they are just the remains of what once were bigger limbs. A snake's vestigial hind legs provide evidence that snakes evolved from animals with legs. In the same way, many blind cave species lack functional eyes in their light-less habitats but retain vestigial eyes (**Figure 9.13**). These vestigial organs suggest that cave species evolved from animals with eyes.

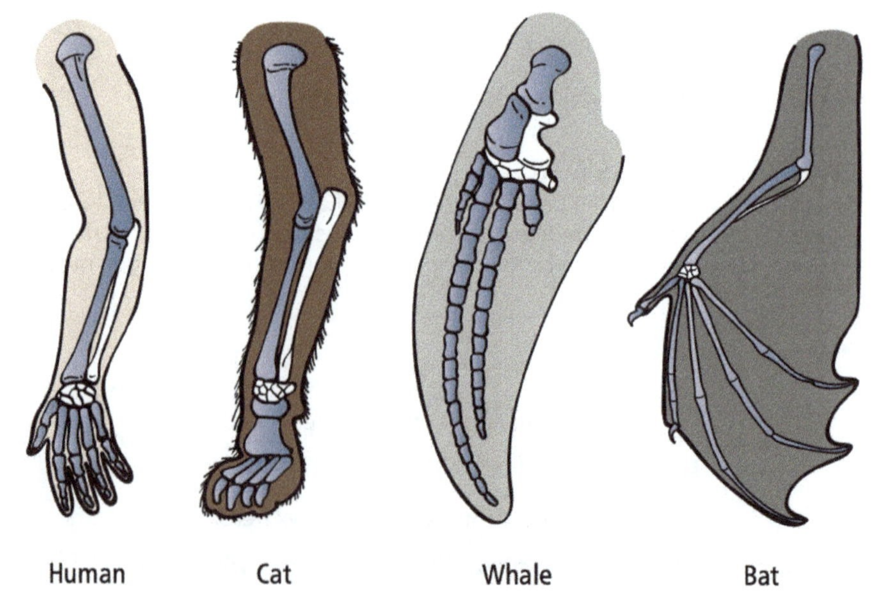

Figure 9.12 >

Although these mammalian limbs are each used for different activities, they are composed of the same set of bones, evidence that they were inherited from a common ancestor.

< **Figure 9.13**

The Texas blind salamander lives in lightless caves. It has tiny vestigial eyes (the dark dots in the photo) that are covered by skin.

Patterns of Development

Related species develop in similar ways. If each species on Earth had originated independently, we wouldn't expect these similarities in development. For example, even though humans have no tails, we go through a tailed stage during development, just like other vertebrates (**Figure 9.14**).

DNA and Molecular Evidence

The DNA of related species has similar nucleotide (ACGT) sequences. In fact, the more closely related two species are, the more similar their DNA sequences tend to be. This is true not only for DNA sequences that code for proteins but even for sequences with no known function. If each species on Earth had originated independently, would we expect to see similar non-coding DNA in related species? DNA similarity suggests that DNA did not originate independently in each species but was inherited from a common ancestor and then modified during evolution.

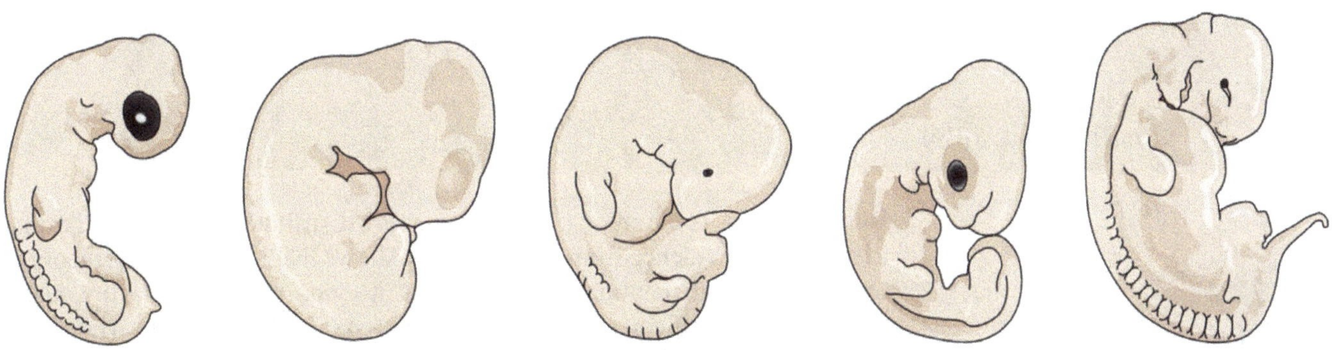

∧ **Figure 9.14**

Related species go through similar stages in their development. The human embryo goes through a tailed stage just like the other vertebrates, even though humans don't have tails.

Hierarchical Organization of Living Things

Darwin's theory of evolution through natural selection explains Earth's diversity of species as originating through numerous speciation events. If this is the case, then we expect living things to be organized into hierarchical sets of "nested groups"—that is, "groups within groups."

Each living species should have fewer traits in common with more distant relatives, and more traits in common with species that it split off from more recently. This is, in fact, how living things on Earth are organized. Humans, for example, share a backbone with other vertebrates such as fishes, amphibians, reptiles, and mammals; they share four limbs with terrestrial vertebrates such as amphibians, reptiles, and mammals but not with fish, which are more distantly related; they share a waterproof skin with reptiles and mammals but not with amphibians, which are more distantly related; and they share the trait of nursing their young with milk with other mammals but not with the more distantly related reptiles.

Living things fit into a hierarchical organization, as predicted by evolution. We do not see traits scattered across living things. For example, we do not see a backbone in vertebrates plus some worms and some insects and some snails. The characteristics that organisms have make sense based on their evolutionary history and relationships.

CONCEPT **CHECK**

How does DNA provide evidence for evolution?

CHECK YOUR ANSWER

When scientists look at the genomes of related species, they find similarities in the species' DNA sequences. Closely related species tend to have more similar DNA sequences than more distantly related species. This is true not only for DNA sequences that code for proteins—in which case you could argue that the functional requirements of the proteins require certain DNA sequences—but even for DNA sequences with no known function. If each species on Earth had originated independently, you would not expect to see similar non-coding DNA in related species. These similarities suggest that DNA did not originate independently in each species but was inherited from a common ancestor and then modified during evolution.

More to Explore

Do humans have vestigial organs? You can read about some here:

https://www.livescience.com/21513-vestigial-organs.html

9.6 Biogeography

READING CHECK

What does the evidence from biogeography tell us about why members of a particular species can be observed at various locations around the globe?

Biogeography is the study of how species are distributed on Earth. An examination of the evidence from biogeography shows that it supports the theory of evolution through natural selection. That is, it supports the idea that organisms evolved in a specific place and then left descendants in the places where they were able to spread. Biogeography does not support the idea that organisms were specially designed to fit into a specific habitat type and then distributed where these habitats occur on Earth. For example, although the Arctic and Antarctic have similar environments, they are occupied by entirely different species (**Figure 9.15**).

Biogeography shows that the ranges of many species are bounded by geographic barriers such as oceans or mountain ranges. For example, many organisms are restricted to a single continent. In addition, closely related species tend to be found close together, suggesting that they evolved in one place and then spread. For example, all of Darwin's finches are found in or near the Galápagos, and all the honeycreepers are found in Hawaii. Similarly, island species are usually most closely related to species found on the closest mainland. Islands also tend to have fewer species than an area of equal size on the mainland, and many island species are *endemic*, meaning they are found nowhere else on Earth.

Finally, islands tend to be occupied by many flying animals but few terrestrial ones (**Figure 9.16**). All these points suggest that organisms were not dispersed purposefully around Earth; instead, they evolved in one place and then left descendants where they could spread.

^ Figure 9.15

The Arctic and Antarctic, which have similar habitats, are occupied by very different species. Polar bears are found in the Arctic but not in the Antarctic. Penguins are found in the Antarctic but not in the Arctic.

< Figure 9.16

Why are terrestrial vertebrates rare or absent from islands, whereas flying species are common? This is the Hawaiian hoary bat, the only mammal found in Hawaii prior to human colonization of the islands.

CONCEPT CHECK

Many species found on islands resemble species found on the nearest mainland. How does this provide evidence for evolution?

CHECK YOUR ANSWER

This pattern suggests that island species evolved when some mainland individuals colonized the island and then evolved in isolation, rather than that species were distributed purposefully around Earth.

More to Explore

You can read more about biogeography and evolution here:

https://evolution.berkeley.edu/evolibrary/article/0_0_0/history_16

9.7 The Evolution of Humans

Humans are primates, a group of mammals including monkeys and apes. This does not mean we are descended from any modern species of monkey or ape, just that we share a common ancestor with these species more recently than we do with a dog, a lizard, or a plant. Humans are also *hominids*, the group within the primates that includes modern *Homo sapiens* (our species) as well as some of our extinct relatives. Although humans are the only hominids in existence today, fossil hominids provide clues as to how humans evolved. A timeline of human evolution is shown in **Figure 9.17**.

Some of the earliest hominids known belong to the group *Australopithecus*. Fossil *Australopithecus* have been found at multiple sites in Africa, where hominids originated. "Lucy," the famous *Australopithecus afarensis* fossil shown in **Figure 9.18**, dates from about 3.2 million years ago. When she was alive, Lucy stood 3 feet 8 inches tall and had a brain about the size of a chimpanzee's. However, the bones of Lucy's pelvis make it clear that she walked upright on two legs. In fact, older *Australopithecus* fossils show that an upright posture dates to at least 4 million years ago and therefore evolved long before increased brain size and intelligence.

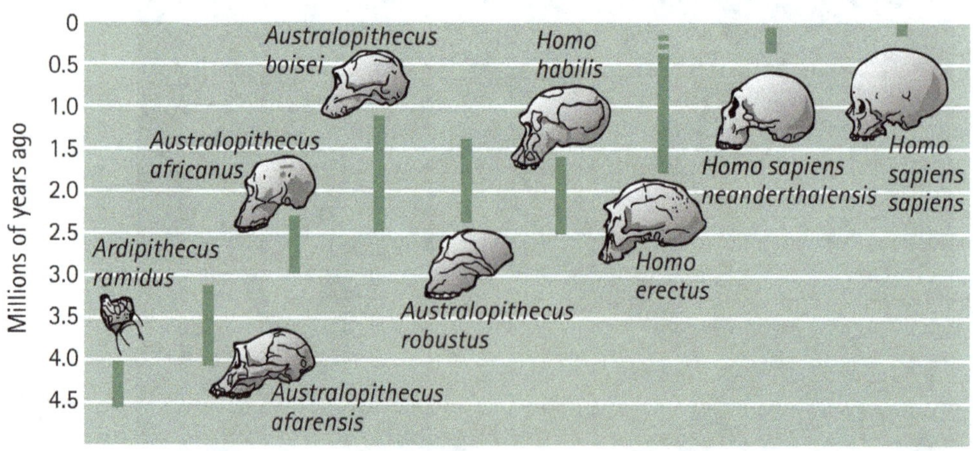

Figure 9.17 >

This timeline shows when certain hominid species existed on Earth. The skulls (or skull fragments) are all drawn to the same scale to show relative brain sizes.

Homo habilis is the earliest known species in the group *Homo*, which includes the species most closely related to modern humans. Some Homo habilis fossils are 2.2 million years old. *Homo habilis* had a larger brain than *Australopithecus*. *Homo habilis* also made stone tools—in fact, its scientific name means "handyman." In addition, male *Homo habilis* were much larger than females. This is interesting because in other primates, such as gorillas and baboons, a big size difference between males and females is a sign that males fight each other for female mates.

Homo erectus lived from about 2 million to about 400,000 years ago. *Homo erectus* had an even larger brain than *Homo habilis*. In fact, the brain of *Homo erectus* was only a little smaller than that of modern humans.

Homo erectus was a skilled toolmaker as well as the first hominid species to migrate out of Africa and spread into what is now Europe and Asia. Like *Homo habilis*, older *Homo erectus* fossils show that males were much larger than females. However, later fossils of the same species show a male–female size difference closer to that present in modern humans, suggesting the development of a more human-like social system.

The Neanderthals—*Homo neanderthalensis*—are closely related to modern humans (**Figure 9.19**). They lived from about 400,000 years ago to about 30,000 years ago.

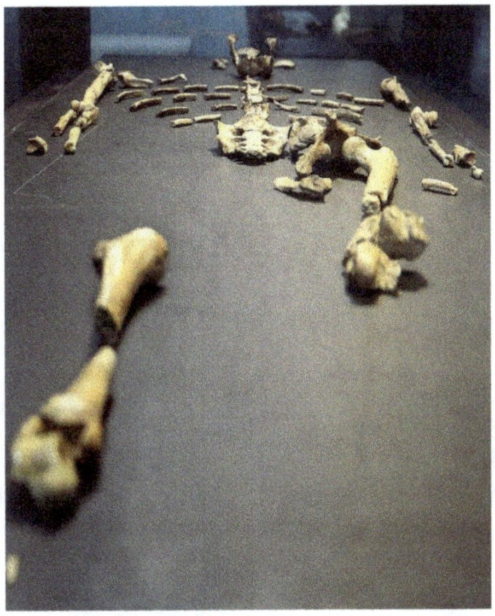

∧ Figure 9.18

"Lucy," a fossil Australopithecus afarensis, stood upright and walked on two feet.

Neanderthals had very thick arms and legs, and their brains were as large as those of modern humans. Archaeological finds show that Neanderthals were effective hunters, had complex burial rituals, and made use of medicinal plants. One question that remains unanswered is whether the Neanderthals had language. For thousands of years, modern humans coexisted with Neanderthals. However, Neanderthal populations disappeared as modern humans spread. Scientists are not sure why, although it seems possible that modern humans out-competed the Neanderthals and drove them to extinction. The development of modern genetic techniques has allowed scientists to collect information about the Neanderthals from a new source—DNA (**Figure 9.20**). Using fossil remains, scientists were able to sequence the Neanderthal genome.

READING CHECK

How old are the oldest fossils of Homo habilis, the oldest species of the group that includes modern humans?

This resulted in a huge surprise—most modern humans have some Neanderthal DNA! Neanderthal DNA accounts for about 1 to 4 percent of the genome of most humans. The genetic data indicate that Neanderthals and modern humans interbred at some point in the past.

The earliest fossils of modern humans, *Homo sapiens*, were found in Ethiopia and are 195,000 years old.

Although anatomically modern humans are quite old, the cultural traits we associate with humans—things like art, music, and religion—are more recent, appearing only about 50,000 years ago. The reason for this gap between modern anatomy and modern behavior is the subject of continued debate.

Figure 9.19 >

Neanderthals coexisted with modern humans and interbred with them. This reconstruction appear at the Neanderthal Museum in Mettmann, Germany.

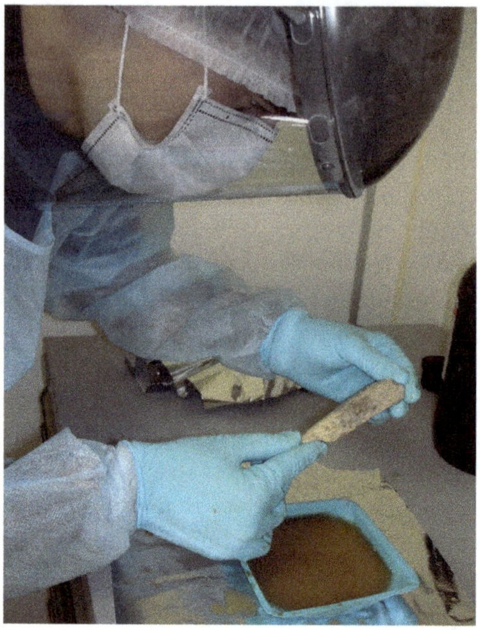

Figure 9.20 >

A scientist prepares a sample of bone from a Neanderthal fossil in order to extract and study its DNA.

CONCEPT CHECK

1. Have multiple species of hominids ever coexisted on Earth? Do any hominids other than humans survive to this day?

2. What is the significance of the transition from a large male–female size difference in early Homo erectus fossils to a size difference closer to that of modern humans in later fossils of the same species?

3. What is the result of trillions and trillions of living things passing genetic traits to their offspring, here and there making an adaptive change, and surviving to today?

CHECK YOUR ANSWERS

1. The timeline of hominid evolution shows that multiple species of hominids coexisted during much of hominid history. Today, however, humans are the only species of hominids in existence. The others have all died out.

2. A large size difference between males and females is a sign that males fought each other for female mates. This may have been true in early Homo erectus. More equal body sizes in later Homo erectus suggests that males and females may have formed longer-term bonds, perhaps as they raised offspring together.

3. We and Earth's other living things are the results of this long and astounding journey!

More to Explore

You can read more about human evolution at the Smithsonian Institution's:

https://humanorigins.si.edu/education/introduction-human-evolution

But wait! If humans and Neanderthals once interbred, why do we consider them different species? Here's an expert's take on this interesting question:

https://www.nhm.ac.uk/discover/are-neanderthals-same-species-as-us.html

Concept Review

Chapter 9

Detailed Chapter Summary

At the genetic level, evolution can be described as changes in the allele frequencies of genes over time. Four mechanisms of evolution are:

1.) Natural selection: the frequency of an advantageous allele increases over time due to its positive effects on survival and reproduction.

2.) Mutation pressure: the tendency of alleles to mutate in one direction more than the other direction, causing a change in allele frequencies.

3.) Genetic drift: the evolution of populations due to chance. Genetic drift is especially significant in small populations.

4.) Gene flow: the evolution of a population due to the movement of alleles into or out of the population.

A species is a group of organisms whose members can breed with one another but not with members of other species. Speciation describes the formation of new species. Speciation relies on the evolution of reproductive barriers between groups. A prezygotic reproductive barrier prevents mating or prevents the formation of a fertilized egg. A postzygotic reproductive barrier acts after fertilization has taken place. There are two types of speciation, allopatric and sympatric. In allopatric speciation, new species are formed after a geographic barrier divides a population. In sympatric speciation, there is no such barrier. Allopatric speciation is much more common than sympatric speciation, but hybridization and polyploidy are examples of the latter.

Many types of evidence support evolution, which is inherited changes in populations of organisms over time. These include:

1.) Observations of natural selection in action, as seen in resistance to myxoma virus in Australian rabbits, peppered moth coloration, evolution in Darwin's finches, and other examples.

2.) Artificial selection, as seen in many domesticated animals and crops, including dogs and corn.

3.) Fossils that show the evolution of organisms over time, such as fossil whales.

4.) Similarities in body structures, as seen in the forelimbs of mammals.

5.) Vestigial organs, such as the eyes of blind species and tiny limbs in some snakes.

6.) Patterns of development, as seen in the tails of human embryos, among other examples.

7.) DNA evidence, such as sequence similarities among species in parts of the genome with no known function.

8.) Hierarchical organization of living things.

9.) Biogeography, as seen in Arctic vs. Antarctic species and the distribution of species on islands, among other examples.

Humans are hominids. Many hominid fossils have been found. "Lucy," an early hominid of the genus *Australopithecus*, had an upright stance. *Homo habilis*, a toolmaker, is the earliest known member of the genus *Homo*. *Homo erectus* evolved from large male-female size differences to more equal sizes, suggesting a more human-like social system. *Homo erectus* was also the first hominid to move out of Africa. Neanderthals were larger and more muscular than modern humans, but had the same brain size. Neanderthals were larger and more muscular than modern humans but had the same brain size. Neanderthals coexisted for a while with modern humans and interbred with them but ultimately went extinct. Most modern humans have some Neanderthal DNA. The earliest fossils of our species, Homo sapiens sapiens, were found in Ethiopia and are 195,000 years old.

Key Terms

Biogeography The study of how species are distributed on Earth.

Artificial selection The selective breeding of organisms with desirable traits in order to produce offspring with similar desirable traits.

Gene flow The evolution of a population due to the movement of alleles into or out of the population.

Genetic drift The evolution of a population due to chance.

Speciation The formation of new species.

Species A group of organisms whose members can breed with one another but not with members of other species.

Review Questions

Simple Review Questions

9.1 Mechanisms of Evolution

1. Does natural selection act on genotype or phenotype?
2. Define genetic drift and provide an example of how genetic drift can cause a population to evolve.
3. What is gene flow? Use an example to explain how gene flow can cause a population to evolve.
4. Why are genetic mutations and sexual reproduction important to creating and maintaining genetic variation in populations?

9.2 How New Species Form

5. What is a species?
6. What is the difference between a prezygotic reproductive barrier and a postzygotic reproductive barrier? Give an example of each.
7. Explain the difference between allopatric speciation and sympatric speciation. Which is more common?
8. What is adaptive radiation?

9.3 Evidence of Evolution: Natural Selection in Action

9. What is artificial selection? Why does artificial selection provide evidence for evolution?

9.4 Evidence of Evolution: Fossils

10. How do fossils provide evidence for evolution?
11. Explain how the fossil whales that have been discovered support Darwin's theory of evolution.
12. What does the fossil Archaeopteryx tell us about bird evolution?

9.5 Evidence of Evolution: Body Structures and Genetics

13. Why does the similarity of the mammalian limb in all different species of mammals provide evidence for evolution?

9.6 Evidence of Evolution: Biogeography

14. How does biogeography provide evidence for evolution?

9.7 The Evolution of Humans

15. What important feature of modern humans can already be seen in 4-million-year-old Australopithecus fossils?
16. What was the first species of hominid to leave Africa and spread into Europe and Asia?
17. How old is our species, the modern humans known as Homo sapiens?

Challenging Review Questions

9.1 Mechanisms of Evolution

18. Can chance cause a population to evolve?
19. You are studying a population of beetles that includes some red individuals and some yellow individuals. You know that color is an inherited trait in the population. By counting the number of red and yellow beetles over a period of 5 years, you notice that the proportion of red individuals is increasing over time while the proportion of yellow individuals is decreasing over time. How could you determine whether this is a result of natural selection? Are there other potential explanations?

20. In a population of mice that you are studying, tail length appears to be increasing over time. However, you find no evidence that natural selection is acting on tail length. What are two alternative explanations for your observation?

9.2 How New Species Form

21. Why is speciation often associated with the introduction of a geographic barrier?

22. Individuals of two different fish species sometimes mate, but their offspring die soon after hatching. Is this an example of a prezygotic or postzygotic reproductive barrier?

23. Finches on two closely situated islands look different; on one island, they have brown tail feathers, and on the other island, they have black tail feathers. Can you conclude that these are two different species? How could you determine whether they are, in fact, distinct species?

24. At your field site, there are butterflies with yellow wings and butterflies with orange wings. After observing them carefully, you notice that the yellow butterflies always mate in shady areas under trees, whereas the orange butterflies always mate in sunny meadows. Can you conclude that they are different species?

9.3 Evidence of Evolution: Natural Selection in Action

25. What are some examples of artificial selection?

26. How are artificial selection and natural selection similar? How are they different?

9.4 Evidence of Evolution: Fossils

27. Why do fossil whales have legs?

28. This photo shows the leg bone of a fossil horse (upper) next to the leg bone of a modern horse (lower). What do these fossils tell us about horse evolution?

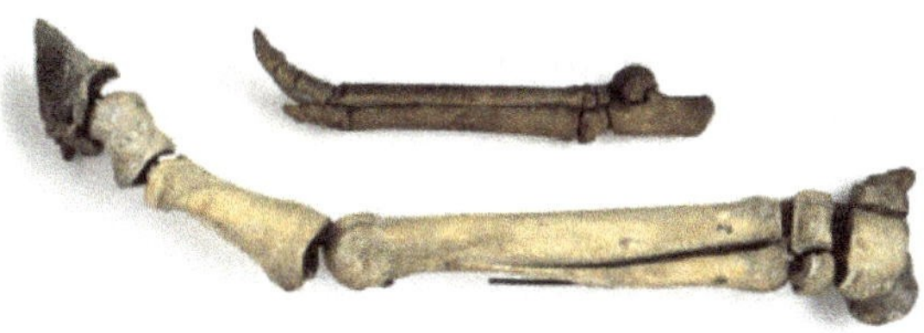

29. Scientists have found a number of fossil turtles. One early fossil turtle did not have a "shell" but did have wide, expanded ribs along its back ("top") side and expanded belly coverings on the belly ("bottom") side. A second, more recent turtle fossil had a full bottom shell. Although this second fossil turtle did not have a full top shell, it did have a partial shell extending from its backbone as well as wide, expanded ribs. What do these two fossils tell us about turtle evolution? Does it support the hypothesis that turtles evolved in water rather than on land? (Hint: If the bottom shell evolved first, what kinds of predators did it help defend turtles from?) Explain your answer.

9.5 Evidence of Evolution: Body Structures and Genetics

30. Many dogs have a dewclaw, a toenail located on the inside of the leg. Although some dogs can use their dewclaws to help grip objects, in many, it serves no purpose. In fact, if the nail is not trimmed, it can become overgrown and cause issues for the dog. Is the dewclaw a vestigial organ? How does a dog's dewclaw provide evidence for evolution?

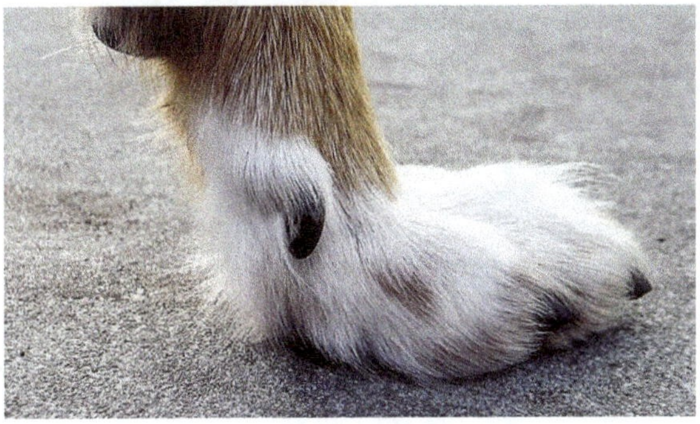

31. How do the bones of the human hand provide evidence for evolution?

9.6 Evidence of Evolution: Biogeography

32. Islands tend to have fewer species than the nearby mainland. Furthermore, island species often include many flying organisms and few terrestrial ones. Do these biogeographic patterns support evolution? Why?

33. Many of the living organisms in Hawaii are found nowhere else on Earth. Hawaii has numerous unique species of plants, birds, insects, mammals, mushrooms, and other living things. What is the best explanation for this?

9.7 The Evolution of Humans

34. Can Homo sapiens and Neanderthals (Homo neanderthalensis) be classified as different species if genetic evidence shows they interbred in the past?

35. Laura says she doesn't believe that humans were at one time chimpanzees or gorillas. Jeff says he doesn't believe it either. Explain why biologists also don't believe that humans are descended from chimps or gorillas.

36. In one of the most spectacular archaeological finds in centuries, the skeletons of a tiny human relative were discovered on a remote Indonesian island in 2004. Nicknamed "hobbits," Homo floresiensis adults had skulls the size of grapefruits and were no bigger than 3-year-old modern children. Homo floresiensis lived alongside pygmy elephants, giant rodents, and Komodo dragons. Why might Homo floresiensis have evolved to be a very small size in its environment?

Apply & Discuss Questions

37. A beetle population that includes sandy and green individuals is introduced into a grassy environment. How do you expect the population to evolve due to natural selection? Now suppose that another beetle population lives in a nearby sandbank. Most of the individuals in this population are sandy-colored. If beetles regularly migrate from one population to the other, what will be the effect on each population? Does gene flow make it easier or harder for these beetle populations to adapt to their environments?

38. Islands tend to have fewer species than an area of equal size on the mainland. Is this consistent with the idea that species were spread around Earth purposefully? Is it consistent with evolution?

39. Scientific theories must be falsifiable. Is evolution falsifiable? For example, can you imagine some biogeographic evidence that would not be consistent with evolution? Has any evidence of this sort been found?

40. Scientists who are searching for new fossils of early hominids usually look in Africa. Does this make sense, or should they expand their search?

Solutions to Odd-Numbered Questions

Simple Review Solutions

1. Natural selection acts on an organism's phenotype (traits), not its genotype (genes).

3. Gene flow describes changes in allele frequencies resulting from a net movement of alleles into or out of a population. For example, a half-polluted town may be next to a pristine woodland that is home to a population of light moths. If a few of these light moths migrated from the woodland into town, the frequency of the light allele in the town would increase.

5. A species is a group of organisms whose members can breed with one another but not with members of other species.

7. In allopatric speciation, new species are formed after a geographic barrier divides a single population into two isolated populations. Sympatric speciation occurs without geographic isolation. As a result, sympatric speciation is less common than allopatric speciation.

9. Artificial selection is the selective breeding of organisms with desirable traits in order to obtain organisms with similar traits. In artificial selection, humans control the reproductive success of different organisms and bring about distinct evolutionary changes in populations over time

11. Fossil whales show that whales are descended from hoofed mammals and tell us how many key whale traits evolved. Whale fossils show how over time, whale nostrils moved from the front of the skull to the top, forming a blowhole. Whale fossils also show how whales gradually lost their hind legs as they adapted more and more to an aquatic existence.

13. If each of these animals had originated independently, we would expect their limbs to look completely different. Each limb would have been designed from scratch to best perform its function. But, despite the different functions of human hands, cat legs, whale flippers, and bat wings, all these limbs show the same arrangement of bones. This suggests that the limbs were inherited from a common ancestor and then modified through natural selection for different functions.

15. The bones of Lucy's pelvis make it clear that she walked upright on two legs. In fact, older Australopithecus fossils show that an upright posture dates to at least 4 million years ago.

17. The earliest fossils of modern humans, Homo sapiens, were found in Ethiopia and are 195,000 years old.

Challenging Review Solutions

19. To determine whether natural selection is responsible for the shift, you could compare the fitness (number of offspring left) of red individuals versus yellow individuals. If this turned out to be difficult, you could also compare their survival or ability to acquire mates in an attempt to identify other potential causes of fitness differences. If it's not natural selection, evolution in the population could also be due to genetic drift, migration into or out of the population, or mutation pressure.

21. A geographic barrier causes the two separated populations to evolve in isolation, making a reproductive barrier more likely to evolve.

23. No, you cannot conclude they are distinct species merely because they are distinguishable. However, you can determine whether they are distinct species by figuring out whether they interbreed.

25. We saw artificial selection in the breeding of dogs as well as in the breeding of corn from teosinte. There are numerous other examples of domesticated animals and crops that humans have bred through artificial selection, including most of the food we eat and most of the animals we regularly keep as pets or livestock.

27. Because whales are descended from animals that had legs. Some fossil whales show an intermediate state where legs were present but no longer fully functional.

29. These fossils suggest that the turtle's bottom shell evolved before the top shell. This sequence does support the hypothesis that turtles evolved in water rather than on land. A bottom shell is more likely to protect you from predators attacking from below – something that could happen in water but not on land.

31. The pattern of bones in the human hand is similar to that found in other mammals. If each mammal had originated independently, we would expect their limbs to look completely different. Each limb would have been designed from scratch to best perform its function. But, despite the different functions of human hands, cat legs, whale flippers, and bat wings, all these limbs show the same arrangement of bones. This suggests that the limbs were inherited from a common ancestor and then modified through natural selection for different functions.

33. Hawaii is extremely isolated from all mainlands, so organisms that arrived there then had plenty of time to evolve in isolation and speciate from mainland species.

35. Humans are primates. We belong to a group of mammals that also includes monkeys and apes. This doesn't mean we're descended from any existing species of monkey or ape. It just means we're more closely related to monkeys and apes than we are to other living things, such as dogs, lizards, or plants.

Apply & Discuss Solutions

37. In the grassy environment, the green individuals are likely to have higher fitness because they are better camouflaged. The population is likely to evolve to have more green individuals. Because of migration, sandy genes will constantly enter the grassy population, and green genes will constantly enter the sandy population. Migration and gene flow make it harder for each population to adapt to its local environment.

39. Evolution is falsifiable. In the specific case of biogeography, it is possible to imagine a situation where
1.) the same species is found in many different isolated places on Earth,
2.) islands have more biodiversity than mainlands,
3.) islands are full of terrestrial animals that would have had trouble crossing over from the mainland,
4.) closely related species are scattered over the Earth, etc.
However, few or no examples of these can be found, whereas there is abundant evidence of the opposite.

Living Earth Essay

Life on Earth Connections

Living Under the Ice

When the temperature of a substance is increased, its molecules vibrate faster and move farther apart on average. The result is that the substance expands. With few exceptions, all phases of matter—solids, liquids, and gases—expand when heated and contract when cooled. In many cases, these changes in volume are not very noticeable, but with careful observation, you can usually detect them. Telephone wires, for instance, are longer and sag more on a hot summer day than on a cold winter day. Metal lids on glass jars can often be loosened by heating them under hot water. If one part of a piece of glass is heated or cooled more rapidly than adjacent parts, the resulting expansion or contraction may break the glass.

Within any given phase, water also expands with increasing temperature and contracts with decreasing temperature. This is true of all three phases—ice, liquid water, and water vapor. Liquid water at near-freezing temperatures, however, is an exception.

Liquid water at 0°C can flow just like any other liquid, but at 0°C, the temperature of this liquid water is cold enough that nanoscopic ice crystals can form. These crystals slightly "bloat" the liquid water's volume, as shown in **Figure A**. As the temperature increases to above 0°C, more and more of these crystals collapse, and as a result, the volume of the liquid water decreases.

Figure B shows that liquid water *contracts* between 0°C and 4°C as its temperature is raised. This contraction, however, continues only up to 4°C. As near-freezing liquid water is heated, there is a simultaneous tendency for the water to expand due to greater molecular motion. Between 0°C and 4°C, the decrease in volume caused by collapsing ice crystals is greater than the increase in volume caused by the faster-moving molecules. As a result, the water volume continues to decrease. At temperatures just above 4°C, expansion overrides contraction because most of the ice crystals have collapsed.

Figure A
Within a few degrees of 0°C, liquid water contains ice crystals. The open structure of these crystals makes the volume of the water slightly greater than it would be without the crystals.

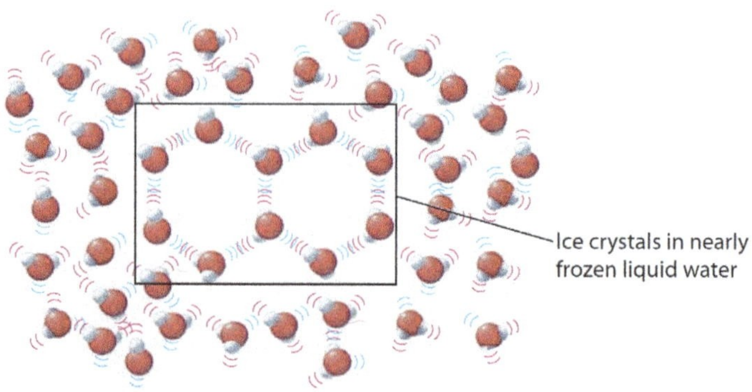

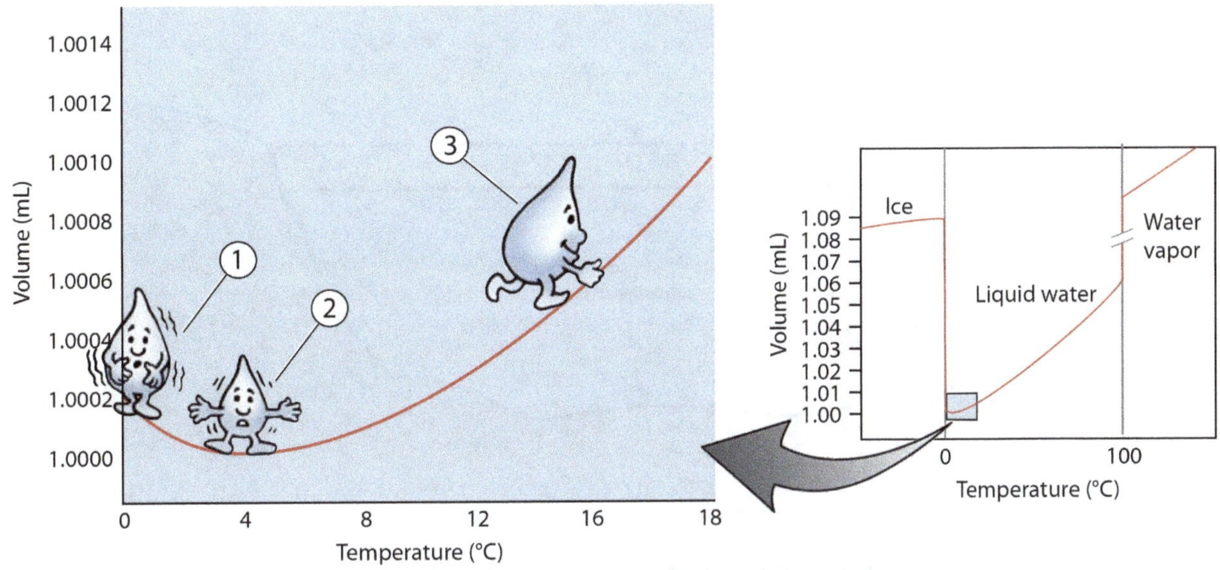

(1) Liquid water below 4°C is bloated with ice crystals.

(2) Upon warming, the crystals collapse, resulting in a smaller volume for the liquid water.

(3) Above 4°C, liquid water expands as it is heated because of greater molecular motion.

Figure B
Between 0°C and 4°C, the volume of liquid water decreases as the temperature increases. Above 4°C, water behaves the way all other substances do: its volume increases as its temperature increases. The volumes shown here are for a 1-gram sample.

Because of the effect of collapsing nanoscopic ice crystals, liquid water has its smallest volume and thus its greatest density at 4°C. By definition, 1 gram of pure water at this temperature has a volume of 1.0000 milliliter. As Figure B shows, 1 gram of liquid water at 0°C has only a slightly larger volume, 1.0002 milliliters. By comparison, 1 gram of ice at 0°C has a volume of 1.0870 milliliters. As can be seen in the small graph to the upper right in Figure B, the volume of 1 gram of ice stays above 1.08 milliliters even below 0°C, meaning that even when ice is cooled to temperatures well below freezing, it is still less dense than liquid water—ice will always nicely float in liquid water, even when that water is boiling!

READING CHECK

What forms in 0°C liquid water?

Water Freezes Top Down

Although liquid water at 4°C is only slightly denser than liquid water at 0°C, this small difference is of great importance in nature. Consider that if water were densest at its freezing point, as is true of most other liquids, the coldest water in a pond would settle to the bottom, and the pond would freeze from the bottom up, destroying living organisms in winter months.

Fortunately, this does not happen. As winter comes on and the temperature of the water drops, its density also drops. But the entire volume of water in the pond does not cool all at once. Instead, surface water cools first because it is in direct contact with the cold air. Being cooler than the underlying water, this surface water is denser and so sinks, with warmer water rising to replace it. That new batch of surface water then cools to the air temperature, gets denser as it does so, and sinks, only to be replaced by warmer water that cools.

This process continues until the entire body of water has been cooled to 4°C. Then if the air temperature remains below 4°C, the surface water also cools to below 4°C. This colder surface water does not sink. Why not? Because it is less dense than the 4°C water beneath it. Thus, cooler, less dense water stays on the surface, where it can cool further, eventually reach 0°C, and turn to ice. While this ice formation develops at the surface, the organisms that require a liquid environment happily swim below the ice in liquid water at a "warm" 4°C, as **Figure C** illustrates.

An important effect of the vertical movement of water is the creation of vertical currents that, to the benefit of organisms living in the water, transport oxygen-rich surface

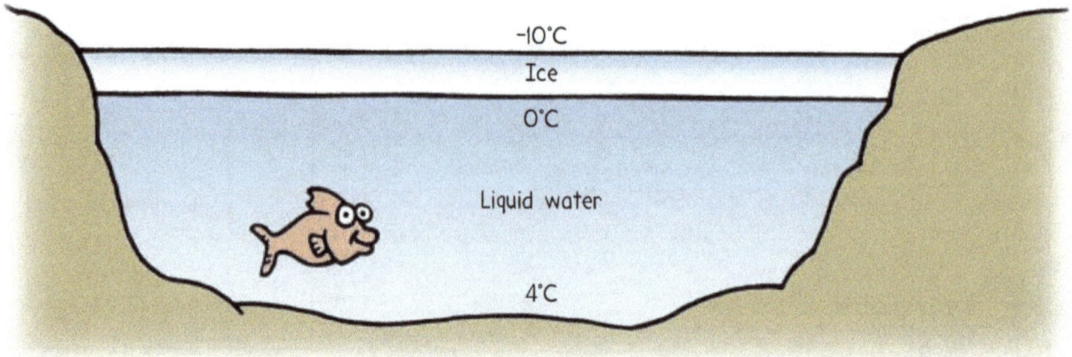

Figure C
As water cools to 4°C, it sinks. Then as water at the surface is cooled to below 4°C, it floats on top and can freeze. Only after surface ice forms can temperatures lower than 4°C extend down into the pond. This does not happen very readily, however, because surface ice insulates the liquid water from the cold air.

water to the bottom and nutrient-rich bottom water to the surface. Marine biologists refer to this vertical cycling of water and nutrients as *upwelling*.

Very deep bodies of fresh water are not ice-covered even in the coldest of winters. This is because, as noted previously, all the water must be cooled to 4°C before the surface water temperature can drop below 4°C. For deep water, the winter is not long enough for this to occur.

Consider that if only some of the water in a pond or lake is 4°C, this water lies on the bottom. Because of water's ability to resist changes in temperature and its poor ability to conduct heat, the bottom of deep bodies of water in cold regions is a constant 4°C year-round, which works to the advantage of the many living organisms living there.

To Life!

CONCEPT **CHECK**
What was the precise temperature at the bottom of Lake Michigan on New Year's Eve in 1901?

CHECK YOUR ANSWER
If a body of water has 4°C water, then the temperature at the bottom of that body of water is 4°C, for the same reason that rocks are at the bottom. Both 4°C water and rocks are denser than water at any other temperature. If the body of water is deep and in a region of short summers, as is the case for Lake Michigan, the water at the bottom is 4°C year-round.

More to Explore
From NASA and the National Snow & Ice Data Center, learn about the diminishing sea ice covering of the Arctic Ocean:

https://nsidc.org/arcticseaicenews/

Chapter 10

Diversity of Life 1

THE MAIN IDEA

Life on Earth can be classified into three domains: Bacteria, Archaea, and Eukarya.

10.1 Classifying Life
10.2 Evolutionary Trees
10.3 Three Domains of Life
10.4 Bacteria
10.5 Archaea
10.6 Protists
10.7 Plants

There is an amazing diversity of life on Earth. Yet, all known organisms are related—some more closely and others more distantly. These relationships form the basis of the Linnaean classification system, which segregates living organisms by domain, kingdom, phylum, class, order, family, genus, and species. Where do humans fall within this system? And how are we both different and related to plants and bacteria?

10.1 Classifying Life

Imagine having 1.5 million books. If you want to identify the different types of books in your library, you need some way to organize them. You'll likely want each book classified based on such things as the content, the author, and the date of publication. Consider that there are some 1.5 million known species in biology and many more yet to be identified. To help us identify the many different kinds of species and how they might be related, we can likewise organize these species using a classification system.

The most familiar biological classification system used today began with the Swedish naturalist Carolus Linnaeus in the 18th century. The Linnaean system divides species into groups based on their physical similarities. Under the Linnaean system, all living things are divided into different domains. Each domain is divided into different kingdoms. Each kingdom is divided into different phyla (plural for "phylum"). Each phylum is divided into different classes. Each class is divided into different orders. Each order is divided into different families. Each family is divided into different genera (plural for "genus"). Finally, each genus includes one or more different species.

Figure 10.1 shows the classification of humans. Humans are in the domain Eukarya, which includes all eukaryotes. This domain includes animals, plants, fungi, and many other living things.

Humans are in the kingdom Animalia, which includes all animals. We are in the phylum Chordata, which includes vertebrates and a few other organisms. We are in the class Mammalia, which includes all mammals. We are in the order Primates, along with monkeys, apes, lemurs, and other species. We are in the family Hominidae, which includes humans and many

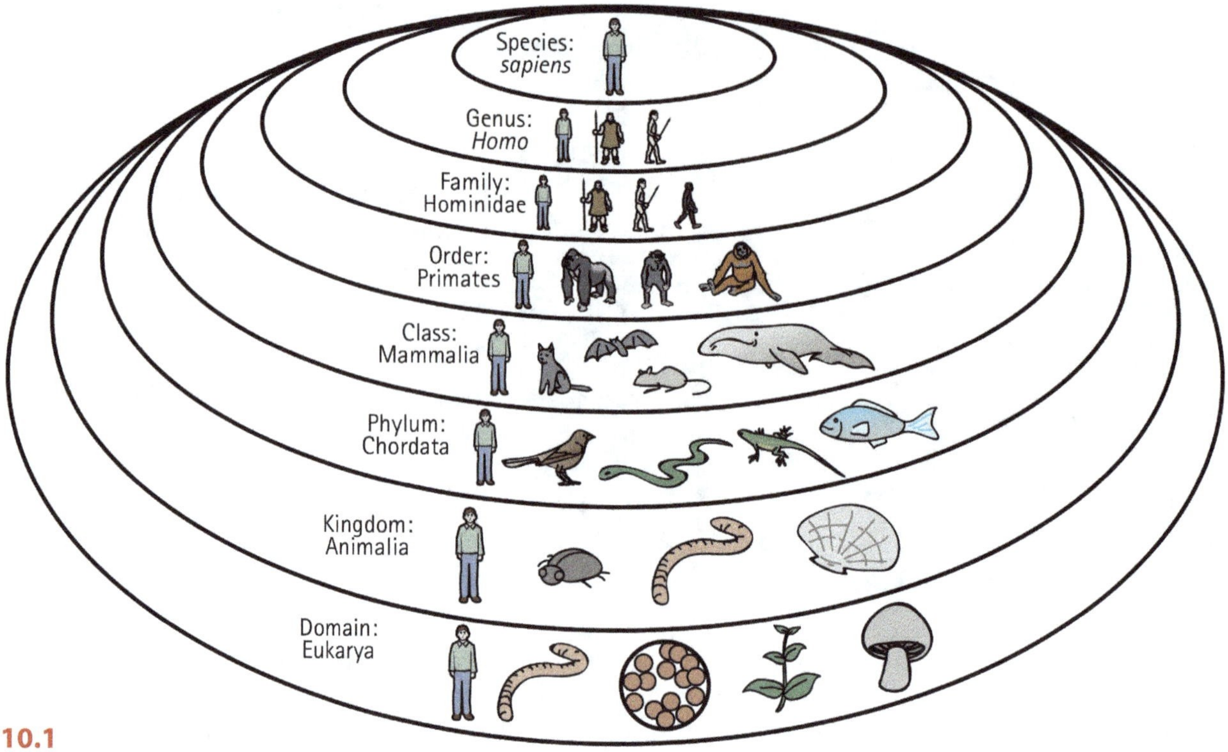

^ Figure 10.1

In the Linnaean classification system, humans belong to the domain Eukarya (eukaryotes), kingdom Animalia (animals), phylum Chordata (chordates), class Mammalia (mammals), order Primates (monkeys and apes), family Hominidae (humans and some of our extinct relatives), genus Homo, and species sapiens.

species of extinct human relatives. We are in the genus *Homo*, which includes humans and our closest fossil relatives, such as the Neanderthals, who went extinct about 30,000 years ago. Finally, our species is *sapiens*.

Do the levels of Linnaean classification seem hard to remember? Mnemonics are little sayings that help people remember things. You can use this mnemonic to remember the levels of classification in the Linnaean system: In his domain, *King Phillip* called the class to *order*—the *family genius* will now speak. In addition to this classification system, Linnaeus also devised rules about how species are named. All species have a two-part scientific name made up of their genus name and species name. For example, humans are *Homo sapiens* ("wise human"), and dogs are *Canis familiaris* ("intimate dog"). Genus and species names are Latin; by convention, the names are in italics with the genus name capitalized. Sometimes the genus name is abbreviated as a single letter, as in *H. sapiens* or *C. familiaris*.

READING CHECK

What family do we humans belong to in the Linnean classification system?

CONCEPT CHECK

The scientific name of the endangered orangutan of Sumatra is Pongo abelii. What is its genus name? Its species name?

CHECK YOUR ANSWER

The genus name is Pongo; the species name is If abelii.

More to Explore

For more about Linnaean classification:

https://www.britannica.com/science/taxonomy/The-objectives-of-biological-classification

10.2 Evolutionary Trees

Since Linnaeus's time, science has expanded our understanding of the history of life on Earth. Darwin's theory of evolution showed that the wealth of species on Earth is the result of numerous instances of speciation followed by the independent evolution of new lineages. While maintaining the framework of the Linnaean classification system, biologists now classify living things based on this evolutionary history. Specifically, they group species based on how closely related they are to one another rather than on how physically "similar" they are. How does this work?

Biologists use a diagram called an **evolutionary tree** to show the evolutionary relationships among a group of species. The tree of all living things is sometimes called the "tree of life." A smaller and simpler evolutionary tree is shown in **Figure 10.2**. This tree includes three species—daisies, elephants, and humans. Notice the dot labeled Speciation Event 1. Up until the time specified by this dot, daisies, elephants, and humans had a shared history. Then a speciation event occurred, producing two separate lineages. A *lineage* is a line of descendants from an ancestor. One of the lineages eventually gave rise to daisies (and many other species, which are not shown).

The other lineage eventually gave rise to humans and elephants (as well as other species not shown). At a later point in time, Speciation Event 2 caused the lineage that led to elephants to split from the lineage that led to humans. This evolutionary tree tells us that because humans and elephants split from each other more recently than they split from daisies, humans and elephants are more closely related to each other than daisies.

READING CHECK

Why are humans more closely related to elephants than daisies according to Figure 10.2

So, humans and elephants should be classified together, and daisies should be classified separately. This is precisely what we do when we say humans and elephants are *animals*, but daisies are *plants*.

Many traditional biological groups already satisfy these criteria. They already divide living things in a way consistent with their evolutionary relationships. Familiar groups such as animals, plants, fungi, mammals, and amphibians remained the same when the classification was changed to incorporate evolutionary relationships. However, other traditional biological groups have required some rethinking.

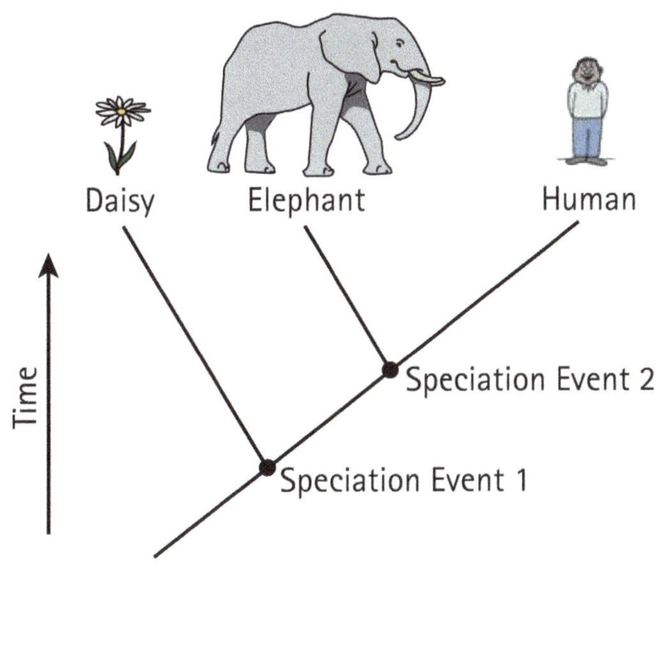

< Figure 10.2

An evolutionary tree shows how species or other biological groups are related. For example, this evolutionary tree for humans, elephants, and daisies shows that humans and elephants are more closely related to each other than either is related to daisies.

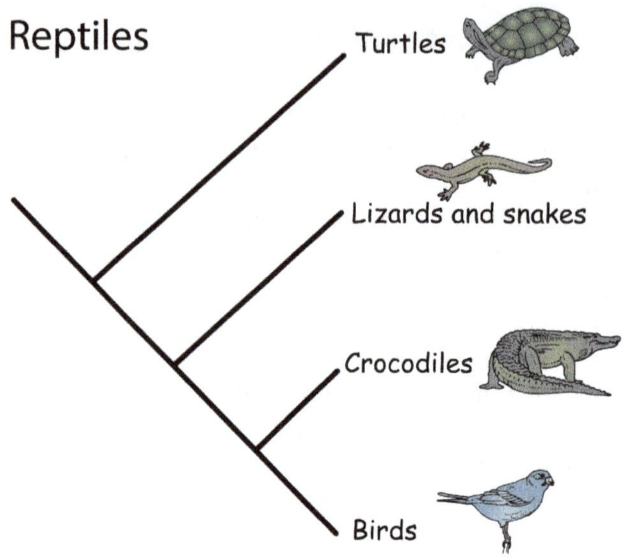

Figure 10.3 >

This evolutionary tree shows that birds are nested within the reptile tree. This means that they must be considered reptiles also.

One such group is the reptiles. The reptiles traditionally included turtles, lizards, snakes, and crocodiles. Birds were considered a separate group, not part of the reptiles. Turtles, lizards, snakes, and crocodiles were grouped together based on shared features such as "cold-bloodedness" and the possession of scales. Birds were placed in a different group because they are "warm-blooded" and have feathers. However, the evolutionary tree of the reptiles in **Figure 10.3** shows that birds actually represent one branch within the reptiles. This means that birds must be considered reptiles too. As a matter of fact, birds aren't just reptiles—they're dinosaurs! This is because birds are nested within the dinosaurs also, which makes them dinosaurs too. This means that dinosaurs didn't *all* go extinct: Birds survived and are certainly alive and well today.

But how do scientists construct the evolutionary trees that classification is based on? Biologists use fossils as well as information about the anatomy, behavior, and genetics of existing species to reconstruct evolutionary relationships. DNA sequences have proved to be a particularly valuable source of information.

It is also important to remember that biological classification is always a work in progress. As biologists learn more about the evolution of species, they sometimes need to draw new evolutionary trees. Classification then changes to reflect this new understanding of species relationships.

More to Explore

Evolutionary trees and how they related to classification

https://evolution.berkeley.edu/evo101/IIDClassification.shtml

CONCEPT**CHECK**

1. Why are birds reptiles? Explain your answer.
2. Which of the following are the closest relatives of birds—turtles, lizards, or crocodiles? (Hint: Look at the evolutionary tree in Figure 10.3)

CHECK YOUR ANSWERS

1. Birds are reptiles because they are nested within the reptile tree. A classification based on evolutionary relationships would have to include birds among the reptiles.
2. Based on the evolutionary tree of reptiles, birds are most closely related to crocodiles.

10.3 Three Domains of Life

Life can be classified into three domains: Bacteria, Archaea, and Eukarya. Early in the history of life—probably 2.5 to 3.5 billion years ago—living organisms split into two separate lineages: one that produced Bacteria and one that produced Archaea and Eukarya (**Figure 10.4**).

Bacteria and Archaea are prokaryotic organisms whose cells do not have a nucleus. **Eukarya** includes all eukaryotic organisms, living things whose cells have a nucleus. The domain Eukarya is further divided into four kingdoms: Protists, Plants, Fungi, and Animals.

Table 10.1 shows some key characteristics of each group, including whether species are prokaryotic or eukaryotic, single-celled or multicellular, and how species obtain nutrition. **Heterotrophs** obtain energy and organic molecules from other living things, as humans and other animals do today. **Autotrophs** convert inorganic molecules into food and organic molecules. Plants are examples of autotrophs.

The Protist kingdom is problematic because it includes all eukaryotes that aren't plants, animals, or fungi—that is, it's a hodgepodge of species that represent many separate lineages. For example, Amoebas, kelp, and diatoms are all protists, although they have little in common besides the fact that they are eukaryotes. Until a more accurate classification emerges, however, we are stuck with the group *Protists*.

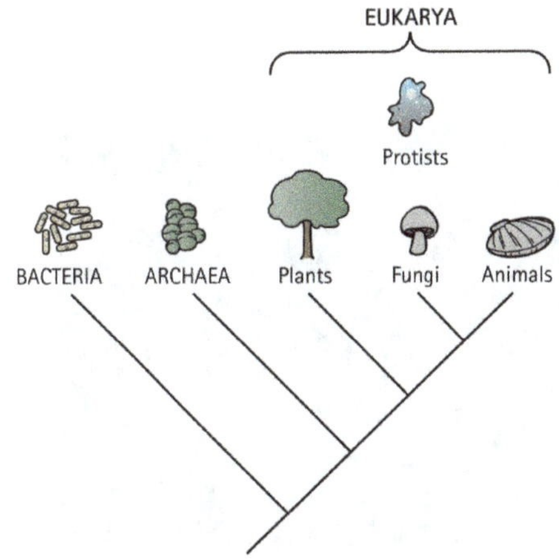

∧ Figure 10.4

The three domains of life are Bacteria, Archaea, and Eukarya. Eukarya includes Plants, Fungi, Animals, and Protists. Protists are not shown in the tree because the group includes many separate lineages and does not have a clear place to go. (Protists would have to be drawn as multiple "twigs" in various spots within Eukarya.)

Table 10.1 key characteristics of the major groups of living things.

Group	Type of Cells	Single-Celled or Multicellular	Way of Obtaining Nutrition
Bacteria	Prokaryotic	Mostly single-celled	Heterotrophs or autotrophs
Archaea	Prokaryotic	Single-celled	Heterotrophs or autotrophs
Eukarya			
Protists	Eukaryotic	Single-celled or multicellular	Heterotrophs or autotrophs
Plants	Eukaryotic	Multicellular	Autotrophs
Fungi	Eukaryotic	Single-celled or multicellular	Heterotrophs
Animals	Eukaryotic	Multicellular	Heterotrophs

READING CHECK

Name the four kingdoms included in the domain Eukarya.

CONCEPT **CHECK**

Are humans more closely related to bread mold or cherries?

CHECK YOUR ANSWER

Humans are animals, bread mold is a fungus, and cherries are plants. In the evolutionary tree shown in Figure 10.4, we see that animals and fungi are more closely related to each other than either is to plants. So, we are more closely related to bread mold!

More to Explore

For more on the three domains of life, check out the website here:
https://ucmp.berkeley.edu/alllife/threedomains.html

10.4 Bacteria

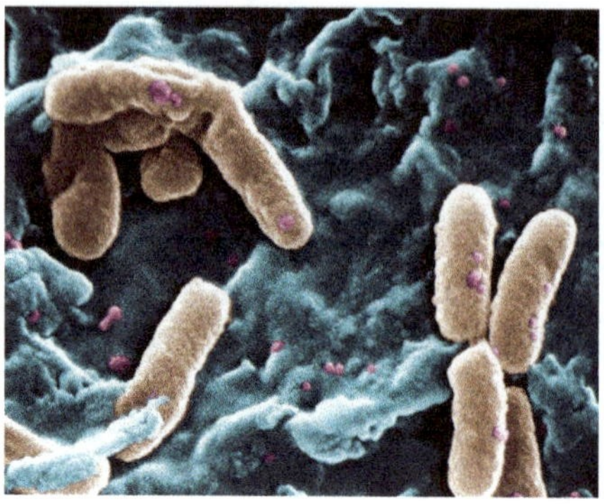

^ Figure 10.5

Bacteria can live in habitats where no other organisms can survive. Some were found in a 120,000-year-old ice core 3000 meters beneath a glacier in Greenland. They live in a habitat with subzero temperatures, high pressure, and very little oxygen or nutrients.

They live on your body by the trillions, occupy habitats where no other organisms can survive (**Figure 10.5**), and devastate human populations with diseases such as plague and tuberculosis. Yet life on Earth would quickly end without them. *They* are bacteria, one of the most ancient lineages on Earth. Earth's oldest fossils, 3.5 billion years old, are of bacteria.

Bacteria are prokaryotes so diverse that it is hard to make generalizations about them. Some bacteria are autotrophs that make their own food through photosynthesis, like plants. Others are chemoautotrophs that make food using chemical energy rather than energy from sunlight. Still, others are heterotrophs that obtain food from organic matter. Heterotrophic bacteria are so diverse in what they eat that just about any organic molecule is food for some species of bacteria. Most bacteria are single-celled, but others gather in multicellular clusters. Bacteria come in varied shapes, including spheres, rods, and spirals. Many can move by using whip-like structures called *flagella.* Most bacteria are microscopic. However, a few are actually big enough to see with the naked eye! The largest bacteria known are *Thiomargarita namibiensis*, whose name means "sulfur pearls of Namibia." (Discovered in Namibia, they grow in long strands like strings of pearls.) These giant bacteria are about the size of the period at the end of this sentence.

Bacteria typically reproduce asexually by dividing. However, most species exchange genetic material at least occasionally—when they take up small pieces of naked DNA from the environment when bacterial viruses inadvertently transfer DNA between organisms, or when two bacteria join together, and one passes DNA to the other. Under favorable conditions, bacteria can divide very quickly, as often as every 20 minutes.

READING CHECK

How quickly can bacteria divide?

Bacterial populations grow rapidly when food is plentiful, and conditions are good. In poor conditions, many bacteria form spores—tough, thick-walled structures that stay dormant until conditions improve.

Life on Earth would be impossible without bacteria because they play an important role in *decomposition*, breaking down organic matter. Without bacterial decomposition, carbon would stay trapped in dead organic matter, all the carbon dioxide in the atmosphere would eventually be used up, and photosynthesis would stop. Bacteria also help cycle other nutrients on Earth; some bacteria fix nitrogen, transforming it from its inorganic atmospheric form to varieties that living organisms can use.

Countless bacteria live in and on our bodies, particularly on the skin, mouth, respiratory tract, and intestines. A few of these are potentially harmful, but others benefit us by producing vitamins and keeping more dangerous bacteria from invading our bodies. Bacteria are used to make foods such as cheese and yogurt, and some genetically engineered strains produce human insulin and other medically important molecules. Of course, other bacteria cause diseases, including tuberculosis, syphilis, and Lyme disease. The development of *antibiotics*, which kill bacteria, was a huge step forward in medicine.

CONCEPT **CHECK**

Some people get yeast (fungal) infections after taking antibiotics. Why?

CHECK YOUR ANSWER

Antibiotics kill "bad" bacteria as well as the normal "good" bacteria that live in our bodies. The "good" bacteria help keep yeast in check. With the good bacteria out of the way, yeast has a chance to grow.

More to Explore

You can learn more about bacteria here:

https://www.livescience.com/51641-bacteria.html

If you'd like to see photos of many different kinds of bacteria, check out this website here:

https://www.atsu.edu/faculty/chamberlain/Website/gallery.htm

One biologist finds bacteria so beautiful, he grows them in unique ways to create colorful art:

https://ideas.ted.com/gallery-the-most-beautiful-bacteria-youll-ever-see/

10.5 Archaea

Once considered a group of funny-looking bacteria, **archaea** ("OUR-kee-uh") are now recognized as a distinct domain of prokaryotic organisms more closely related to eukaryotes than to bacteria. Some features of archaean genetics in, particular link, archaea to eukaryotes—their ribosomes are like those of eukaryotes, their genes contain introns like those of eukaryotes, and their DNA is associated with histone proteins, like that of eukaryotes.

Many archaea are adapted to extreme environments, such as very salty ponds or the scalding waters of hot springs and hydrothermal vents (**Figure 10.6**). These archaea are called "extremophiles"—lovers of the extreme. Biologists are interested in extremophiles because they live in conditions that are similar, in some ways, to those found on the young Earth. Because of this, extremophiles may provide clues about what the earliest living organisms were like. For example, certain archaea thrive in hydrothermal vent habitats deep in the oceans, where life may have first evolved. These archaea obtain energy from an abundant chemical, hydrogen sulfide, and form the basis of remarkable vent communities entirely independent of sunlight.

Not all archaea are extremophiles, though—many live in more familiar places. For example, some are found in the open ocean, and others live in the digestive tracts of termites, cows, and other herbivores. These archaea help their hosts digest plant material, and they release methane as a waste product. In fact, because methane is a powerful greenhouse gas, raising large numbers of cows, sheep, and other livestock contributes significantly to global warming.

READING CHECK

Are all archea extremophiles?

Figure 10.6 >

Large colonies of extremophile archaea—the orange and yellow layers—live in the scalding waters of this Nevada geyser.

CONCEPT **CHECK**

Why are some archaea described as "extremophiles"?

CHECK YOUR ANSWER

They live in extreme environments, such as very salty or very hot habitats.

More to Explore

More about Archaea here:

https://ucmp.berkeley.edu/archaea/archaea.html

10.6 Protists

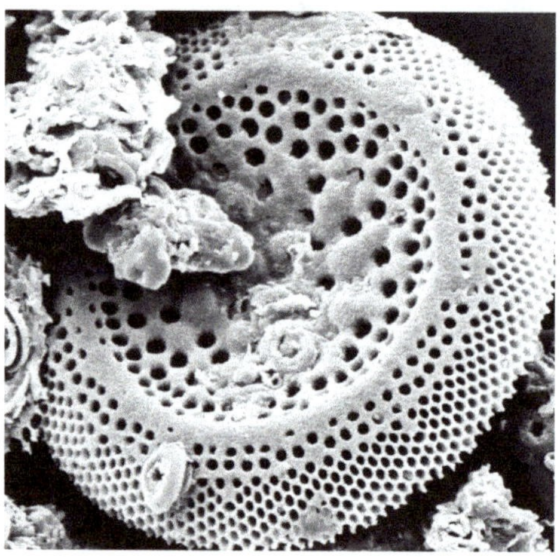

^ Figure 10.7

This microscopic view of a diatom shows its silica shell.

Eukaryotes that are not plants, animals, or fungi are lumped together in a group called **Protists**. This group includes autotrophs, heterotrophs, and even species that use both strategies to obtain nutrition. Protists may be single-celled or multicellular. Certain protists, the slime molds, are actually somewhere between single-celled and multicellular—they go from one condition to the other during the course of their lives. Many protists reproduce asexually, but others use sexual reproduction. Because Protists currently includes many separate lineages, some biologists are in the process of splitting it into separate groups.

Many protists are autotrophs that get their food from photosynthesis. *Diatoms* are single-celled protists that float in the open ocean. They perform most oceanic photosynthesis and are critical to many marine food chains. Diatoms have elaborate shells made of silica (**Figure 10.7**). These shells are sometimes used in human-made products—for example, they provide the gritty texture of some toothpastes. *Dinoflagellates* are another group of single-celled marine protists. Some dinoflagellates are autotrophs, and others are heterotrophs.

When sunlight and nutrients are plentiful, dinoflagellate populations can explode, producing "red tides." The discoloration that gives red tides their name is caused by the huge number of dinoflagellates in the water. Red tides aren't always red, though. Instead, the water may be pink, purple, green, orange, brown, or blue, depending on the dinoflagellate that is responsible. In addition, some red tides are toxic; shellfish that eat the dinoflagellates become contaminated and poisonous to humans.

READING CHECK

What group of single-celled marine protists produces red tides?

Some photosynthetic protists are multicellular and can grow quite large. For example, all the different kinds of seaweeds are protists. Kelp forms huge oceanic forests that are home to many unique species (**Figure 10.8**). Red algae are the source of some of the seaweed we eat, including Japanese nori. Green algae are a group of multicellular protists that likely gave rise to terrestrial plants.

Heterotrophic protists are typically active, single-celled hunters with special cell vacuoles for digesting prey. *Amoebas* move by extending part of their body forward and then pulling the rest of the body behind (**Figure 10.9**). The extensions are called pseudopodia ("false feet"). Amoebas surround and engulf their prey.

Ciliates move by beating numerous hairlike projections called cilia. *Flagellates* move by whipping a single long flagellum. Both ciliates and flagellates have openings that function as "mouths." One group of flagellates called the *choanoflagellates*, probably gave rise to animals.

Protists cause a number of serious human diseases, including malaria, African sleeping sickness, and amoebic dysentery. Malaria is caused by *Plasmodium* protists that divide their life cycle between mosquitoes and humans. Humans contract the disease when infected mosquitoes bite them. The protists then move into human red blood cells (**Figure 10.10**) and reproduce in huge numbers. The synchronized emergence of protists from host red blood cells causes chills, fever, and vomiting.

^ Figure 10.8

Kelp forms marine "forests" that are home to diverse species. This kelp forest is off the West Coast of the United States.

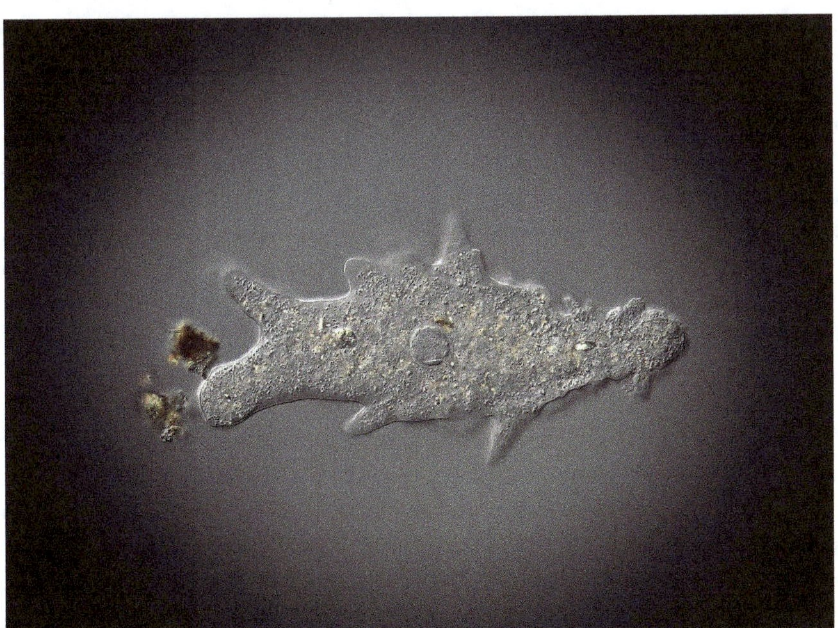

< Figure 10.9

This freshwater amoeba (left) feeds on bacteria and smaller protists. It uses extensions of its body to move as well as to engulf food.

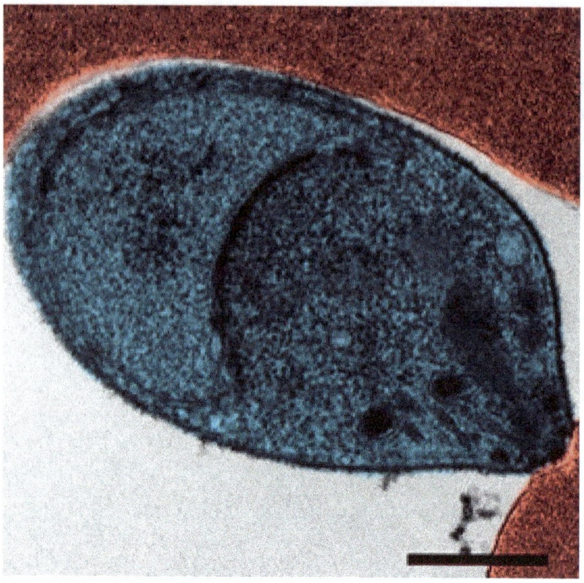

< **Figure 10.10**

The malaria-causing protist Plasmodium (shown in blue) latching onto a red blood cell (shown in red).

CONCEPT CHECK

Why is the group of Protists currently being divided into separate groups?

CHECK YOUR ANSWER

Currently, the group Protists basically consists of all eukaryotes that aren't plants, fungi, or animals. "Protists" include many separate lineages that are lumped together for no good reason.

More to Explore

More about protists, and see more photos:

https://www.livescience.com/54242-protists.html

10.7 Plants

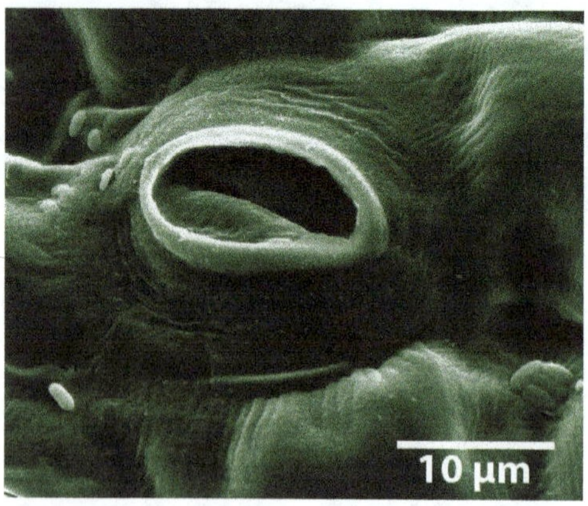

∧ **Figure 10.11**

Stomata are tiny pores in plant leaves that allow carbon dioxide to enter. You can see several stomata in this magnified image of a rose plant leaf.

Photos of Earth from space show large green patches stretching across vast areas of the continents. Much of Earth's land surface is green because it is covered with plants. **Plants** are terrestrial, multicellular, autotrophic eukaryotes that obtain energy through photosynthesis. Plants are green because they contain chlorophyll, a pigment used in photosynthesis.

Plants have a variety of adaptations for living in terrestrial environments. Roots anchor them to the ground and absorb water and nutrients from the soil. Shoots, the stems, and leaves of a plant conduct photosynthesis. The leaves of plants have a large surface area for catching sunlight, which powers photosynthesis. Carbon dioxide, which is also needed for photosynthesis, diffuses from the air into leaves through tiny pores called *stomata* (**Figure 10.11**).

Most plants also have a *vascular system*, a "circulatory system" that distributes water and other resources. The plant vascular system consists of two types of tissue: xylem and phloem. The *xylem* is made up of dead, tube-shaped cells through which water and nutrients move up from the roots. The *phloem* consists of living cells that pass the sugars produced during photosynthesis down from the leaves. The liquid that flows in a plant's vascular system is known as *sap*. One sap you may be familiar with is maple syrup, which is made by boiling down liquid collected from the vascular systems of maple trees.

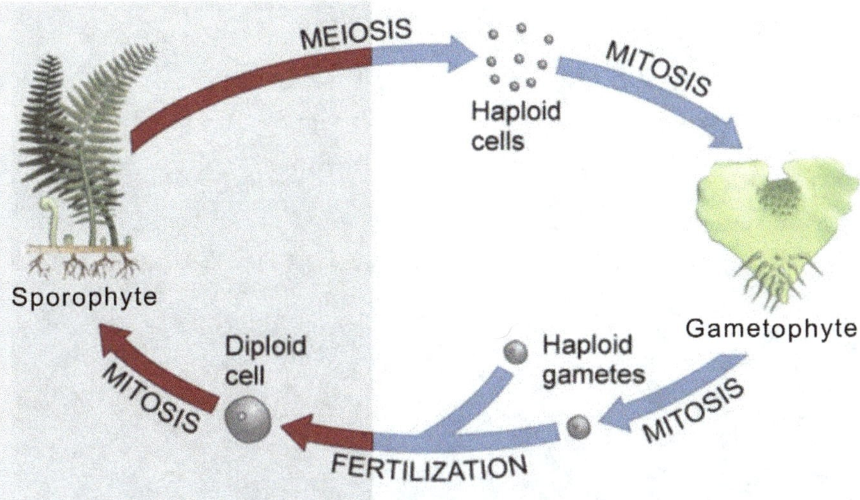

< Figure 10.12

The plant life cycle is characterized by the "alternation of generations." Plants alternate between a haploid gametophyte (shown on the right) and a diploid sporophyte (shown on the left). Different groups of plants differ in whether the gametophye or sporophyte dominates the life cycle.

Plant reproduction occurs through an *alternation of generations*, in which the life cycle alternates between a haploid stage called a *gametophyte* and a diploid stage called a *sporophyte*. You may remember that haploid cells contain a single set of chromosomes, whereas diploid cells contain two sets of chromosomes. The details of this life cycle vary among the three major groups of plants—mosses, ferns, and seed plants. A diagram of the alternation of generations in plants is shown in **Figure 10.12**.

Mosses

Mosses are small plants with simple leaves and no vascular system. They often grow in dense clumps in bogs, forests, and other moist, shady habitats (**Figure 10.13**). Mosses are restricted to moist habitats for several reasons. First, because mosses do not have a vascular system, every part of a moss plant must obtain water directly from the environment through diffusion. Second, male and female moss plants are usually separate, and during reproduction, sperm must swim from the male plant through a film of water in the environment to fertilize eggs in the female plant.

Mosses are unique among plants in that the gametophyte is much larger than the sporophyte. When you see moss in the forest, you are looking at a haploid gametophyte.

< Figure 10.13

Mosses growing in the forest understory in Great Smoky Mountains National Park, Tennessee.

Figure 10.14 >

Ferns grow in Muir Woods National Monument in California.

Ferns

Ferns are seedless plants with distinctive feathery leaves. They are often found in the forest understory, where they thrive in the shade of large trees (**Figure 10.14**). Unlike mosses, ferns have a vascular system for transporting water and nutrients.

However, ferns are similar to mosses in that their sperm swim through the environment to fertilize eggs. Because of this, ferns can also live only in moist habitats. In ferns, the diploid sporophyte is much larger than the haploid gametophyte—when you see a fern in the forest, you are looking at a diploid sporophyte.

Seed Plants

Seed plants are the largest group of plants by far. Two key features of seed plant reproduction have made these plants successful in a wide variety of land habitats: pollen and seeds. **Pollen** consists of many tiny grains, each of which is a male gametophyte wrapped in a protective coating. The male gametophyte produces the sperm used in reproduction.

Pollen can be transported to female gametophytes, which produce eggs, by wind or (as we will see) by animals. Because the sperm of seed plants do not have to swim through the environment to fertilize eggs, seed plants are not restricted to moist environments.

Seed plants also make seeds. The fertilized eggs of seed plants grow into small embryos that are encased in a tough outer coating along with a food supply—this entire structure is a **seed**. Seeds can survive in a dormant state until environmental conditions are appropriate for growth. This is why many seeds do not sprout until you plant them in soil and water them. All the seed plants you see are diploid sporophytes.

The two largest groups of seed plants are conifers and flowering plants. *Conifers* include plants such as redwoods, pines, cedars, and firs. Conifers have waxy, needle-like leaves and reproductive structures called *cones* (**Figure 10.15**). Male cones release pollen, and then the wind carries the pollen to female cones. Note that most conifers have both male and female cones. Because the wind blows pollen all over the place, conifers make large amounts of pollen—this makes it more likely that some of the pollen will reach female cones. As a result, fertilization occurs in the female cones, which eventually drop the mature seeds.

< Figure 10.15

Conifers are seed plants with reproductive structures called cones. (a) This is a pine branch showing male and female cones. (b) This is a cone and seeds from a Sequoia redwood.

Flowering plants are the largest and most successful group of seed plants. Flowering plants have two important features absent in conifers: flowers and fruit. A **flower** functions in reproduction. Flowers may contain the male structures that produce pollen, the female structures that produce eggs, or both (**Figure 10.16**). The *stamen* is the male reproductive structure. It consists of a stalk capped with an *anther* where pollen develops. The *carpel* is the female reproductive structure. It includes an *ovary* where eggs develop and a stalk capped by the *stigma*, a sticky structure that traps pollen. As in conifers, a single flowering plant usually has both male and female reproductive structures, whether in separate flowers or the same flowers. However, there are a small number of plants that only have male reproductive structures, or only female reproductive structures.

In many flowering plants, insects or other animals transport pollen from one flower to another. The petals, scent, and nectar of many flowers have evolved to attract specific animal pollinators. In fact, flowers and their pollinators can be so perfectly suited to each other that you can often look at one and predict the features of the other. For example, after Charles Darwin studied a night-blooming orchid in Madagascar, he predicted that there should exist a nocturnal moth with a tongue 30 centimeters long.

Forty years later, that moth was finally discovered! Not all flowers smell sweet, however. Flowers of the "dead horse arum lily" smell like rotting meat. Who are their pollinators? Flies that lay their eggs in rotting meat. **Figure 10.17** shows more examples of flowers and their animal pollinators.

Flowering plants surround their seeds with a structure called a **fruit**. A fruit is an adaptation for spreading seeds. When an animal eats a fruit, the seeds pass through its digestive system and eventually are deposited far from the parent plant. Tasty fruits evolved in certain plants because animals were more likely to eat them. But not all fruits have evolved to be eaten. For example, the burrs that catch on your socks during a hike are also fruits. These fruits hitch a ride until you pull them off and drop them on the ground—again, far from the parent plant.

Seed plants are very important to human societies. Humans use many different trees for wood, and much of our food comes from the roots, stems, leaves, and fruits of flowering plants.

Also, did you know that the most massive living organism on Earth is a seed plant? In Utah, a quaking aspen has more than 47,000 separate tree trunks covering over 100 acres. The aspen is nicknamed "Pando"—Latin for "I spread." Although Pando looks like a forest of separate trees, its trunks are all connected to a single enormous root system—and they all have the same DNA. Pando might be 80,000 years old or more.

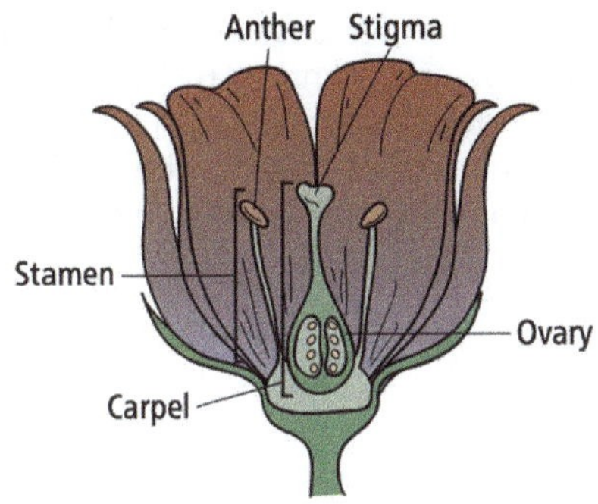

Λ Figure 10.16

Flowers contain the reproductive structures of flowering plants.

READING CHECK

How old is Pando, the most massive living organism on Earth?

(a) (b) (c)

Figure 10.17

Many flowering plants are pollinated by insects or other animals. (a) Flowers pollinated by bees are often blue or yellow, the colors bees see best. Bee-pollinated flowers may also give off a pleasant scent and provide suitable landing spots for the bees. (b) Flowers pollinated by hummingbirds are often red. Hummingbirds have a poor sense of smell, and the flowers they pollinate are usually odorless. Hummingbirds hover while feeding, so hummingbird-pollinated flowers may be trumpet-shaped. (c) Flowers pollinated by bats, which are active at night, tend to be white, which is easy to see at night. Bat-pollinated flowers also have strong odors and grow at the tops of plants for easy access.

How Plant Structures Respond and Work Together

Plants have various systems that work together to respond to external stimuli such as light, temperature, and gravity. These systems include the below-ground root, above-ground shoot, and vascular systems that carry fluids throughout. Each system has a specific structure that facilitates its function and interaction with other systems.

The root system, for example, is responsible for absorbing water and nutrients from the soil. The roots have specialized structures, such as root hairs, which increase the surface area available to collect water and nutrients from the soil. The roots also have root caps that protect the growing tip. Meanwhile, the shoot system, which consists of stems, leaves, flowers, and fruit, is responsible for photosynthesis and the production of plant hormones. The leaves have specialized structures, such as stomata (Figure 10.11), which facilitate gas exchange with the environment.

The vascular system is responsible for transporting water, nutrients, and hormones throughout the plant. It consists of xylem and phloem tissues, which have specialized structures such as vessels and sieve tubes that facilitate the movement of substances.

The interactions among these systems are facilitated by their structures. For example, the vascular system connects the root and shoot systems, transporting water and nutrients from the roots to the leaves. Plant hormones allow communication between different systems, triggering responses such as growth and development. The specialized structures in each system allow for efficient functioning and coordination with other systems.

More to Explore

More about plants, including the different groups of plants:

https://basicbiology.net/plants

CONCEPT CHECK

Fruits are adaptations that help flowering plants spread their seeds. Why is it adaptive for a plant to spread its seeds?

CHECK YOUR ANSWER

If plants are able to spread their seeds to a variety of environments, their offspring have a better chance of encountering environments that are well suited to their survival and reproduction.

Concept Review

Chapter 10

Detailed Chapter Summary

Under Linnaean classification, living species are divided into separate groups based on their characteristics and overall "similarity." The levels of classification are: domain, kingdom, phylum, class, order, family, genus, and species. A species' scientific name consists of its genus name and species name.

However, biologists now aim to classify species based on their evolutionary relationships -- that is, how closely related they are to one another. These relationships are not always easy to work out, but they are not subjective the way "similarity" is. Relationships among species are shown with a diagram called an evolutionary tree.

The three domains of life are Bacteria, Archaea, and Eukarya. Archaea and Eukarya are more closely related to each other than either is to Bacteria. Bacteria and Archaea consist of prokaryotes, whereas Eukarya consists of eukaryotes.

Bacteria represent one of the oldest lineages on Earth. Bacteria are extremely diverse. Some bacteria are photosynthetic autotrophs, some are chemoautotrophs, and some are heterotrophs. Most bacteria are single-celled, but some gather in multicellular clusters. Some bacteria are able to move using flagella. Bacteria reproduce asexually by dividing, and many form hardy spores when conditions are poor. Bacteria play an essential role in decomposition, the breaking down of organic matter. They also help cycle important nutrients such as nitrogen. A large number of bacteria live within our bodies, performing essential functions. Others are used in the production of food. Some bacteria cause important human diseases.

Archaea are a domain of single-celled prokaryotic organisms. Some archaean species are extremophiles that live in unusual, extreme environments such as very salty ponds or hot hydrothermal vents.

Protists are a diverse group that includes all the eukaryotes that are not plants, animals, or fungi. The group Protists includes a variety of species that have been lumped together while scientists continue to work out their evolutionary relationships. It is no surprise, then, that the group includes many different kinds of species. Autotrophic protists include single-celled marine species such as diatoms and (some) dinoflagellates, as well as multicellular seaweeds such as kelp, red algae, and green algae. Heterotrophic protists include single-celled amoebas, ciliates, and flagellates. Slime molds are unusual among living things in that they may move from single-celled to multicellular during their life cycle.

Plants are terrestrial, multicellular, autotrophic eukaryotes. They obtain energy and organic molecules through photosynthesis. Plant adaptations for living on land include roots, shoots, and leaves. Leaves typically have a large surface area for catching sunlight and stomata for obtaining carbon dioxide. Some plants also have a vascular system for transporting materials. The xylem transports water and soil nutrients from the roots and is made of dead cells. The phloem transports sugars down from the leaves and is made of living cells. Plant reproduction occurs through an "alternation of generations" that differs in its details among the main groups of plants. During the alternation of generations, plants alternate between a haploid gametophyte and a diploid sporophyte.

The major groups of plants are mosses, ferns, and seed plants. Mosses are small plants with no vascular systems. In mosses, the gametophyte is much larger than the sporophyte and dominates the life cycle. Moss sperm swim directly through the environment to reach the eggs. Because of this feature and the absence of a vascular system, mosses are restricted to moist environments.

Ferns have a vascular system, but their sperm also swim through the environment to fertilize the eggs. The fern sporophyte is much larger than the gametophyte.

The seed plant life cycle is also dominated by the sporophyte. Seed plants have two key features, pollen and seeds, that have enabled them to live in a wider range of habitats, including drier ones. Pollen grains are male gametophytes that can be carried to female structures during reproduction, making swimming sperm (and moist habitats) less necessary. Seeds are embryonic sporophytes encased in a tough outer coating along with a food supply.

The two main groups of seed plants are conifers and flowering plants. Conifers have reproductive structures called cones, and they are pollinated by wind. Flowering plants are the largest and most diverse group of plants. Their reproductive structures are flowers, which include male parts such as the stamen and anther, and female parts such as the carpel, ovary, and stigma. Insects and other animals are involved in the pollination of many flowering plants. Flowering plants surround their seeds with a structure called a fruit, which functions in spreading seeds.

Key Terms

Archaea One of the three domains of life, consisting of many different kinds of prokaryotic organisms that are more closely related to eukaryotes than bacteria; some archaea are adapted to extreme environments.

Autotrophs Living organisms that convert inorganic molecules into food and organic molecules.

Bacteria One of the three domains of life, consisting of an extremely diverse array of prokaryotic organisms.

Eukarya One of the three domains of life, consisting of eukaryotic organisms whose cells have a nucleus: animals, fungi, plants, and protists.

Evolutionary tree A diagram that shows the relationships among a set of organisms.

Flower The reproductive structure of flowering plants.

Fruit In flowering plants, a structure surrounding the seeds that typically helps to spread the seeds.

Heterotrophs Living organisms that obtain energy and organic molecules from other living organisms or other organic materials.

Plants A group of autotrophic, multicellular, terrestrial eukaryotes that make food through photosynthesis.

Pollen In seed plants, immature male gametophytes wrapped in protective coatings; it ultimately produces sperm in flowering plants.

Protists A miscellaneous group of eukaryotic organisms that includes all the eukaryotes that are not plants, animals, or fungi.

Seed In seed plants, a structure consisting of a plant embryo, a food supply, and a tough outer coating.

Review Questions

Simple Review Questions

10.1 Classifying Life

1. What criteria are used to classify species in the Linnaean classification system?

10.2 Evolutionary Trees

2. How did scientific ideas about evolution change the way biologists classify organisms?
3. What information do scientists use to construct evolutionary trees?

10.3 Three Domains of Life

4. What are the three domains of life?
5. To which domain of life do eukaryotes belong?

10.4 Bacteria

6. Explain how bacteria that are autotrophs and chemoautotrophs obtain food.
7. How do bacteria reproduce? Do bacteria ever exchange genetic material?
8. Why is bacterial decomposition important to life?

10.5 Archaea

9. Which features of archaeans suggest they are more closely related to eukaryotes than to bacteria?
10. Why are some archaea described as “extremophiles”?

10.6 Protists

11. What are protists?
12. What is a diatom, and why are diatoms important to many other marine organisms?
13. Name three kinds of multicellular photosynthetic protists.
14. Heterotrophic protists move to capture prey. Describe some of the different ways these protists move.
15. Describe how *Plasmodium*, the protist that causes malaria, infects humans. Also, describe the symptoms of the disease.

10.7 Plants

16. What are the two components of the plant vascular system? What is the function of each?
17. Why are mosses restricted to living in moist environments?

18. How is pollen transferred from one plant to another in conifers? In flowering plants?

Challenging Review Questions

10.1 Classifying Life

19. If two species belong to the same order, do they have to belong to the same class? Do they have to belong to the same genus?

10.2 Evolutionary Trees

20. Which is more arbitrary: classifying organisms based on their evolutionary relationships or based on their "similarities"? Explain your answer.

21. Is an eagle a reptile? Defend your answer.

10.3 Three Domains of Life

22. How are the three domains of life—Bacteria, Archaea, and Eukarya—related to one another? To which domain do humans belong?

23. Scientists are working on separating the Protists into separate groups. Explain why.

24. What is the difference between a heterotroph and an autotroph? Name a group of living things that consists exclusively of heterotrophs and one that consists exclusively of autotrophs. Name a group that includes both heterotrophs and autotrophs.

25. What is a chemoautotroph? What does a chemoautotroph have in common with a plant? How does a chemoautotroph differ from a plant in how it obtains food?

10.4 Bacteria

26. Why would life on Earth be impossible without bacteria?

27. Bacteria reproduce asexually. However, they sometimes exchange DNA with other bacteria. Explain some of the ways in which this can happen.

28. The photo below shows soil bacteria formed spores. What is the advantage of being able to produce spores, as many bacteria do?

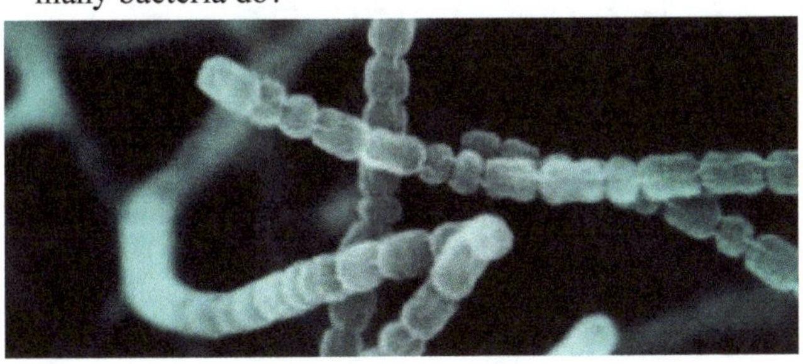

29. What are some ways in which bacteria are helpful in human lives?

10.5 Archaea

30. You are studying some microbes that were collected from the Pacific Ocean. So far, you can tell that the microbes are prokaryotes rather than eukaryotes. How can you tell whether they are bacteria or archaea?

31. How do certain archaea benefit animals such as sheep and cows?

32. What are some extreme environments in which certain archaea thrive?

10.6 Protists

33. We saw that life on Earth would be impossible without bacteria. Would life on Earth be impossible without eukaryotes?

34. Are protists single-celled or multicellular? Use examples to defend your answer.

35. Are protists autotrophs, heterotrophs, or both? Use examples to defend your answer.

36. Which protists likely gave rise to plants? Which protists likely gave rise to animals?

10.7 Plants

37. What are the three major groups of plants? Identify an example of each in the photo below

38. Of the three major plant groups, which is most dependent on living in a moist habitat? Why? Which is least dependent?

39. Some plants, including many grasses, have small green flowers with no petals (see photo). How do you think these flowers are pollinated?

40. Some people are allergic to pollen. Do you think bee-pollinated plants or wind-pollinated plants are more likely to cause allergies? Why?

Apply & Discuss Questions

41. Of the three domains of life, Bacteria and Archaea both consist of prokaryotes, whereas Eukarya consists of eukaryotes. Why can't we lump Bacteria and Archaea together and call them all Bacteria?

42. Bacteria generally reproduce asexually. What are some advantages of asexual reproduction?

43. Most living organisms reproduce sexually sometimes or have some other mechanism for exchanging genetic material. What is the advantage of sexual reproduction or genetic exchange?

44. Do plants have to use energy to obtain the carbon dioxide they need for photosynthesis?

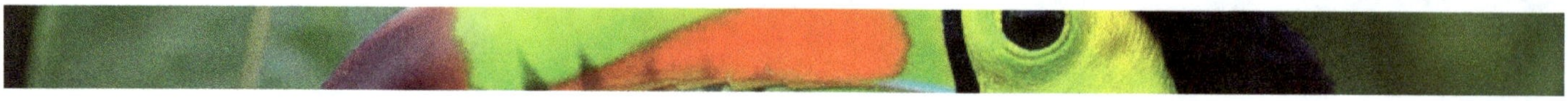

Solutions to Odd-Numbered Questions

Simple Review Solutions

1. The Linnaean system divides species into groups based on their similarities.

3. Biologists use fossils as well as information about the anatomy, behavior, and genetics of existing species to reconstruct evolutionary relationships. DNA sequences have proved to be a particularly valuable source of information.

5. Eukarya includes all eukaryotic organisms, living things whose cells have a nucleus.

7. Bacteria typically reproduce asexually by dividing. However, most species exchange genetic material at least occasionally—when they take up small pieces of naked DNA from the environment, when bacterial viruses inadvertently transfer DNA between organisms, or when two bacteria join together and one passes DNA to the other.

9. Features of archaean genetics in particular link archaea to eukaryotes—their ribosomes are like those of eukaryotes, their genes contain introns like those of eukaryotes, and their DNA is associated with histone proteins, like that of eukaryotes.

11. Eukaryotes that are not plants, animals, or fungi are lumped together in a group called Protists.

13. Some photosynthetic protists are multicellular and can grow quite large. For example, all the different kinds of seaweeds are protists. Kelp forms huge oceanic forests that are home to many unique species. Red algae are the source of some of the seaweed we eat, including Japanese nori. Green algae are a group of multicellular protists that likely gave rise to terrestrial plants.

15. Malaria is caused by *Plasmodium* protists that divide their life cycle between mosquitoes and humans. Humans contract the disease when infected mosquitoes bite them. The protists then move into our red blood cells and reproduce in huge numbers. The synchronized emergence of protists from host red blood cells causes chills, fever, and vomiting.

17. Mosses are restricted to moist habitats for several reasons. First, because mosses do not have a vascular system, every part of a moss plant must obtain water directly from the environment through diffusion. Second, male and female moss plants are usually separate, and during reproduction, sperm must swim from the male plant through a film of water in the environment to fertilize eggs in the female plant.

Challenging Review Solutions

19. Two species that belong to the same order must also belong to the same class. However, these two species do not have to belong to the same genus – this is because an order can contain one or more families, and each family can contain one or more genera. So, two species that are in the same order could be in different genera. For example, in the order Primates, there are many different genera.

21. Birds are reptiles because the evolutionary tree of the reptiles includes the birds as one branch within the rest of the reptile tree. In other words, birds are descended from reptiles, and so they are reptiles too. This means that eagles are reptiles.

23. The Protist kingdom is problematic because it includes all eukaryotes that aren't plants, animals, or fungi—that is, the group is basically a hodgepodge of species that represent many separate lineages. Scientists are working on separating the different lineages into separate groups.

25. A chemoautotroph is an autotroph that makes food and organic molecules using chemical energy. Like plants, chemoautotrophs are autotrophs—they convert inorganic molecules into food and organic molecules. However, plants use energy from sunlight to do this, and chemoautotrophs use chemical energy.

27. Bacteria may exchange DNA when they take up small pieces of naked DNA from the environment, when bacterial viruses inadvertently transfer DNA between organisms, or when two bacteria join together and one passes DNA to the other.

29. Countless bacteria live in and on our bodies, particularly on the skin and in the mouth, respiratory tract, and intestines. A few of these are potentially harmful, but others benefit us by producing vitamins and by keeping more dangerous bacteria from invading our bodies. Bacteria are used to make foods such as cheese and yogurt, and some genetically engineered strains produce human insulin and other medically important molecules. You also learned about bacteria that clean up oil spills and hazardous waste.

31. They live in the digestive tracts of sheep, cows, and other animals. There, the archaea help their hosts digest plant material.

33. Life would go on just fine without eukaryotes. In fact, life did just fine for several billion years before the first eukaryotes evolved.

35. Some protists are autotrophs, and others are heterotrophs. Examples of autotrophic protists include diatoms and kelp. Examples of heterotrophic protists include ciliates and amoebas.

37. The three major groups of plants are mosses, ferns, and seed plants. The trees in the photo are seed plants. Mosses can be seen covering the trunks of the trees and rocks. The plants in the bottom left corner are ferns.

39. Green flowers are not particularly visible, so they likely are not attracting vision-oriented pollinators. In fact, many are wind pollinated.

Apply & Discuss Solutions

41. Because archaea are more closely related to eukaryotes than either is to bacteria, classifying archaea and bacteria together while excluding eukaryotes obscures the evolutionary history of the three groups. In terms of a classification based on evolutionary history, the fact that archaea and bacteria are both prokaryotes is not relevant – only the evolutionary relationships among the three groups matter in constructing biological groups.

43. The advantage of genetic exchange is genetic diversity among the offspring you produce. This way, you don't put all your eggs in one (genetic) basket. Since future environmental conditions are impossible to predict, at least some of your offspring are likely to do well no matter what the environmental conditions turn out to be.

Living Earth Essay

Life on Earth Connections

The Wonder of Wind

If you have ever witnessed a tire as it goes flat, you probably noticed a hissing sound of air escaping. If your hand was in front of the leak, you also felt air rushing out of the tire. Air moves naturally from a region of high pressure (inside the tire) to a region of low pressure (outside the tire). Any wind we feel while outdoors is an example of the same phenomenon.

Consider a region of the atmosphere with a high pressure, which means many air molecules are scrunched together. Consider this high-pressure region being in contact with a lower-pressure region, which means relatively few air molecules are scrunched together. What happens next? Air molecules naturally flow from where they are very scrunched (high pressure) to where they are less scrunched (low pressure). We experience the air moving between those two regions as wind. Understandably, the bigger the difference in air pressure, the stronger and faster the winds.

Fun fact: Winds are named according to the direction from which they blow. Face the wind and consider the direction you are then facing. For example, if you find yourself facing the northwest, then the winds can be described as "northwesterly." Of course, the direction of that northwesterly wind will be to the southeast.

fyi FOR YOUR INFORMATION

Before humans arrived some 1500 years ago, the Hawaiian Islands were well colonized by many plant and animal species. Centered in the middle of the Pacific Ocean and thousands of miles from any mainland, how did these species get there? The common answer is wind, wings, and waves. The wings of birds and the waves of water are inseparable from the action of moving air. The role of wind in spreading life's seeds around the globe cannot be underestimated.

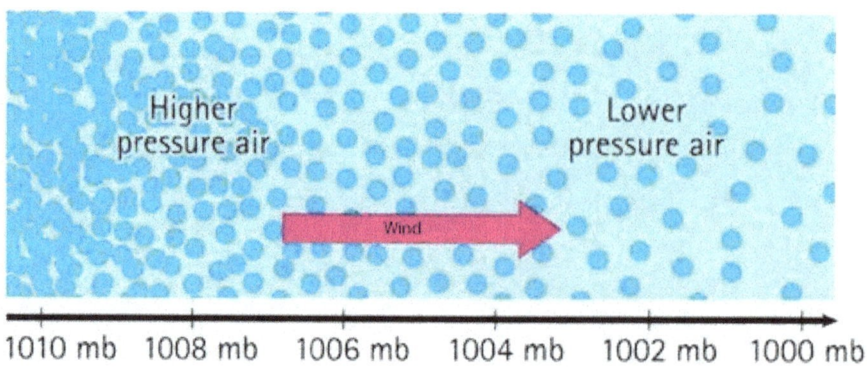

Figure A
Air moves from a region of higher pressure to a region of lower pressure

Wind Chill Chart

Wind (mph) \ Temperature (°F)	40	35	30	25	20	15	10	5	0	-5	-10	-15	-20	-25	-30	-35	-40	-45
5	36	31	25	19	13	7	1	-5	-11	-16	-22	-28	-34	-40	-46	-52	-57	-63
10	34	27	21	15	9	3	-4	-10	-16	-22	-28	-35	-41	-47	-53	-59	-66	-72
15	32	25	19	13	6	0	-7	-13	-19	-26	-32	-39	-45	-51	-58	-64	-71	-77
20	30	24	17	11	4	-2	-9	-15	-22	-29	-35	-42	-48	-55	-61	-68	-74	-81
25	29	23	16	9	3	-4	-11	-17	-24	-31	-37	-44	-51	-58	-64	-71	-78	-84
30	28	22	15	8	1	-5	-12	-19	-26	-33	-39	-46	-53	-60	-67	-73	-80	-87
35	28	21	14	7	0	-7	-14	-21	-27	-34	-41	-48	-55	-62	-69	-76	-82	-89
40	27	20	13	6	-1	-8	-15	-22	-29	-36	-43	-50	-57	-64	-71	-78	-84	-91
45	26	29	12	5	-2	-9	-16	-23	-30	-37	-44	-51	-58	-65	-72	-79	-86	-93
50	26	19	12	4	-3	-10	-17	-24	-31	-38	-45	-52	-60	-67	-74	-81	-88	-95
55	25	18	11	4	-3	-11	-18	-25	-32	-39	-46	-54	-61	-68	-75	-82	-89	-97
60	25	17	10	3	-4	-11	-19	-26	-33	-40	-48	-55	-62	-69	-76	-84	-91	-98

Frostbite Times: 30 minutes; 10 minutes; 5 minutes

Figure B
A wind chill chart. Wind speed and air temperature combine to make the air feel colder to our bodies.

Wind Chill

Warm air heated by our bodies is held in place by our hair and clothes. This layer of warmed air keeps us warm. Any wind, however, will blow this layer of warm air away. In the summer, this wind keeps us cool, and this is a good thing. But if it's cold outside, losing body heat can be dangerous. Extreme cold can bring on frostbite and hypothermia. Death from freezing occurs when the body does not have enough heat to allow the chemical reactions that sustain life.

The National Weather Service publishes a Wind Chill Index. This table shows how air temperature and wind combine to approximate a temperature that your body actually feels, called a "wind chill." As the wind increases for a given temperature, the wind chill decreases. The chart's background color also helps identify how long it might take for a given wind chill to produce frostbite.

FOR YOUR INFORMATION

Wind is a form of solar energy. Modern wind turbines that generate electricity from this energy source have become more numerous across many parts of the world. By 2020, wind power met about 8 percent of the electricity needs of the United States, surpassing hydroelectric generation. By 2030 this is projected to exceed 20 percent. Presently, about 85,000 people are employed by the U.S. wind power industry and related fields. To learn more about this clean power, visit CleanPower.org.

A World of Wind

There are two different types of winds: local and global. Local winds arise from pressure differences over a small region. Local winds are also markedly affected by local conditions such as the terrain, nearby bodies of water, and humidity.

On a planetary scale we have global winds, which occur because of unequal heating of Earth at the equator and poles. At equatorial latitudes, Earth absorbs the Sun's rays directly. Air is warmed efficiently here, and it rises leaving behind a low-pressure region. The air that rises flows toward the poles until it sinks back to Earth upon becoming too cold and dense to stay aloft. The result is a series of six convection cells as shown in **Figure C**.

READING CHECK

Why does the air close to the equator tend to lower pressures?

If Earth did not rotate, these global winds would blow straight in a north-south direction. But Earth does rotate, which shifts the direction of the global winds, as shown in **Figure D** (*next page*).

The global winds that flow between 0° and 30° latitude are called the *trade winds*. Trade winds are named for their role in propelling trading ships centuries ago. North of the equator, they are known as the northeast trade winds, which were used by early European sailing ships traveling to the American continent, typically landing them in

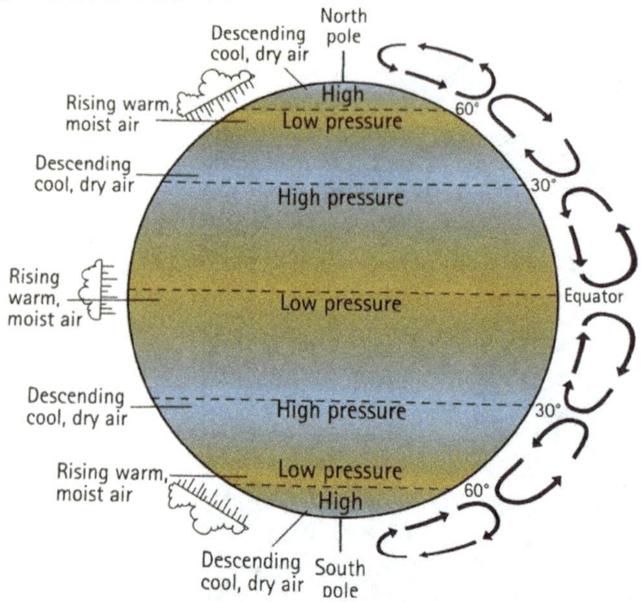

Figure C
Warm air rising at Earth's equator generates a set of six major convection cells running from the equator to the poles. Where the air rises, it leaves room behind it, creating a low-pressure area. Conversely, where the air comes downward, it concentrates, and the air pressure increases.

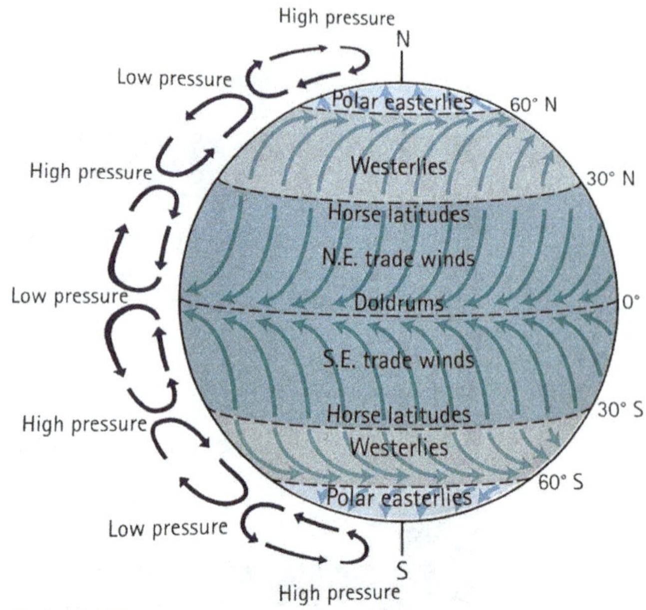

Figure D
Note the locations of the major global winds: the trade winds, the temperate westerlies, and the polar easterlies. Also note that the convection cells tend to be taller closer to the equator due to Earth's spin, which essentially tosses the atmosphere at the equator to higher altitudes. (Fun fact: Earth rotates eastward, which explains why the Sun rises in the east.)

the Caribbean. Then these sailors caught the global winds called the "westerlies" back to Europe.

Near the equator, the trade winds tend to die. How so? This is a zone in between the two equatorial cells. The result is a zone of relatively still air. Sailors of long ago cursed the equatorial seas as their ships stalled for lack of wind, and they named the area the *doldrums*. Sailors were also frequently stalled in the calm air where convection cells meet at 30° latitude. According to legend, as food and water supplies dwindled, horses on board were either eaten or cast overboard to conserve fresh water. As a result, these regions become known as the *horse latitudes*.

Jet Streams

There is another type of global wind—*jet streams*. Jet streams are rivers of high-speed wind that blow in the upper troposphere (where jets fly) within the area where the convection cells meet, as shown in **Figure E**. Many complex interactions among many variables result in the jet stream frequently bending in a north-south direction. This results in a looping behavior that can significantly impact weather conditions on the surface. For example, this can help transport warm tropical air to high latitudes and cold polar air to temperate latitudes. To Life!

CONCEPT CHECK

A migrating bird flies west to east across the United States, then back again. Which trip is easier for the bird and why?

CHECK YOUR ANSWER

The westerlies blow from west to east across the United States. This helps the bird when it flies from west to east but hinders the bird when it flies from east to west. This also explains why a commercial flight from Denver to Boston is always quicker than a flight from Boston to Denver.

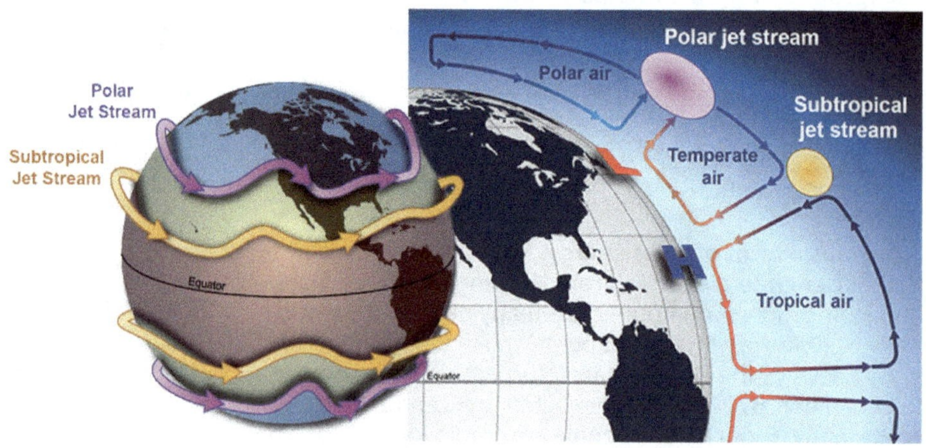

Figure E
Jet streams occur in both the northern and southern hemispheres. Because of Earth's rotation, westerly winds flow to the east, reaching speeds of more than 275 mph.

Chapter 11

Diversity of Life 2

THE MAIN IDEA

Earth is home to millions of living species, including tiny bacteria, giant redwood trees, and human beings.

11.1 Fungi
11.2 Animals Part 1
11.3 Animals Part 2
11.4 Animals Part 3
11.5 Viruses and Prions
11.6 Life is Interconnected

More than 1.5 million known species live on Earth. Many more—probably between 10 million and 100 million—have yet to be discovered. What are the major kinds of living things? How are they related to one another? Is it true that we humans are more closely related to mold than to cherries? Why do certain flowers smell like dead horses? Why do some salamanders have no lungs? And are there really thousands of species of dinosaurs flying around today? We'll answer these questions in the chapter.

11.1 Fungi

When you keep a loaf of bread too long, you often end up with some fuzzy stuff called *mold*. Mold is a fungus, a living organism that belongs to the group Fungi. And it is doing what you wanted to do—it is eating your food!

Fungi were once grouped with plants because of their stationary way of life, but they are actually more closely related to animals. Like animals, fungi are heterotrophs that obtain food from other organisms. Fungi release digestive enzymes over organic matter and then absorb the nutrients. This distinguishes them from animals, which digest food inside their bodies. Many fungi are decomposers that obtain the bulk of their nutrients from dead organic material, as the fungus in **Figure 11.1** is doing. Fungi and bacteria are the most essential decomposers in terrestrial ecosystems.

READING CHECK

In what important way are fungi like animals?

Some fungi, such as yeast, are single-celled organisms, but most fungi are multicellular. Multicellular fungi are composed of bunches of small threadlike filaments. Fungi may reproduce either sexually or asexually; most species actually use both strategies at some point in their life cycles.

Reproduction occurs through the formation of spores, tiny reproductive bodies that remain in a dormant state until conditions are favorable for growth. Fungal spores spread far and wide mostly by floating through air or water—this explains why mold finds your leftovers no matter where you hide them. Mushrooms are the spore-producing structures of certain fungi. Notice that what we think of as a "mushroom" is only a small part of the entire organism. Most of the fungus actually lives underground, as you can see from the illustration in **Figure 11.2**.

^ Figure 11.1

Many fungi are decomposers and obtain their nutrients from dead organic material. The fungus to the left is growing in and on dead wood and obtaining nutrients from it. The colorful and nutritious mushrooms seen on the table are being prepared for sale. All mushrooms you can see above the surface are reproductive structures.

Fungi are essential to the survival and growth of many, probably most, plants. This is because, in most plant species, the roots form close associations with fungi. These associations, called *mycorrhizae* ("my-kuh-RYE-zuh"), benefit both fungi and plants. The fungus receives nutrients from the plant while helping the plant roots absorb water and minerals from the soil. Mycorrhizae are shown in **Figure 11.3**.

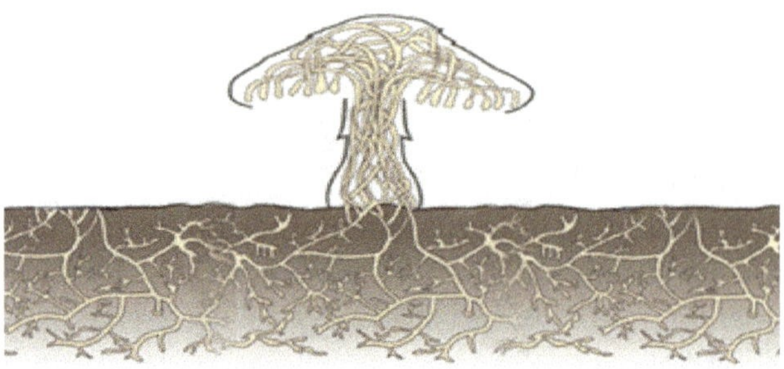

^ Figure 11.2

Mushrooms are spore-producing structures in certain fungi. Most of the organism lives underground.

^ Figure 11.3

Mycorrhizae are close associations between fungi and the roots of plants. This photo shows a fungus (white) encasing the roots (brown). This association is beneficial to both species.

Fungi important to humans include yeast, which is used in baking and brewing, and edible mushrooms. Fungi are also used to make blue cheeses such as Roquefort and Gorgonzola (the blue stuff is actually zillions of tiny fungal spores—enjoy!). Penicillin, the first antibiotic, was originally found in a fungus. Finally, human fungal diseases include yeast infections, ringworm, and athlete's foot.

Finally, did you know that the living organism with the largest overall size in the world is a fungus? The heaviest, most *massive* organism is the aspen described in the previous chapter, but a fungus wins in terms of overall size!

It is an underground honey mushroom in Oregon that measures 5.6 kilometers (3.5 miles) across. The fungus covers more than 2200 acres! This honey mushroom was discovered when scientists tried to figure out why large groves of trees were dying. The answer? The fungus was eating them.

CONCEPT **CHECK**

If fungi can't move around the way an animal can, how do they find new sources of food?

CHECK YOUR ANSWER

Fungi release large numbers of spores, some of which may settle on a new food source and begin to grow there.

More to Explore

To learn more about fungi:

https://www.livescience.com/53618-fungus.html

11.2 Animals Part 1: Sponges and Cnidarians

Animals include organisms as varied as starfish, beetles, coral, and antelope. **Animals** are multicellular, heterotrophic eukaryotes that obtain nutrients by eating other organisms. What are some other characteristics of animals? Animals *ingest* food, taking it into their bodies for digestion. Most animals reproduce sexually and are diploid during most of their life cycle. The gametes—sperm and eggs—are the only haploid stage. Many animals go through a juvenile period as a *larva* that is markedly different from the adult in form and ecology; examples are butterfly caterpillars and frog tadpoles. Most animals also have muscles for moving, sense organs for collecting information from their environments, and nervous systems for controlling their actions. The evolutionary tree in **Figure 11.4** shows a tentative hypothesis of the relationships among the major groups of animals.

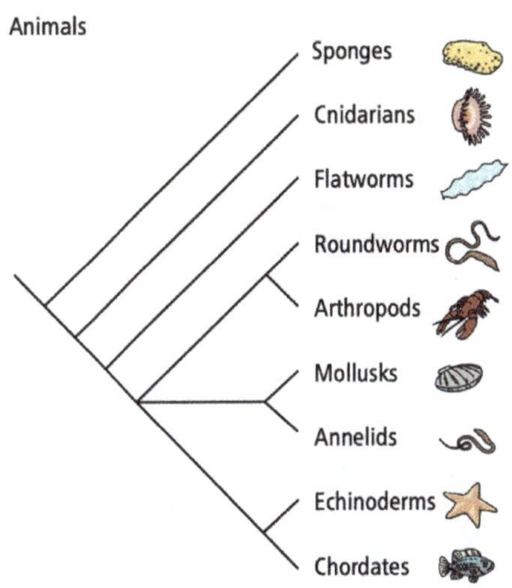

Figure 11.4

This evolutionary tree shows how the major groups of animals are related to one another.

Sponges

Sponges are sedentary marine animals (**Figure 11.5**). Most sponges have a tubelike shape with a large central cavity. Special cells in the sponge beat their flagella to produce a constant flow of water through the animal. Water enters through numerous pores, flows into the sponge's central cavity, and goes out the top. This constant current allows the sponge to catch food. Sponge cells trap bacteria from the water, digest them, and then distribute the nutrients to other cells.

Figure 11.5 >

These yellow tube sponges filter water for nutrients.

Sponges are the only animals that lack tissues, groups of similar cells that perform a specific function. This allows sponges to do unusual things—for example, if you separate a sponge's cells by passing it through a sieve, the cells will reassemble on the other side, forming a new sponge. No other animals can do that.

Cnidarians

Cnidarians ("nye-DARE-ee-uhns") include animals such as jellyfish, sea anemones, and corals. Unlike sponges, cnidarians have two distinct tissue layers: an outer layer that protects the body and an inner layer that digests food. These layers are separated by a jellylike middle layer.

READING CHECK

How do cnidarians catch prey?

Cnidarians use tentacles armed with barbed stinging cells to catch prey. In many species, the stinging cells release powerful toxins. (This is why jellyfish can be a danger to ocean swimmers.)

Prey is digested in a *gastrovascular cavity* that has a single opening that serves as both mouth and anus. Many cnidarians alternate between a sedentary polyp stage and a mobile, bell-shaped medusa stage. Cnidarians, such as sea anemones and corals, spend most of their lives as polyps. Cnidarians such as jellyfish spend most of their lives as medusas (**Figure 11.6**). One important group of cnidarians are the corals.

^ Figure 11.6

These are moon jellyfish, a species found in coastal waters worldwide. The moon jellyfish is a cnidarian in its medusa form.

Coral reefs are found in tropical oceans, in clear, shallow waters with temperatures between 20°C and 28°C (68°F and 82°F). Coral reefs are among the most diverse ecosystems in the world. Numerous marine species, including commercially important fish, spend all or part of their lives in coral reefs. Reefs also help protect shorelines from ocean waves. But, as humans burn more and more fossil fuels, the amount of carbon dioxide in Earth's atmosphere has increased, resulting in global warming and more acidic oceans. How has this affected corals and other marine species?

Corals live in colonies of tiny polyps, each wrapped in a calcium carbonate skeleton. Unlike many other cnidarians, however, corals are not hunters. They obtain most of their nutrients from photosynthesizing dinoflagellates known as *zooxanthellae* that live within their cells, as shown in **Figure 11.7a**. (We discussed dinoflagellates, a group of Protists, in Chapter 10.)

Coral bleaching occurs when environmental conditions cause corals to kick out their zooxanthellae. The corals literally turn white because it is their dinoflagellates that give them their colors (**Figure 11.7b**). Coral bleaching is most often triggered by an increase in seawater temperature. High temperatures interfere with photosynthesis in dinoflagellates, causing toxic molecules to build up in them. This, in turn, causes the corals to eject them.

Corals can survive for a short time without their dinoflagellates. If water temperatures decrease again, dinoflagellates move back into bleached corals, and the corals survive. But, if warm temperatures continue for too long, the corals starve to death.

Mass coral bleaching has become common in recent years due to the high water temperatures associated with global warming. In 2016, an estimated 29% of corals in Australia's Great Barrier Reef—the largest coral reef in the world—died. Bleaching was worst in the northern part of the reef, where over 70% of corals died. A second bleaching event occurred in 2017, this time striking the middle part of the reef. This event caused another 20% of the corals to die. Overall, about 90% of the reef has been affected. Unfortunately, as temperatures continue to rise, bleaching is likely to become more widespread and more severe.

Will coral reefs disappear? Corals do vary in their preferred temperatures and in their susceptibility to bleaching. For example, certain corals use special fluorescent pigments, like sunscreen, to shield their dinoflagellates. These fluorescent corals have survived mass bleaching episodes better than non-fluorescent corals. But, it is uncertain how much warming coral reefs can stand.

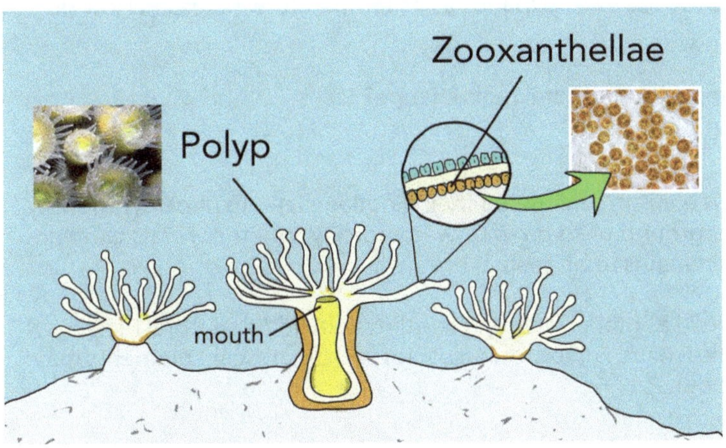

(a)

< Figure 11.7

(a) Coral polyps tend to be tiny but live in huge colonies. Corals receive as much as 98% of their nutrients from photosynthesis within the colorful zooxanthellae living within their tentacles. (b) These bleached corals have kicked out their zooxanthellae dinoflagellates. (c) A rare moment caught on camera when corals under heat stress turn vibrant colors usually preceding full coral bleaching and death.

(b)

(c)

High carbon dioxide levels also lead to ocean acidification. Atmospheric carbon dioxide (CO_2) is absorbed by the ocean, where it reacts with water to form carbonic acid (H_2CO_3). This lowers the pH of the ocean while also transforming carbonate minerals, such as calcium carbonate ($CaCO_3$), into bicarbonate compounds:

$$\underset{\text{Carbon dioxide}}{CO_2} + \underset{\text{Water}}{H_2O} \rightarrow \underset{\text{Carbonic acid}}{H_2CO_3}$$

$$\underset{\text{Carbonic acid}}{H_2CO_3} + \underset{\text{Calcium carbonate}}{CaCO_3} \rightarrow \underset{\text{Calcium bicarbonate}}{Ca(HCO_3)_2}$$

Many marine animals—including corals, echinoderms, crustaceans, and mollusks—need calcium carbonate to build their shells. In addition, acidified seawater increases the rate of dissolution of their shells. For these reasons, ocean acidification causes shelled animals to grow more slowly and to have weaker shells. Scientists have observed that coral growth slows in acidified waters and that the larvae of bivalves such as oysters and mussels are smaller. Some plankton, including certain photosynthetic species, also have calcium carbonate shells. As seawater becomes more acidified, plankton mass decreases. Because photosynthetic plankton forms a crucial part of marine food chains, the consequences could extend across entire marine communities.

CONCEPT CHECK

1. Global warming also causes sea levels to rise, mainly because seawater expands as its temperature increases. How do rising sea levels affect corals?

2. As global warming continues, how might the species composition of coral reefs change?

CHECK YOUR ANSWER

1. Corals are found only in clear, shallow waters, where there is enough sunlight for their dinoflagellates to photosynthesize. Rising sea levels will require corals to shift to the shallower ground or to grow upward quickly. Whether corals can find appropriate habitats and keep pace with sea-level changes remains to be seen.

2. Corals that are adapted to warmer temperatures may spread as global warming continues. In addition, coral reefs may come to be dominated by corals that are more resistant to bleaching. For example, species with fluorescent pigments may survive, whereas species without these sunscreens may die out.

More to Explore

Visit these sites to learn more about the dire problem of coral bleaching and what can be done about it:

https://www.chasingcoral.com/

https://www.worldwildlife.org/pages/everything-you-need-to-know-about-coral-bleaching-and-how-we-can-stop-it

For more on sponges

http://tolweb.org/treehouses/?treehouse_id=4291

To read more about cnidarians

https://ucmp.berkeley.edu/cnidaria/cnidaria.html

11.3 Flatworms, Roundworms, Arthropods, Mollusks, and Annelids

Flatworms

Flatworms are, as their name suggests, flat. They have distinct "head" and "tail" ends as well as "back" and "belly" sides. A single body opening serves as both mouth and anus, and an elaborately branched digestive tract transports nutrients to the entire body. The flat shape of flatworms allows oxygen to be absorbed efficiently across the skin via diffusion. Flatworms include many parasites as well as some non-parasitic species. The most familiar flatworms are tapeworms (**Figure 11.8**)—long, ribbon-like worms that live as parasites in the intestines of humans and other animals.

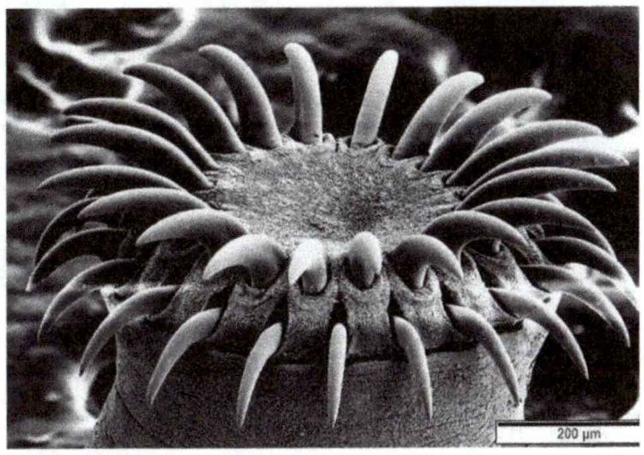

^ Figure 11.8

This tapeworm has hooks at its head end that help it attach to its host's intestines.

Roundworms

Roundworms (not to be confused with the more familiar earthworms) have small, slender bodies with tapered ends and a round cross-section (**Figure 11.9**). They have both a mouth and an anus. Roundworms have a tough outer cuticle that is shed periodically during growth. They eat a variety of things—bacteria, plants, fungi, and other animals—but many specialize in decaying organic material. In many habitats, roundworms are important decomposers. Roundworms have muscles that run from head to tail. As a result, these worms move like flailing whips as muscles on alternate sides of the body contract. Roundworms are responsible for several human diseases, including hookworm, pinworm, elephantiasis, and trichinosis.

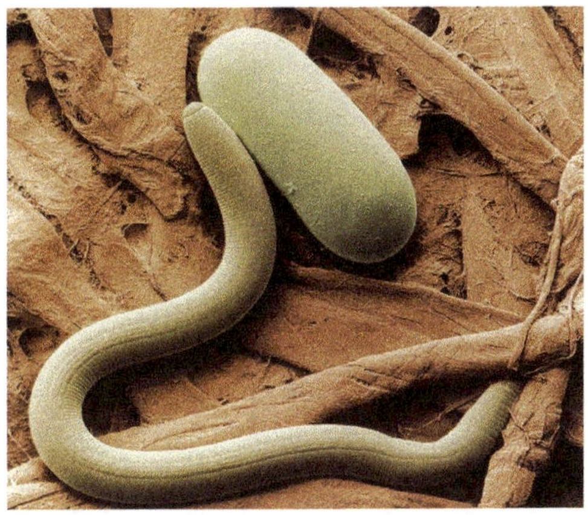

^ Figure 11.9

A roundworm moves through rotting vegetables in a compost heap.

Arthropods

Arthropods include lobsters, barnacles, spiders, scorpions, ticks, centipedes, insects, and many other species. They are found in just about every known habitat on Earth. All arthropods have an external skeleton called an **exoskeleton** that protects and supports the body. The exoskeleton cannot grow and must be shed periodically as animals grow. The bodies of arthropods are divided into different segments, and their legs have bendable joints. Some arthropod legs have been modified during evolution to function as mouth parts, antennae, and reproductive organs. Arthropods have a brain and well-developed sense organs. Many species pass through a distinct larval stage during their growth and development.

The major groups of arthropods are the crustaceans, the chelicerates, and the uniramians (**Figure 11.10**). *Crustaceans* are primarily aquatic and include species such as lobsters, crabs, shrimp, krill, and barnacles. *Chelicerates* are eight-legged animals such as horseshoe crabs, spiders, scorpions, ticks, and mites. *Uniramians* include centipedes, millipedes, and insects. Insects are the most diverse group of living organisms on Earth, with more than a million known species and perhaps as many as ten times that number waiting to be discovered. All insects have three body parts—a head, thorax, and abdomen—and three pairs of legs. Most also have two pairs of wings. Many insects are important to humans as plant pollinators. Others affect us because they transmit disease (mosquitoes carry malaria and West Nile virus) or are agricultural pests.

Despite their diversity and success, the one thing insects have not achieved is large size: Why are insects so small? The answer has to do with how they obtain oxygen. Insects obtain oxygen via branched tubules connected to the outside air. Oxygen must diffuse through the tubules to reach

READING CHECK

How do insects obtain the oxygen they need?

Figure 11.10

Arthropods have segmented bodies and jointed legs. (a) This barnacle is a crustacean. Its shell may make you think of mollusks, but its feathery jointed legs, which it uses to filter food from the water, are an arthropod feature. (b) This jumping spider is a chelicerate. (c) This damselfly is an insect and a uniramian. Here, it is shedding its exoskeleton.

the tissues, a strategy that works only in extremely small bodies. However, there were much bigger insects during the Carboniferous period, 300 million years ago, when atmospheric oxygen levels were much higher than they are today. One Carboniferous dragonfly had a 2.5-foot wingspan!

Today, scientists are concerned about the huge numbers of insects that have disappeared. Decades ago, taking a long road trip by car meant stopping every hundred miles or so to clean countless squashed insects off the windshield. Now, people can drive hundreds of miles and barely hit any insects at all. Scientists sometimes refer to this as the "windshield effect." But what happened to all the insects?

A recent study suggests that huge numbers of insects have indeed disappeared. At a German nature preserve, scientists counted insects using "malaise traps" like the one shown in **Figure 11.11**. Insects fly into the bottom of the tent-like structure and get funneled to the top of the tent, where they are captured. The results were astounding—insect numbers in the preserve had decreased by 76% between 1989 and 2016. It was especially surprising that such a steep decline had occurred in a nature preserve, where habitats are protected, and the environment is relatively unpolluted. Insects could be doing far worse in areas where the effect of humans is greater, such as cities, suburbs, and rural areas.

Figure 11.11

Researchers use a Malaise trap to collect insects. Invented by the Swedish entomologist, René Malaise, the tent-like structure funnels insects to the top for collection.

The drastic decline of insect populations has been a shock to many scientists. Insects were thought to be less vulnerable to human disturbances than most other species.

Compared to larger animals, insects need less habitat and can breed more quickly. What could be responsible for insect declines? In the German study, researchers were able to eliminate two important candidates—habitat destruction (since the study occurred in protected preserves) and global warming. They now think pesticide use in adjacent agricultural areas is responsible for the huge die-offs.

Insects are critical to most ecosystems. They are important as plant pollinators and as prey for other species. Scientists estimate that 80% of plants rely on insects to pollinate them. Insects are also the primary prey for many species, especially birds, bats, amphibians, and reptiles. The loss of insects would impact not only those species but also entire food chains. If insects disappear, many ecosystems will surely collapse.

Mollusks

Mollusks are soft-bodied animals such as clams, oysters, squids, octopuses, snails, and slugs. Most mollusks have a protective shell, although the shell is tiny in some species (squids) and entirely absent in others (octopuses and slugs). All mollusks have a muscular "foot" responsible for locomotion, a visceral mass that holds the digestive and reproductive organs, and a mantle that secretes the shell. There are three main groups of mollusks. *Bivalves* have two-hinged shells and include species such as clams, oysters, mussels, and scallops. Most bivalves are sedentary and feed by filtering small particles from the water.

Cephalopods such as squids and octopuses (**Figure 11.12**) are active predators that use arms (eight in octopuses and ten in squids) to capture prey. Cephalopods also have well-developed brains and eyes. *Gastropods* have a single spiral shell and include species such as snails, abalone, and limpets. Most gastropods are herbivores.

< Figure 11.12

This common octopus is a cephalopod, a type of mollusk.

^ **Figure 11.13**

The leech is an annelid that feeds on blood. The leech shown here is sometimes used for medical purposes.

Annelids

Annelids are a group of segmented worms that include earthworms, leeches (**Figure 11.13**), and other species. Annelids have some muscles that go around the body and others that go head-to-tail. Their muscles allow for great flexibility of motion— for example, annelids can contract one part of their body while keeping the rest still. Earthworms are important decomposers that feed by ingesting large amounts of soil and absorbing the available nutrients. Their burrowing activity also helps to aerate soil, supplying it with oxygen. Leeches are parasites that feed on blood. They have blade-like teeth for cutting through their host's skin and secrete anticoagulants that prevent blood from clotting while they feed. Leeches are sometimes used in medicine today. During certain delicate surgeries, leeches are used to keep blood from clotting. The anticoagulants they make are still more effective than anything humans have invented.

CONCEPT CHECK

Of the animal groups considered in this section—flatworms, roundworms, arthropods, mollusks, and annelids—which has the greatest number of species? What are some of the characteristics of the organisms in this group?

CHECK YOUR ANSWER

The arthropods have the greatest number of species. Arthropods include over one million species of insects. Characteristics of arthropods include an exoskeleton, a body divided into segments, legs with bendable joints, a brain, and well-developed sense organs. Many species also pass through a distinct larval stage during growth and development.

More to Explore

You can read more about these living things at the following:

Flatworms:

https://animaldiversity.org/accounts/Platyhelminthes/

Roundworms:

https://ucmp.berkeley.edu/phyla/ecdysozoa/nematoda.html

Arthropods:

https://ucmp.berkeley.edu/arthropoda/arthropoda.html

Mollusks:

http://www.oceanicresearch.org/education/wonders/mollusk.html

Annelids:

http://tolweb.org/Annelida

11.4 Animals Part 3: Echinoderms and Chordates

Echinoderms

Echinoderms include starfish, sea urchins, and sea cucumbers (**Figure 11.14**). All echinoderms have an internal skeleton, or **endoskeleton**, made of small, interlocking plates. They use small, sucker-like appendages called *tube feet* to move and stick to rocks. Starfish also use their tube feet to pry open the shells of bivalves. Most echinoderms live on the ocean floor and move very ... very ... slowly.

Chordates

The **Chordates** are the animal group that includes tunicates, lancelets, and vertebrates, the group to which humans belong. Chordates share four key features: a brain and a spinal cord that runs along the back of the body, a notochord—a stiff but bendable rod that supports the back, gill slits, and a tail that extends beyond the anus.

READING CHECK

What four features do chordates share?

These features are not apparent in all adult chordates (for example, humans don't have tails) but are generally present at some stage of development (human embryos go through a tailed stage during development). Let's look at the major groups of chordates.

Tunicates are sedentary marine species, also known as "sea squirts" (**Figure 11.15**). They feed by filtering small particles from the water. *Lancelets* are small, blade-shaped, swimming marine species that often bury themselves in the sand. Like tunicates, lancelets filter food from the water.

Vertebrates are animals with backbones. The vertebrates include several groups of fish, amphibians, reptiles, and mammals. Did you know that the earliest vertebrates had mouths but not hinged jaws? Only a few jawless vertebrates are still in existence. Lampreys, parasites that suck blood from other fish, are shown in **Figure 11.16**.

Cartilaginous fishes include sharks, skates, and rays. These fishes have skeletons made of cartilage rather than bone. Sharks are famous predators. Their unusual *electroreceptive organs* help them detect the tiny electric currents in the muscles and nerves of nearby prey.

< **Figure 11.14**

This photo shows interaction between echinoderms in the Adriatic Sea.

∧ **Figure 11.15**

It may not be obvious at first glance, but these sea squirts are marine chordates and are close relatives of the vertebrates.

∧ **Figure 11.16**

The earliest vertebrates had mouths but did not have hinged jaws. Modern lampreys retain this characteristic. These individuals are clinging to glass with their circular mouths.

Ray-finned fishes include most of the animals we think of as "fish"—tuna, salmon, bass, and so on. Ray-finned fishes have a sac in their bodies called a *swim bladder*. The swim bladder is filled with just enough gas to make the fish's density the same as the density of water. This means that a ray-finned fish doesn't sink or float, giving it great water mobility. Cartilaginous fishes, which do not have swim bladders, will sink if they stop swimming.

At first glance, *lungfishes* and *coelacanths* (**Figure 11.17**) may look like ray-finned fishes. However, they are more closely related to terrestrial vertebrates such as amphibians, reptiles, and mammals. The bones in their fins are arranged serially, like those in the limbs of terrestrial vertebrates, rather than as rays emanating from the base of the fin. Although only a few species remain, these species are of great interest to biologists because they provide clues about how the limbs of terrestrial vertebrates evolved.

Amphibians include salamanders, frogs, and caecilians, a group of limbless species. The name *amphibian* refers to the fact that many species use both aquatic and terrestrial habitats. In many amphibians, aquatic larvae metamorphose into terrestrial adults. For example, aquatic tadpoles may metamorphose into terrestrial frogs.

However, some amphibians are entirely aquatic, and many are entirely terrestrial. Amphibians are restricted to moist habitats because their skins are made of living cells that dry out. Amphibian eggs do not have shells and also require moisture (**Figure 11.18**). In the last few decades, dozens of amphibian species have suddenly gone extinct, and hundreds have experienced huge population declines. What is especially worrisome is that many species, such as the remarkable golden toad (**Figure 11.19**), have disappeared from protected nature reserves.

(a)

(b)

Figure 11.17 >

(a) This lungfish may look like a ray-finned fish but is more closely related to terrestrial vertebrates. Note the lungfish's unusual fins, which have features in common with the legs of terrestrial vertebrates. (b) The coelacanth is another species that, despite looking like a ray-finned fish, is more closely related to terrestrial vertebrates.

< **Figure 11.18**

Amphibian eggs have no shells, and so they require moisture to keep from drying out. These are developing frog eggs of the species *Raorchestes tuberohumerus*.

Reptiles (including the feathered ones—birds) and mammals are *amniotes*. Amniotes have two traits that enable them to live in diverse terrestrial habitats. First, their skin is made of dead cells, which helps to prevent water loss. Second, their eggs have shells, which keep them from drying out.

Reptiles include turtles, lizards and snakes, crocodilians, and birds (**Figure 11.20**). Turtles have protective shells that are modified ribs. (Imagine squeezing your entire body inside your rib cage!)

Lizards are a large and diverse group found in a wide array of terrestrial environments. Snakes are a group of lizards that have lost their legs and evolved adaptations for subduing large prey and swallowing them whole. Birds are exceptional in that they have evolved the ability to fly. In birds, adaptations for flight include wings, feathers, and hollow bones (for lightness).

All reptiles except birds are ectotherms. An **ectotherm** uses behavior to regulate its body temperature. For example, lizards bask in the sunlight to warm up and retreat to the shade to cool down. The body temperatures of ectotherms vary, to some extent, depending on environmental conditions. Birds and mammals are endotherms. An **endotherm** maintains a constant,

< **Figure 11.19**

The golden toad is one of many amphibian species that has disappeared in recent decades.

Figure 11.20 >

Snakes and birds are reptiles, a group that also includes turtles, lizards, and crocodilians. (a) Many snakes have adaptations for swallowing large prey or food items. This snake is eating eggs. (b) Birds are reptiles that have evolved the ability to fly. This is a great blue heron.

(a)

(b)

high body temperature by breaking down food; this process releases heat, which warms the body. Because an endotherm warms itself by breaking down food, it has to eat more than an ectotherm of the same size. This is why you can feed a pet snake once a week, but most dogs eat daily. Ectotherms and endotherms were once called "cold-blooded" and "warm-blooded," but these terms just aren't accurate. For example, some desert lizards have body temperatures higher than 38°C (100°F), hotter than the body temperatures of most birds and mammals.

Mammals have hair and feed their young milk. The majority of mammals live on land, but bats fly, and seals and whales are partly or fully aquatic. There are three major groups of mammals. *Monotremes*, such as the platypus (**Figure 11.21**) and spiny echidna, lay eggs. *Marsupials*, such as possums, koalas, and kangaroos, give birth to very immature live young. Newborn marsupials crawl up into their mother's pouch and attach to a nipple, where they eat and continue to develop (**Figure 11.22**). *Placentals*, which include most living mammal species, also give birth to live young; however, newborn placentals are more mature than newborn marsupials.

Figure 11.21 >

The platypus is an egg-laying mammal.

< **Figure 11.22**

A baby kangaroo peeks out of its mother's pouch.

CONCEPT **CHECK**

Which terrestrial vertebrates are least dependent on living in a moist environment? What adaptations help them live in dry terrestrial environments?

CHECK YOUR ANSWER

Amniotes, the group that includes reptiles and mammals, are the least dependent on living in a moist environment. Two traits found in amniotes enable them to live in drier habitats than amphibians. First, their skin is made of dead cells, which helps to prevent water loss. Second, their eggs have shells, which keep them from drying out.

More to Explore

You can learn more about these animal groups through the following:

Echinoderms:

http://www.oceanicresearch.org/education/wonders/echinoderm.html

Vertebrates and other chordates:

http://www.oceanicresearch.org/education/wonders/echinoderm.html-https://ucmp.berkeley.edu/chordata/chordata.html

11.5 Viruses and Prions

A **virus** is a small piece of genetic material, either DNA or RNA, wrapped in a protein coat called a capsid, which encloses the nucleic acid. Some viruses also have an outer envelope of lipids derived from the host cell membrane. Viruses lack organelles and cannot perform metabolic functions on their own. Therefore, viruses exist on the border between living and nonliving. They are not made of cells, and they can reproduce only within a host cell. However, they have genes, and they evolve. Where did viruses come from? Scientists believe viruses originate when little pieces of host DNA or RNA somehow evolve the ability to move from one cell to another. This means that viruses have probably originated many times. A typical virus is shown in **Figure 11.23**, along with a photo of Human immunodeficiency virus (HIV), the virus that causes acquired immunodeficiency syndrome (AIDS).

READING CHECK

What types of organisms are infected by viruses?

Many viruses have normal double-stranded DNA, but others have genes made of single-stranded DNA, single-stranded RNA, or double-stranded RNA. Viruses reproduce by infecting a host cell and then using the cell's resources to copy their genetic material and build viral proteins. These are then assembled to form new viruses that exit the host to propagate within other organisms, which is the spread of the viral disease. All forms of life—from bacteria to plants and animals—can be infected by viruses. Viruses can cause disease by damaging or killing host cells, disrupting normal cellular functions, and triggering an immune response. Symptoms of viral infections can range from mild, such as the common cold, to severe, such as Ebola or HIV/AIDS.

Viruses on Earth can be found nearly everywhere, from the soil to the air to the water we drink. The adult human body harbors viruses as well. Some of these viruses are of the sort that infects not human cells but bacterial cells, which have evolved various defenses. Thus, in many ways, your body is but a battleground between these two biological agents. When there

More to Explore

More about viruses and prions:

https://www.news-medical.net/health/What-is-a-Virus.aspx

https://www.cdc.gov/prions/index.html

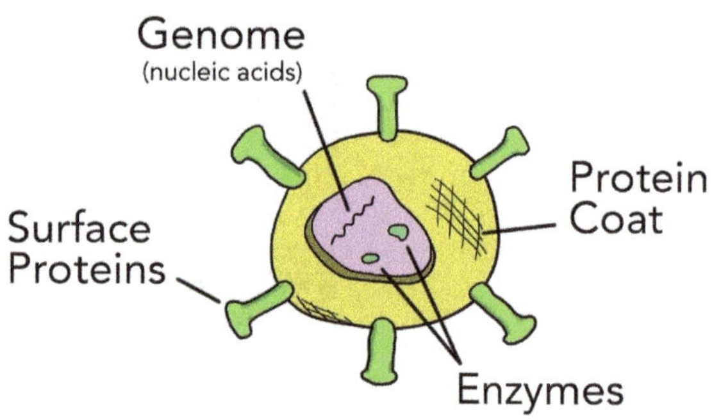

(a)

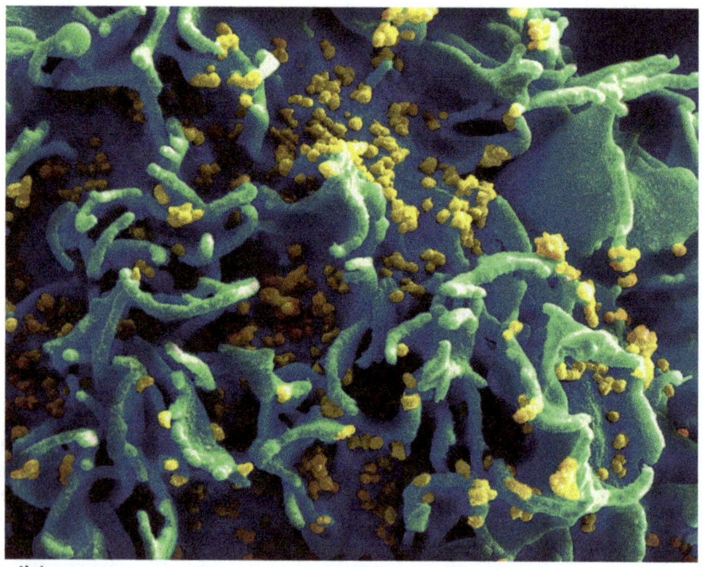

(b)

Figure 11.23

(a) Most viruses consist of a protein coat encapsulating an inner chamber containing genetic material and essential enzymes. The protein coat can also feature surface structures that play a role in the virus's ability to penetrate the host cell. (b) This human immune cell is infected with HIV (yellow), the virus that causes AIDS. The cell is covered with viral particles that will soon spread and infect other cells.

is a balance between these opposing forces, the result can be your good health. However, any upset with this balance can lead to poor health.

Some of the more notorious viruses include those responsible for human diseases such as the common cold, influenza (the "flu"), AIDS, smallpox, and Covid. One feature of viruses is that their genes may mutate very quickly. This explains why there are always plenty of colds to catch and why the flu comes back in a different form year after year.

In summary, viruses and cells have different structures and functions, with viruses being much simpler than cells. Viruses require host cells to reproduce and can cause disease by damaging or disrupting normal cellular functions. Understanding these differences is important for understanding how viruses spread and cause disease.

Prions are incorrectly folded proteins that cause disease. (Remember that proteins are folded strings of amino acids. In prions, the folding has somehow gone wrong.) Prions cause mad cow disease and the related Creutzfeldt–Jakob disease in humans. Both these diseases cause severe damage to the brain. Prions infect cells and then "reproduce" by converting normal proteins into the incorrectly folded form. Prions are spread when we eat infected meat. Cooking, which kills bacteria and viruses, has no effect on prions.

CONCEPT **CHECK**

How are prions different from the viruses and bacteria that cause many human diseases?

CHECK YOUR ANSWER

Unlike viruses and bacteria, prions are disease-causing agents that don't have any genetic material at all and are certainly not living things. Prions are incorrectly folded proteins that have the ability to convert the proteins near them to the misfolded variety. Large numbers of misfolded protein result in diseases such as Creutzfeldt–Jakob disease.

11.6 Life Is Interconnected

Placing living organisms into distinct branches of an evolutionary tree provides a broad overview of how these living organisms, ourselves included, are related. This has value. Probing deeper, however, we find that all living organisms here on Earth are completely interconnected, even those within separate branches.

For example, vital components of eukaryotic cells have a bacterial or archaeal origin. Specifically, the mitochondria and chloroplasts found in eukaryotic cells are remnants of bacteria or archaea that were captured by an early eukaryotic cell that later evolved a mutually beneficial relationship with them. Mitochondria and chloroplasts now provide eukaryotic cells with the means to carry out cellular respiration and photosynthesis, helping eukaryotic cells obtain energy.

Similarly, soon after the human genome was sequenced in the early 2000s, it was discovered that about 8 percent of the nucleotide sequences in our DNA were of viral origin. This viral DNA passes down from one generation to the next in our human lineage. Our very genetic identity is about 8 percent viral. But it gets even more profound: This viral DNA within us has been found to play a key role in the development of the placenta. This suggests that a viral infection many years ago eventually allowed for the evolution of pregnancy as we know it in placental mammals. If true, we can thank a virus for allowing us to give birth to mature young.

READING CHECK

How can bacteria of one species spread resistance to bacteria of another species quickly?

Horizontal gene transfer (HGT) describes the movement of genetic material from one organism to another through a path other than the usual transmission of DNA from parent to offspring. Both the examples above describe important past instances of HGT.

HGT also explains one way bacteria can become resistant to antibiotics fairly quickly. Answer: One bacterium that, through mutations, has become resistant to the drug can transfer its drug-resisting genes to a neighboring bacterium, even if it is of a different species. Horizontal gene transfer occurs in nature regularly and can play an important role in evolution.

The CRISPR technology described in Chapter 7 shows how humans can use technology to transfer genes horizontally from one organism to another as well. This technology is revolutionary as it allows us to manipulate our genome within a single generation to a variety of ends we might desire. It may allow us to cure genetic diseases such as sickle cell anemia, cystic fibrosis, or muscular dystrophy. It can also potentially include designing new traits for future generations—from higher IQ to greater athleticism. Clearly, this application is fraught with ethical issues.

So, zoom closer into any evolutionary tree, and you'll find innumerable tangled relationships among seemingly unrelated branches. We may seek simplicity when it comes to classifying life. But when it comes to life itself, the name of the game is complexity with a heavy dose of interconnectedness. There is still much we have yet to learn about the fundamental nature of life. But what we have discovered so far is beyond fascinating.

More to Explore

The history of the evolutionary tree and the development of our understanding of Horizontal Gene Transfer:

https://www.simonandschuster.com/books/The-Tangled-Tree/David-Quammen/9781476776637

CONCEPT **CHECK**

Zooming into the branches of any evolutionary tree, what would one find?

CHECK YOUR ANSWER

One would find that those branches have connections between them, especially given the ability of DNA to transfer between different life forms.

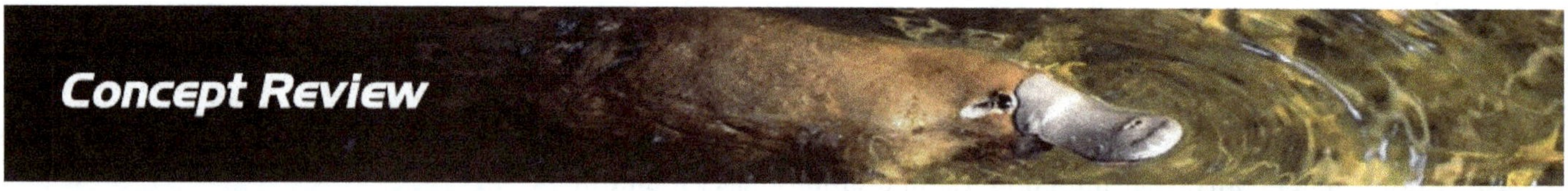

Concept Review

Chapter 11

Detailed Chapter Summary

Fungi are heterotrophs. They obtain food by releasing digestive enzymes over organic matter and absorbing the nutrients. Many fungi are important decomposers. Some fungi are single-celled, and others are multicellular. Fungi may reproduce sexually or asexually, and many species use both strategies at some point in their life cycles. Reproduction often occurs through the formation of tiny spores that are able to spread far and wide. Mushrooms are the spore-producing structures of some fungi. Mycorrhizae are associations between fungi and plant roots that benefit both species. These associations are essential to many plants.

Animals are multicellular, heterotrophic eukaryotes that obtain nutrients by ingesting organic matter. Animals reproduce sexually, and many species go through a juvenile larval period. Most animals also have muscles, sense organs, and nervous systems.

Sponges are sedentary marine species. They are the only animals that lack tissues.

Cnidarians, such as jellyfish and sea anemones, have two tissue layers separated by a jellylike middle layer. They use tentacles armed with barbed stinging cells to catch prey. A single opening serves as both mouth and anus. Global warming and ocean acidification threaten cnidarians such as corals as well as many other marine species. Coral bleaching could damage coral reefs, the most diverse marine ecosystems.

Flatworms such as tapeworms have distinct head and tail ends and back and belly sides.

Roundworms are a diverse group of small worms that move like flailing whips. Many roundworms are important decomposers.

Arthropods are a very diverse animal group that includes crustaceans, spiders, insects, and many other species. Arthropods have an exoskeleton that is repeatedly shed during growth, segmented bodies, and jointed legs. Insects are the most species-rich group of living organisms on Earth. All insects have three body parts and six legs, and most also have two pairs of wings.

Mollusks are soft-bodied animals with shells (lost in some species). Mollusks include bivalves, cephalopods such as octopuses and squids, and gastropods such as snails. Annelids are segmented worms such as earthworms and leeches.

Echinoderms include species such as starfish and sea urchins. They have an endoskeleton and move using tube feet.

Chordates include a few small animal groups as well as the vertebrates, animals with backbones. Important vertebrate groups include jawless fishes, cartilaginous fishes, ray-finned fishes, and terrestrial species such as amphibians, reptiles (including birds), and mammals. Reptiles and mammals are amniotes that are characterized by shelled eggs and a skin made of dead cells. These two features allow them to live in dry terrestrial habitats. The three main groups of mammals are the egg-laying monotremes, marsupials, and placentals. Ectotherms use behavior to regulate body temperature, whereas endotherms maintain a constant, high body temperature by breaking down large amounts of food, a process that releases heat. Birds and mammals are endotherms.

Viruses are small pieces of genetic material, either DNA or RNA, wrapped in a protein coat. Viruses are not made of cells, but they have genes, and they evolve. Viruses can only reproduce within a host cell. They infect a host cell and then use the cell's resources to copy their genetic material and build new viral proteins. Viruses infect many different species and are responsible for many important human diseases.

Prions are incorrectly folded proteins that "reproduce" by converting normal proteins to the misfolded variety. Prions cause mad cow disease and the related Creutzfeldt-Jakob disease in humans.

Horizontal gene transfer describes the transfer of DNA from organism to organism through mechanisms other than the usual parent-offspring transmission. The origin of eukaryotic mitochondria and chloroplasts from prokaryotes (bacteria or archaea) is one example. Horizontal gene transfer is also significant in human history—the human genome consists of 8% viral DNA. The prominence of horizontal gene transfer in the history of life means that life is more interconnected than the separate branches of an evolutionary tree might suggest.

Key Terms

Animals A group of multicellular, heterotrophic eukaryotes that take food into their bodies for digestion.

Chordates The animal group that includes tunicates, lancelets, and vertebrates that is characterized by four main features: a brain and spinal cord that run along the back of the body, a notochord, gill slits, and a tail that extends beyond the anus.

Ectotherm An organism that uses behavior to regulate its body temperature, which may vary to some extent.

Endoskeleton An internal skeleton, such as that found in echinoderms and chordates.

Endotherm An organism that maintains a constant, high body temperature by breaking down food, releasing heat.

Exoskeleton An external skeleton, such as that found in arthropods.

Fungi A group of heterotrophic eukaryotes that obtain food by secreting digestive enzymes over organic matter and then absorbing the nutrients.

Prions Incorrectly folded proteins that cause diseases, including mad cow disease and Creutzfeldt-Jacob disease.

Vertebrates The animal group within the Chordates that include animals with backbones, such as several groups of fishes, as well as amphibians, reptiles (including birds), and mammals.

Virus A small piece of genetic material wrapped in a protein coat that infects and reproduces within a host cell.

Review Questions

Simple Review Questions

11.1 Fungi

1. How do fungi obtain food?
2. What are fungal spores?
3. Why are fungi essential to the growth and survival of most plants?

11.2 Animals Part 1

4. How do animals obtain nutrients?
5. What single factor is most frequently responsible for coral bleaching?
6. Why do corals turn white during bleaching episodes?
7. Explain why ocean acidification threatens shelled marine species.

11.3 Animals Part 2:

8. What are some features of arthropods?
9. Describe the three major groups of mollusks.

11.4 Animals Part 3:

10. Why must amphibians live in moist habitats?
11. What is the difference between an ectotherm and an endotherm? Which vertebrates are ectotherms, and which are endotherms?
12. How do monotremes differ from other mammals?

11.5 Viruses and Prions

13. What is a virus? How do viruses reproduce?
14. What are prions? How do they "reproduce"?

11.6 Life Is Interconnected

15. How did the mitochondria and chloroplasts of eukaryotic cells originate?
16. What is horizontal gene transfer?

Challenging Review Questions

11.1 Fungi

17. What do fungi and animals have in common? How do they differ?
18. What are you eating when you eat a mushroom?
19. Truffles, prized mushrooms for high-end cuisine, remain elusive and costly. Due to the inability to farm them, people resort to gathering them from the wild, often with the aid of trained dogs. Truffle hunters have discovered that the prime truffle grounds lie close to specific oak tree species' roots. Why are they found in association with only certain plants?

11.2 Animals Part 1

20. The bluefire jellyfish in the photo has caught a fish. Name two different strategies used by cnidarians to obtain food.
21. What kinds of living things are corals? How do corals obtain nutrients?
22. How does global warming threaten corals and other marine life?

11.3 Animals Part 2

23. How do the muscles of roundworms and earthworms differ? What does this mean about the way each animal moves?

11.4 Animals Part 3

24. Why is a salamander more dependent on living in a moist habitat than a lizard?
25. Many snakes can survive by eating just once every few weeks. Why can't birds do this?

11.5 Viruses and Prions

26. Viruses straddle the line between living and nonliving. How do viruses resemble living things? What features of living things do they lack?
27. What are some examples of viruses that make us sick? Does the fact that viral genes mutate very quickly have consequences for humans?
28. What are prions? What kinds of diseases do prions cause?

11.6 Life Is Interconnected

29. What is the relevance of horizontal gene transfer to evolution?

Apply & Discuss Questions

30. What are some fungi you have encountered in the past month? Have you eaten any? Used any in baking? Seen any on your food or in the neighborhood?
31. Compare endoskeletons and exoskeletons. What do they have in common? What are the advantages and disadvantages of each?
32. Why are certain new viral diseases, such as COVID-19, so devastating? What characteristics of the disease made the pandemic so harmful to human populations?
33. Classify the following animals: A sand dollar, a barnacle, an oyster, a conch, a lobster, a seal, a squid, a cow, a koala bear, a bat. Feel free to research the features of these living things online in order to help determine your answers.

Solutions to Odd-Numbered Questions

Simple Review Solutions

1. Like animals, fungi are heterotrophs that obtain food from other organisms. Fungi release digestive enzymes over organic matter and then absorb the nutrients. This distinguishes them from animals, which digest food inside their bodies.

3. Fungi are essential to the survival and growth of many, perhaps most, plants. This is because, in most plant species, roots form close associations with fungi called mycorrhizae. The fungus receives nutrients from the plant while helping roots absorb water and minerals from the soil.

5. Coral bleaching is most often triggered by increased seawater temperature.

7. The extra hydrogen ions resulting from acidification react with carbonate, reducing the amount of carbonate in the water. Many marine animals—including corals, echinoderms, crustaceans, and mollusks—need carbonate to build their calcium carbonate shells. More acidic seawater also increases the rate of dissolution of their shells.

9. Bivalves have two-hinged shells and include species such as clams, oysters, mussels, and scallops. Most bivalves are sedentary and feed by filtering small particles from the water. Cephalopods such as squids and octopuses are active predators that use arms (eight in octopuses and ten in squids) to capture prey. Cephalopods also have well-developed brains and eyes. Gastropods have a single spiral shell and include species such as snails, abalone, and limpets. Most gastropods are herbivores.

11. All reptiles except birds are ectotherms. An ectotherm uses behavior to regulate its body temperature. For example, lizards bask in the sun to warm up and retreat to the shade to cool down. The body temperature of ectotherms fluctuates, depending on environmental conditions. Birds and mammals are endotherms. An endotherm maintains a constant, high body temperature by breaking down food—this process releases heat, which warms the body.

13. A virus is a small piece of genetic material wrapped in a protein coat. Viruses reproduce by infecting a host cell and then using the cell's resources to copy their genetic material and build viral proteins. These are then assembled to form new viruses.

15. The mitochondria and chloroplasts found in eukaryotic cells are remnants of bacteria or archaea that were captured by an early eukaryotic cell that later evolved a mutually beneficial relationship with them.

Challenging Review Solutions

17. Fungi and animals are both heterotrophs. However, fungi tend to release digestive enzymes out over their food and absorb the nutrients, whereas animals tend to ingest their food, taking it into their bodies for digestion.

19. The fungi associated with truffles form mycorrhizae with the roots of these oak trees.

21. Corals are animals in the group Cnidaria. Unlike other cnidarians, corals house dinoflagellates in their bodies and obtain the bulk of their nutrients from these photosynthesizers.

23. The muscles of roundworms all run longitudinally (from head to tail) down the body. As a result, roundworms move like flailing whips as muscles on alternate sides of the body contract. The muscles of annelids are arranged in both circular (around the body) and longitudinal (head-to-tail) orientations, allowing for great flexibility of motion. Unlike roundworms, for example, annelids can contract one part of the body while keeping the rest still.

25. Birds are endotherms and need to break down a lot of food molecules to help maintain their high, stable body temperatures. Snakes are also unusual in that they have adaptations for eating very large prey – this is part of the reason why they don't need to eat as frequently.

27. Many answers are possible, as many viruses make us sick. For example, the cold and flu are both caused by viruses, as is Covid-19. Because viral genes mutate very quickly, the immunity we develop as a result of catching one cold will not prevent us from also catching the next cold. So, we continue to catch many viral diseases over and over again.

29. Darwin's theory of evolution focuses on offspring inheriting traits from their parents ("vertical" gene transfer). Horizontal gene transfer describes the transmission of DNA from one organism to another in a way other than this standard parent-to-offspring transmission. Horizontal gene transfer can cause an organism to very quickly gain a trait, such as when antibiotic resistance is passed from one species of bacteria to another. This is not the gradual shaping of a trait through small-scale changes that was the focus of much of Darwin's work.

Apply & Discuss Solutions

31. Both endoskeletons and exoskeletons support an animal's body and provide places for muscles to attach to move the body. Exoskeletons also help protect animals, which endoskeletons can also do, though they don't tend to provide the complete casing of, for example, a crab's shell. On the other hand, endoskeletons can grow with the animal, whereas exoskeletons have to be shed.

33. Sand dollars are echinoderms. Barnacles are arthropods (crustaceans, specifically). Oysters are mollusks (bivalves). Conchs are mollusks (gastropods). Lobsters are arthropods (crustaceans). A seal is a mammal (placental). A squid is a mollusk (cephalopod). A cow is a mammal (placental). A koala is a mammal (marsupial). A bat is a mammal (placental).

Living Earth Essay

Life on Earth Connections

Interconnectedness

Ubuntu

In referring to humanity, the Zulu word "ubuntu" roughly translates to "I am because we are." It points to an underlying truth that any living being cannot come into existence, survive, or let alone thrive on its own. Rather, any individual is necessarily part of a larger relational, communal, societal, environmental, and even spiritual world.

A fish spends its life surrounded by water, which becomes invisible to the fish because the water is just so ever-present. In a similar way, we spend our lives within an intricate 3-dimensional web of interdependence, which can also become invisible to us as we navigate through our daily tasks. Not mindful of this web, we may see ourselves as separate entities with a strong sense of self-identity and self-determination. But consider the food you eat. Where did that food come from? How did it end up before you? What of your shelter? Your employment? Your teachers? Your family and friends? Your clothing? Your entertainment? Your transportation? When you get sick, how is care provided to you? We survive and thrive by the kindness and actions of others.

The concept of ubuntu tells us it is impossible for each of us to make any movement without creating propagating jiggles within that 3-dimensional web surrounding us. The reality is that each living being's identity is defined not so much by the living being but by the fabric within which that living being is found. This is what we mean when we say that all life is interconnected.

Figure A
Newton's Third Law tells us we cannot touch without being touched—for every action, there is a reaction. On a broader scale, we see this law underpins the interconnectedness of life—no one can remain an island unto themself.

Figure B
Trees depend upon each other and their surrounding community for the survival of not just themselves but of the forest at large.

Forest For the Trees

Much of biology is an exploration of the interconnectedness of life forms. A wonderful example comes from forests. Our old image of a forest as a group of trees competing for water, nutrients, and sunlight has been replaced by a much deeper understanding. Through scientific research, we are learning that any native forest is a complex system of many parts that can work in parallel to optimize the living conditions for its members.

Perhaps most fascinating is what research has revealed about the fungal mycelium that threads itself within a healthy forest floor. Researchers, such as Suzanne Simard at the University of British Columbia, have found that trees share sugar with each other using the underground mycelium network as a conduit. They discovered this using a technique known as *radioactive tracing*. By injecting radioactive sugar into one tree, they could follow that sugar as it made its way through the mycelium network to another tree. This explains how the stump of a cut tree can continue to survive: through the mycelium network, the sugar it needs can come from its healthy neighbors. This system thus acts as a social security safety net where if one tree gets sick, it will be supported by its neighbors for the benefit of the entire forest.

This system has also been found to work between different species of trees. For example, Fir trees and birch trees produce sugar optimally at different times of the year. When optimal for the fir trees, sugar is transferred through the mycelium network to the birch trees, who return the favor later in the year.

A tree infested by damaging insects or fungi can alert neighboring trees of its condition by releasing specific molecules through the mycelium network and directly through the air. The neighboring trees will then turn on their self-defense mechanisms. This same sort of communication helps the forest to manage itself in times of stress, such as during a drought. It also explains how certain trees are able to coordinate the release of their seeds, not by some external signal, such as the length of the day, but through internal signals that they share. In releasing their seeds all at once, helps attract animals that act to spread those seeds far and wide.

This communication among the trees and within their community is key to the health of a forest. Thus, there is a significant difference between a long-standing diverse native forest and a cultivated mono-crop forest where all trees are of the same age and are harvested the moment they reach a given height. A truly healthy forest is one where new and old-growth trees care for each other while responding to a dynamic environment.

Figure C
Within a healthy forest, underground fungal mycelia form an expansive network through which chemicals can pass. Trees connect to this network through their root system.

Figure D
The transpiration occurring within forests plays a significant role within the hydrologic cycle.

Forest and the Water Cycle

Taking a step back from the forest to a collection of many forests, we see an even bigger picture. On a planetary scale, we know that trees absorb large amounts of carbon dioxide, CO_2, while also releasing oxygen, O_2. Both these actions are very good things supporting oxygen-breathing organisms such as ourselves. Perhaps less well appreciated is the world's forests' role in the water cycle. Through *transpiration*, each tree absorbs water from the soil and releases it through its leaves into the atmosphere. Because of local geography, the area of the Amazon River of Brazil easily tends to be a grassland ecosystem. However, transpiration from the expansive forests in the area creates the moist conditions indicative of the Amazon Rainforest—where the rain is literally coming from the forest.

When you think of the water cycle, it's easy to think of water evaporating from our oceans, creating clouds that precipitate over the land, thereby creating rivers and lakes of fresh water. However, any water evaporating from the ocean is typically precipitated within 500 kilometers of the coast. Beyond that 500 kilometers, where does the atmospheric humidity come from? A good part of this humidity comes from forests. This explains a virtuous cycle: There will be trees and other photosynthesizing plants where there is just enough moisture. Where there are trees, there will be more moisture, which allows for more life to prosper, including more trees.

Lastly, the water cycle is but one of many cycles operating on this living Earth. There is also the carbon cycle, the nitrogen cycle, the rock cycle, as well as others. What all these cycles have in common is the movement of chemical elements and compounds through the biosphere, geosphere, hydrosphere, and atmosphere. While it's helpful to focus on a particular cycle, it should be understood that no cycle operates independently of the others. Ubuntu and the ideas of interconnectedness—applied to both the living and non-living—tell us there can't be one without the other. To Life!

CONCEPT CHECK

What effect might there be upon cutting down trees around and within the Amazon rainforest?

CHECK YOUR ANSWER

Cutting down trees pushes toward the reverse of a virtuous cycle. Cut down the tree, and there will be less transpiration, meaning less moisture for any other tree or living organism. Cutting down a tree is not just cutting down a tree—it's cutting down a rainmaker.

READING CHECK

How is it possible to tell that trees are sharing sugar through the mycelium network?

fyi FOR YOUR INFORMATION

Life is interconnected down to the level of atoms and molecules. Consider that a living organism's atoms are the same ones that were once within other living organisms. We don't own our atoms; we are their temporary caretakers. Consider amino acids come in left-handed and right-handed forms, yet all organisms on Earth rely on only the left-handed amino acids. Why? Because an organism made of right-handed amino acids could not digest the plentiful left-handed amino acids within its environment. Nor would that organism be digestible by any predator. Such an organism could not participate in the interconnected living world as we know it.

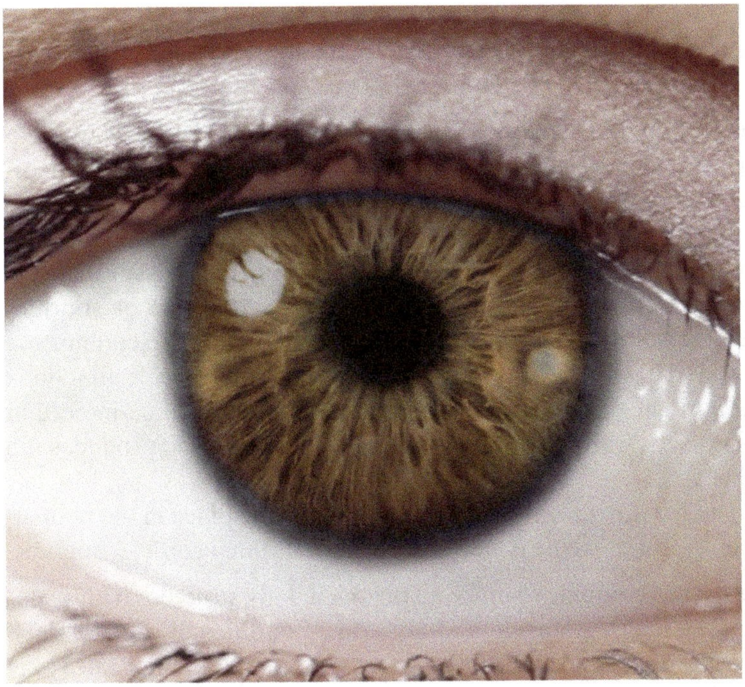

Chapter 12

The Nervous System

THE MAIN IDEA

The nervous system controls and coordinates bodily movement and functions.

12.1 Organization of the Human Body
12.2 Homeostasis
12.3 The Brain
12.4 The Nervous System
12.5 How Neurons Fire
12.6 How Neurons Communicate
12.7 The Senses

Cells combine to make tissues, which combine to make organs. Related organs work together within an organ system. The coordination of multiple organ systems is what makes for a living organism. In this chapter, you will learn about the nervous system and the senses. How can we ride a bike or fold laundry without even thinking about it? What part of the brain is working when we write a poem or do a math problem? How can we tell that a strawberry is tasty but a rotten egg isn't? You will learn all this and more.

12.1 Organization of the Human Body

You look at the traffic light and step off the curb. As you cross the street, you consider whether you want tacos or a sandwich for lunch. You breathe in, bringing oxygen into your body, and then breathe out, removing carbon dioxide. Blood, pumped by your heart, moves through your blood vessels. At the same time, immune cells crowd around a small cut on your finger, dispatching the invading bacteria. At any moment, your body is involved in numerous different activities. To do their jobs, the cells in your body are organized into tissues, organs, and organ systems (**Figure 12.1**).

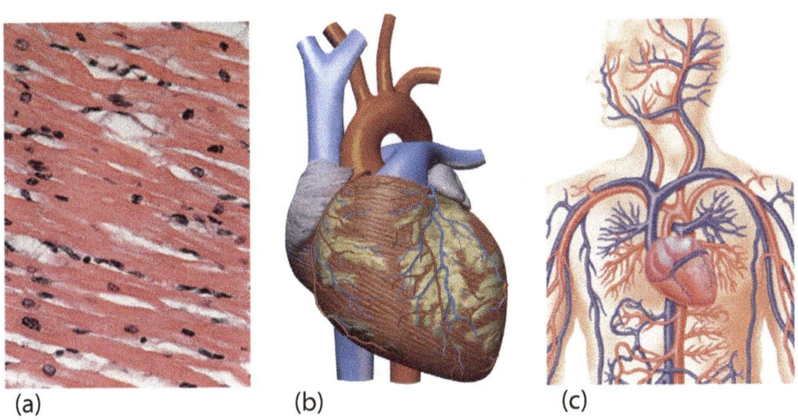

< Figure 12.1

The human body organizes cells into tissues, organs, and organ systems. (a) Many muscle cells make up muscle tissue. This photo shows muscle tissue from a human heart. (b) The heart is an organ. (c) The circulatory system comprises the heart, blood, and blood vessels. The circulatory system transports nutrients, gases, and wastes to different parts of the body.

A **tissue** is a group of similar cells that performs a distinctive function. There are four main types of tissue in the body.

- *Epithelial tissue* consists of sheets of tightly packed cells covering the body's internal and external surfaces. Skin is an example of epithelial tissue.
- *Connective tissue* consists of cells scattered within an external matrix. Bone, cartilage, and blood are all connective tissues.
- *Muscle tissue* is made up of cells that can contract or shorten. Three types of muscle tissue are found in the body. Skeletal muscle is responsible for voluntary movements. Smooth muscle functions in the internal organs of the digestive system as well as in certain blood vessels. Cardiac muscle is found in the heart and produces the heartbeat.
- *Nervous tissue* transmits information from one place in the body to another. Nervous tissue is found in the brain, spinal cord, and nerves.

READING CHECK

Where can nervous tissue be found in the body?

Multiple tissues combine to make an **organ**, a body structure with a specific function. The heart, stomach, and brain are examples of organs. Each of these organs is made up of multiple tissues. For example, the heart (1) is surrounded by epithelial tissue on the outside and lined with it on the inside; (2) contains blood vessels carrying blood, a connective tissue; (3) has walls made up mainly of cardiac muscle; and (4) contains nerves, made of nervous tissue, that help to control its activity.

An **organ system** is a set of organs that work together to perform a particular bodily function. For example, the circulatory system moves nutrients, gases, and wastes throughout the body. The human body has ten major organ systems: nervous, sensory, endocrine, reproductive, muscular and skeletal, circulatory, respiratory, digestive, excretory, and immune. In the following chapters, we'll examine what these organ systems do.

CONCEPT CHECK

You are at the gym lifting weights. You flex your arm, using your biceps muscle to lift a barbell. Is muscle tissue involved in this action? Are the other types of tissues (epithelial, connective, and nervous tissues) involved in this action?

CHECK YOUR ANSWER

Muscle tissue is definitely involved. Your biceps muscle is made up mainly of skeletal muscle cells that are able to contract. The muscle itself is surrounded by epithelial tissue. Blood, which is a connective tissue, supplies the muscle with the oxygen it needs to contract. In addition, muscle contraction can make your arm bend and lift the weight because the muscle is attached to bones, another connective tissue, in your arm. Finally, nerves, made of nervous tissue, direct the biceps muscle contraction as you lift the barbell.

More to Explore

How cells are organized into tissues, organs, and organ systems

https://www.bbc.co.uk/bitesize/guides/z9hyvcw/revision/6

12.2 Homeostasis

Whether swimming in icy waters or hiking through scorching heat, your body temperature stays close to 37°C (98.6°F). This consistency in body temperature is an example of homeostasis, the maintenance of a relatively stable internal environment. Homeostasis is a characteristic of all living organisms, and a considerable amount of the body's activity goes toward maintaining it.

As another example of homeostasis, your cells need a certain amount of oxygen to function. Your lungs and heart maintain a normal level of activity to supply this oxygen. If your activity level increases—say, because you run to catch a bus—your cells use more oxygen. What happens? Your body responds by breathing harder to take in more oxygen and increasing your heart rate to move that oxygen to your cells. Once your activity level returns to normal and your oxygen use decreases, your breathing and heart rate slow again.

To go back to body temperature, when it's cold outside, you feel cold and pile on more clothes or wrap your arms around your body to reduce heat loss. In addition, less blood is sent to your limbs and extremities, which lose heat faster than your body's core. (This explains why your fingers and toes often feel most cold when you're cold.) You may also shiver to generate heat (**Figure 12.2**). On the other hand, when it's hot outside, you take off clothes, look for shade, and sweat to cool off. Also, more blood goes to the extremities and the face, which are good at shedding heat. (This explains why your face turns red when you're hot.)

READING CHECK

How does the flow of blood through your body change when it's hot outside?

Controlling body temperature is an example of *feedback regulation*. In feedback regulation, changes in one variable affect a second variable, and changes in the second variable, in turn, affect the first variable. We say the variables are related in a *feedback loop*. In this example, changes in body temperature trigger specific responses that, in turn, affect body temperature. Homeostasis is often maintained by a type of feedback regulation called *negative feedback*. In negative feedback, a stimulus causes a response that counters the stimulus. So, if body temperature decreases, the response will cause body temperature to increase back to normal. On the other hand, if body temperature increases, the response will cause body temperature to decrease back to normal.

Oxygen supply and body temperature are only two of the many variables the body carefully regulates. The amount of water in the body, the concentration of nutrients and waste products in the blood, the concentrations of essential ions inside and outside cells, and blood pH—these and many other variables are carefully regulated. When all this regulation is working well, we say that the body is maintaining homeostasis.

(a) (b)

< Figure 12.2

Your body responds to changing conditions in ways that help maintain homeostasis. (a) When you're cold, you shiver to warm up. (b) When you're hot, you sweat to cool down.

CONCEPT CHECK

Your activity level increases and your cells begin to use more oxygen. Then, you begin to breathe harder and take in more oxygen. Your heart rate also increases to help move that oxygen to your cells. Is this an example of negative feedback? Why or why not?

CHECK YOUR ANSWER

Negative feedback describes a situation in which a stimulus causes a response that counters the stimulus. In this example, the stimulus is that your cells begin to use more oxygen which causes the oxygen levels in your tissues to decrease. The response is that you breathe harder, and your heart beats faster, transporting more oxygen to your tissues and increasing oxygen levels there. In other words, this response (increased oxygen) counters the stimulus (decreased oxygen), so yes, this is an example of negative feedback.

More to Explore

More about homeostasis at this website:

https://www.livescience.com/65938-homeostasis.html

12.3 The Brain

Your brain strongly influences who you are. All of your thoughts, feelings, and desires come from your brain. Your brain also controls all of your activities—conscious ones, such as choosing a shirt or kicking a soccer ball, and unconscious ones, such as breathing, blood circulation, and digestion. Your brain "heads up" your nervous system. It is made up of five main parts (**Figure 12.3**).

1. **Brainstem** Can you imagine how much less you would get done if you had to remember to breathe every few seconds? It's a good thing the brainstem does this for you. The brainstem connects the spinal cord to the rest of the brain. It controls many of the body's primary involuntary activities, such as heartbeat, respiration, and digestion. The brainstem also wakes you up every morning, bringing your body from sleep to wakefulness (**Figure 12.4**).

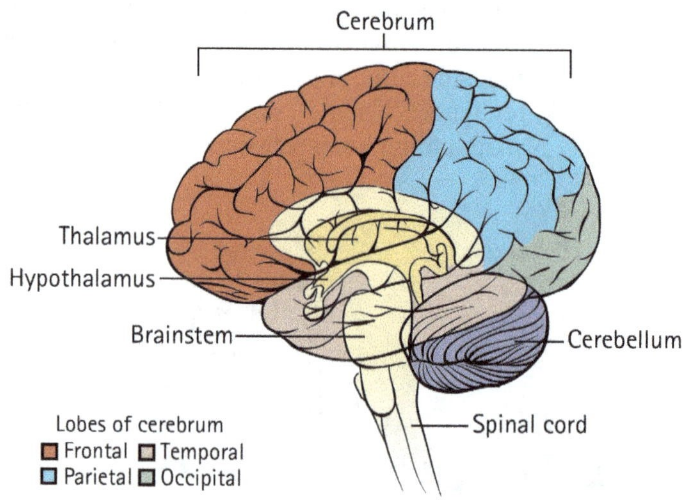

Figure 12.3 >

The major parts of the human brain are the brainstem, cerebellum, cerebrum, thalamus, and hypothalamus.

2. **Cerebellum** The cerebellum controls balance, posture, coordination, and fine movements. It also controls all the motions you perform "without thinking." Consider bicycle riding. When you first learn how to ride a bike, you think carefully about what your arms and legs are doing, and your motion is very awkward. With enough practice, though, you can ride a bike easily—"without thinking." When you do something "without thinking," it means the cerebellum has taken over.

3. **Cerebrum** The cerebrum is the largest part of the brain. It contains more than 10 billion neurons! The cerebrum is responsible for high-level functions such as reasoning, problem-solving, language, and creativity. Your personality also resides in your cerebrum. The cerebrum collects information from the senses and controls all the conscious, voluntary activities of the body. The cerebrum has a right hemisphere (right side) and a left hemisphere (left side). The right hemisphere controls the left side of the body, and vice versa. This means that when you move your right hand, the left hemisphere of your cerebrum controls that action.

   Most of the information processing in the cerebrum occurs in the cerebral cortex, the thin layer covering the surface of the cerebrum. "Wrinkles" in the cerebral cortex give the brain its familiar convoluted appearance and increase the surface area available for information processing.

   Each cerebral hemisphere consists of four lobes that are responsible for different activities. The *frontal lobes* deal with reasoning, voluntary movements, and speech. The *parietal lobes* take in sensory information about temperature, touch, taste, and pain. The *occipital lobes* process what you see—that is, visual information. The *temporal lobes* deal with sound and help you comprehend language.

   The control of certain cognitive functions is dominated by either the right or left cerebral hemisphere. The left hemisphere is more adept at math, reasoning, language, and detail-oriented activities. The right hemisphere is more adept at spatial relationships, emotional processing, and music. This distinction has led to the popular conception of "left-brained" people who are organized, analytical, and attentive to detail and "right-brained" people who are intuitive, flexible, and creative.

   Incidentally, this description of the different functions of the two sides of the brain applies to most right-handed and left-handed people. However, where 5% of right-handed people may have some of the functions "flipped" or reversed, up to 30% of left-handed people have brains with reversed functions.

4. **Thalamus** The thalamus receives information from many different parts of the brain. It sorts and filters this information and then passes it on to the cerebrum.

5. **Hypothalamus** The hypothalamus is responsible for emotions such as pleasure and rage. It also controls bodily drives such as hunger, thirst, and the sex drive and regulates body temperature and blood pressure. Another function of the hypothalamus is to control your body's internal clock, which tells you when it is day and when it is night. The hypothalamus performs some of its activities using molecules called *hormones*, which we will discuss in the next chapter.

Figure 12.4

The brainstem controls the transition between sleep and wakefulness.

READING CHECK

What part of your brain controls emotions such as pleasure and rage?

CONCEPT CHECK

1. Io is reading a book. Which lobes of the cerebrum is she using?

2. In sports, it's called choking. It's a tense, decisive moment in a game, and the pressure is on. The ball comes toward you—it's an easy play. But suddenly, you become highly aware of all your movements, and then it happens—the ball rolls right between your legs. What part of the brain normally controls practiced, automated movements? What part of your brain takes over when you "choke?"

CHECK YOUR ANSWERS

1. Io is using her occipital lobes to process visual information (the letters on the pages of her book) and her temporal lobes to comprehend language.

2. Automated movements that you can perform "without thinking" are normally controlled by the cerebellum. You use the cerebrum's frontal lobes when you consciously control your movements and "choke."

More to Explore

To read more about the parts of the brain and see a cool 3-D animation of the brain:

https://www.visiblebody.com/learn/nervous/brain

12.4 The Nervous System

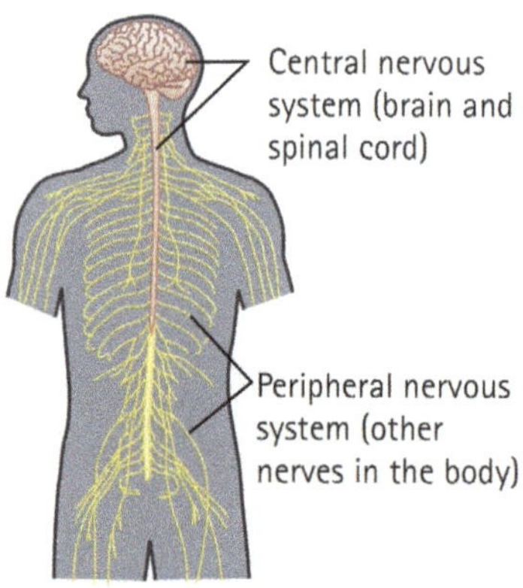

Λ Figure 12.5

The central nervous system consists of the brain and spinal cord. The peripheral nervous system includes all the other nerves in the body.

You're eating cornflakes and reading the sports page. You glance at the clock and see that you're late for class. Immediately, your body goes into overdrive. Your heart starts to pound, and you begin to sweat. You dash out the door, run down the steps—and trip. Pain shoots through your knee, but you pull yourself up and continue your race to class. There, you drop into your seat. Your heart rate gradually slows, and your body relaxes. This entire sequence of actions and responses is controlled by your nervous system.

The *nervous system* collects information about the body's internal and external environments and controls the body's activities. It can be divided into two parts. The **central nervous system** includes the brain and spinal cord, and the **peripheral nervous system** includes all the other nerves in the body (**Figure 12.5**).

The nervous system includes two different kinds of cells: neurons and glial cells. A **neuron** is a specialized cell that receives and transmits electrical signals from one part of the body to another (**Figure 12.6**). A typical neuron has three parts: (1) dendrites receive information from other neurons or cells, (2) the cell body contains the neuron's nucleus and organelles, and (3) the axon transmits information to other neurons or cells. *Glial cells* support, insulate, and protect neurons. The human brain has 10 to 50 times more glial cells than neurons.

READING CHECK

Name the three parts of a neuron.

Kinds of Neurons

There are three types of neurons: sensory neurons, interneurons, and motor neurons (**Figure 12.7**). **Sensory neurons** carry information from the senses to the central nervous system. The neurons that transmit visual information from your eyes to your brain are sensory neurons. In the story about your

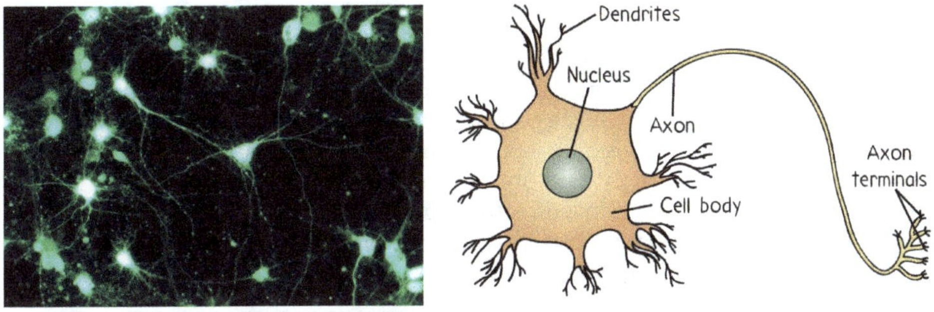

(a) (b)

< Figure 12.6

(a) Neurons transmit electrical signals from one part of the body to another. The green-blue cells shown here are neurons. (b) Neurons have a cell body, dendrites that receive information from other cells, and an axon that transmits information to other cells.

run to class, sensory neurons told your brain about the time on the clock and the pain of hitting your knee on the walkway. **Interneurons** connect neurons to other neurons. Interneurons are found exclusively within the central nervous system. **Motor neurons** carry instructions from the central nervous system to muscles or organs. During your run to class, motor neurons directed the motion of your legs. A motor neuron also made your heart beat faster. Motor neurons belong to either the somatic nervous system or the autonomic nervous system. Motor neurons in the **somatic nervous system** control voluntary actions, such as running to class. These neurons provide instructions to your voluntary muscles, such as the muscles of your arms, legs, and fingers. Motor neurons in the **autonomic nervous system** control involuntary actions. These neurons control organs and involuntary muscles (such as the heart or stomach muscle).

The neuron that made your heart beat faster is part of your autonomic nervous system. The autonomic nervous system includes a sympathetic and a parasympathetic division (Figure 12.8). The **sympathetic nervous system** prepares your body for danger. It triggers a "fight or flight" response. "Fight or flight" refers to the fact that when you're threatened, you need to either run away ("flight") or stay and "fight." The sympathetic nervous system sped up your heartbeat when you realized you were late for class, speeding the transport of oxygen to your muscles. The sympathetic nervous system also slows down digestion and other activities that are unimportant during emergencies. The **parasympathetic nervous system** works during calmer times. Its effects are the opposite of those of the sympathetic nervous system—it calms you down. Your parasympathetic nervous system took over after you got to class and were able to relax.

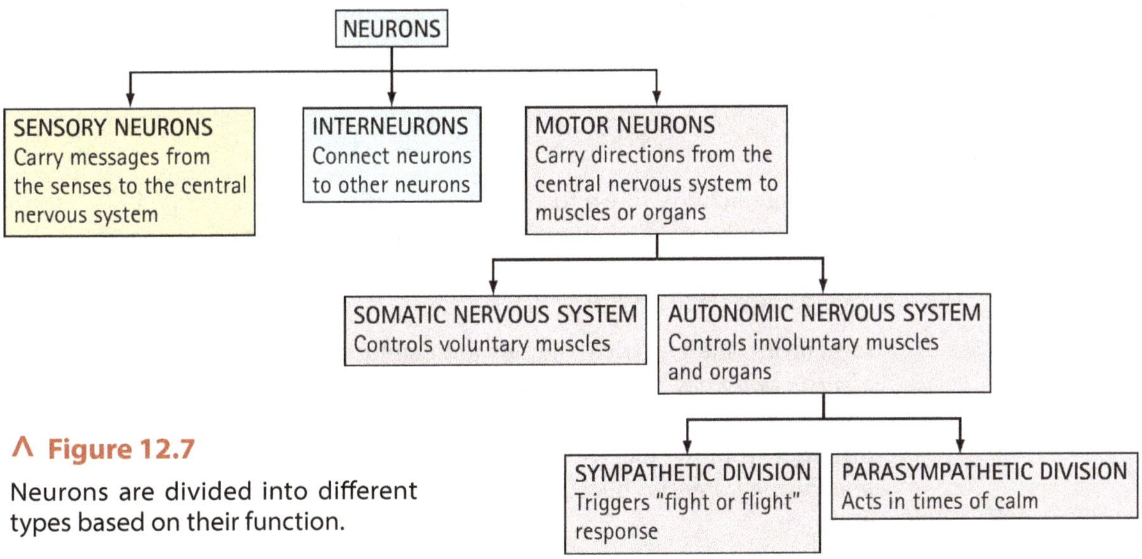

Λ Figure 12.7

Neurons are divided into different types based on their function.

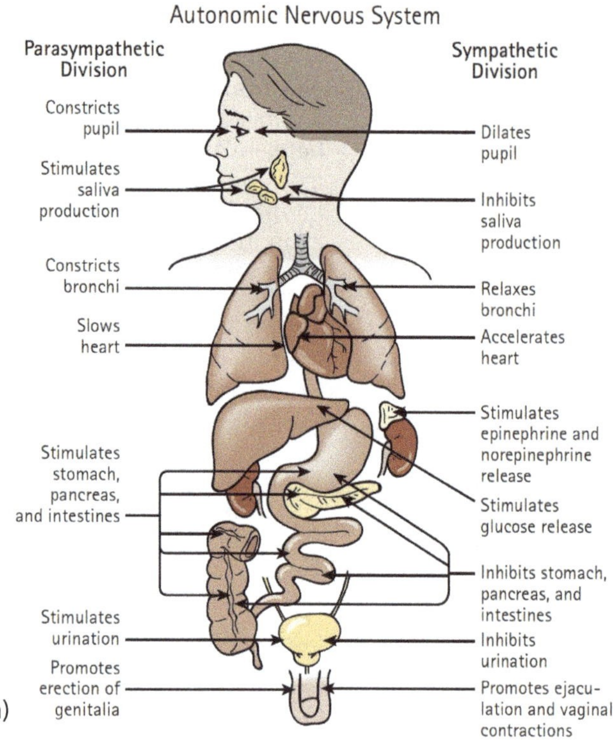

(a)

< Figure 12.8

The autonomic nervous system has a sympathetic division and a parasympathetic division. (a) The sympathetic division prepares the body for danger—it initiates the "fight or flight" response. The parasympathetic division acts in a directly opposed way and operates in times of calm. (b) Is this man's sympathetic nervous system working hard right now?

(b)

CONCEPT CHECK

1. You stand at the edge of a pool and wonder whether the water is too cold for a swim. What kinds of neurons do you use to find out?
2. What kind of motor neuron did you use to test the water—somatic or autonomic?

CHECK YOUR ANSWERS

1. You use a motor neuron to dip your toe into the pool. Then a sensory neuron in the skin of your toe (one that senses temperature) tells you how cold the water is.
2. Somatic, because dipping your toe in the pool is a voluntary action.

More to Explore

To learn more about the nervous system::

https://www.news-medical.net/health/What-is-the-Nervous-System.aspx

12.5 How Neurons Fire

Is the human body like a toaster? Well, it is in one small way. Like a toaster or a computer, your body relies on electrical signals to do its work.

The Action Potential

In neurons, the electrical signals are changes in the voltage, also known as *electric potential*, across the cell membrane. A neuron has an electric potential across its cell membrane because the electric charge inside a neuron is different from the electric charge outside. This electric potential is called a *membrane potential* because the cell membrane keeps the charged particles separate.

When a neuron is not signaling, it is at its *resting potential*. The resting potential of a neuron is negative. Why? First, like other cells, neurons have more potassium ions inside the cell than outside and more sodium ions outside the cell than inside. The overall effect of these concentration gradients is to help establish a negative resting potential. (You may remember that the sodium-potassium pump maintains these concentration gradients by using active transport to move sodium ions out of cells and potassium ions into cells.) Second, neurons contain many negatively charged ions, including proteins and other organic molecules. As a result, the inside of a neuron is normally negatively charged, and the outside is normally positively charged, creating a resting potential of about -70 millivolts (mV) across the cell membrane.

A neuron sends information by signaling or "firing." How does this happen? A neuron is stimulated when its membrane potential is increased. If a neuron's membrane potential reaches a threshold value of about -55 mV, sodium channels in the neuron's cell membrane suddenly open, allowing positively charged sodium ions to move into the neuron. (Sodium ions move into the neuron because more sodium ions are outside the cell than inside.) This influx of positively charged ions causes the membrane potential to spike and become positive. This spike is called an **action potential**. The action potential is the neuron's way of signaling or "firing." Once the spike occurs, the sodium channels quickly close, and the potassium channels open. Positively charged potassium ions flow out of the neuron (they flow out because there are more potassium ions inside the cell than outside), causing the membrane potential to return to resting potential. The sequence of events in an action potential is shown in **Figure 12.9**.

An action potential is an all-or-nothing event—a neuron either fires or doesn't. A neuron cannot fire "harder" when a

READING CHECK

What happens when a neuron's membrane potential reaches its threshold value?

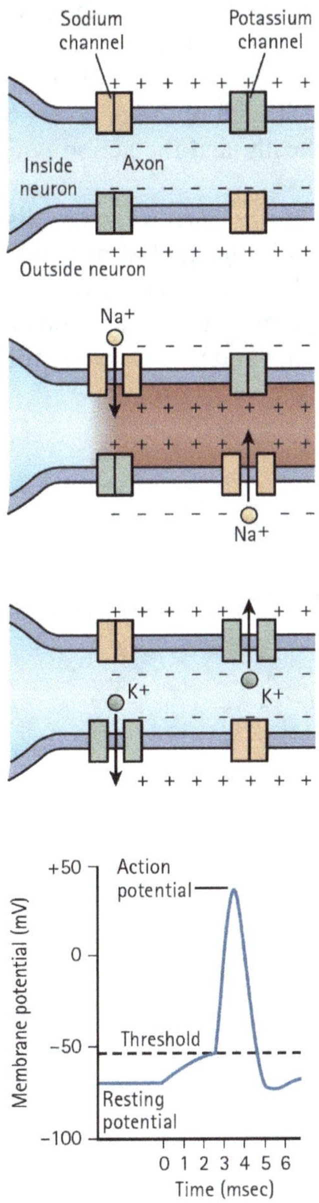

(a) The inside of the neuron is negatively charged, and the outside is positively charged.

(b) During an action potential, sodium channels in the neuron's cell membrane open. Sodium ions move into the neuron.

(c) Sodium channels close, and potassium channels open. Potassium ions move out of the neuron, causing the membrane potential to return to resting potential.

(d) This graph shows how the membrane potential changes during an action potential. Resting potential is negative. When the neuron is stimulated, the membrane potential increases slowly until it reaches the threshold. Then the action potential happens—sodium channels open, causing the membrane potential to spike. The membrane potential decreases when potassium channels open. It returns to resting potential.

Figure 12.9 >

The action potential is the neuron's way of signaling.

stimulus is more intense. However, it does fire more often. For example, touch sensors in your skin fire slowly when they feel a little pressure (tick ... tick ... tick ...) and quickly when they feel lots of pressure (tick-tick-tick-tick-tick).

How an Action Potential Is Propagated

An action potential doesn't spike everywhere on a neuron's cell membrane at once. It begins at one end of the axon, near the neuron's cell body, and then travels down the axon. How does this happen?

When an action potential begins, sodium ions enter the end of the axon that is closest to the cell body. These ions diffuse into adjacent areas along the axon. Because the sodium ions are positively charged, they cause the local membrane potential to increase. When the local membrane potential reaches a threshold, a new action potential, farther along the axon, begins. The process is similar to the way a row of dominoes falls: The first domino knocks down the next one, which knocks down the next one, and so forth. Similarly, an action potential sets off an action potential farther down the axon, and the action potential travels down the entire axon (**Figure 12.10**).

Some axons, like the one shown in **Figure 12.11**, are surrounded by a *myelin sheath* that allows the neuron's signal to be transmitted more quickly. The sheath consists of glial cells wrapped around and around the axon, insulating it with multiple layers of cell membrane. There are periodic gaps in the sheath, and the action potential jumps down the axon from one gap to the next.

In *multiple sclerosis* disease, the body's immune system destroys myelin in the central nervous system. The symptoms of multiple sclerosis vary depending on the parts of the brain and spinal cord that are affected. Symptoms can include problems with fatigue, dizziness, bladder and bowel control, vision, muscle control, and balance.

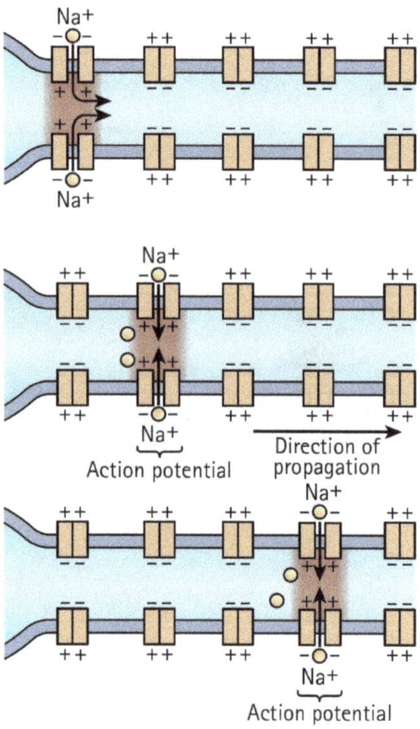

Λ Figure 12.10

Action potentials are propagated down an axon. The action potential starts near the cell body. Sodium ions that enter the neuron diffuse down the axon, initiating an action potential farther along the axon. The action potential continues to move down the axon until it reaches the end of the axon.

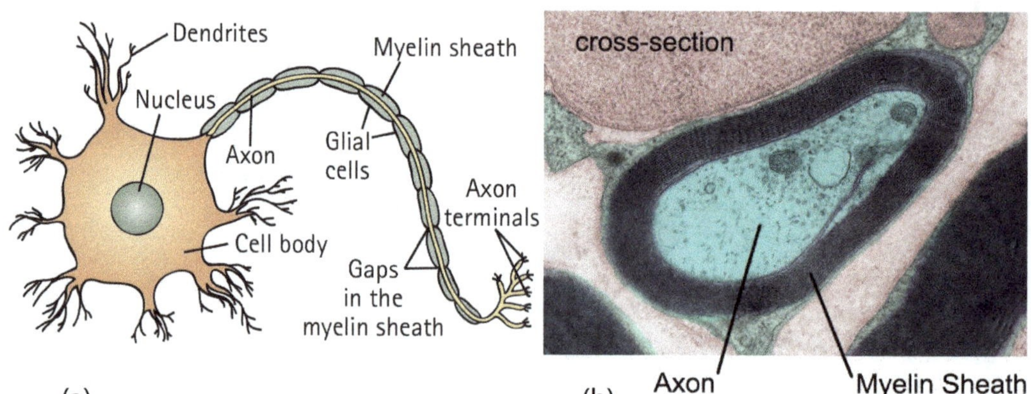

(a) (b)

Λ Figure 12.11

(a) A myelin sheath surrounds some axons, enabling the signals they transmit to travel faster. (b) The myelin sheath consists of glial cells wrapped around and around an axon, as shown in this cross-section. The dark ring is the myelin. (The blue is the cytoplasm of the glial cell.)

CONCEPT CHECK

1. What causes sodium channels to open, initiating an action potential?
2. What causes an action potential to end?

CHECK YOUR ANSWERS

1. A stimulus that increases the membrane potential to its threshold value.
2. The opening of potassium channels returns the membrane potential to its resting potential.

More to Explore

For more on the action potential:

https://faculty.washington.edu/chudler/ap.html

12.6 How Neurons Communicate

So far, we have seen what happens during an action potential and how an action potential travels down an axon. What happens when an action potential arrives at the end of an axon? At the end of an axon, a neuron connects with one or more target cells and passes information to them. A target cell can be another neuron or a cell that does something, such as a muscle or organ cell. A connection between a neuron and a target cell is called a **synapse**. There are two types of synapses: electrical synapses and chemical synapses.

In an *electrical synapse*, tiny channels in the cell membrane called *gap junctions* connect a neuron to a target cell. When the neuron fires, ions move through these channels into the target cell, starting an action potential in the target cell (**Figure 12.12a**). Because the ions move very quickly, electrical synapses are very fast. Electrical synapses are found in places where speed is essential. For example, in the heart, electrical synapses allow many muscle cells to contract nearly simultaneously to produce the heartbeat. Despite their speed, though, electrical synapses are fairly rare in the body.

Most synapses are *chemical synapses*. Chemical synapses are not as fast as electrical synapses, but they allow for a finer degree of control. In a chemical synapse, a neuron communicates with a target cell using **neurotransmitter** molecules. When an action potential arrives at the end of an axon, vesicles containing neurotransmitter fuse to the neuron's cell membrane (**Figure 12.12b**). Neurotransmitter molecules are released into the narrow space between the axon and the target cell and diffuse to the target cell. There, the neurotransmitter molecules bind to receptors on the target

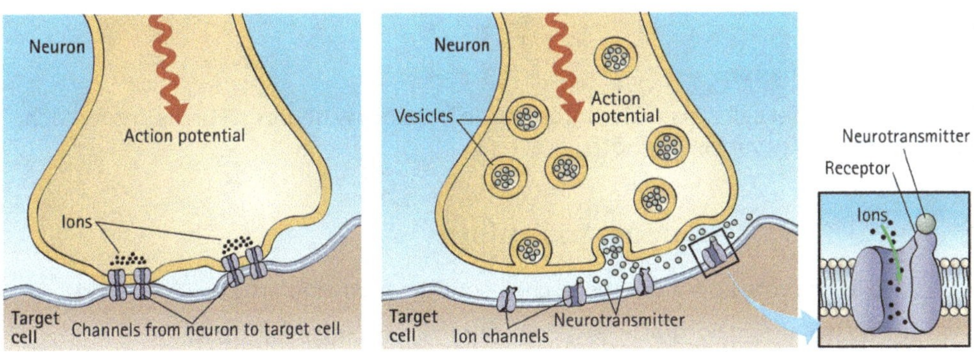

(a) (b)

< Figure 12.12

(a) In an electrical synapse, ions move through the channels that connect a neuron to a target cell and start an action potential in the target cell. (b) In a chemical synapse, a neuron uses a neurotransmitter to communicate with a target cell. The neurotransmitter is released when an action potential arrives at the end of the axon. The neurotransmitter diffuses to the target cell and binds to receptors on the target cell's membrane, causing ion channels to open. Ions enter the target cell and change its membrane potential.

cell's membrane. The binding causes ion channels to open, and ions enter the target cell.

The membrane potential of the target cell may increase or decrease depending on the type of ion that enters. An *excitatory signal* increases the target cell's membrane potential and makes the target cell more likely to fire. An *inhibitory signal* decreases the target cell's membrane potential and makes it less likely to fire. A target cell may receive signals from many other neurons (**Figure 12.13**). A typical neuron has between 1,000 and 10,000 synapses with other neurons. Depending on the sum total of their effects, the target will either reach its threshold and fire or not.

READING CHECK

How many synapses does a typical neuronhave with other neurons?

The effect of a neurotransmitter on a target cell ends when the neurotransmitter is removed. Neurotransmitter molecules can be degraded by enzymes, taken up by the original neuron and repackaged into vesicles, or collected and broken down by glial cells. Some medications affect neurotransmitter function. For example, certain anti-depressants are selective serotonin re-uptake inhibitors (SSRIs). SSRIs prevent neurons from taking up the neurotransmitter serotonin after it has been released at synapses. As a result, more serotonin remains to bind to and stimulate target neurons.

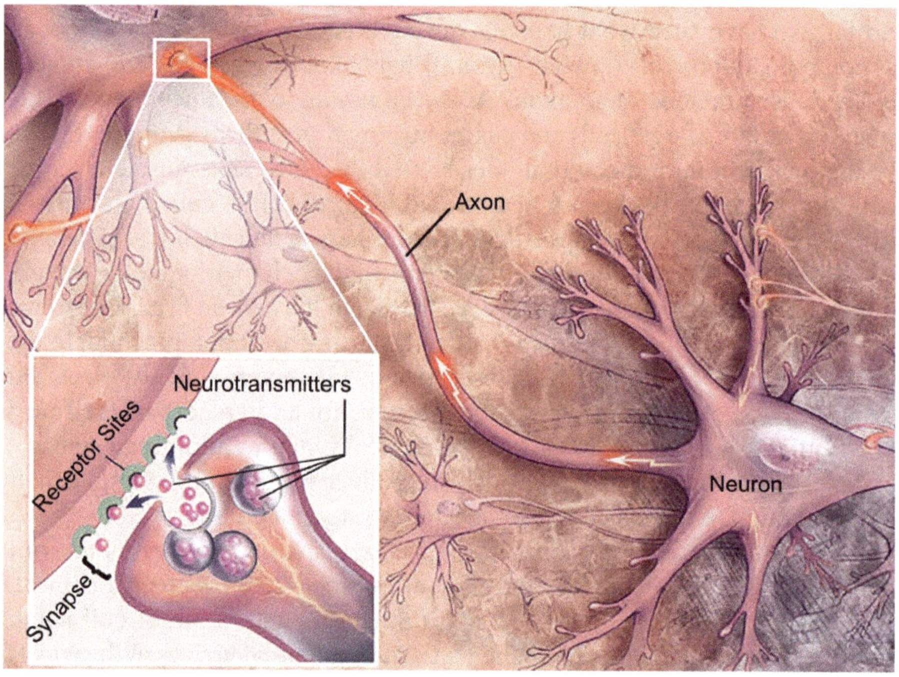

Figure 12.13 >

Axons from many different neurons have synapses with this target cell. Some of these signals make the target cell more likely to fire, whereas others make it less likely to fire.

CONCEPT CHECK

1. What is the advantage of an electrical synapse over a chemical synapse?
2. What is the advantage of a chemical synapse over an electrical synapse?

CHECK YOUR ANSWERS

1. An electrical synapse is faster than a chemical synapse.
2. A chemical synapse allows for finer control. A target cell can receive information from many neurons. Furthermore, each signal may be either excitatory or inhibitory and may vary in strength.

More to Explore

To learn more about synapses, check out the following article:
https://www.dana.org/article/qa-neurotransmission-the-synapse/

12.7 The Senses

Your senses are your connection to the world. Each of your senses takes information from the environment—light, sound, touch, or molecules—and converts it into action potentials that are sent to the brain. As you will see, each sense accomplishes this in its own way.

Vision

The light enters your eyes (**Figure 12.14a**) through a tough, transparent layer called the *cornea*, which is continuous with the "whites" of your eyes. Light then passes through a small hole, the *pupil*. The *iris*, the part of the eye that gives you your eye color, surrounds the pupil and controls its size. In bright light, the pupil is small. In dim light, the pupil expands to let in more light. From the pupil, light passes through the lens. The *lens* focuses light on the **retina** at the back of the eyeball. The retina is covered with light-sensitive cells called *rods* and *cones* (**Figure 12.14b**). Light hitting the rods and cones changes the action potentials they transmit to the brain. The *optic nerve* is the bundle of neurons that takes visual information from the retina to the brain.

The two types of light-sensitive cells in your eyes, rods and cones, have different functions. *Rods* are very sensitive to light and are especially important for seeing in dim light. Rods cannot discriminate colors and allow you to see only black, white, and shades of gray. This is why, in a dark room, you cannot see the difference between a navy-blue shirt and a maroon shirt. (Most people are so used to this that they don't even realize they've lost their color vision!) Rods are also not very good at making out fine details, which is why your night vision is grainy and not very sharp. *Cones* detect color. Your eyes have three types of cones that respond most strongly to red, green, and blue light. All the shades you see are made up of different combinations of these three colors. Color blindness results from having a non-functioning version of one or more cone types. Rods and cones make up 70% of the sense cells in the body—evidence that vision is our best-developed sense.

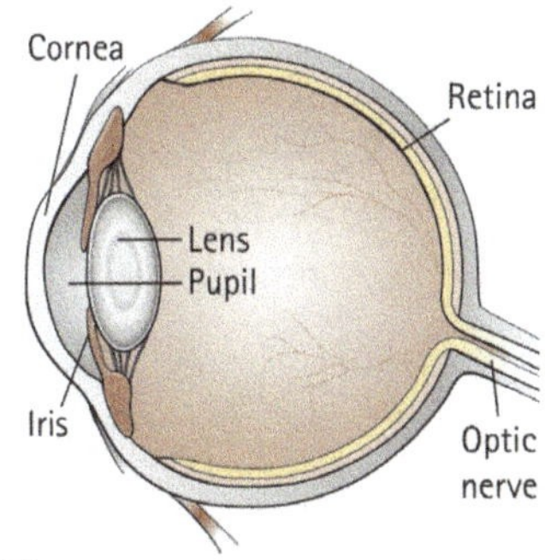

(a)

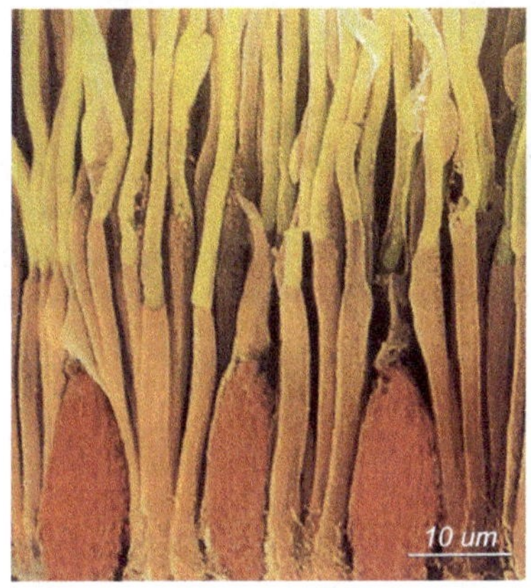

(b)

Figure 12.14

(a) The eyes convert light into action potentials. (b) Rods and cones are the light-sensitive cells in the eyes. As you can see, both rods and cones are named for their shape.

Hearing

The ear consists of three parts: the outer ear, middle ear, and inner ear (**Figure 12.15a**). Sound waves move through the air to the *pinna*, the cartilaginous flap on the side of the head. The pinna funnels the waves in, and they move toward a thin membrane of skin—the *eardrum*. Sound waves make the eardrum vibrate, similar to how blowing on a piece of paper makes the paper vibrate. The eardrum's vibrations move three middle ear bones—the hammer, the anvil, and the stirrup—in sequence. These bones amplify the vibrations, making them more powerful. The stirrup then transfers the vibrations to the fluid-filled inner ear.

In the inner ear, sound vibrations enter the cochlea, a coiled tube that holds the organ of Corti. The **organ of Corti** contains the sensory cells responsible for hearing. Fluid vibrations in the inner ear move the organ of Corti's basilar membrane, causing sensory "hairs" embedded in it to brush against an overlying membrane and bend (**Figure 12.15b**). This bending causes ion channels to open, initiating action potentials transmitted by the auditory nerve to the brain. Note that the "hairs" of hearing cells are not like the hairs on your head; they are long extensions of sensory cells that happen to be shaped like hairs (**Figure 12.15c**). You can distinguish different noises—the

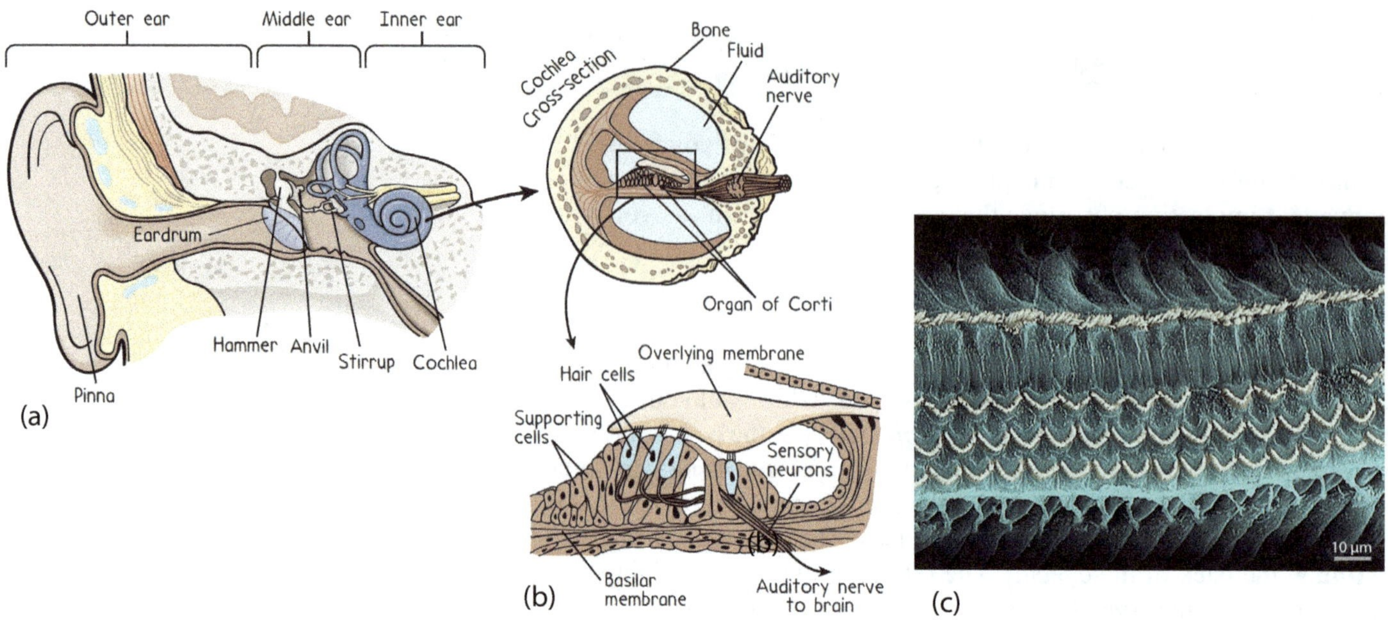

Figure 12.15

(a) The ear includes the structures of the outer ear, middle ear, and inner ear. (b) Sound-sensitive cells are contained in the organ of Corti. Fluid vibrations in the inner ear vibrate the basilar membrane of the organ of Corti, causing sensory "hairs" to brush against the overlying membrane and bend. The bending opens ion channels, starting action potentials. (c) This microscope photo shows the "hairs" (yellow) in the organ of Corti.

high pitch of a siren versus the low pitch of a jackhammer—because different parts of the organ of Corti respond to different pitches.

Smell and Taste

The senses of smell and taste work through chemoreception. In **chemoreception**, chemicals bind to receptors in the cell membrane of chemosensory cells. The binding causes ion channels to open, initiating action potentials (**Figure 12.16**).

The sensory cells for smell lie in two patches at the top of your nasal passages. Each patch is about the size of a dime. Humans have more than 1000 different kinds of chemosensory cells for smell, each containing only one or a few different kinds of receptors. Because different chemicals trigger different combinations of sensory cells, you can distinguish well over 10,000 distinct odors.

One interesting thing about the sense of smell is that smells are sometimes linked to emotionally powerful memories. Have you ever had a smell bring back a memory? The part of the brain that processes smells is very close to two important parts of the brain: the hippocampus and the amygdala. The hippocampus is responsible for memory. The amygdala is involved with emotion. Using fMRI experiments, scientists have shown that smells can trigger memories and experiences in extraordinary ways. This is not the case for our other senses.

Taste cells cluster in small bumps called *taste buds*. Taste buds are located on your tongue, the insides of your cheeks, and the roof of your mouth. Humans distinguish five basic tastes: sweet, salty, sour, bitter, and umami. *Umami* is the Japanese word for "delicious," and it describes the flavor found in foods such as meat, mushrooms, cheese, and asparagus. Monosodium glutamate, or MSG, has a strong umami taste.

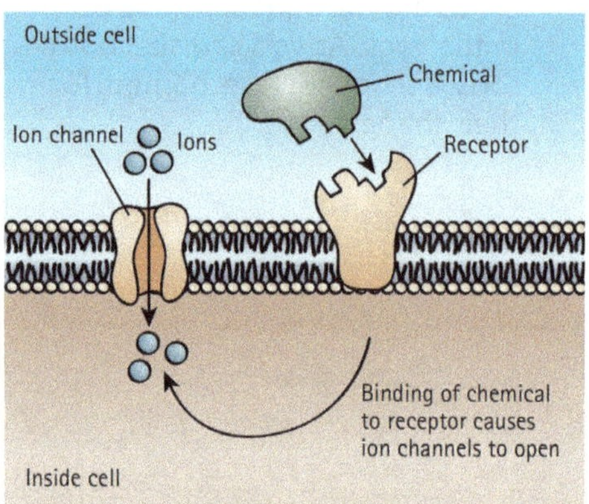

Figure 12.16

Smell and taste are examples of chemoreception. Chemicals bind to receptors on the cell membrane of chemosensory cells, causing ion channels to open. The ions initiate action potentials.

It is interesting that when you are hungry, your taste buds become more sensitive to salt and sugar, making food taste better. Hunger does not affect your ability to taste bitterness, which helps you identify toxic foods. In addition, your experience of food—what you think of as "taste"—comes partly from your sense of smell. This is why food doesn't have nearly as much "taste" when you have a stuffy nose.

Touch

Your sense of touch is several different senses. These senses tell you about stimuli such as pressure, temperature, and pain. Your skin's sensory cells for touch are shown in **Figure 12.17**. Pressure causes the "hairs" on sensory cells to bend, opening ion channels and initiating action potentials. You have separate sensory cells for detecting light touch and heavy pressure. Temperature-sensing cells have ion channels that are affected by heat or cold. Some temperature sensors respond to heat, others to cold. The chemical menthol (found in peppermint) also stimulates cold receptors—it is this coincidence, not an actual cold temperature, that explains the cool feeling you get from eating a mint.

Pain sensors respond to stimuli that damage the body. These sensory cells require strong stimulation before they respond. However, damaged tissues release chemicals called *prostaglandins* that increase the sensitivity of pain receptors. Aspirin and similar drugs provide pain relief by interfering with prostaglandin production. Pain sensors also become more sensitive with continued stimulation. This makes pain receptors different from most other sensory cells, which become less sensitive with repeated stimulation. (This is why you stop noticing the smells in your house or feeling the weight of your backpack as you walk.) Scientists believe that some types of chronic pain result from pain receptors that have become abnormally sensitive.

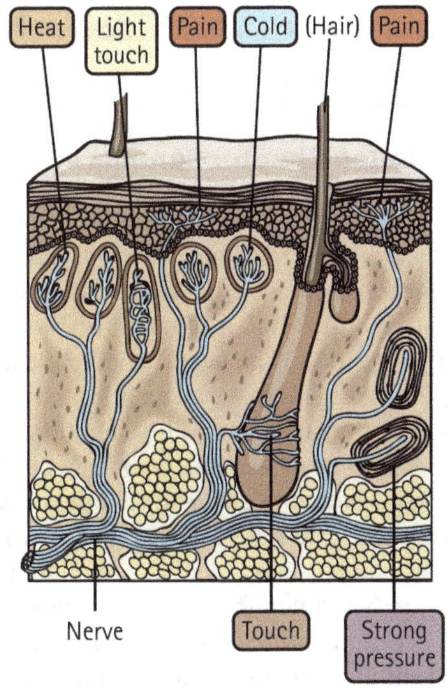

^ Figure 12.17

Sensory cells in the skin are responsible for the various senses of touch.

Other Senses

You have other senses in addition to the big five. *Proprioceptors* in your muscles, tendons, and joints tell you where different parts of your body are. You can try out this sense by closing your eyes and touching your nose with your finger. (You may be too used to this ability to be impressed, but consider the fact that you can't easily touch *another* person's nose with your eyes closed!) Finally, the *vestibular senses* tell you about body rotation and movement as well as which way is up.

READING CHECK

What information do your vestibular senses provide for you?

CONCEPT CHECK

1. When you are looking at a starry sky outside at night, are you using your rods or cones?

2. Deafness can result from a number of problems, including a ruptured eardrum or damaged sensory cells in the organ of Corti. Why would each of these problems make hearing difficult?

CHECK YOUR ANSWERS

1. You primarily use your rods, allowing you to see in dim light. However, when you see some color in a star, your cones are also activated. But this color can be hard to discern because the light is so dim.

2. If the eardrum is ruptured, sound waves cannot be conducted to the middle ear, and no vibrations reach the organ of Corti. If sensory cells in the organ of Corti are damaged, they are unable to send signals to the brain.

More to Explore

Check out the great visualizations:

https://www.visiblebody.com/learn/nervous/five-senses

Concept Review

Chapter 12

Detailed Chapter Summary

A tissue is a group of similar cells that performs a particular function. The four major body tissue types are epithelial, connective, muscle, and nervous tissue. Multiple tissues combine to form an organ, a body structure with a specific function. The heart, stomach, and brain are organs. Multiple organs make up an organ system, which is responsible for a particular bodily function.

Homeostasis is the maintenance of a relatively stable internal environment. Body temperature and blood oxygen levels are examples of variables for which the body maintains homeostasis. Feedback regulation, in which changes in one variable affect a second variable, and changes in the second variable, in turn, affect the first variable, is an essential mechanism for maintaining homeostasis.

The brain controls most all bodily activities. The brainstem connects to the spinal cord and is responsible for involuntary activities such as heartbeat, respiration, and digestion. The cerebellum controls balance, posture, coordination, and fine motor movements. The cerebrum is the largest part of the brain and is responsible for high-level functions such as reasoning, problem-solving, language, and creativity. The cerebrum also receives information from the senses and controls voluntary actions. Each cerebral hemisphere consists of four lobes. The frontal lobes deal with reasoning, voluntary movements, and speech. The parietal lobes deal with sensory information such as temperature, touch, taste, and pain. The occipital lobes deal with sight. The temporal lobes deal with sound and language comprehension. The thalamus sorts and filters information headed to the cerebrum. The hypothalamus is responsible for emotions such as pleasure, rage, and bodily drives.

The central nervous system (CNS) includes the brain and spinal cord, and the peripheral nervous system includes all other nerves in the body. The nervous system has two types of cells, neurons and glial cells. Neurons are signaling cells and have dendrites, a cell body, and an axon. Glial cells support and protect neurons. The three types of neurons are sensory neurons, interneurons, and motor neurons. Sensory neurons carry information from the senses to the CNS. Interneurons connect neurons to other neurons. Motor neurons carry directions from the CNS to muscles or organs. Motor neurons belong to either the somatic nervous system (controlling voluntary actions) or the autonomic nervous system (controlling involuntary actions). The autonomic nervous system has two divisions. The sympathetic division is responsible for the "fight or flight" response and is active during danger. The parasympathetic division works during times of calm.

A neuron fires during a sequence of events called an action potential. A neuron's membrane potential is normally negative -- this is called its resting potential. Stimulation increases the membrane potential. If the membrane potential reaches a value known as the threshold, an action potential occurs -- sodium channels in the membrane open, sodium ions enter the neuron, and the membrane potential spikes. Potassium channels in the membrane open, potassium ions exit the neuron, and the membrane potential returns to resting potential. An action potential is an all-or-nothing event.

Once a neuron fires, the action potential is propagated down its axon. This happens as sodium ions entering the neuron diffuse down the axon and cause successive parts of the axon to reach the threshold, initiating action potentials in a domino effect. Some axons are enclosed by a myelin sheath that allows an action potential to jump from one gap in the sheath to the next, greatly increasing the speed of propagation.

At the end of an axon, a neuron connects with one or more target cells. The connection is called a synapse. In an electrical synapse, ions move directly from the signaling neuron to the target cell, allowing for rapid signal transmission. Electrical synapses are responsible for the simultaneous muscle contractions that produce the heartbeat. In a chemical synapse, a signaling neuron releases neurotransmitter molecules that bind to receptors in the cell membrane of the target cell. This leads to changes in the membrane potential of the target cell that can be either excitatory or inhibitory. Target cells may receive signals from many neurons, and the sum of their effects determines whether the target cell fires. Chemical synapses allow for a finer degree of control and are much more common than electrical synapses.

Our senses allow us to obtain information about our environment. Our eyes allow us to see. Important structures of the eye include the cornea, pupil, iris, lens, and retina. Light enters the eyes and stimulates the sensory cells of the retina—rods and cones. Rods and cones send information to the brain via the optic nerve. Rods are more sensitive to light but offer grainy black-and-white vision. Cones detect color.

Sound enters through the pinna of the ear and reaches the eardrum, where waves are transmitted via the middle ear bones (hammer, anvil, and stirrup) to the fluid-filled inner ear. The sensory cells for sound are found in the organ of Corti in the cochlea. The bending of "hairs" found on these sensory cells causes action potentials to be transmitted to the brain via the auditory nerve.

Smell and taste occur through chemoreception. Action potentials are initiated when chemicals bind to receptors in the cell membrane of chemosensory cells. Humans have more than 1,000 different kinds of chemosensory cells for smell. Chemosensory cells in the taste buds distinguish five basic tastes – sweet, sour, salty, bitter, and umami. Much of our sense of "taste" relies on our sense of smell.

The sense of touch is multiple senses that respond to pressure, temperature, and pain. Other senses include proprioceptors, which tell us where different body parts are, and vestibular senses, which tell us about body rotation and movement.

Key Terms

Action potential A signal from a neuron or other cell that occurs when the cell's membrane potential becomes positive.

Axon The part of a neuron that transmits information to other neurons or cells.

Autonomic nervous system The motor neurons that control involuntary actions in the body.

Brainstem The part of the brain that connects the spinal cord to the rest of the brain and controls many of the body's basic involuntary activities, such as heartbeat, respiration, and digestion.

Central nervous system The brain and spinal cord.

Cerebellum The part of the brain that controls balance, posture, coordination, and fine movements.

Cerebrum The largest part of the brain, which is responsible for high-level functions such as reasoning, problem-solving, and language, as well as collecting information from the senses and controlling the voluntary activities of the body.

Chemoreception A way of sensing that occurs when chemicals bind to receptors on sensory cells, initiating action potentials.

Dendrites The parts of a neuron that receive information from other neurons or cells.

Homeostasis The maintenance of a relatively stable internal environment.

Hypothalamus The part of the brain that is responsible for emotions and bodily drives as well as regulating the hormones of the endocrine system.

Interneurons Neurons that connect one neuron to another neuron.

Motor neurons Neurons that carry instructions from the central nervous system to muscles or organs.

Neuron A cell that receives and transmits electrical signals from one part of the body to another.

Neurotransmitter Molecules used by a neuron to communicate with another neuron at a chemical synapse.

Organ A structure in the body that has a specific function.

Organ of Corti The structure in the ear that contains the sensory cells responsible for hearing.

Organ system A set of organs that work together to perform a particular bodily function.

Parasympathetic nervous system The division of the autonomic nervous system that calms your body; its effects are the opposite of those of the sympathetic nervous system.

Peripheral nervous system All the nerves in the body that are not found in the brain or spinal cord.

Retina The structure at the back of the eyeball that contains the rods and cones, the light-sensitive cells of the eye.

Sensory neurons Neurons that carry information from the senses to the central nervous system.

Somatic nervous system The motor neurons that control the voluntary actions of the body.

Sympathetic nervous system The division of the autonomic nervous system that prepares your body for danger, also known as the "fight or flight" response.

Synapse A connection between a neuron and a target cell.

Thalamus A major part of the brain that functions in collecting information from other parts of the brain, which it then sorts and filters to pass to the cerebrum

Tissue A group of similar cells that performs a certain function.

Review Questions

Simple Review Questions

12.1 Organization of the Human Body

1. What are the four main types of tissues in the human body? Give an example of each.
2. Multiple tissues combine to make a(n)____, a structure in the body that has a specific function.
3. What is an organ system?

12.2 Homeostasis

4. What is homeostasis?
5. Give some examples of variables for which the body maintains homeostasis.

12.3 The Brain

6. What are some of the functions of the brainstem?
7. Which part of the brain provides balance and posture?
8. Why can damage to one side of the cerebrum affect the functioning of the opposite side of the body?
9. Describe the functions of each of the four lobes of the cerebrum.

12.4 Organization of the Nervous System

10. Which structures make up the central nervous system?
11. A typical neuron has dendrites, a cell body, and an axon. What is the function of each of these parts?
12. What are the functions of sensory neurons, interneurons, and motor neurons?
13. What kinds of functions are controlled by the somatic nervous system and the autonomic nervous system?

12.5 How Neurons Fire

14. What is a membrane potential?
15. What happens to the membrane potential of a neuron during an action potential?
16. Why is an action potential described as an all-or-nothing event?
17. What causes an action potential to travel down a neuron's axon?

12.6 How Neurons Communicate With Other Cells

18. How does an electrical synapse work?
19. How does a chemical synapse work?

12.7 The Senses

20. What are the two types of light-sensitive cells in the eyes? How do they differ from each other?
21. Describe how sound waves enter the ear and ultimately cause you to hear.
22. How are action potentials generated in chemosensory cells?
23. What are proprioceptors?

Challenging Review Questions

12.1 Organization of the Human Body

24. Is the brain a tissue, an organ, or an organ system? Defend your answer.
25. The stomach is an organ. Describe some of the different tissues that make up the stomach.

12.2 Homeostasis

26. Why do you shiver when you are cold?
27. One aspect of homeostasis is maintaining the appropriate amount of water in the body. How does the body regulate this? Is feeling thirsty an important aspect of maintaining homeostasis?

28. When you exercise, your cells use more oxygen, and you breathe harder. Is this an example of negative feedback? Defend your answer.

29. This man is cooling off after an intense run. He will probably drink some of the water in his water bottle too. Explain how sweating, pouring water on yourself, and drinking water relate to homeostasis.

12.3 The Brain

30. When you move your body, is your cerebrum in complete control of your motions? What other parts of your brain are involved? Explain.

31. Why is the surface of your brain wrinkled?

32. The figure shows a map of the motor control area of the brain, found in the frontal lobe of the cerebrum. Why is such a large part of the brain responsible for controlling the actions of the hands and lips? Why is only a small part of the brain responsible for controlling the trunk?

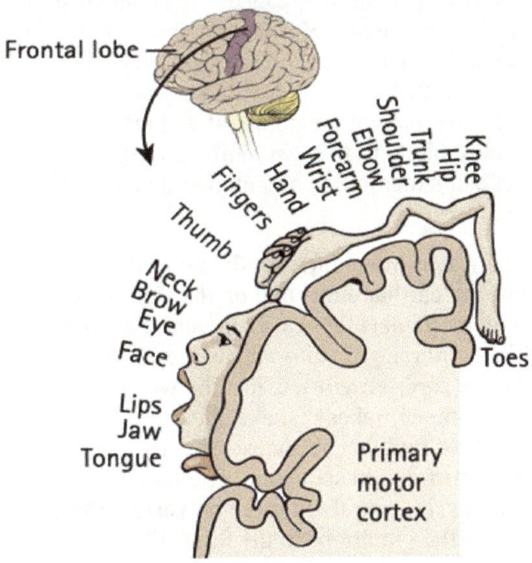

12.4 Organization of the Nervous System

33. Describe the structure of a typical neuron.

34. Of the three types of neurons—sensory neurons, motor neurons, and interneurons—which type goes to your biceps muscle and tells you to bend your elbow? Which type tells you whether your feet feel cold?

35. What happens during the "fight or flight" response?

36. Is a neuron that slows your heartbeat part of the somatic or autonomic nervous system? Is this neuron part of the sympathetic or parasympathetic division?

12.5 How Neurons Fire

37. What is an action potential? Describe how the membrane potential changes during an action potential.

38. What role do sodium channels and potassium channels play in an action potential?

39. What would be the effect of removing the myelin sheath from the axon of a neuron?

12.6 How Neurons Communicate With Other Cells

40. Do neurotransmitters enter the target cell? If not, how do they have an effect on the target cell?

41. What determines whether a target cell responds to a specific type of neurotransmitter?

12.7 The Senses

42. Many nocturnal animals have only rods in their retinas. In addition, many have a reflective layer of tissue behind the retina that bounces light back toward the retina. You can see this in the "eyeshine" of the wolf in the photo. How do these two features help animals see at night?

43. Are your rods or cones more important for reading a book? Defend your answer.

44. If you accidentally hit your eye, you may "see" a flash of light. Why?

45. In some people, the bones of the middle ear stiffen with age. This can result in deafness. Why?

46. You eat an apple. What senses are involved? Explain.

Apply & Discuss Questions

47. Does maintaining homeostasis of body temperature involve feedback regulation? Explain.

48. Animals vary in how "wrinkled" their brains are. Some animals have very wrinkled brains, while others have smooth brains. What would you predict about an animal that has a wrinkled brain rather than a smooth one?

49. You talk with a friend on the telephone. Which parts of your brain do you use?

50. If a signaling neuron has an excitatory effect on a target cell, does it increase or decrease the membrane potential of the target cell? Defend your answer.

51. Stars come in different colors depending on their surface temperatures. But when you look up at a starry night sky, all the stars look like they are the same color. Why?

Solutions to Odd-Numbered Questions

Simple Review Solutions

1. Epithelial tissue consists of sheets of tightly packed cells covering the body's internal and external surfaces. Skin is an example of epithelial tissue. Connective tissue consists of cells scattered within an external matrix. Bone, cartilage, and blood are all connective tissues. Muscle tissue consists of cells that can contract or shorten. Three types of muscle tissue are found in the body. Skeletal muscle is responsible for all voluntary movements. Smooth muscle functions in the internal organs of the digestive system as well as in certain blood vessels. The cardiac muscle produces the heartbeat. Nervous tissue transmits information from one place in the body to the other. Nervous tissue is found in the brain, spinal cord, and nerves.

3. An organ system is a set of organs that work together to perform a particular bodily function.

5. Oxygen supply and body temperature are two of the many variables the body carefully regulates. The amount of water in the body, the concentration of nutrients and waste products in the blood, the concentrations of important ions inside and outside cells, and blood pH—all these variables are carefully regulated as part of maintaining homeostasis.

7. The cerebellum controls balance, posture, coordination, and fine movements.

9. The frontal lobes deal with reasoning, voluntary movements, and speech. The parietal lobes take in sensory information about temperature, touch, taste, and pain. The occipital lobes process what you see—that is, visual information. The temporal lobes deal with sound and help you comprehend language.

11. A typical neuron has three parts: (1) dendrites receive information from other neurons or cells, (2) the cell body contains the neuron's nucleus and organelles, and (3) the axon transmits information to other neurons or cells.

13. Motor neurons in the somatic nervous system control voluntary actions, such as running to class. These neurons provide instructions to your voluntary muscles, such as the muscles of your arms, legs, and fingers. Motor neurons in the autonomic nervous system control involuntary actions. These neurons control organs and involuntary muscles (such as the heart or stomach muscles).

15. An influx of positively charged ions causes the membrane potential to spike and become positive. This spike is called an action potential.

17. When an action potential begins, sodium ions enter the end of the axon that is closest to the cell body. These ions diffuse into adjacent areas along the axon. Because the sodium ions are positively charged, they cause the local membrane potential to increase. When the local membrane potential reaches a threshold, a new action potential, further along the axon, begins. The process is similar to the way a row of dominoes falls—the first domino knocks down the next one, which knocks down the next one, and so forth. In this way, the action potential travels down the entire axon.

19. In a chemical synapse, a neuron communicates with a target cell using neurotransmitter molecules. When an action potential arrives at the end of an axon, vesicles containing neurotransmitter fuse to the neuron's cell membrane. Neurotransmitter molecules are released into the narrow space between the axon and the target cell, and they diffuse to the target cell. There, the neurotransmitter molecules bind to receptors on the target cell's membrane. The binding causes ion channels to open, and ions enter the target cell. The membrane potential of the target cell may increase or decrease, depending on the type of ion that enters. An excitatory signal increases the target cell's membrane potential and makes the target cell more likely to fire. An inhibitory signal decreases the target cell's membrane potential and makes it less likely to fire.

21. Sound waves move through the air to the outer ear, or pinna, the cartilaginous flap on the side of your head. The pinna funnels the waves in, and they move towards a thin membrane of skin—the eardrum. Sound waves make the eardrum vibrate, just the way blowing on a piece of paper makes it shake. The eardrum's vibrations move three middle ear bones—the hammer, the anvil, and the stirrup—in sequence. These bones amplify the vibrations, making them more powerful. The stirrup then transfers the vibrations to the fluid-filled inner ear. In the inner ear, sound vibrations enter the cochlea, a coiled tube containing the organ of Corti. The organ of Corti contains the sensory cells responsible for hearing. Fluid vibrations in the inner ear move the organ of Corti's basilar membrane, causing sensory "hairs" embedded in it to brush against an overlying membrane and bend. This bending causes ion channels to open, initiating action potentials that are transmitted to the brain.

23. Proprioceptors in your muscles, tendons, and joints tell you where different parts of your body are.

Challenging Review Solutions

25. The stomach is lined on both the inside and outside with epithelial tissue. The stomach has a lot of smooth muscle tissue that functions in churning and digesting food. The neurons that go to the stomach to regulate its activities are made of nervous tissue. The blood vessels that supply it are filled with blood, a type of connective tissue.

27. The body regulates water levels largely by controlling

whether the urine that is produced is concentrated or dilute. Antidiuretic hormone causes urine to become more concentrated. However, behavior is important too. When water levels in the body are low, you become thirsty, and you seek out and drink water.

29. Sweating and pouring water on the body both help with heat regulation, or maintaining homeostasis of body temperature. Since the man has become overheated due to running (all the work the body does while running releases a lot of heat), sweating and covering himself with water helps cool his body. Drinking water, on the other hand, helps with regulating water levels in the body. The man has lost water through sweating, and his feeling of thirst causes him to drink water. This helps maintain homeostasis of water levels in the body.

31. The cerebral cortex is the thin layer that covers the surface of the cerebrum. It performs most of the information processing that occurs in the cerebrum. "Wrinkles" in the cerebral cortex increase the surface area available for information processing without increasing the overall size of the brain and head.

33. A typical neuron has dendrites that receive information from other neurons or cells, a cell body that performs the cellular processes required by all cells, and an axon that transmits information to other neurons or cells.

35. The "fight or flight" response occurs when you are in danger and need to either run away ("flight") or stay and "fight." The sympathetic nervous system promotes this response, speeding up your heartbeat to speed the transport of oxygen to your muscles. The sympathetic nervous system also slows down digestion and other activities that are unimportant during emergencies.

37. An action potential is a neuron's way of "firing" or signaling. During an action potential, the neuron's membrane potential increases slowly until it reaches a certain threshold value—at that point, it suddenly spikes. The spike is brief—the membrane potential quickly decreases again and returns to threshold value.

39. The action potential would travel much more slowly along the axon, and not "jump" from one gap in the myelin sheath to the next.

41. A target cell responds to a neurotransmitter if it has receptors for that neurotransmitter.

43. Your cones – cones are responsible for making out fine details, such as the letters in a book.

44. Hitting your eye accidentally stimulates the rods and cones in the retina and causes them to send signals to the brain. However, the brain interprets any signal that comes from the retina as something visual, so you "see" a flash of light.

45. If the middle ear bones stiffen, they are less effective at transferring and amplifying sound vibrations from the outer ear to the inner ear. Reduced vibrations in the fluid of the inner ear results in fewer signals being sent from the organ of Corti to the brain. As a result, you hear less.

Apply & Discuss Solutions

47. Yes. You learned that feedback regulation occurs when the level of a body variable initiates a process that comes back to affect that variable. In the case of body temperature, when body temperature is low, the body initiates activities such as shivering or putting on more clothes that help to increase body temperature. When body temperature is high, sweating and other activities help to lower it. These are both examples of feedback regulation.

49. You use the temporal lobes of your cerebrum to understand spoken language (what your friend says to you on the phone). You use the frontal lobes of your cerebrum to talk to your friend—to speak.

51. When you are looking at the night sky, there is very little light, so you are using your rods to see. Rods see only in black and white, so you cannot see the colors of the stars. This is why stars look much more colorful in photographs than to the naked eye.

Living Earth Essay

Plant Nutrients

Plants have a significant impact on the geosphere, hydrosphere, and atmosphere, especially through the process of photosynthesis as well as transpiration. For this Living Earth Essay, we turn the table to show how chemical compounds from the soil (geosphere), water (hydrosphere), and air (atmosphere) are utilized by plants. After exploring this chemistry, let there be no doubt that plants and this planet are deeply interconnected.

Plants Require Nutrients

Plant material consists primarily of carbohydrates, which are made of the elements carbon, oxygen, and hydrogen. Plants get these three elements from carbon dioxide and water, but the soil they live in also provides many other elements vital to their survival and good health. **Table A** lists these nutrients as *macronutrients*, those nutrients needed in large quantities, and *micronutrients*, those nutrients needed only in trace quantities. Some micronutrients are needed in such trace quantities that a plant's lifetime supply is provided by the seed from which the plant grew.

TABLE A Essential Elements for Most Plants

ELEMENT	FORM AVAILABLE TO PLANTS
Macronutrients	
Nitrogen, N	NO_3^-, NH_4^+
Potassium, K	K^+
Calcium, Ca	Ca^{2+}
Magnesium, Mg	Mg^{2+}
Phosphorus, P	$H_2PO_4^{2-}$, HPO_4^{2-}
Sulfur, S	SO_4^{2-}
Micronutrients	
Chlorine, Cl	Cl^-
Iron, Fe	Fe^{3+}, Fe^{2+}
Boron, B	$H_2BO_3^-$
Manganese, Mn	Mn^{2+}
Zinc, Zn	Zn^{2+}
Copper, Cu	Cu^+, Cu^{2+}
Molybdenum, Mo	MoO_4^{2-}

Nitrogen, Phosphorus, and Potassium

Plants need nitrogen to build proteins and a variety of other biomolecules, such as *chlorophyll*, the green pigment responsible for photosynthesis. As Table A shows, plants are able to absorb nitrogen from the soil in the form of ammonium ions, NH4+, and nitrate ions, NO3-. **Figure A**, shows the natural sources of nitrogen for a plant. The two reactions shown here are examples of nitrogen fixation, which is defined as any chemical reaction that converts atmospheric nitrogen to a form of nitrogen that plants can use. The two most common forms are ammonium ions and nitrate ions.

Most of the ammonium ions in soil come from nitrogen fixation carried out either by bacteria living in the soil or by microorganisms living in root nodules of certain plants, especially those of the legume family, including clover, alfalfa, beans, and peas (plants generally called *nitrogen fixers*). These microorganisms possess the enzyme *nitrogenase*, which catalyzes the formation of ammonium ions from atmospheric nitrogen and soil-bound hydrogen ions, as Figure A shows.

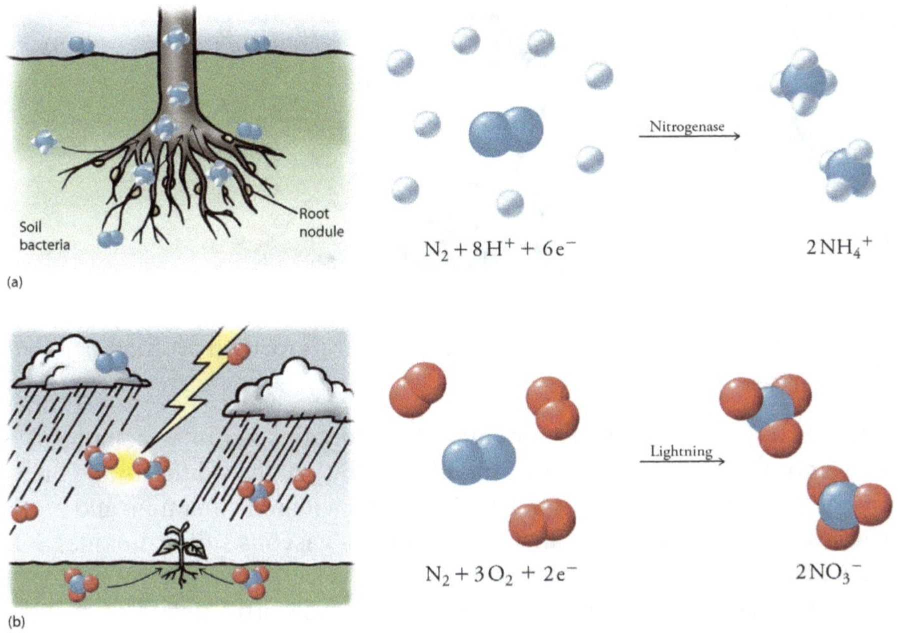

Figure A
Two pathways for nitrogen fixation, a source of nitrogen for plants. (a) Both free-living bacteria in the soil and microorganisms in root nodules produce ammonium ions. (b) Lightning provides the energy needed to form nitrate ions from atmospheric nitrogen.

Plants need phosphorus to build nucleic acids, phospholipids, and a variety of energy-carrying biomolecules such as ATP. All phosphorus comes to plants in the form of phosphate ions. The major natural source of these ions is eroded phosphate-containing rock. Significant amounts of phosphates are also recycled as organisms die and become incorporated into the soil. After nitrogen, phosphorus is most often the limiting element in soil. Phosphorus-deficient plants are stunted.

Potassium ions activate many of the enzymes essential to photosynthesis and respiration. As with phosphates, the major natural sources of potassium ions are eroded rock and the recycling of potassium ions from decomposing plant material. After nitrogen and phosphorus, soils are usually most deficient in potassium. Symptoms of potassium deficiency include small areas of dead tissue, usually along leaf tips or edges. As with nitrogen and phosphorus, potassium ions are easily redistributed from mature to younger parts of the plant, so deficiency symptoms first appear on older leaves. When cereal grains, such as wheat or corn, are potassium-deficient, they develop weak stalks, and their roots become more easily infected with root-rotting organisms. These two factors cause potassium-deficient plants to be easily bent to the ground by wind, rain, or snow.

Nitrogen fixation also results to a lesser extent from the action of lightning on atmospheric nitrogen, as Figure A also shows. The high energy of the lightning is sufficient to initiate the oxidization of atmospheric nitrogen to nitrate ions, which are then washed into the soil by rain.

In a natural setting, nitrogen fixation is the soil's source of ammonium and nitrate ions. However, most of this fixed nitrogen is recycled from one organism to the next. After a plant dies, for example, bacterial decomposition of the plant releases ammonium and nitrate ions back into the soil, where these ions become available to plants that are still living.

Plants deficient in nitrogen have stunted growth. Because nitrogen is needed for making chlorophyll, yellow leaves are another symptom of nitrogen deficiency in plants, as **Figure B** on the following page shows. The yellowing is most pronounced in older leaves—younger ones remain green longer because soluble forms of nitrogen are transported to them from older dying leaves.

READING CHECK
What macronutrients do plants need for making chlorophyll?

Calcium, Magnesium, and Sulfur

Plants absorb both calcium and magnesium as the positively charged calcium and magnesium ions, Ca2+ and Mg2+, respectively, and sulfur is absorbed as the negatively charged sulfate ion, SO42-. Most topsoil contains enough of these ions for adequate plant growth.

Calcium ions are essential for building cell walls. Once absorbed by the plant, calcium ions are relatively immobile; that is, they do not travel well from one part of the plant to another. Therefore, the plant is incapable of reallocating calcium supplies in times of need. This is why new-growth zones, such as the tips of roots and stems, are most susceptible to calcium deficiencies. The results are twisted and deformed growth patterns.

Magnesium ions are essential for the formation of chlorophyll, which is the green-pigmented molecule

Figure B
(a) The leaves of nitrogen-deficient plants turn yellow prematurely. (b) Phosphorus deficiencies are marked by stunted growth. (c) The leaves of potassium-deficient plants develop dead areas.

essential for photosynthesis. Chlorophyll houses a magnesium ion at the center of a structure called a porphyrin ring, as shown in **Figure C**. Besides its presence in chlorophyll, magnesium is essential because it activates many metabolic enzymes. Magnesium deficiency, which is rare, produces yellow leaves because the plant cannot generate chlorophyll. Most of the sulfur in plants occurs in proteins, especially in the amino acids cysteine and methionine. Another compound that contains sulfur is coenzyme A, which is essential for respiration, the synthesis and breakdown of fatty acids, and the vitamins thiamine and biotin. Leaves can absorb sulfur as gaseous sulfur dioxide, SO2, an environmental pollutant released from active volcanoes and burning wood or fossil fuels. To Life!

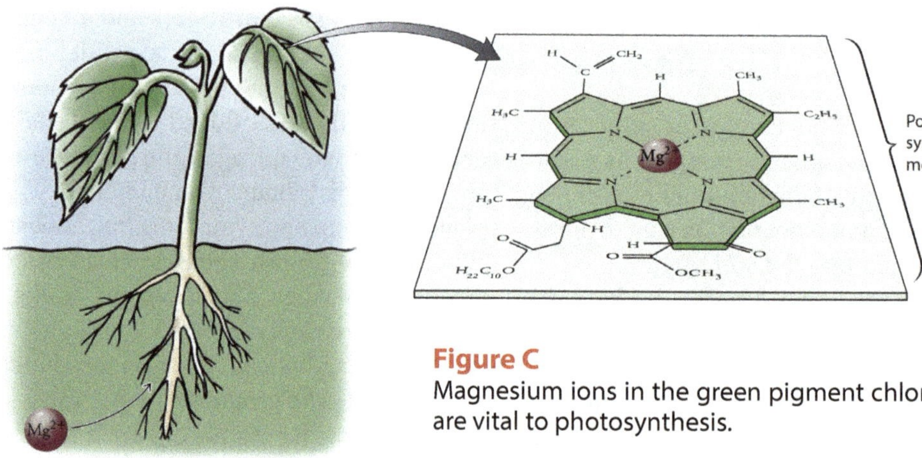

Figure C
Magnesium ions in the green pigment chlorophyll are vital to photosynthesis.

CONCEPT CHECK

Plants require more calcium and magnesium than phosphorus and potassium, yet deficiencies of calcium and magnesium are much more infrequent than deficiencies of phosphorus and potassium. How can this be?

CHECK YOUR ANSWER The potential for a nutrient deficiency depends on the amount of nutrients needed and the nutrient's availability. Calcium and magnesium are abundant in most soils, but phosphorus and potassium are not. Deficiencies of phosphorus and potassium, therefore, tend to be more frequent.

FOR YOUR INFORMATION

Plants readily absorb potassium ions but have a low tolerance for sodium ions, which they actively pump out of their systems. Interestingly, all living cells, not just plant cells, have a low tolerance for sodium ions and therefore have evolved mechanisms for pumping them out. A neuron pumping out sodium ions sets the stage for a nerve impulse. Multiple nerve impulses ultimately lead to conscious thought. From a macroscopic perspective, plants and animals are significantly different. However, their chemistries are much the same on a deeper submicroscopic level. In other words, plants and animals are more closely related than one might initially think.

Chapter 13

Control and Development

THE MAIN IDEA

Multiple organ systems work together to control the body's function as we develop, grow, and reproduce.

13.1 Hormones
13.2 Reproduction
13.3 Development
13.4 The Skeleton
13.5 Muscles

You've probably heard of hormones, but what exactly are hormones, and what do they do? Do they help control our growth, development, and reproduction? How do these critical human functions happen? How does the skeleton protect our body and allow us to move? What happens when we flex our muscles or use them to pick up a cup or run a marathon? It's time to take a good look at these wonders of the human body.

13.1 Hormones

While the nervous system handles rapid actions and reactions for the human body, the hormones of the endocrine system regulate activities that take place over longer time periods. For example, hormones regulate your growth and development, prepare you for reproduction, determine how quickly you metabolize food, and tell you whether it is time to be awake or asleep. Hormones also play an essential role in maintaining homeostasis in the body. So, what exactly are hormones, and how do they work?

A **hormone** is a molecule that gives instructions to the body. Hormones are produced in one place in the body, released into the bloodstream, and then received by target cells elsewhere in the body. Hormones come in two types. **Protein hormones** are, as their name suggests, proteins or modified amino acids. **Steroid hormones** are made from cholesterol.

Protein hormones and steroid hormones work in different ways. Protein hormones bind to receptors on the cell membranes of their target cells. This binding starts a series of chemical reactions that result in the target cells' response to the hormone (**Figure 13.1a**). Steroid hormones cross the cell membrane and bind to receptors inside target cells—a steroid hormone's receptors may be in a target cell's cytoplasm or nucleus. The steroid hormone-receptor complex then binds to DNA in the nucleus, directly affecting gene transcription (**Figure 13.1b**).

Figure 13.1 >

(a) Protein hormones bind to receptors in the cell membranes of target cells. This starts a sequence of events that result in the target cells' response to the hormone. (b) Steroid hormones enter target cells and bind to receptors in the cytoplasm or nucleus. Together, the steroid hormone and receptor directly affect gene transcription.

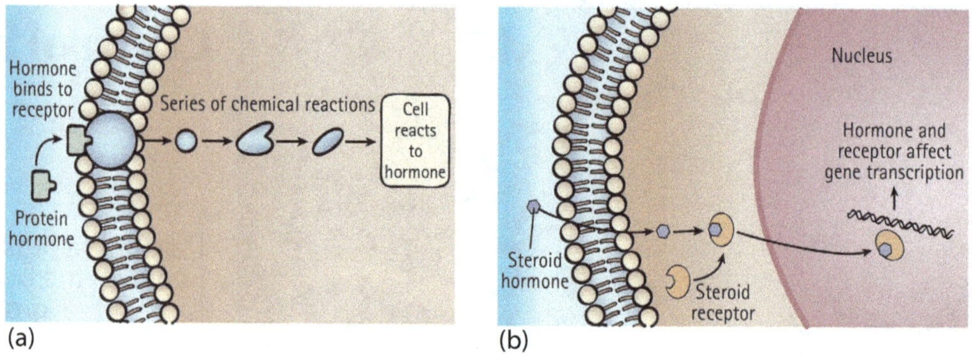

Endocrine Organs and Their Hormones

Let's look at the human endocrine organs (**Figure 13.2**) and the hormones they make.

The *hypothalamus* is a part of the brain, the endocrine system's control center. Many of the hypothalamus's hormones regulate the activity of another nearby endocrine organ, the pituitary gland. The pituitary gland has two major parts: the anterior pituitary and the posterior pituitary.

Hormones released by the hypothalamus cause the anterior pituitary to act in specific ways, such as producing and releasing its hormones. Meanwhile, hormones released by the hypothalamus can also be stored for later release by the posterior pituitary. Let's take a closer look.

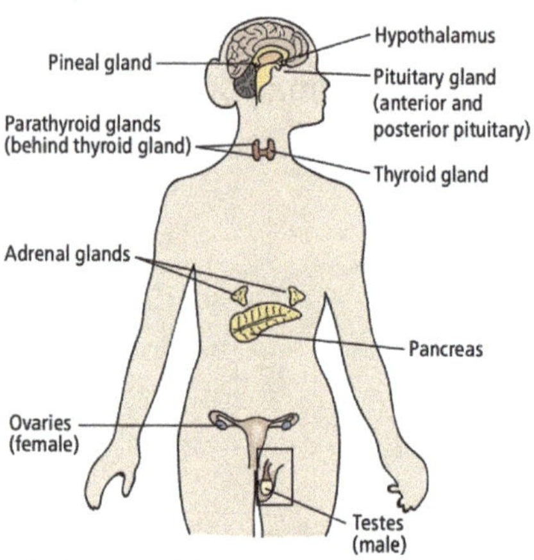

∧ Figure 13.2

Human endocrine organs include the hypothalamus, pituitary gland (anterior and posterior pituitary), thyroid gland, parathyroid glands, adrenal glands, pancreas, ovaries and testes, and pineal gland.

The **anterior pituitary** is sometimes called the "master gland." This is because many of its hormones regulate the activity of other endocrine organs. For example, anterior pituitary hormones tell the thyroid, sex organs, and adrenal glands to release their hormones. The anterior pituitary also makes growth hormone and prolactin. **Growth hormone** does what its name suggests—it promotes growth. Too little growth hormone results in dwarfism, and too much results in gigantism (**Figure 13.3**). *Prolactin* stimulates milk production in nursing mothers.

The *posterior pituitary* stores and controls the release of hormones made by the hypothalamus. **Antidiuretic hormone** helps regulate the amount of water in the body. Specifically, it helps the body conserve water by instructing the kidneys to produce more concentrated urine. Alcohol inhibits the release of antidiuretic hormone. This is why people produce more urine—and sometimes become dehydrated—when they consume alcohol. *Oxytocin* stimulates contraction of the uterus during childbirth. Women whose labor does not progress or need to have labor induced may be given pitocin, a synthetic form of oxytocin, to stimulate contractions.

The *thyroid gland* makes thyroid hormones. *Thyroid hormones* are involved in metabolism, growth, and development. For example, thyroid hormones are necessary for proper brain development during childhood. Thyroid hormones contain iodine, and getting too little iodine can result in developmental problems. This is why iodine is added to many brands of table salt. The thyroid gland also makes *calcitonin*, which decreases calcium levels in the blood.

The *parathyroid glands*, which lie behind the thyroid gland, make parathyroid hormone. *Parathyroid hormone* increases calcium levels in the blood. It does this in three ways: It causes calcium to be released from bones, increases calcium absorption in the intestine, and decreases calcium excretion in urine. Maintaining appropriate calcium levels in the blood is vital for many reasons. For example, you will see later in this chapter that muscles require just the right amount of calcium to function properly.

The *adrenal glands*, located above the kidneys, make epinephrine (also called adrenaline) and norepinephrine. *Epinephrine* and *norepinephrine* are involved in the "fight or flight" response. Signals from the sympathetic nervous system trigger the release of these hormones. The adrenal glands also produce glucocorticoids and mineralocorticoids. *Glucocorticoids* increase glucose levels in the blood; at certain times, your blood needs to deliver more glucose, the molecule that gives you energy, to your body's cells. *Mineralocorticoids* help regulate water and salt levels in the body.

The *pancreas* makes insulin and glucagon. These hormones regulate the amount of glucose in the blood. **Insulin** decreases blood glucose levels in two ways: It tells body cells to take in glucose, and it tells the liver to convert glucose into the storage substance glycogen. Glucagon increases blood glucose levels. *Glucagon* tells the liver to break down glycogen and release glucose. The control of blood glucose levels is an example of negative feedback and homeostasis: High blood glucose levels cause insulin to be released, and insulin causes blood glucose to decrease back to normal. Low blood glucose levels cause glucagon to be released, and glucagon causes blood glucose to increase back to normal (Figure 13.4).

The disease *diabetes* results when the pancreas doesn't make enough insulin or when the body's cells do not respond to insulin. In either case, blood glucose levels become abnormally high. Symptoms of diabetes include thirst, fatigue, weight loss, blurred vision, and nerve damage in the hands and feet. People who have diabetes have to control their diets and monitor their blood glucose levels. Some require regular injections of insulin.

The sex organs—*ovaries* in women and *testes* in men—make sex hormones. Both women and men make all three types of sex hormones: estrogens, progestins, and androgens. Women make more **estrogens** and **progestins**. These hormones regulate ovulation and the menstrual cycle and are involved in pregnancy. Estrogen also promotes breast development and fat storage in the hips and thighs. Men produce more **androgens**. Androgens such as testosterone are required for sperm production. Androgens also promote the development of male secondary sexual characteristics such as facial hair and increased muscle mass. It is this last effect of androgens that tempts some athletes to use anabolic steroids—synthetic versions of testosterone—to

^ Figure 13.3

This composite picture shows dwarfism (upper left) and gigantism (center) relative to a person of normal size (right). Dwarfism and gigantism result from too little or too much growth hormone.

READING CHECK

What kinds of sex hormones do men make? What types of hormones do women make?

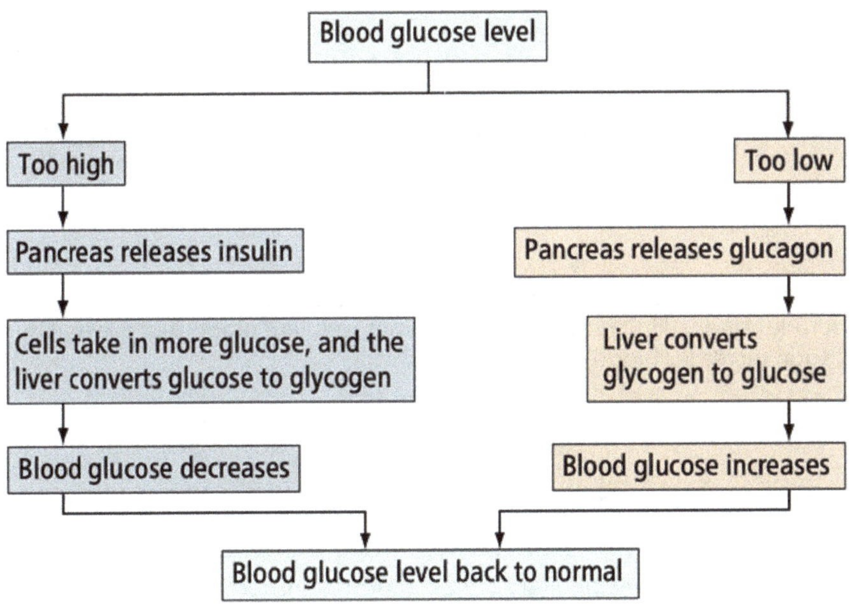

< Figure 13.4

Negative feedback enables the body to maintain appropriate blood glucose levels. If blood glucose levels are too high or too low, the endocrine system responds and brings the levels back to normal.

improve athletic performance. Unfortunately, steroids also have many adverse side effects. They can cause aggressive behavior, mood swings, and irritability. In men, steroids also cause shrinking of the testicles, decreased sperm count, baldness, and breast development. In women, steroids disrupt the menstrual cycle, deepen the voice, and promote facial hair growth. In adolescents, steroids stunt growth and accelerate puberty.

The *pineal gland* produces the hormone melatonin. **Melatonin** regulates the body's internal clock, telling you when it is day and when it is night. The pineal gland releases melatonin during the night using light cues from the eyes. This is why some people use melatonin as a sleeping pill. The major endocrine organs and the hormones they produce are summarized in **Table 13.1**.

Table 13.1 Major endocrine organs

Endocrine Organ	Hormone	Hormone Type	Effect
Hypothalamus	Hormones released by the posterior pituitary and hormones that regulate the anterior pituitary (see below)		
Pituitary gland			
Anterior pituitary	Growth hormone	Protein	Stimulates growth
	Prolactin	Protein	Stimulates milk production
	Follicle-stimulating hormone	Protein	Stimulates production of eggs and sperm
	Luteinizing hormone	Protein	Stimulates ovaries and testes
	Thyroid-stimulating hormone	Protein	Stimulates thyroid gland
	Adrenocorticotropic hormone	Protein	Stimulates adrenal glands to secrete glucocorticoids
Posterior pituitary (releases hormones made by hypothalamus)	Antidiuretic hormone	Protein	Promotes retention of water by kidneys
	Oxytocin	Protein	Stimulates contraction of uterus
Thyroid gland	Thyroid hormones	Protein	Promote growth and development
	Calcitonin	Protein	Lowers blood calcium level
Parathyroid glands	Parathyroid hormone	Protein	Raises blood calcium level
Adrenal glands	Epinephrine (adrenaline) and norepinephrine	Protein	Promote "fight or flight" response
	Glucocorticoids	Steroid	Raise blood glucose level
	Mineralocorticoids	Steroid	Control water and salt levels
Pancreas	Insulin	Protein	Lowers blood glucose level
	Glucagon	Protein	Raises blood glucose level
Sex organs			
Testes	Androgens (testosterone)	Steroid	Support sperm formation; promote development and maintenance of male characteristics
Ovaries	Estrogens	Steroid	Stimulate uterine lining growth; promote development and maintenance of female characteristics
	Progesterone	Steroid	Promotes uterine lining growth
Pineal gland	Melatonin	Protein	Regulates internal clock

CONCEPT CHECK

Some hormones come in pairs with opposing effects. Each pair helps maintain homeostasis in the body by regulating the amount of an important substance or molecule—often through feedback regulation. Name some pairs of hormones with opposing effects.

CHECK YOUR ANSWER

Calcitonin and parathyroid hormone together regulate blood calcium levels. Calcitonin decreases blood calcium levels, and parathyroid hormone increases blood calcium levels. Insulin and glucagon together regulate blood glucose levels. Insulin decreases blood glucose levels, and glucagon increases blood glucose levels.

More to Explore

To read more about hormones:

http://www.biologyreference.com/Ho-La/Hormones.html

https://www.pituitary.org.uk/information/hormones/

13.2 Reproduction

Where do babies come from? Without reproduction, humans—or any other species—would quickly go extinct.

Human reproduction begins with the production of **gametes**, or sex cells—eggs in women and sperm in men. Eggs and sperm are produced through meiosis. They are haploid cells with only half the usual number of chromosomes. At **fertilization**, the egg and sperm join to form a diploid cell that develops into a new human being. Let's start by looking at the female and male reproductive systems.

In females, eggs are made in the **ovaries** (**Figure 13.5**). Each ovary is made up of about 400,000 follicles, which are developing eggs surrounded by support cells. (This is the number at puberty, but it decreases over time.) During each menstrual cycle, a single follicle matures and releases an egg in a process called **ovulation**. The egg is a large cell with lots of nutrients stored in its cytoplasm. Eggs are relatively large because they are the result of unequal meiosis: During cell division, the future egg receives almost all the cytoplasm, while the other cells (which quickly degenerate) receive very little cytoplasm. After ovulation, the egg enters the oviduct, where cilia sweep it toward the uterus. Fertilization typically takes place while the egg is still in the oviduct. The fertilized egg then continues to the **uterus**, where it implants and continues development.

In males, sperm are made in the **testes** located in the scrotum (**Figure 13.6**). The scrotum hangs away from the body to keep the testes at a temperature lower than the body temperature. This is essential for sperm production. From the testes, sperm enter the *epididymis*, where they complete development and become mobile. Each mature sperm cell has a head that contains DNA, mitochondria, and enzymes for penetrating the egg. Sperm also have a tail for swimming.

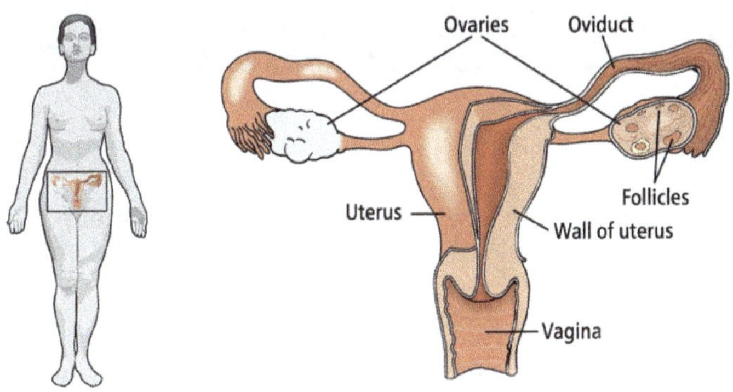

< **Figure 13.5**

The female reproductive system.

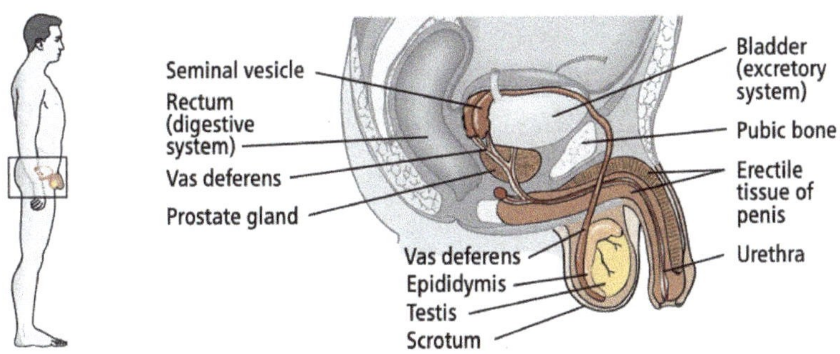

< **Figure 13.6**

The male reproductive system.

During sexual intercourse, sperm travel along the *vas deferens* to the *urethra*, a tube inside the penis. Sperm are ejaculated from the urethra in *semen*. In addition to sperm, semen contains fluids from the *seminal vesicles* and *prostate gland* that nourish sperm and protect them from the acidic environment of the vagina. There are about half a billion sperm in each ejaculate, each vying for the chance of encountering the egg.

READING CHECK

What is the jellylike layer that covers an egg cell called?

After sexual intercourse, sperm swim up the oviduct toward the egg (**Figure 13.7**). The egg is covered by a jellylike layer called the *zona pellucida*. Enzymes released from the heads of many sperm eat away at this cover. A single sperm finally reaches the egg's cell membrane, and the cell membranes of the egg and sperm fuse. At this point, the zona pellucida quickly undergoes a change in thickness that makes it impenetrable to additional sperm, assuring that the fertilized egg doesn't end up with too many chromosomes.

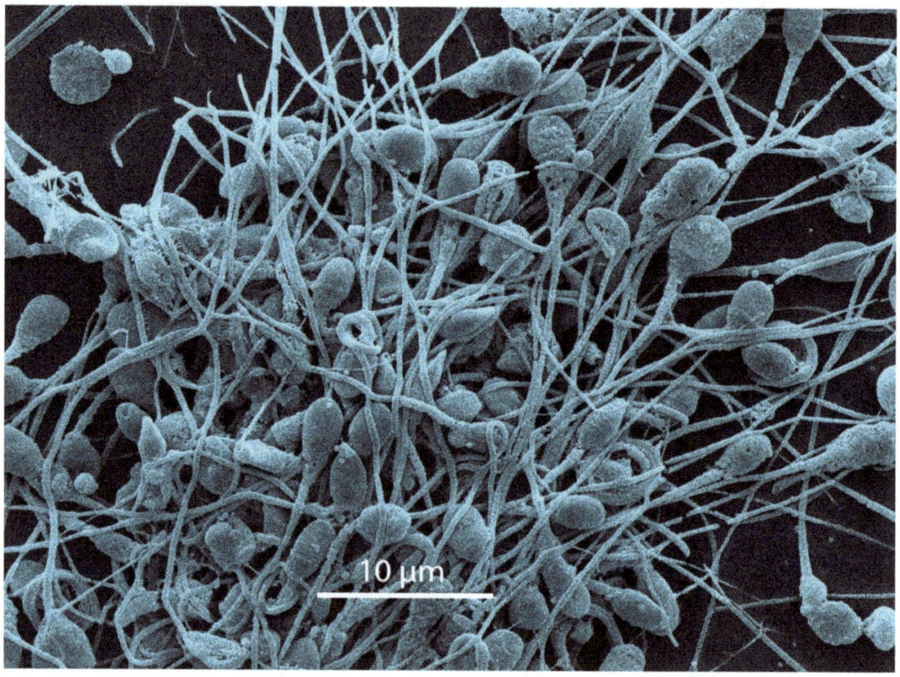

Figure 13.7 >

Human sperm as viewed through an electron microscope.

CONCEPT **CHECK**

If you place an unfertilized egg with its zona pellucida removed into a petri dish containing many sperm, what is the likely result?

CHECK YOUR ANSWER

Multiple sperm will probably fertilize the egg. Normally, the zona pellucida undergoes changes at fertilization to prevent additional sperm from entering the egg.

More to Explore

You can read more about the female and male reproductive systems here:
https://www.livescience.com/26741-reproductive-system.html

13.3 Development

After an egg is fertilized, it begins to divide through the process of cell division, or mitosis. By the time implantation occurs in the uterus, about 6 days after fertilization, the developing egg has become a hollow ball of cells called a *blastocyst*. Part of the blastocyst forms the embryo, the future baby. The rest of the blastocyst forms structures that protect and nourish the embryo, such as the amnion and the embryonic portion of the placenta.

The **amnion** is a membrane that surrounds the embryo. It is filled with amniotic fluid, a liquid that cushions and protects the developing embryo. The amnion is what ruptures when a pregnant woman's "water breaks" during labor. The **placenta** provides oxygen and nutrients to the developing embryo and carries away wastes. The placenta consists of both embryonic and maternal tissues. Maternal blood and embryonic blood do not come into direct contact in the placenta; however, they are close enough to allow for the exchange of nutrients and wastes. The placenta also produces estrogen and progesterone (a progestin). These hormones prevent further ovulation and maintain the uterus's nurturing condition throughout pregnancy.

The 9 months of pregnancy are divided into three 3-month trimesters. All of the embryo's major organs and body parts develop during the first trimester. Further development and most of the fetus's growth occurs in the second and third trimesters. Three stages in human development are shown in **Figure 13.8**.

READING CHECK

When does a fetus grow the most during its development?

EMBRYONIC DEVELOPMENT

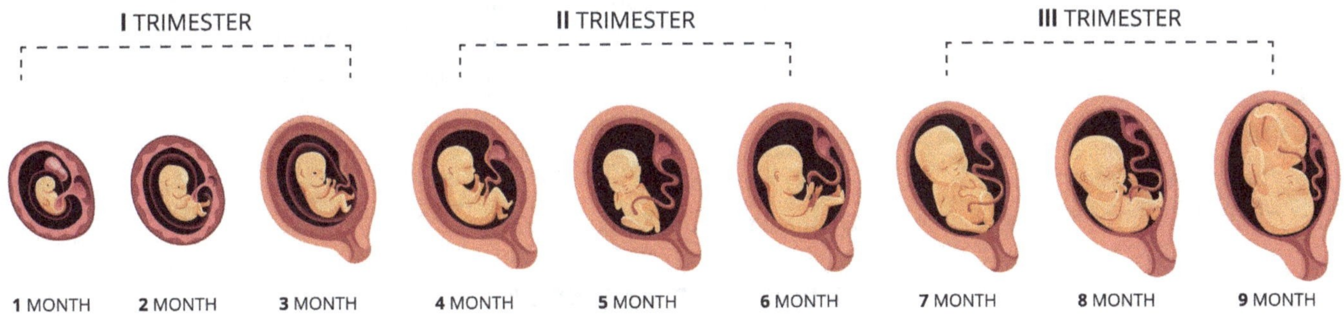

∧ Figure 13.8

By the end of the first trimester, all major organs and body parts are ready to start maturing. During the second trimester the bones are hardening and eye movement begins. During the third trimester the fetus can see and hear while the brain and organs continue to develop.

CONCEPT **CHECK**

What is the function of the placenta?

CHECK YOUR ANSWER

The placenta consists of embryonic and maternal tissues found in close proximity to one another. It allows oxygen and nutrients to pass from the mother to the developing embryo and remove wastes.

More to Explore

Additional information on human development:

https://embryology.med.unsw.edu.au/embryology/index.php/Embryonic_Development?source=post_page

13.4 The Skeleton

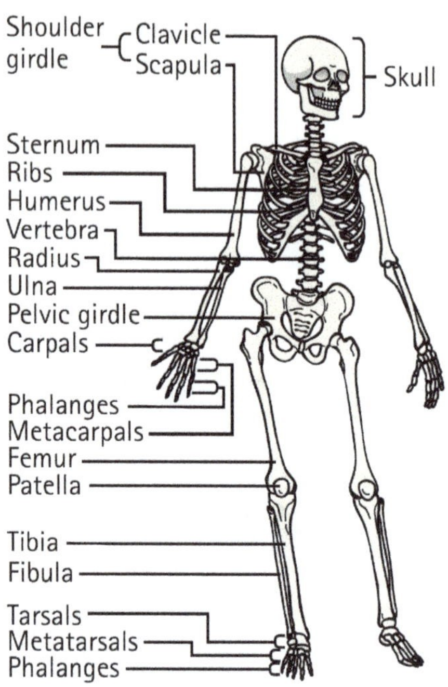

Figure 13.9

The human skeleton includes 206 bones (in adults) as well as a number of cartilages.

The human skeleton, shown in **Figure 13.9**, is made up of the bones and cartilage that protect and support the body. Adults have 206 bones in all. Babies are born with more, but many of these bones fuse during growth. The largest bone in the body is the femur, or thigh bone. The smallest is the stirrup, a bone in the middle ear that is about a quarter of a centimeter long. Besides bones, the skeleton includes several cartilages, including your external ears (pinnas) and the tip of your nose.

One function of the skeleton is to protect the body. For example, the skull protects the brain, the vertebrae protect the spinal cord, and the ribs protect the heart and lungs. A second function of the skeleton is to support the body and move it with the help of the muscles. Joints are movable connections between bones. Some joints, like the elbow and knee, act like hinges, bending in only one direction. Other joints, like the one between the hip and thigh, resemble a ball and socket, allowing for a greater range of motion. At a joint, the ends of connecting bones are covered with smooth cartilage and enclosed in a fluid-filled capsule. The fluid lubricates the joint so the bones can move smoothly without rubbing against each other. *Arthritis* is a condition where the tissues of the joint become inflamed and produce too much fluid. Both the inflammation and excess fluid cause painful bone damage.

Bones are made up of three layers. First, there is a strong, hard outer layer of *compact bone*. Compact bone surrounds a lighter layer of *spongy bone*. Spongy bone surrounds a jellylike substance called **bone marrow** (**Figure 13.10**). *Red bone marrow* makes red and white blood cells for the circulatory system. *Yellow bone marrow* stores fat. Like other parts of the body, bones are made of living cells. These cells secrete the hard calcium-containing matrix that gives bones both strength and flexibility.

READING CHECK

What happens inside the body when arthritis occurs?

CONCEPT **CHECK**

What are two functions of the skeleton?

CHECK YOUR ANSWER

To protect organs in the body, such as the brain, heart, and other internal organs, and to support and move the body.

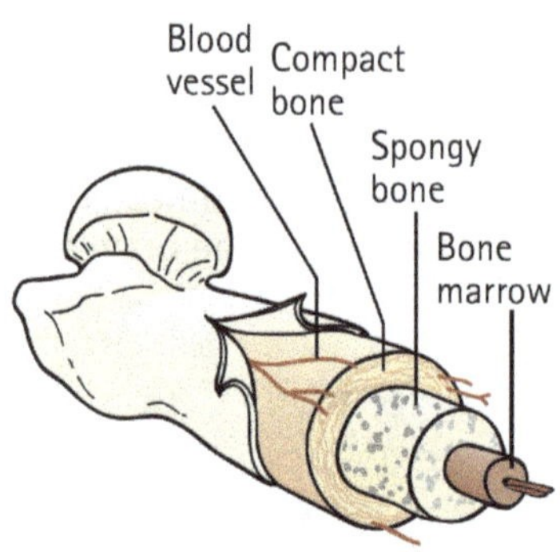

Figure 13.10

In bones, compact bone surrounds a layer of spongy bone, which surrounds the bone marrow.

More to Explore

You can read more about the human skeleton here:

https://www.livescience.com/22537-skeletal-system.html

And more about joints here:

https://www.visiblebody.com/learn/skeleton/joints-and-ligaments

13.5 Muscles

You have more than 600 muscles in your body, including 30 in your face alone. You use these muscles for everything from smiling to making sandwiches. In fact, almost any time you say that you are "doing" something, you are doing it with your muscles.

Muscles work by contracting, or shortening. Many of your muscles are connected to bones via *tendons*. When these muscles contract, they pull at your bones, moving you. Because muscles can only pull, not push, you have many pairs of muscles with opposing effects. For example, your biceps muscle pulls on the inner part of your forearm, bending your forearm up toward your shoulder (**Figure 13.11**). Your triceps muscle pulls on the back end of your forearm (it attaches to the "funny bone" at the end of your elbow) and straightens the arm.

How does a muscle contract? Let's start by looking at its structure. A muscle (**Figure 13.12**) consists of bundles of long muscle fibers. Each muscle fiber is a single cell with multiple nuclei. When you lift weights, you don't increase the number of muscle fibers in your muscles. That number stays the same. Instead, your muscles get bigger because the muscle fibers become thicker.

A muscle fiber contains bundles of smaller elements called *myofibrils*. Each myofibril is made up of a series of contractile units called **sarcomeres**. Sarcomeres are made up of carefully arranged fibers of two proteins: thin filaments of **actin** and thick filaments of **myosin**. When a muscle contracts, the actin and myosin filaments slide past each other, shortening the length of each sarcomere. Let's look at how this happens.

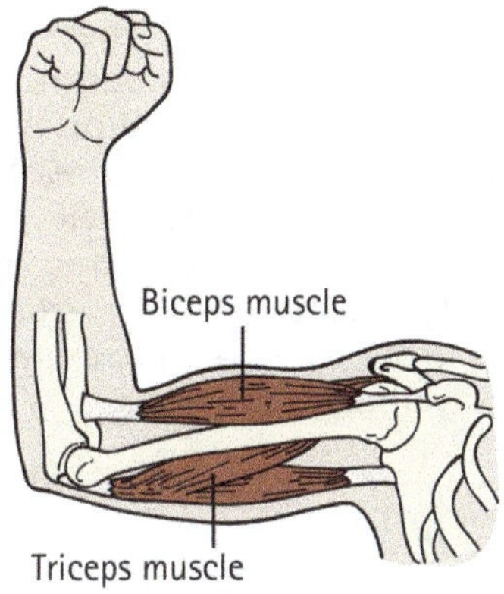

Figure 13.11

The biceps and triceps muscles move the forearm in opposite directions. The biceps bends the forearm, and the triceps straightens it.

READING CHECK

How many muscle cells make up a muscle fiber?

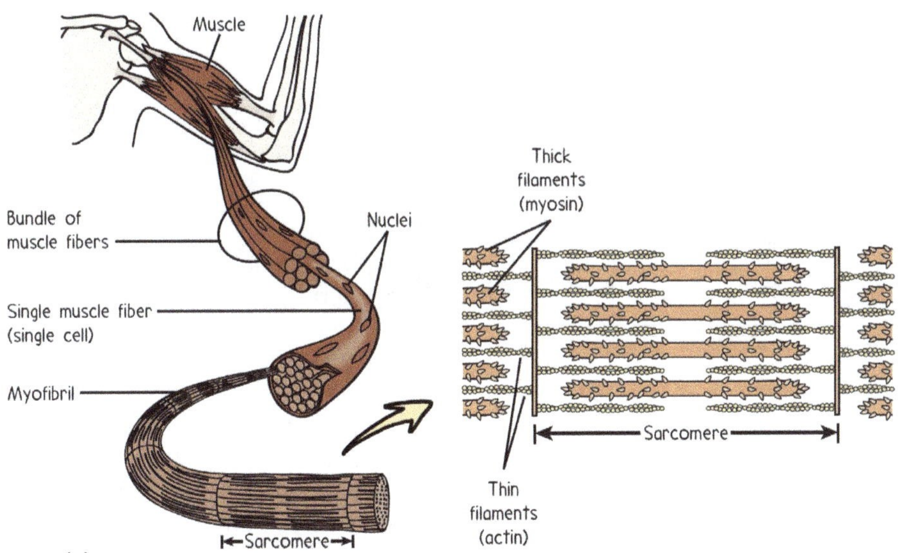

(a)

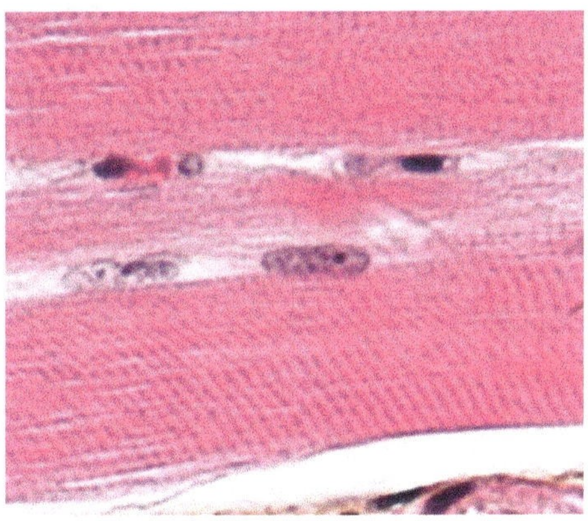

(b)

Figure 13.12

(a) A muscle is made up of bundles of muscle fibers. A muscle fiber contains smaller fibers called myofibrils. A myofibril consists of sarcomeres arranged end to end. Two proteins in the sarcomere, actin and myosin, allow sarcomeres—and consequently muscles—to contract. (b) This photo shows the "striated" or striped appearance of skeletal muscle due to the appearance of the sarcomeres. The darker oval shapes are cell nuclei.

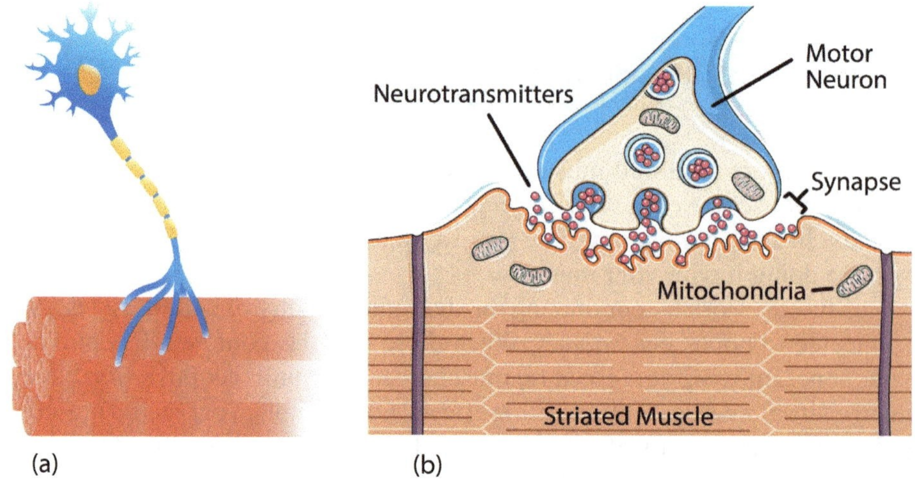

Figure 13.13 >
(a) This graphic shows a motor neuron (blue) and its connection with muscle cells (red). (b) This is a close-up view of the synapse between a motor neuron and a muscle cell. The motor neuron is shown in blue and the muscle cell is shown in red. Small vesicles containing the neurotransmitter acetylcholine (red) are clustered near the muscle cell where they activated receptor sites.

Muscle Contraction

A muscle contracts when it receives a signal from a motor neuron. An action potential in the motor neuron arrives at a chemical synapse connecting the neuron to a muscle cell. The action potential triggers the release of the neurotransmitter **acetylcholine** at the synapse (**Figure 13.13**). Acetylcholine then binds to receptors on the muscle cell's membrane, starting an action potential in the muscle cell.

An action potential in a muscle cell causes calcium ions to be released from the muscle cell's endoplasmic reticulum (a cell organelle). Calcium ions enable a series of "heads" on the myosin fibers to attach to actin (**Figure 13.14**). Specifically, calcium ions bind to proteins that normally block the myosin-binding spots on actin and essentially move them out of the way so

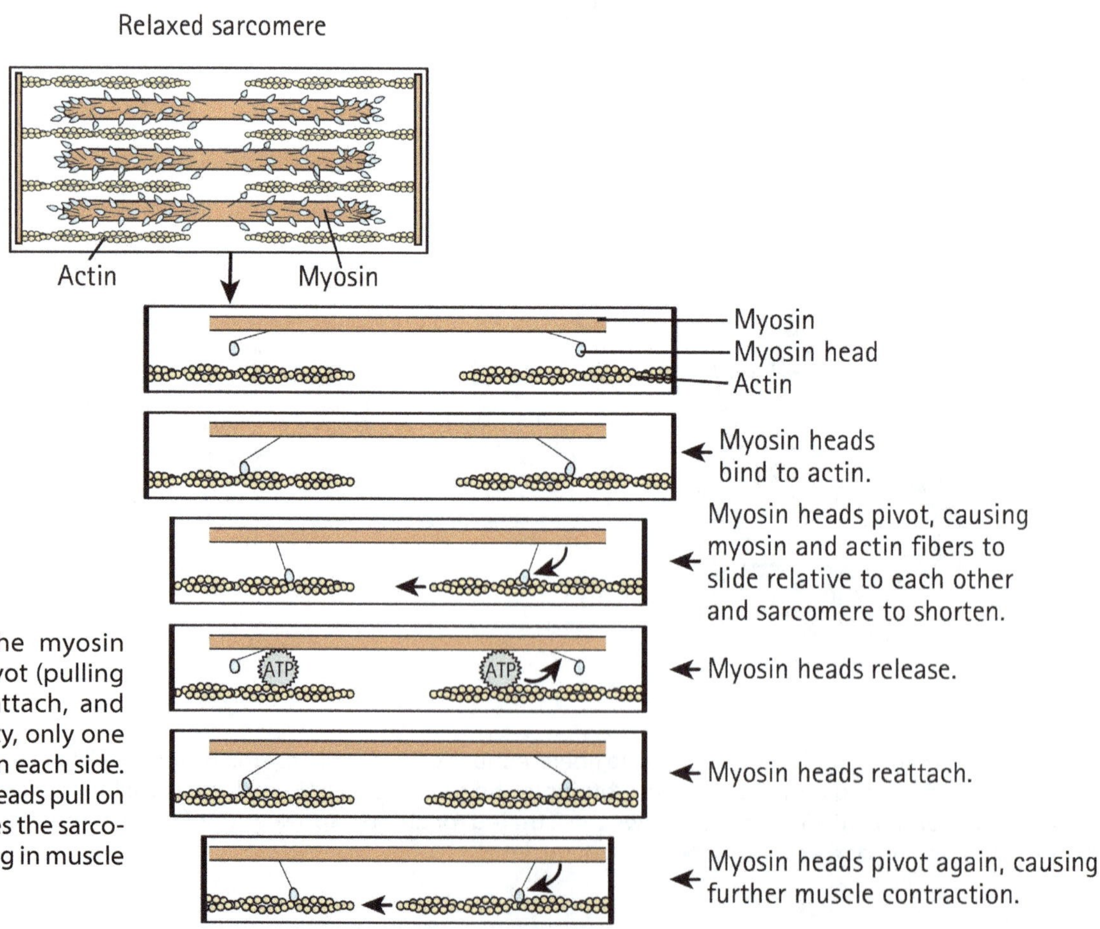

Figure 13.14 >
During contraction, the myosin heads bind to actin, pivot (pulling the actin), release, reattach, and pull again. (For simplicity, only one myosin head is shown on each side. In reality, many myosin heads pull on actin.) This process causes the sarcomere to shorten, resulting in muscle contraction.

myosin can bind to actin. The myosin heads attach and pivot due to intermolecular forces, such as dipole-dipole attractions. This results in a pulling on the actin filaments. Each pull shortens the length of the sarcomere a tiny bit—about 10 nanometers—and, consequently, the length of the muscle as a whole. After pulling, the myosin heads release, extend, attach, and pull again. This cycle repeats until the signal to contract ends or until the muscle has fully contracted.

Muscle contraction requires energy: ATP is needed for the myosin heads to release actin, an essential step in the contraction cycle. Once the motor neuron stops signaling, acetylcholine stops binding to the muscle cell, and the muscle cell's endoplasmic reticulum stops releasing calcium. Without calcium, the myosin heads are unable to bind to actin, and the muscle relaxes. Several well-known toxins work by interfering with the connection between neurons and muscles. Curare, an arrow poison used in the South American tropics for hunting, binds to acetylcholine receptors on muscle cells, preventing acetylcholine itself from binding. Curare causes paralysis and then death as the respiratory muscles become paralyzed. The powerful nerve gas sarin prevents acetylcholine from breaking down after muscles contract. Muscles are stimulated continuously and soon become exhausted. Again, death occurs through asphyxiation as the respiratory muscles stop working.

Muscle contraction also explains rigor mortis, the stiffness of the body that sets in after death. After death, calcium ions leak from the endoplasmic reticulum of muscle cells. Myosin binds to actin, and muscles contract. Once the available ATP is used up, the myosin heads are unable to disengage, and the muscles remain contracted.

CONCEPT CHECK

1. Why is calcium necessary for muscle contraction?
2. Why is ATP necessary for muscle contraction?

CHECK YOUR ANSWERS

1. Calcium enables the myosin heads to attach to actin.
2. ATP is required for the myosin heads to release actin, an essential step in the contraction cycle.

More to Explore

You can read more about muscle contraction here:

https://www.visiblebody.com/learn/muscular/muscle-contractions

Chapter 13

Detailed Chapter Summary

Hormones are molecules that give instructions to the body. Compared to the nervous system, the hormones of the endocrine system generally control activities that take place over longer time scales.

Hormones are produced in one place in the body, released into the blood, and received by target cells elsewhere in the body. Protein hormones bind to receptors in the cell membrane of target cells, initiating a series of chemical reactions. Steroid hormones, which are made from cholesterol, cross the cell membrane and bind to receptors inside target cells. The steroid hormone-receptor complex then binds to DNA in the nucleus and impacts gene transcription.

The hypothalamus, a part of the brain, is the endocrine system's control center. Many of its hormones regulate the anterior pituitary. Hypothalamus hormones are also stored and released by the posterior pituitary, including antidiuretic hormone, which regulates the amount of water in the body, and oxytocin, which stimulates uterine contraction. The anterior pituitary is known as the master gland because many of its hormones regulate other endocrine glands, including the thyroid gland, sex organs, and adrenal glands. The anterior pituitary also makes growth hormone, which promotes growth, and prolactin, which stimulates milk production. The thyroid gland makes thyroid hormones involved in metabolism, growth, and development, as well as calcitonin, which regulates calcium levels in the blood. The parathyroid gland makes parathyroid hormone, which also helps control calcium levels in the blood. The adrenal glands make epinephrine and norepinephrine, which are involved in the "fight or flight" response. They also make glucocorticoids and mineralocorticoids, which regulate levels of glucose, water, and salt in the body. The pancreas makes insulin and glucagon, which together regulate blood glucose levels. The ovaries and testes make the sex hormones: estrogens, progestins, and androgens. The pineal gland produces melatonin, which regulates the body's internal clock. Many of these hormones function in pairs through feedback regulation.

Human reproduction begins with the fertilization of an egg by a sperm. The fertilized egg then develops into a new human during the nine months of pregnancy. Eggs are made in the ovaries and are released during ovulation. Sperm are made in the testes. Sperm must work to penetrate the zona pellucida surrounding the egg before fertilization can occur. Fertilization occurs in the oviduct. Changes in the zona pellucida help to ensure that only one sperm fertilizes the egg. The fertilized egg implants in the uterus and continues development. The fertilized egg forms the embryo as well as the protective, fluid-filled amnion and the embryonic portion of the placenta. The placenta, which provides oxygen and nutrients to the developing embryo and removes wastes, is made up of both maternal and fetal tissues.

The skeleton includes the bones and cartilage of the body. These function in protection, support, and movement. Joints are movable connections between bones. Bones are made up of multiple layers, including compact bone, spongy bone, and bone marrow. Red bone marrow makes blood cells and yellow bone marrow stores fats.

Muscles work by contracting. Many of our muscles are found in pairs with opposing effects. A muscle is made up of muscle fibers, each of which is a single long cell. A muscle fiber contains bundles of myofibrils. Each myofibril is made up of a series of sarcomeres. Sarcomeres include filaments of actin and myosin. A sarcomere contracts when the actin and myosin filaments slide past each other, shortening the length of the sarcomere and hence the muscle as a whole.

Muscles contract when they receive signals from motor neurons. These signals produce action potentials in the muscle cells, which cause calcium ions to be released from the endoplasmic reticulum of the muscle cell. Calcium, in turn, enables myosin heads to bind to actin and pull. The myosin heads continue to release, extend, attach, and pull until the muscle is fully contracted or the signal from the motor neuron stops. ATP is required for the myosin heads to release actin, an essential step in the contraction cycle.

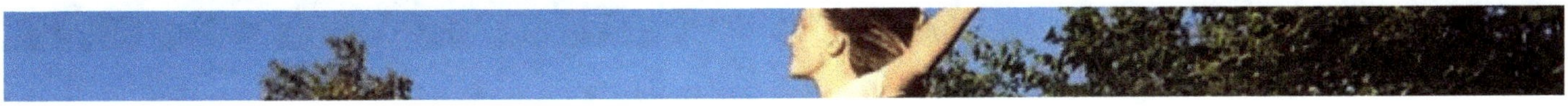

Key Terms

Acetylcholine The neurotransmitter used by motor neurons to communicate with muscle cells at the synapses that connect the two.

Actin Along with myosin, actin is one of the two types of filaments in sarcomeres that work together to cause muscle contraction.

Amnion A membrane that surrounds a developing embryo. It is filled with amniotic fluid, which cushions and protects the embryo.

Androgens Sex hormones involved in sperm production and the development of some male secondary sexual characteristics.

Anterior pituitary The "master gland," many of whose hormones regulate the activity of other endocrine organs.

Antidiuretic hormone A hormone made by the posterior pituitary that helps the body conserve water by producing a more concentrated urine.

Bone marrow The jellylike substance at the center of bones that makes red and white blood cells (red bone marrow) or stores fat (yellow bone marrow).

Estrogens Sex hormones that, together with progestins, regulate ovulation, the menstrual cycle, and pregnancy. Estrogens also promote certain female secondary sexual characteristics.

Gametes Haploid sex cells, such as eggs and sperm.

Fertilization The joining of an egg and a sperm to form a diploid cell that can develop into a new organism.

Growth hormone A hormone made by the anterior pituitary that promotes growth.

Hormones Molecules that give instructions to the body. They are produced in one place in the body, released into the bloodstream, and received by target cells elsewhere in the body.

Insulin A hormone made by the pancreas that acts to decrease blood glucose levels.

Melatonin A hormone produced by the pineal gland that regulates the body's internal clock.

Myosin Along with actin, one of the two types of filaments in sarcomeres that work together to cause muscle contraction.

Ovaries The organs in females where eggs are produced.

Ovulation The release of a mature egg cell that occurs once during each menstrual cycle.

Placenta The organ that allows nutrients and wastes to be exchanged between a pregnant woman and a developing embryo.

Progestins Sex hormones that, together with estrogens, regulate ovulation, the menstrual cycle, and pregnancy.

Protein hormones Hormones that are proteins or modified amino acids. They have their effect by binding to receptors on the cell membrane, initiating a series of chemical reactions.

Sarcomeres The contractile units of muscle cells.

Steroid hormones Hormones made from cholesterol that bind to receptors inside target cells. The steroid hormone-receptor complex then enters the nucleus where it directly affects gene transcription.

Testes The organs in males in which sperm are made.

Uterus The organ in females where a fertilized egg implants and the embryo develops during pregnancy.

Review Questions

Simple Review Questions

13.1 Hormones

1. What are the two types of hormones? How does each type of hormone produce an effect in a target cell?
2. Why is the anterior pituitary called the "master gland"?
3. What hormone made by the hypothalamus helps regulate the amount of water in the body? What is its effect?
4. Which endocrine organ produces the hormones associated with the "fight or flight" response?

13.2 Reproduction

5. What is unusual about the process of meiosis that produces the egg?
6. How do sperm get past the zona pellucida?

13.3 Development

7. What is the function of the placenta?
8. When do an embryo's major organs and body parts develop during pregnancy?

13.4 The Skeleton

9. What are the functions of the skeleton?
10. Describe the structure of a typical bone.
11. What are the functions of red bone marrow and yellow bone marrow?

13.5 Muscles

12. How does a signal from a motor neuron result in the contraction of a muscle?
13. When during muscle contraction is ATP required?

Challenging Review Questions

13.1 Hormones

14. Suppose you know that the receptor for a hormone you are studying is found in the target cell's cytoplasm. Are you studying a protein hormone or a steroid hormone?
15. On a brilliant, sunny day, you take a long hike through open country. You sweat a lot, losing a lot of water. What hormone does your body release? Why?
16. Why does your body release insulin after a large meal?
17. What is diabetes? What are the symptoms of diabetes?

13.2 Reproduction

18. What is the role of meiosis in human reproduction?
19. Vasectomy is a form of male sterilization in which a section of each vas deferens is removed. How does this cause sterility, preventing a man from fathering children?

13.3 Development

20. Does a fertilized human egg make anything other than the embryo?
21. Does a mother's blood mix with her baby's during pregnancy? Explain your answer.

13.4 The Skeleton

22. How do the skull, vertebrae, and ribs provide protection to the body?
23. Where in your body do you have a hinge joint? Where do you have a ball-and-socket joint?

13.5 Muscles

24. If muscles can only pull, not push, how is it possible for you to both bend your arm as well as straighten it?
25. Curare is a dart poison used by certain South American tribes in hunting. The darts in the photo shown here were poisoned with curare. Curare binds to acetylcholine receptors in muscle cells. How does curare cause death?

26. What causes rigor mortis, the stiffening of the body after death?

Apply & Discuss Questions

27. Osteoporosis is a disease that primarily affects post-menopausal women, causing decreased bone density and brittle bones that are vulnerable to fracture. The hormone calcitonin is sometimes used to treat osteoporosis. Why?
28. Jet lag describes the fatigue and disorientation that result from flying across many time zones. What causes jet lag, and how does the body eventually adjust to a new time zone? Why is there no such thing as "train lag" or "bicycle lag"?
29. Only one sperm fertilizes an egg. If this is the case, why can low sperm count be a factor in infertility?
30. Each time myosin heads pull on actin, the sarcomere contracts only about 10 nm (10-9 meter). Given that, how are you able to produce large motions with your muscles?
31. Explain what happens when you wiggle your toe. Start with the decision to wiggle your toe, which occurs in your brain, and end with a description of the activity of your muscles and bones.

Solutions to Odd-Numbered Questions

Simple Review Solutions

1. Protein hormones are, as their name suggests, proteins or modified amino acids. Steroid hormones are made from cholesterol. Protein hormones and steroid hormones work in different ways. Protein hormones bind to receptors on the cell membrane of their target cells. This binding starts a series of chemical reactions that result in the cell's response to the hormone. Steroid hormones cross the cell membrane and bind to receptors inside target cells—their receptors may be in the target cell's cytoplasm or nucleus. The hormone-receptor complex then binds to DNA in the nucleus and directly affects gene transcription.

3. Antidiuretic hormone helps regulate the amount of water in the body. Specifically, it helps the body conserve water by instructing the kidneys to produce more concentrated urine.

5. Eggs are large because they are the result of unequal meiosis—during cell division, the future egg receives almost all the cytoplasm, while the other cells (which quickly degenerate) receive very little cytoplasm.

7. The placenta provides oxygen and nutrients to the developing embryo and carries away wastes.

9. One function of the skeleton is to protect the body. For example, the skull protects the brain, the vertebrae protect the spinal cord, and the ribs protect the heart and lungs. The skeleton also supports the body and, working with the muscles, moves it.

11. Red bone marrow makes red and white blood cells for the circulatory system. Yellow bone marrow stores fat.

13. ATP is needed for the myosin heads to release actin, an essential step in the contraction cycle.

Challenging Review Solutions

15. Antidiuretic hormone – this hormone helps you conserve water by producing a more concentrated urine.

17. The disease diabetes results when the pancreas doesn't make enough insulin or when the body's cells do not respond to insulin. In either case, blood glucose levels become abnormally high. Symptoms of diabetes include thirst, fatigue, weight loss, blurred vision, and nerve damage in the hands and feet. Diabetes is controlled through diet and medication.

19. If a section of each vas deferens is removed, sperm do not enter the semen.

21. The placenta consists of both embryonic and maternal tissues that are in close contact with one another. Maternal blood and embryonic blood do not come into direct contact in the placenta; however, they are close enough to allow for the exchange of nutrients and wastes between the mother's blood supply and the embryo's blood supply.

23. Hinge joints are found at your elbow, knee, ankle, and between the segments of your fingers. Ball-and-socket joints are found at your shoulder and hip.

25. When curare binds to acetylcholine receptors in muscle cells, acetylcholine itself is unable to bind to those receptors. This means that when motor neurons try to tell muscles to contract, they will not respond. Curare causes paralysis and then death as the respiratory muscles become paralyzed.

Apply & Discuss Solutions

27. Two hormones, parathyroid hormone and calcitonin, work together to regulate blood calcium levels. Parathyroid hormone raises calcium levels in the blood, partly by causing calcium to be released from bones. Calcitonin has the opposite effect – it lowers blood calcium levels by causing bones to take up calcium. This is why it is useful for treating osteoporosis – calcitonin helps increase bone strength by increasing bone calcium.

29. Only one sperm fertilizes an egg, but many sperm are involved in the process of getting through the zona pellucida that surrounds the egg. Enzymes released from the heads of many sperm help to penetrate the zona pellucida.

31. You decide to wiggle your toe. Wiggling your toe is a voluntary movement, so the instruction to do so originates in the frontal lobe of your cerebrum (the left frontal lobe if you wiggle your right toe, the right frontal lobe if you wiggle your left toe). An action potential in a motor neuron that goes to your toe signals muscles there to contract. The action potential arrives at a chemical synapse connecting the neuron to the muscle cells. The action potential triggers the release of the neurotransmitter acetylcholine at the synapse. Acetylcholine then binds to receptors on the muscle cell's membrane, initiating an action potential in the muscle cell. This action potential causes calcium ions to be released from the muscle cell's endoplasmic reticulum. Calcium ions allow a series of "heads" on the myosin fibers to attach to actin. The myosin heads attach and pivot, pulling on the actin filaments. Each pull shortens the length of the sarcomere a tiny bit—about 10 nanometers—and, consequently, the length of the muscle as a whole. After pulling, the myosin heads release, extend, attach, and pull again. This cycle repeats, and the muscle becomes shorter and shorter. The shortened muscle pulls on the toe bone it is attached to, and your toe bends one way. A similar process causes opposing muscles to contract, bending your toe the other way. And there you are, wiggling your toe!

Living Earth Essay

Life on Earth Connections

Soil Fertility

Soil is a mixture containing sand, silt, and clay. All three of these components are ground-up rock; the differences are in how finely the particles are ground. Sand particles are the largest, and clay particles are the smallest.

Soil often occurs as a series of horizontal layers called **soil horizons**, shown in **Figure A**. The deepest horizon, which lies just above solid rock, is the *substratum*, which is rock just beginning to be disintegrated into soil by the action of water that has seeped down to this level. No growing plant material is found in the substratum. Above the substratum is the *subsoil*, which consists mostly of clay. Only the deepest roots penetrate into the subsoil, which may be up to 1 meter thick. Above the subsoil is the *topsoil*, which lies on the surface and varies in thickness from a few centimeters to 2 meters. The topsoil usually contains equal amounts of sand, silt, and clay. This is the horizon where the roots of plants absorb most of their nutrients.

Fertile topsoil is a mixture of at least four components—mineral particles, water, air, and organic matter. The mineral particles are the particles of sand, silt, and clay. Many of the nutrients that plants need are released as these particles are formed from rock erosion. The size of the particles greatly affects soil fertility. Large particles result in porous soil with many pockets of space that collect water and air—up to 25 percent of the volume of fertile topsoil consists of pockets. Roots absorb water and oxygen from the air in these pockets. Small rock particles pack tightly together so that no or only very few air pockets are present, and as a result, little or no oxygen or water is available to roots. This is why plants do not grow well in clay soils. **Figure B** compares these two extremes of soil.

The organic matter in topsoil is a mixture of fallen plant material, the remains of dead animals, and such decomposers as bacteria and fungi, as **Figure C** illustrates. This organic matter is called **humus**, and it is rich in a variety of plant nutrients. Humus tends to be porous, giving roots access to subterranean water and oxygen. It also binds the soil, helping to prevent erosion.

The flow of water through soil is called *percolation*. The more porous the soil, the greater the rate of percolation. With excessive percolation, flowing water removes many water-soluble nutrients needed to make the soil productive. This process is known as *leaching*. With too little percolation, topsoil becomes waterlogged, choking off a plant's supply of oxygen. Soils with optimal percolation drain water from all but the smallest air pockets.

Figure A
The vertical structure of soil is a series of layers called soil horizons.

READING CHECK
Fertile topsoil is a mixture of what four components?

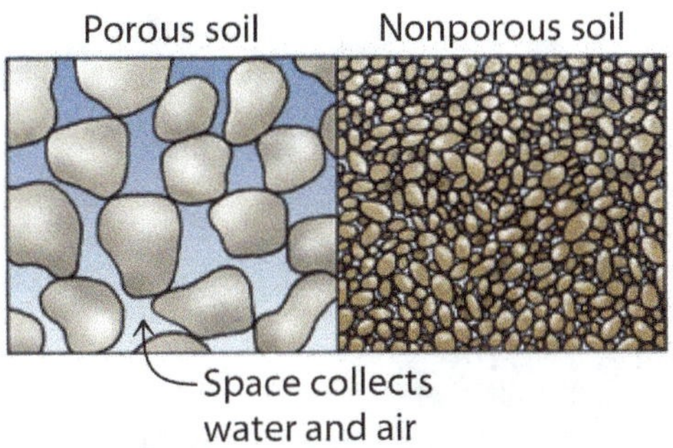

Figure B
Large soil particles create larger pockets of space than do smaller soil particles.

Soil Readily Retains Positively Charged Ions

Mineral particles play an important role in keeping nutrients in the soil. Many plant nutrients are positively charged ions. The surfaces of most mineral particles, however, are negatively charged. **Figure D** illustrates how the resulting ionic attractions help keep nutrients from being washed away. The degree of nutrient retention is most pronounced in clay soils because its mineral particles are the smallest and thus have the largest surface area relative to volume.

Decaying matter in humus contains many carboxylic acid and phenolic groups, which are ionized to negatively charged carboxylate and phenolate ions at a typical soil pH. So, like mineral particles, the humus helps retain positively charged nutrients.

The pH of the soil is largely a function of the amount of carbon dioxide present. Recall from chemistry that carbon dioxide reacts with water to form carbonic acid, which in turn forms hydronium ions:

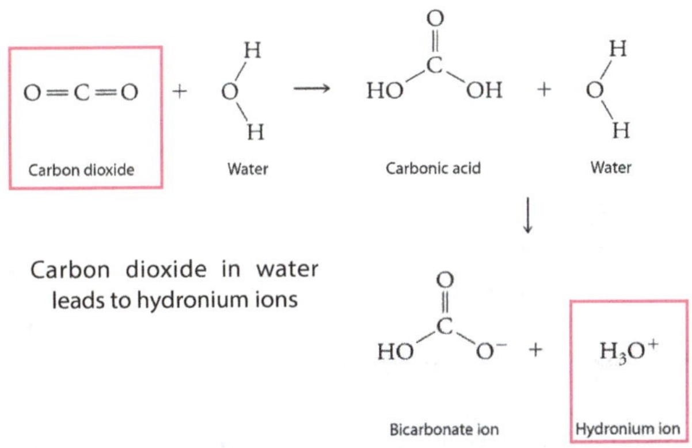

The greater the amount of carbon dioxide in the soil, the more hydronium ions, so the lower the pH. Soil that has a low pH is referred to as sour. (Recall from chemistry that many acidic foods, such as lemons, are characteristically sour.) Two main sources of soil carbon dioxide are humus and plant roots. The humus releases carbon dioxide as it decays, and plant roots release carbon dioxide as a product of cellular respiration. Healthy soil may have enough carbon dioxide released from these processes to give a pH range from about 4 to 7. If the soil within a farm becomes too acidic, a weak base, such as calcium carbonate (known as lime or limestone), can be added.

Hydronium ions are able to displace nutrient ions held to mineral particles and humus. Plants use this fact to great advantage. **Figure E** illustrates how a plant releases

Figure C
Organic matter and living organisms are important components of topsoil.

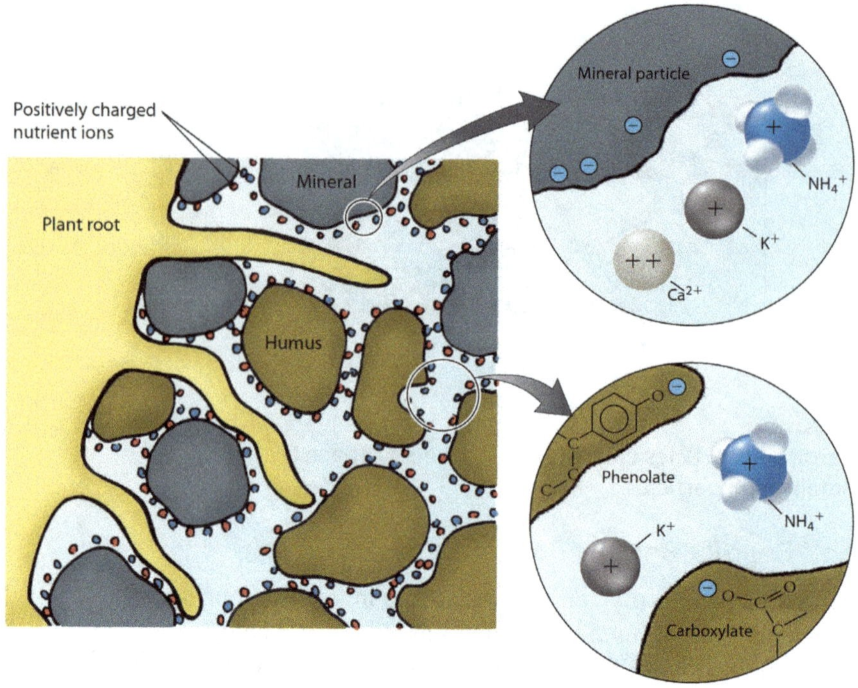

Figure D
The negatively charged surfaces of soil mineral particles and humus help retain positively charged nutrient ions.

CONCEPT CHECK

Soil is able to retain ammonium ions, NH_4^+, better than nitrate ions, NO_3^-. Why?

CHECK YOUR ANSWER

Mineral particles and bits of humus in soil are negatively charged and therefore hold on to positively charged ammonium ions but repel negatively charged nitrate ions.

carbon dioxide through its root system and, in doing so, generates nutrient-displacing hydronium ions. The displaced nutrients, no longer attached to soil particles, thus become available to the plant.

To Life!

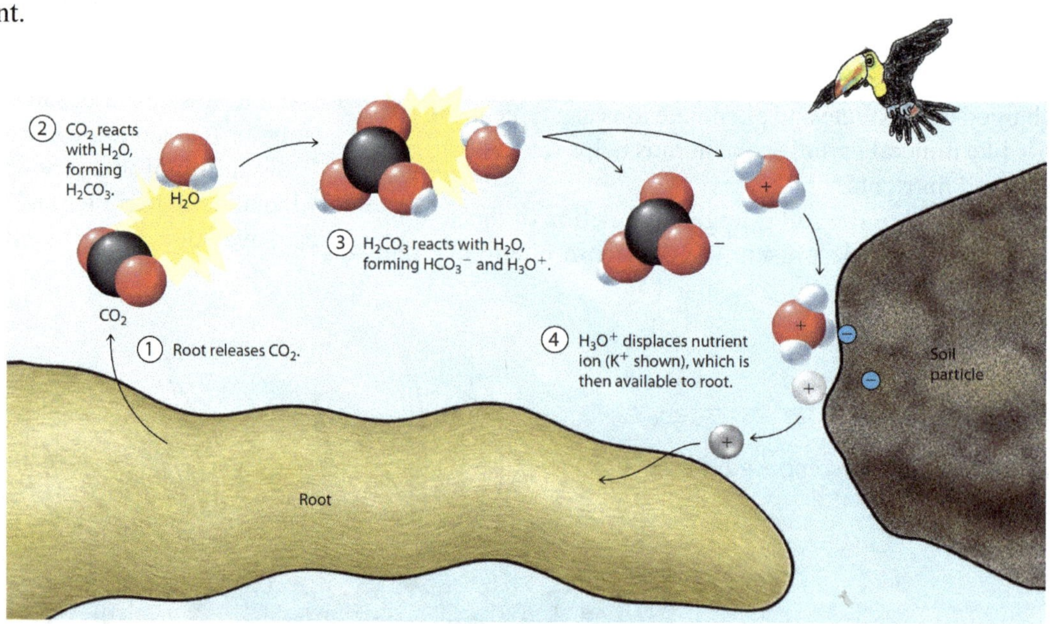

Figure E
By releasing carbon dioxide, a plant guarantees a steady flow of nutrients from the soil to its roots.

CONCEPT CHECK

How might a below-normal pH in soil lead to nutrient deficiencies in plants?

CHECK YOUR ANSWER

When the soil pH is below normal, the water in the soil pockets contains an abundance of hydronium ions, which displace large numbers of nutrient ions from soil and humus particles. Most of these nutrients wash away before the plant is able to absorb them. In a short time, the soils are depleted of nutrients, and the plants become nutrient-deficient

Chapter 14

Maintaining the Body

THE MAIN IDEA

To keep on living, we need nutrients, energy, and protection from disease.

14.1 The Circulatory System
14.2 The Path of Blood Flow
14.3 Blood
14.4 Respiration
14.5 Digestion

Whether you run at world record-breaking speeds—or just sit and read text messages—your body is hard at work. In this chapter, we look at the care and maintenance of the human body. As we explore the body, we'll learn many things. What makes the "lub-dubb" sound of the heartbeat? Is it possible to forget to breathe? What is a hiccup? Is swallowing a voluntary or involuntary action?

14.1 The Circulatory System

Wouldn't it be nice if everything you needed came right to your door? Running low on milk? Here come a couple of gallons now. Hungry for oranges? There should be some along any minute. Broccoli? This is your lucky day. And no need to take the garbage out either. Just drop it, and it will make its own way to the dump.

Is this a couch potato's dream come true? Maybe, but it's also just a day in the life of your cells. Your circulatory system brings your cells everything they need and removes all the garbage they have to dispose of. The circulatory system delivers oxygen and nutrients, removes wastes, and transports special items such as hormones and immune cells. It is like a food, garbage, and mail service rolled into one. In this section, we'll look at the components of the circulatory system. Then, we will examine how blood flows around the body. Finally, we will take a closer look at blood itself.

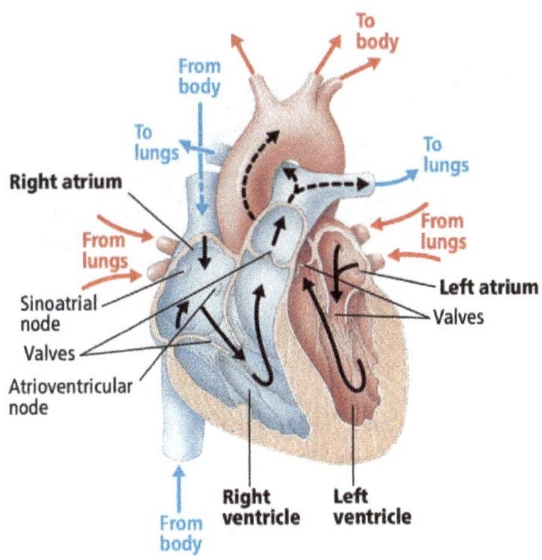

Λ Figure 14.1

The heart has four chambers, the right atrium, right ventricle, left atrium, and left ventricle (labeled in bold). The arrows indicate how blood flows throughout the heart. Heart valves keep blood from flowing backward.

The Heart

The **heart** (Figure 14.1) is a muscular pump that drives blood around the body. The heart is about the size of a clenched fist and has four separate chambers: right atrium, right ventricle, left atrium, and left ventricle. The right side of the heart pumps blood to the lungs, where it picks up oxygen. The left side of the heart pumps blood to the body.

The heart muscle does not need a neuron to tell it to contract, the way voluntary muscles do. Instead, the heart contracts on its own.
Each heartbeat begins in a part of the right atrium called the *sinoatrial node*, or *pacemaker*, as shown in Figure 14.1.

The pacemaker initiates an action potential that sweeps quickly through the right and left atria, which contract simultaneously, driving blood into the ventricles. The pacemaker's signal also travels to the *atrioventricular node*. From the atrioventricular node, the signal passes to the two ventricles, which also contract simultaneously, driving blood out through blood vessels.

Because the atrioventricular node conducts action potentials slowly, there is a delay between the contraction of the atria and the contraction of the ventricles. This is why each heartbeat consists of two separate sounds, the miraculous "lub-dubb ... lub-dubb ... lub-dubb ..." that we associate with life. The "lub" occurs when the two atria contract; the "dubb" occurs when the two ventricles contract. Note, however, that the "lub-dubb" of the heartbeat doesn't come from muscle contractions. (After all, your arm and leg muscles don't thud when they contract!) The noises come from *valves* that snap shut after each contraction. Heart valves are flaps of tissue between the atria and the ventricles, and between the ventricles and blood vessels. These valves prevent blood from flowing backward.

The heart beats about 70 times a minute. This adds up quickly—to more than 100,000 heartbeats a day.

READING CHECK

What are the blood vessels that carry blood away from the heart called?

Blood Vessels

Blood travels throughout the body in rubbery tubes called *blood vessels*. **Arteries** are blood vessels that carry blood away from the heart. Arteries stretch out when blood is pumped, then bounce back. You can feel this stretch and recoil by feeling the pulse at your wrist or temple. From the arteries, blood moves into smaller vessels called *arterioles*.

Each arteriole has a layer of smooth muscle around it that controls its diameter. Adjusting the diameters of different arterioles allows the body to control how much blood different tissues receive. Tissues need more blood when they are working harder. For example, when you've just eaten a meal and are digesting it, arterioles that go to the digestive organs increase in diameter, sending more blood to those organs.

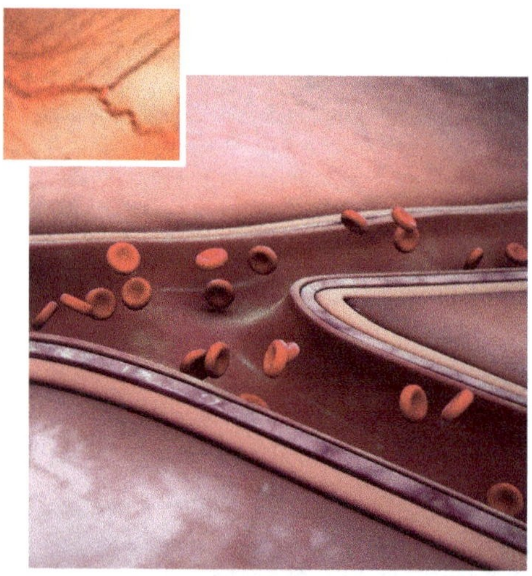

Figure 14.2

Capillaries are tiny, thin-walled blood vessels where molecules are exchanged with body tissues. You can see some red blood cells in this computer-rendered model of a capillary.

From the arterioles, blood flows into capillaries (**Figure 14.2**). **Capillaries** are tiny, thin-walled blood vessels from which molecules are exchanged between blood and body tissues. Oxygen and nutrients move from blood to body tissues, while carbon dioxide and other waste products move from body tissues to blood. Some molecules, including oxygen and carbon dioxide, move via diffusion. Other molecules move by facilitated diffusion or active transport, or they are moved in small vesicles using endocytosis or exocytosis. In order to allow for this exchange, capillaries have very thin walls—often only a single cell thick.

From the capillaries, blood enters small *venules* and then larger **veins**. These are the blood vessels that carry blood toward the heart. The contractions of your voluntary muscles, such as your leg muscles, play an important role in helping to squeeze blood along your veins (**Figure 14.3**). This is why your legs and ankles sometimes swell if you sit or stand for too long without moving. Valves in the veins help make sure blood does not flow backward.

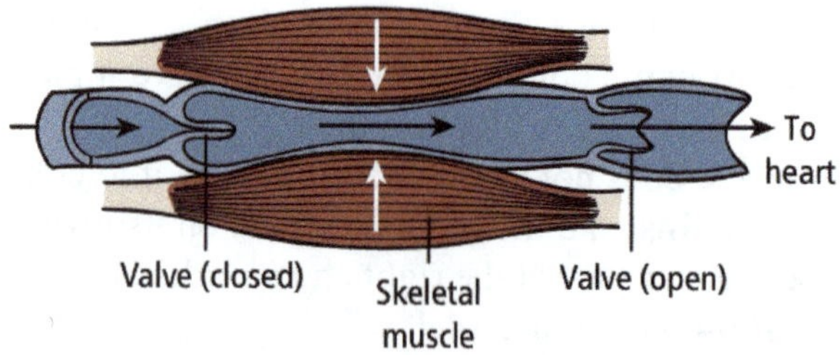

Figure 14.3

The contractions of your voluntary muscles squeeze blood along your veins. Valves make sure the blood doesn't flow backward.

CONCEPT **CHECK**

Explain how the heartbeat occurs.

CHECK YOUR ANSWER

A heartbeat begins on its own in the sinoatrial node, which is part of the right atrium. The sinoatrial node is also known as the pacemaker. An action potential that begins in the pacemaker sweeps quickly through the right and left atria, causing them to contract. This drives blood into the ventricles. The signal from the pacemaker also travels to the atrioventricular node. From the atrioventricular node, the signal passes to the two ventricles, which then contract, driving blood out through blood vessels.

More to Explore

You can read more about the heart and blood vessels, as well as the rest of the circulatory system, here:

https://www.bbc.co.uk/bitesize/guides/zhnk7ty/revision/1

14.2 The Path of Blood Flow

Like all good food, garbage, and mail services, the circulatory system does not move blood haphazardly around the body but follows a set route, shown in **Figure 14.4**. The path of blood flow through the body allows the circulatory system to efficiently carry out one of its primary tasks, delivering oxygen to tissues. How does blood flow?

Let's start with blood returning from the body to the heart. Blood returning from the body contains low levels of oxygen—it is *deoxygenated.* This deoxygenated blood flows from veins into the right atrium of the heart. The right atrium then pumps it to the right ventricle. The right ventricle then pumps it out arteries that go to the lungs. There, in the lungs, blood picks up oxygen and drops off carbon dioxide. The blood is now *oxygenated.*

Oxygenated blood flows back to the heart through veins that go to the left atrium. The left atrium pumps blood to the left ventricle. The left ventricle then pumps it out arteries that go to tissues all over the body.

The oxygen in the blood diffuses into the cells of all your tissues; every one of your cells needs oxygen. After carrying oxygen to the tissues, the blood becomes deoxygenated again and returns to the heart. This entire circuit takes about one minute.

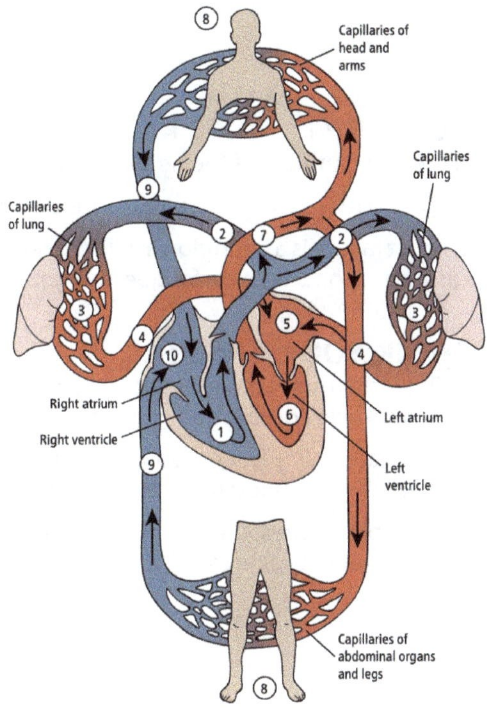

Figure 14.4

This figure shows the path of blood flow around the body. Blue indicates deoxygenated blood and red indicates oxygenated.

Blood Pressure

A common procedure when visiting your doctor is measuring your blood pressure—you may remember the cuff placed around your arm that tightens and then relaxes again (**Figure 14.5**). What exactly is blood pressure, and why do doctors measure it? Pressure is defined as the force exerted over a unit of area. *Blood pressure* is a measure of the pressure your blood exerts against the walls of your arteries.

This pressure is what causes blood to flow through your blood vessels. Blood pressure is highest in the arteries near the heart. It then decreases as blood flows through the arteries to the arterioles, capillaries, venules, and veins. It is lowest just before blood returns to the heart.

When your blood pressure is measured, two separate numbers are reported. The *systolic blood pressure* is the maximum blood pressure in your arteries, which occurs when the ventricles of the heart are contracting. The *diastolic blood pressure* is the minimum blood pressure in your arteries,

READING CHECK

What part of the heart is it that pumps blood out of the arteries that bring oxygen to tissues all over the body?

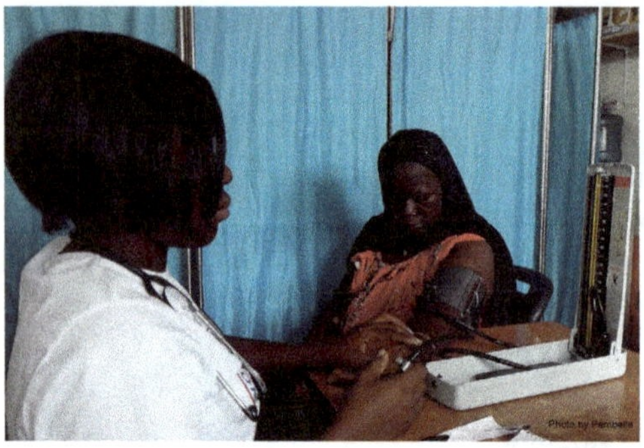

Figure 14.5

Health care provider checks her patient's blood pressure.

which occurs when the ventricles are filling with blood and your heart is resting between beats. Blood pressure is reported as the systolic pressure "over" the diastolic pressure. A normal blood pressure is anything around 120/80 mmHg or lower—that is, a systolic blood pressure that is 120 mmHg or lower, and a diastolic blood pressure that is 80 mmHg or lower. Notice that the units for blood pressure are mmHg (millimeters of mercury). This is the amount of pressure that it takes to push a column of mercury up 1 millimeter. Although mmHg is not a metric system unit, it became standard because mercury was used in the first accurate pressure gauges. This unit of measurement is still widely used in medicine.

A person with high blood pressure has a heart that is working hard. High blood pressure is associated with many health problems, including heart disease, heart attacks, heart failure, kidney disease, and strokes. Unfortunately, high blood pressure is fairly common today. Many people develop the condition with age. A variety of other factors can contribute to high blood pressure, including stress, smoking, lack of exercise, an unhealthy diet, and alcohol consumption. High blood pressure is usually treated with medication.

CONCEPT CHECK

1. Which chambers of the heart contain oxygenated blood? Which chambers contain deoxygenated blood?
2. Most arteries carry oxygenated blood, and most veins carry deoxygenated blood. Are there any exceptions?
3. Heart failure is a condition in which the heart gradually becomes too weak to pump enough blood to the body effectively. Early symptoms include fatigue and getting out of breath easily. Why can high blood pressure contribute to heart failure?

CHECK YOUR ANSWERS

1. The left atrium and left ventricle contain oxygenated blood that has just returned from the lungs. The right atrium and right ventricle contain deoxygenated blood that has just returned from the body.
2. There are only a few exceptions. The arteries that carry blood from the heart to the lungs carry deoxygenated blood. The veins that carry blood from the lungs to the heart carry oxygenated blood.
3. High blood pressure means the heart is working hard to pump blood. Eventually, the heart cannot keep up with its workload and begins to fail.

More to Explore

You can review the path of blood around the body here:

https://www.dummies.com/education/science/biology/the-path-of-blood-through-the-human-body/

14.3 Blood

Now that we know how blood moves around the body, let's look more closely at blood itself. You have about 11 pints of blood in your body, which makes up about 8% of your body weight. A little more than half of this is *plasma.* Plasma is mostly water, but it also contains important molecules such as proteins, hormones, glucose, other nutrients for your cells, and wastes from your cells. The rest of your blood is made up of cells—red blood cells, white blood cells, and platelets. **Red blood cells** transport oxygen to the body's tissues. Red blood cells contain numerous molecules of the protein **hemoglobin**, which binds to and transports oxygen. A single red blood cell has as many as 300 million hemoglobin molecules. Each hemoglobin molecule can carry up to four oxygen molecules. **White blood cells** are part of the immune system and help defend your body against disease. We'll learn more about white blood cells later. **Platelets** are involved in blood clotting. When body tissues are damaged, platelets attach to damaged blood vessels and release special clotting factors. These clotting factors convert proteins in blood plasma into a sticky substance called *fibrin* (**Figure 14.6**). Fibrin prevents more blood from leaking out of the damaged blood vessels. The disease *hemophilia*, associated with a deficiency in blood clotting, results from genetic mutations that affect clotting factors.

READING CHECK

How do the platelets in your blood help you when you cut yourself?

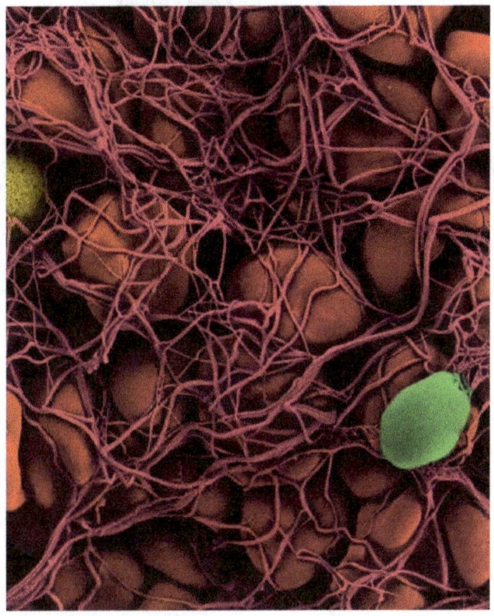

Figure 14.6

Clotting prevents blood from leaking out of damaged blood vessels. This photo shows red blood cells tangled up in threads of fibrin. The small green structures are platelets, and the yellow cell is a white blood cell.

Hemoglobin

Every human red blood cell contains about 300 million molecules of hemoglobin, the oxygen-carrying protein. A hemoglobin molecule is made up of four smaller subunits. Each subunit contains a *heme group* with an iron atom at its center (**Figure 14.7**). It is this iron atom that binds to oxygen. Consequently, each hemoglobin molecule can carry up to four molecules of oxygen.

When oxygen binds to one of the four subunits in a hemoglobin molecule, the other three subunits are altered in such a way that their affinity for oxygen increases— they become more likely to bind oxygen. As a result, most hemoglobin molecules will carry the maximum number of oxygen molecules away from the lungs. This makes oxygen transport more efficient. Similarly, when one subunit unloads an oxygen molecule at a body tissue, the other three subunits become more likely to give up their oxygen molecules, ensuring oxygen is passed efficiently to body tissues. The oxygen affinity of a hemoglobin molecule also changes depending on its local environment.

For example, lower blood pH (a more acidic environment) decreases hemoglobin's oxygen affinity. Why is this adaptive? An active working tissue makes and uses more ATP and so releases more carbon dioxide during cellular respiration. Because carbon dioxide reacts with water in the blood to form carbonic acid, high carbon dioxide levels decrease blood pH. Lower pH decreases the oxygen affinity of local hemoglobin molecules, making it easier for them to unload oxygen to the working tissue.

Hemoglobin also has a high affinity for the toxic carbon monoxide (CO) gas. Carbon monoxide is toxic precisely because it binds to hemoglobin even more strongly than oxygen. Extended exposure to carbon monoxide "fills up" hemoglobin, leaving no place for oxygen to bind. Carbon monoxide is found in car exhaust and cigarette smoke, and some gas appliances also release it. Because the gas is odorless, many people keep a carbon monoxide detector in their homes to warn them of danger.

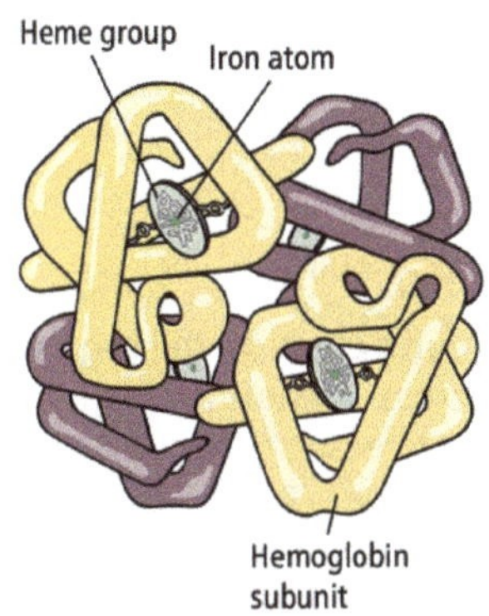

Figure 14.7

Hemoglobin is made up of four subunits. Each subunit includes a heme group with an iron atom at its center.

CONCEPT CHECK

Why is it important to consume enough iron? What happens if you don't consume enough iron?

CHECK YOUR ANSWER

Iron is an essential component of hemoglobin. People who do not consume enough iron may not have enough hemoglobin, resulting in a condition called anemia. In anemia, the blood does not carry enough oxygen. Because the tissues do not receive enough oxygen to support their activities, the result is weakness and fatigue.

More to Explore

You can read more about blood at this website:

https://www.webmd.com/heart/anatomy-picture-of-blood#1

14.4 Respiration

Maybe it's only when you're huffing and puffing that you really think about breathing. Still, breathing in and out is something you do as many as 17,000 times a day! Breathing is your body's way of taking in oxygen, which is needed for cellular respiration, the process cells use to obtain energy in the form of ATP. Through breathing, your respiratory system moves oxygen from the air into your **lungs**, where gas exchange occurs. Oxygen enters the circulatory system from the lungs, delivering it to all your tissues. Breathing also allows your respiratory system to get rid of carbon dioxide, a waste product from an important cellular activity, the production of ATP.

The Path of Air

The respiratory system is shown in **Figure 14.8**. When you inhale, air enters through your nose. (You can also breathe in through your mouth—this is a useful backup because otherwise, a stuffy nose could be fatal.) Hairs in your nostrils trap dust and other particles. Air continues up your nasal passages, where it is moistened by mucus and warmed by a dense network of capillaries. In the nasal passages, your sensory cells for smell also capture odor molecules present in the air. From the nasal passages, air moves through the *pharynx*, the part of the throat above the esophagus and windpipe. It then proceeds through the *larynx*, or voice box, and down the **trachea**, or windpipe. The trachea is a short tube stiffened by rings of cartilage. The rings keep the trachea open. The trachea branches into two tubes called *bronchi* that lead to the right and left lungs. Each bronchus branches into smaller and smaller tubules that finally end at tiny sacs called **alveoli**, where gas exchange occurs.

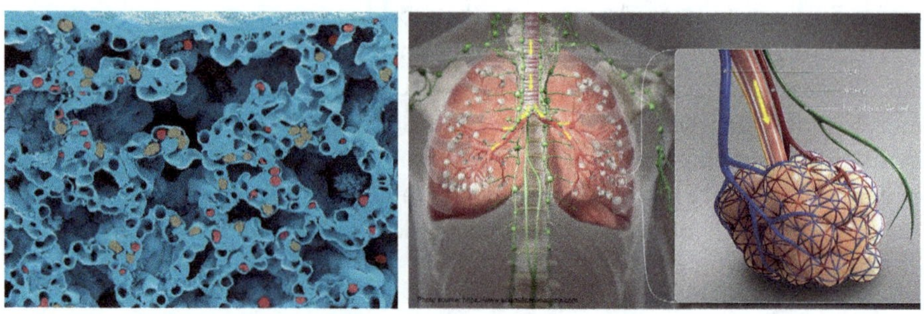

(a) (b)

Figure 14.8 >

(a) A scanning electron micrograph shows a cross-section through the lung. Blood vessels surround alveoli, which appear as round empty spaces in the photo. Red blood cells (red) and white blood cells (yellow) are also visible. (b) The respiratory system brings oxygen into the body and gets rid of carbon dioxide. Gas exchange occurs in the lungs, in tiny sacs called alveoli.

Your lungs contain about 300 million alveoli. Each alveolus is surrounded by a net of capillaries. Both the alveolus and the surrounding capillaries have extremely thin walls, consisting of only a single flattened cell. This allows gas exchange to happen through diffusion. Gas molecules diffuse down their concentration gradients—oxygen diffuses from the alveolus into the blood, and carbon dioxide diffuses from the blood into the alveolus.

When you exhale, air reverses its path. Air moves from the alveoli up through tubules to the bronchi, trachea, and nasal passages, and out through the nose. As air passes the larynx, it may vibrate your vocal cords (**Figure 14.9**). This makes the sound waves that allow you to talk, yell, and sing. In order to make different sounds, you control muscles that stretch your vocal cords, as well as other muscles in your lips, tongue, and cheeks.

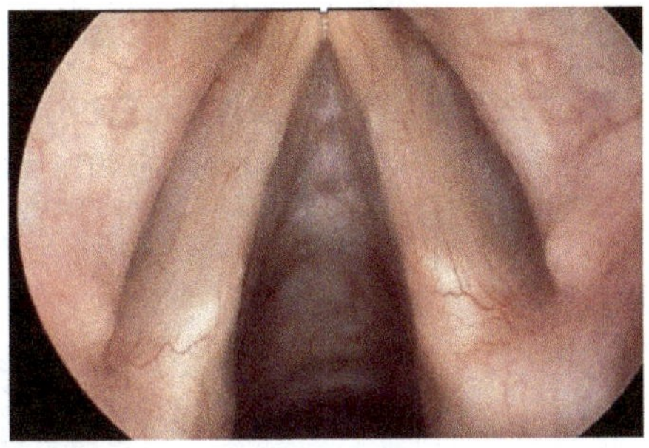

^ Figure 14.9

Vibrations in your vocal cords produce the sounds that makeup speech. You control your sounds by loosening or tightening your vocal cords and moving the muscles of your lips, tongue, and cheeks.

Inhaling and Exhaling

How do you inhale and exhale? Your lungs sit inside your rib cage in an air-filled pocket called the *thoracic cavity*. The bottom of the thoracic cavity is covered by a sheet of muscle called the **diaphragm**. The contraction and relaxation of the diaphragm cause inhalation and exhalation. When the diaphragm is relaxed, it is dome-shaped (**Figure 14.10**). When you inhale, you contract your diaphragm. The muscle fibers shorten, and the diaphragm flattens, increasing the volume of the thoracic cavity. The muscles between your ribs also contract, pulling the rib cage up and out from the body. This further increases the volume of the thoracic cavity. When the volume of the thoracic cavity increases, the air pressure inside it decreases and becomes lower than the pressure of the air outside the body. Since air will move from an area of high pressure to an area of lower pressure, air enters the lungs, filling the alveoli. This process is similar to the way a bicycle pump sucks in air when you pull back its plunger.

READING CHECK

Describe the path of air as it moves from where gas exchange takes place to where it is released from the body when you exhale.

When you exhale, the diaphragm and rib muscles relax, decreasing the volume of the thoracic cavity. This increases the air pressure in the thoracic cavity and pushes air out of the lungs. You breathe about 12 times each minute. During a typical breath, only about 10% of the air in the lungs is exchanged with outside air—this is enough to keep your body tissues supplied with oxygen. Of course, you have some control over how often and how deeply you breathe. You don't have to worry about "forgetting" to breathe, though; respiration is controlled automatically by the brainstem, along with many other involuntary activities.

By the way, if you have ever wondered what a hiccup really is, a hiccup is caused by a sudden spasm of the diaphragm. Each spasm sucks air in, snapping the vocal cords shut and creating a "hic" noise. Hiccups may be caused by any irritation to the diaphragm, including eating too much or too quickly.

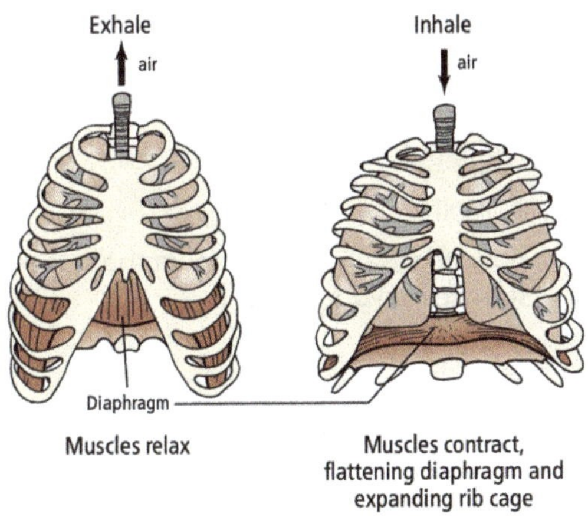

Figure 14.10 >

When you inhale, your diaphragm contracts and flattens, and your rib cage expands. This increases the volume of the thoracic cavity, causing air to flow into your lungs. When you exhale, the diaphragm and rib muscles relax, pushing air out of the lungs.

CONCEPT **CHECK**

Does it take energy to inhale, to exhale, or both?

CHECK YOUR ANSWER

It takes energy to inhale. Energy is needed to contract the diaphragm muscles and the rib cage. It does not take energy to exhale. Exhalation requires only that these muscles relax, a process that requires no energy.

More to Explore

You can read more about the respiratory system here:
https://www.livescience.com/22616-respiratory-system.html

14.5 Digestion

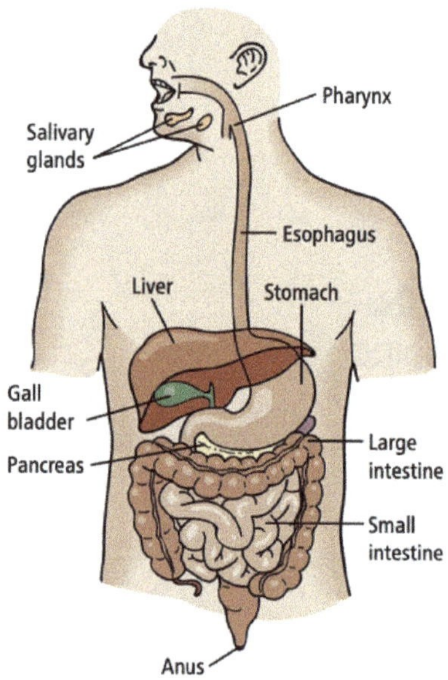

Figure 14.11

The digestive system breaks food down into small molecules that the body can absorb and use.

If you go a few hours without eating, you will likely start thinking about food. Food provides your body with energy-containing organic molecules that are the raw materials for making ATP, the standard energy molecule used by your cells. Food also provides many essential molecules that your body cannot produce on its own. Finally, the breakdown of food releases heat. This heat allows you to maintain a high, stable body temperature, the defining characteristic of mammals and other endotherms.

Although you need food for many reasons, a chunk of food is way too big to travel in your bloodstream or move into your cells. During the process of **digestion**, food is broken down into small organic molecules that can be absorbed and used by the body. You may remember that living things are made up of different kinds of macromolecules—large molecules made up of smaller molecules linked together. During digestion, the macromolecules in your food are broken down into the small molecules that comprise them. For example, the proteins in food are broken down into separate amino acids, and the carbohydrates in food are broken down into simple sugars. These small molecules are then absorbed and used by your body.

Digestion occurs in the digestive system, as shown in **Figure 14.11**. You begin to digest your food as soon as it enters your mouth. You chew food with your teeth, breaking it into little pieces. You also mix your food with *saliva.* Saliva contains a digestive enzyme that breaks down starches. Saliva also moistens food, so that the food can be moved easily around the mouth. Finally, because your taste buds can taste only molecules dissolved in liquid, saliva allows you to taste your food. Eating, or even just thinking about food, causes you to release saliva (**Figure 14.12**).

READING CHECK

How does saliva break down the starches in your food?

After food is chewed and mixed with saliva, you swallow it. Swallowed food is pushed into the pharynx. There are two openings in the pharynx: one to the **esophagus** and one to the trachea. You will choke if food goes down the trachea by mistake. A small flap of cartilage at the back of your tongue—the *epiglottis*—covers the trachea when you swallow so that food will not get into it. This is why you cannot breathe while swallowing, which you may have noticed.

Swallowing begins as a voluntary action. The muscles at the top of the esophagus are voluntary muscles. However, at a certain point, swallowing becomes involuntary. The lower part of the esophagus is made of involuntary muscle, like that found in the rest of the digestive tract. In this part of the esophagus, food is pushed down by a moving wave of involuntary muscle contractions known as **peristalsis**. Peristalsis squeezes food down the esophagus by constricting behind it and pushing it along. You do something similar when you squeeze a LifeSavers® candy from its paper-tube wrapper (**Figure 14.13**). Notice that swallowing does not require gravity—peristalsis does all the work necessary. Astronauts in the zero gravity of space have no trouble swallowing at all. In fact, peristalsis is so effective that you can swallow even while you are upside down.

At the bottom of the esophagus, food moves through a *sphincter*, or ring-shaped muscle, into the **stomach**. Glands in the stomach wall release *gastric juice*, a highly acidic mixture of hydrochloric acid and digestive enzymes (**Figure 14.14**). When you vomit, the acidic nature of your stomach contents becomes immediately apparent, both from the taste and the burning sensation left in your throat. The digestive enzymes in the stomach and the churning of the stomach's muscular walls convert food into a thick liquid called *chyme*. Meanwhile, the acidity of gastric juice kills any bacteria you swallow with your food.

Gastric juice also contains a protective mucus that helps prevent the stomach from digesting its own tissues. Chyme leaves the stomach through a second sphincter and enters the small intestine. Usually, it takes the stomach about 4 hours to process a meal.

The **small intestine** is about 20 feet long. In the first foot of the small intestine, digestion continues with the breakdown of proteins, fats, carbohydrates, and nucleic acids. Some of the digestive enzymes at work in the small intestine are made by the small intestine itself. Others are made by the pancreas and sent to the small intestine. Pancreatic enzymes also play an important role in neutralizing food, which is highly acidic when it arrives from the stomach. The small intestine also receives bile, a substance that is made by the liver and stored in the gall bladder. Bile is an emulsifier—it breaks fats into tiny droplets that are more easily attacked by enzymes.

The rest of the small intestine is responsible for absorbing nutrients into the body. In order to do this job effectively, the small intestine has a huge surface area. The small intestine's inner surface is covered with finger-like projections called *villi*. The villi are covered with even tinier projections called *microvilli* (**Figure 14.15**). Both the villi and microvilli increase the surface area of the small intestine. If your small intestine were flattened, it would fill the area of a tennis court! This impressive surface area enables your small intestine to absorb many digested nutrients at the same time. Most nutrients are absorbed into the body through facilitated diffusion or active transport. Certain bacteria living in your small intestine also play a role in helping you to absorb nutrients. Absorbed nutrients enter capillaries inside each villus and are then carried to all your tissues by the circulatory system.

After the nutrients are absorbed, what is left of your food moves into the **large intestine**. In the large intestine, more water and minerals are absorbed into the body. Huge numbers of bacteria live in the large intestine, where they feed on some of the undigested materials. Some of the bacteria make important vitamins, including vitamin K and some B vitamins. These bacteria are an important part of your microbiome, which includes all the bacteria that live in and on your body. In recent years, scientists have gradually realized how important a healthy microbiome is to human health. From the large intestine, feces are eliminated from the body through the anus. Feces are not exhausted food. Rather, feces are composed primarily of living and dead bacteria as well as indigestible materials such as plant cellulose.

^ Figure 14.12

Just thinking about food can cause the release of saliva.

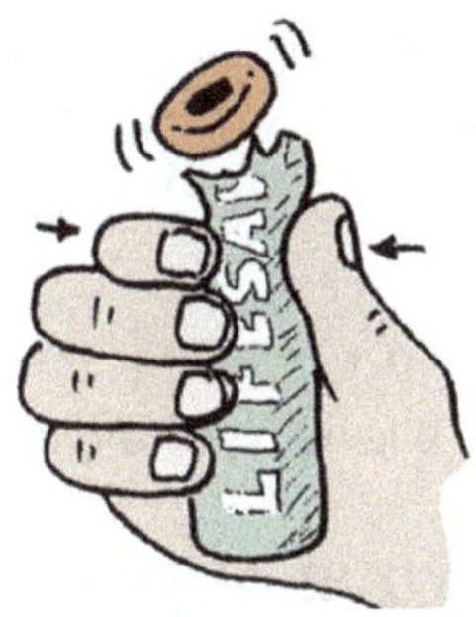

^ Figure 14.13

Peristalsis describes the moving wave of muscle contractions that moves food down the esophagus. It works much like how you squeeze a LifeSavers® candy out of its wrapper.

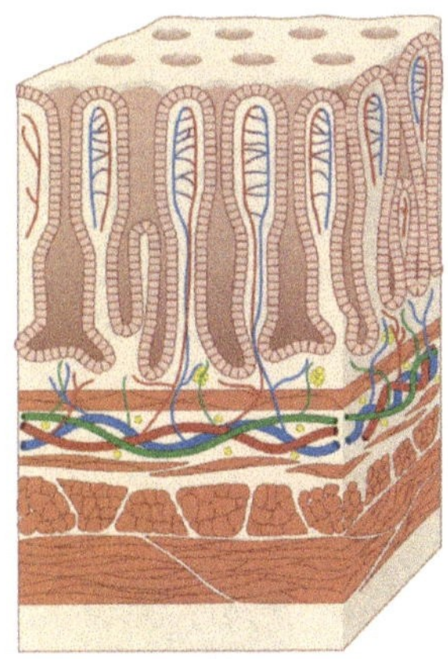

^ Figure 14.14

The stomach has numerous pits on its inner surface. These pits house the glands that release gastric juice.

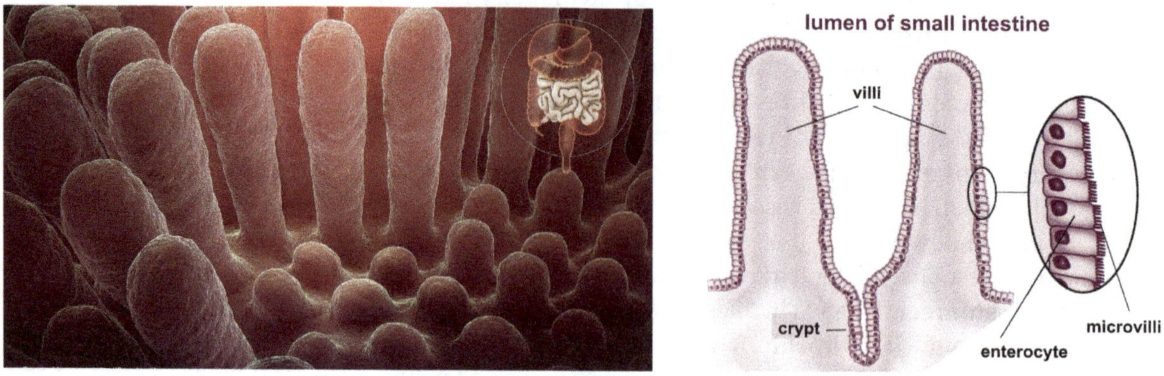

(a) (b)

^ Figure 14.15

(a) Villi are fingerlike projections that increase the surface area of the small intestine. (b) Each villus in the small intestine is covered with even tinier projections called microvilli.

CONCEPT CHECK

How does bile help you digest your food?

CHECK YOUR ANSWER

Bile breaks fats into tiny droplets. Fats, which are hydrophobic, normally stay clumped together. By separating fats into tiny droplets, bile increases the surface area of the fats. This allows digestive enzymes to reach the fat molecules and break them down more quickly.

More to Explore

You can read more about the digestive system at the following websites:

https://www.niddk.nih.gov/health-information/digestive-diseases/digestive-system-how-it-works

https://www.mayoclinic.org/digestive-system/sls-20076373

https://www.uchicagomedicine.org/forefront/gastrointestinal-articles/specific-bacteria-in-the-small-intestine-are-crucial-for-fat-absorption

Concept Review

Chapter 14

Detailed Chapter Summary

The circulatory system is responsible for transporting materials around the body, such as oxygen and nutrients, wastes, and special items such as hormones and immune cells. The circulatory system includes the heart, blood vessels, and blood. The heart is a muscular pump divided into four chambers, the right and left atria and right and left ventricles. The right side of the heart pumps blood to the lungs. The left side pumps blood to body tissues. The heartbeat begins in the sinoatrial node and causes the atria to contract simultaneously. The signal continues to the atrioventricular node and then to the ventricles, which contract simultaneously. Heart valves prevent blood from flowing backward.

The blood vessels include arteries and arterioles that carry blood from the heart, the capillaries, in which molecules are exchanged between blood and body tissues, and venules and veins that return blood to the heart.

Blood follows specific paths around the body. Deoxygenated blood from the body returns to the heart and is pumped to the lungs, where it becomes oxygenated. Oxygenated blood returns from the lungs to the heart and is then pumped to body tissues.

Blood is made up of plasma and three different types of cells, red blood cells, white blood cells, and platelets. Hemoglobin is the protein found in red blood cells that carries oxygen. Each hemoglobin molecule is made up of four subunits. Each subunit includes a heme group with an iron atom that binds to oxygen.

Respiration allows the body to obtain oxygen, which is needed for cellular respiration, and to get rid of carbon dioxide, a waste product of cellular respiration. Air enters through the nostrils, moves through the nasal passages, and continues through the pharynx, larynx, and down the windpipe. Air then enters the two bronchi (left and right) and continues into smaller and smaller tubules until it reaches the alveoli. The lungs contain numerous alveoli, each surrounded by a net of capillaries. Oxygen diffuses from the air in the alveoli into the capillaries, and carbon dioxide diffuses from the capillaries into the alveoli. Inhalation occurs when the muscles of the diaphragm and rib muscles contract. Exhalation occurs when these muscles relax.

During digestion, food is broken down into small molecules that the body can absorb and use. Digestion begins in the mouth, where food is chewed, and digestive enzymes in saliva begin to break down starches. Food is swallowed and moves down the esophagus to the stomach. Food moves through the digestive system through muscular contractions known as peristalsis. The epiglottis closes off the trachea during swallowing to prevent choking. In the stomach, gastric juice includes digestive enzymes as well as hydrochloric acid that helps to kill bacteria or other swallowed organisms. The stomach also churns the food with powerful muscle contractions. Food is converted to chyme and enters the small intestine. In the first foot of the small intestine, proteins, fats, carbohydrates, and nucleic acids continue to be broken down. Digestive enzymes active there are made by the small intestine itself or come from the pancreas. The small intestine also receives bile, an emulsifier that is made by the liver and stored in the gall bladder. The rest of the small intestine is responsible for absorbing nutrients into the body. Villi and microvilli are structures that aid in this process by increasing the small intestine's surface area. In the large intestine, water and minerals are absorbed. Feces, composed of indigestible materials and bacteria, are eliminated from the body through the anus.

Key Terms

Alveoli The tiny sacs in the lungs where gas exchange occurs.

Arteries Blood vessels that carry blood away from the heart.

Capillaries Tiny, thin-walled blood vessels from which molecules are exchanged between blood and body tissues.

Diaphragm The sheet of muscle that covers the bottom of the thoracic cavity; it's contraction and relaxation cause inhalation and exhalation.

Digestion The breakdown of food into small organic molecules that can be absorbed and used by the body.

Esophagus The muscular tube that connects the mouth to the stomach.

Heart The muscular organ that pumps blood through the body; consists of four chambers, the right atrium, right ventricle, left atrium, and left ventricle.

Hemoglobin A protein found in red blood cells that binds to and transports oxygen.

Large intestine A part of the digestive system in which water and minerals are absorbed into the body; also home to large numbers of bacteria that feed on undigested materials and play an important role in human biology.

Lungs The organs in which gas exchange occurs.

Peristalsis A moving wave of muscular contractions that moves food down the digestive tract.

Platelets Blood cells that function in blood clotting.

Red blood cells Blood cells that transport oxygen to body tissues.

Small intestine The long, 20-foot organ in the digestive system that receives chyme from the stomach; here, organic molecules are further broken down and then absorbed into the body.

Stomach The organ in the digestive system that receives food from the esophagus; food is digested by enzymes in gastric juice as well as a muscular churning and then sent to the small intestine.

Trachea A short tube stiffened by rings of cartilage that is a part of the respiratory system and also known as the windpipe; air travels down the trachea, which branches into two bronchi that go to the lungs.

Veins Blood vessels that carry blood toward the heart.

White blood cells Blood cells that are part of the immune system.

Review Questions

Simple Review Questions

14.1 The Circulatory System

1. What stimulates the heart to beat?
2. What makes the "lub-dubb" sound of the heartbeat?
3. Which blood vessels are responsible for nutrient and waste exchange with tissues?

14.2 How Blood Flows Around the Body

4. Trace the path of blood through the body, beginning with blood returning from the tissues to the heart. Be sure to name each of the chambers of the heart.
5. What is blood pressure?
6. What is the difference between systolic blood pressure and diastolic blood pressure?
7. What health issues are associated with high blood pressure?

14.3 Blood

8. What are the three types of blood cells, and what is the function of each?
9. What is the structure of hemoglobin? Which part of the hemoglobin molecule binds to oxygen?
10. How many oxygen molecules can one molecule of hemoglobin carry?
11. Explain how hemoglobin's oxygen affinity is affected by blood pH and how this change in oxygen affinity is adaptive.

14.4 Respiration

12. Trace the path of air as it moves to the alveoli.
13. What structures in the trachea help keep it open?
14. How do the structures of the alveoli and their surrounding capillaries facilitate gas exchange?
15. What is the diaphragm? What is its role in inhalation and exhalation?
16. What happens to the volume of the thoracic cavity when you inhale? What about when you exhale?

14.5 Digestion

17. What does digestion accomplish?
18. What are the functions of saliva?
19. What prevents food from entering the trachea after swallowing?
20. What happens to food while it is in the stomach?
21. What structures increase the surface area available for nutrient absorption in the small intestine?

Challenging Review Questions

14.1 The Circulatory System

22. In a developing embryo, the heart begins to beat long before developing nerves reach it. How is the embryonic heart able to beat without nervous system control?

23. Why are the atria of the heart less muscular than the ventricles? Why is the left ventricle more muscular than the right ventricle?

24. The pumping of the heart does most of the work that is required to move blood around the body. How do your voluntary movements contribute to this process?

14.2 How Blood Flows Around the Body

25. Where in the body is blood most oxygenated?

26. Why does blood pressure decrease as blood flows from arteries to capillaries to veins?

27. High blood pressure can sometimes lead to an enlargement of the left side of the heart. Why?

28. High blood pressure is usually treated with medication. Two kinds of high blood pressure medications are diuretics and ACE inhibitors. Diuretics help the body get rid of excess sodium and water. ACE inhibitors help blood vessels relax and open up. How could each of these medications help reduce blood pressure?

14.3 Blood

29. How does the structure of hemoglobin help ensure that each hemoglobin molecule carries four molecules of oxygen (the maximum number) away from the lungs?

30. How does the structure of hemoglobin help ensure that working tissues, which require more oxygen, get the oxygen they need?

31. Why is carbon monoxide toxic? What effect does it have on the body?

14.4 Respiration

32. Which functions, other than acquiring oxygen for the body, require the work of the respiratory system?

33. Describe the structure of the lungs, as well as how this structure allows the lungs to perform their function in respiration.

34. What role does diffusion play in respiration?

35. How does the shape of the diaphragm change when it contracts? What role does this play in breathing?

36. What role do the rib muscles and rib cage play in inhaling and exhaling?

37. What is the role of air pressure in inhaling and exhaling?

38. Is breathing a voluntary or involuntary action?

14.5 Digestion

39. As food moves down your esophagus, your esophagus bulges. Why doesn't this cause the trachea, which lies adjacent to it, to become closed off?

40. Having the openings to both the trachea and the esophagus in the pharynx is problematic because it can lead to choking. Are there any advantages to this arrangement?

41. What are the functions of gastric juice?

42. What happens to food in the small intestine?

43. Is it true that bacteria live in our large intestines? What do they do there?

Apply & Discuss Questions

44. How do the arterioles react when you are running? When you are doing biceps curls? When you are sitting at your desk thinking and doing homework?

45. You use energy to contract your diaphragm and rib muscles when you breathe. Is energy also required for gas exchange—that is, for the transport of oxygen and carbon dioxide molecules between the air inside the alveoli and the blood in the surrounding capillaries?

46. Why shouldn't you talk with your mouth full (not just because it's impolite)?

47. If you hold a piece of cracker in your mouth without chewing it, the cracker will dissolve. However, this doesn't happen with a piece of meat. Why?

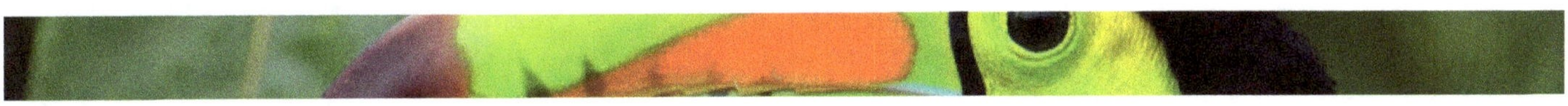

Solutions to Odd-Numbered Questions

Simple Review Solutions

1. Heart muscle does not need a neuron to tell it to contract, the way voluntary muscles do. Instead, the heart contracts on its own. Each heartbeat begins on its own in a part of the right atrium called the sinoatrial node, or pacemaker.

3. Capillaries are tiny, thin-walled blood vessels from which molecules are exchanged between blood and body tissues.

5. Blood pressure is a measure of the pressure your blood exerts against the walls of your arteries. This pressure is what causes blood to flow through your blood vessels.

7. A person with high blood pressure has a heart that is working hard. High blood pressure is associated with many health problems, including heart disease, heart attacks, heart failure, kidney disease, and strokes.

9. A hemoglobin molecule is made up of four smaller subunits. Each subunit contains a heme group that includes

an iron atom at its center. It is this iron atom that binds to oxygen.

11. Lower blood pH (a more acidic environment) decreases hemoglobin's oxygen affinity. Why is this adaptive? An active, working tissue makes and uses more ATP and so releases more carbon dioxide during cellular respiration. Because carbon dioxide reacts with water in the blood to form carbonic acid, the presence of high levels of carbon dioxide decreases blood pH. This acidity decreases the oxygen affinity of local hemoglobin molecules, making it easier for them to unload oxygen to the working tissue.

13. The trachea is a short tube stiffened by rings of cartilage. The rings keep the trachea open.

15. The bottom of the thoracic cavity is covered by a sheet of muscle called the diaphragm. The contraction and relaxation of the diaphragm cause inhalation and exhalation.

17. During the process of digestion, food is broken down into small organic molecules that can be absorbed and used by the body.

19. A small flap of cartilage at the back of your tongue—the epiglottis—covers the trachea when you swallow. That way, food will not get into it.

21. The small intestine's inner surface is covered with fingerlike projections called villi. The villi are covered with even tinier projections called microvilli. Both the villi and microvilli increase the surface area of the small intestine.

Challenging Review Solutions

23. The atria only pump blood to the ventricles, whereas the ventricles have to pump blood to the lungs and body. The left ventricle has to pump blood to the body, so must generate more pressure than the right ventricle, which only has to pump blood to the lungs.

25. Blood is most oxygenated at the lungs, where oxygenation occurs. It stays at this high level of oxygenation as it returns to the heart and is pumped to the body.

27. If a person has high blood pressure, that person's heart is working hard in order to pump blood with a large amount of force. In order to do this, the heart builds more muscle, causing it to become enlarged. It is often the left side of the heart that becomes enlarged because the left side does the harder work of pumping blood to all the body tissues.

29. A hemoglobin molecule is made up of four smaller subunits. Each subunit contains a heme group that includes an iron atom at its center. It is this iron atom that binds to oxygen.

31. Carbon monoxide is toxic because it binds to hemoglobin even more strongly than oxygen binds. Extended exposure to carbon monoxide "fills up" hemoglobin, leaving no place for oxygen to bind.

33. The lungs consist of a set of branching tubules. Each tiny tubule ends in a small sac of air called an alveolus. The wall of each alveolus is made up of only a single flattened cell and is surrounded by a net of capillaries. The large numbers of alveoli and their thin walls allow for effective diffusion of oxygen from the lungs to the blood.

35. The diaphragm flattens when it contracts. This increases the volume of the thoracic cavity, which causes the air pressure inside it to decrease. As a result of this decrease in air pressure, air is pulled from outside the body into the lungs.

37. Changes in the air pressure in the thoracic cavity cause air to move in and out of the lungs. When the diaphragm and rib muscles contract, the volume of the thoracic cavity increases, causing air pressure to decrease. Then, air pressure is greater outside the lungs and moves into the lungs. This is inhalation. When the muscles relax, the volume of the thoracic cavity decreases, increasing air pressure there. Air pressure is then greater inside the lungs than outside, and air is pushed out of the lungs. This is exhalation.

39. The trachea is stiffened by rings of cartilage that keep it open.

41. The digestive enzymes in gastric juice (along with the churning of the stomach's muscular walls) convert food to a thick liquid called chyme. The acidity of gastric juice kills bacteria you swallow with food. Gastric juice also contains protective mucus that helps prevent the stomach from digesting its own tissues.

43. Yes, huge numbers of bacteria live in our large intestines. They eat the food we haven't been able to digest. They also make some key vitamins we need and prevent more dangerous bacteria from being able to survive there.

Apply & Discuss Solutions

45. No -- gas exchange occurs through diffusion, a form of passive transport that does not require energy from your cells.

47. Saliva contains a digestive enzyme that breaks down the starches found in crackers. Meat is largely protein, and the digestive enzymes that break down proteins are not found in your mouth; they are further down the digestive tract.

Living Earth Essay

Life on Earth Connections

Pesticide Bioaccumulation

A high-yield crop needs more than adequate nutrition. It also needs defense against a host of natural enemies, such as the moth larva shown in **Figure A**. To control these pests, farmers can apply substances known as *pesticides*. There are several kinds of pesticides, including insect-killing *insecticides*, weed-killing *herbicides*, and fungus-killing *fungicides*. Despite the promise, pesticides can lead to environmental damage, as exemplified by a process known as *bioaccumulation*. Let's take a closer look focusing only on the insecticides.

Figure A
The larva of the Helicoverpa armigera moth, which is particularly resistant to chemical and biochemical pesticides (Source: David McClenaghan, CSIRO)

Most species of insects are beneficial or even essential to agriculture. Honeybees, for example, are responsible for the pollination of $10 billion worth of produce in the United States. Countless other species take part in nutrient recycling and help maintain soil quality. A small minority of insect species, however, have continually threatened our capacity to grow, harvest, and store crops, and it is against these species that insecticides are used. The most widely used insecticides are chlorinated hydrocarbons, organophosphorus compounds, and carbamates.

The *chlorinated hydrocarbons* are remarkably *persistent*, killing insects for months and years on treated surfaces. There are at least two reasons for this persistence. First, chlorinated hydrocarbons tend to be nonbiodegradable, which means there are no natural pathways to break them down chemically. Second, they are nonpolar compounds, which means they are insoluble in water and thus are not washed away by rainwater.

In 1939, there was a breakthrough in the fight against insect pests with the chemical synthesis of the chlorinated hydrocarbon DDT, shown in **Figure B**. During the 1940s and 1950s, DDT was applied liberally to crops, resulting in markedly greater yields. In addition to protecting plants, DDT was found to protect people from disease. It was applied to rivers, streams, and villages to help control the proliferation of mosquitoes, lice, and tsetse flies, which spread malaria, typhus, and sleeping sickness, respectively. According to the World Health Organization, by protecting against these diseases, DDT has saved an estimated 25 million human lives.

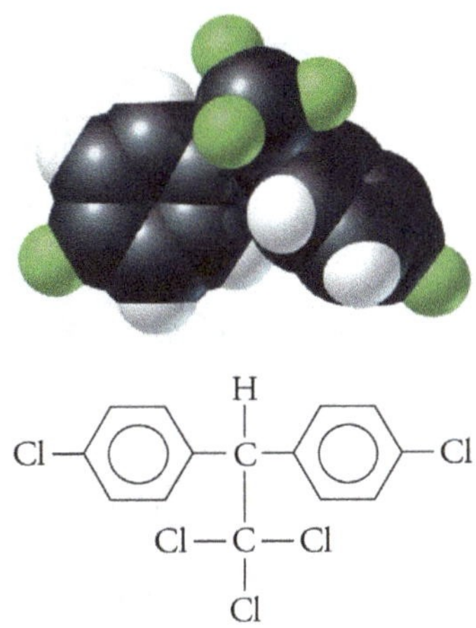

Figure B
The chemical name for DDT is dichlorodiphenyltrichloroethane.

Insect populations, however, began to develop a resistance to DDT within a few years of its first application. Furthermore, DDT was found to be toxic to wildlife, including the natural predators of insects, such as birds. With fewer natural predators, DDT-resistant insects were able to thrive. Therefore, the early increased crop yields resulting from DDT use were not sustainable.

Bioaccumulation

In the 1950s and 1960s, these and other negative aspects of DDT and other pesticides were brought to the public's attention by a number of publications, including the biologist Rachel Carson's book *Silent Spring*. Carson, shown in **Figure C**, described the importance of understanding the dynamics of ecosystems, most of which are highly sensitive to human activities.

Figure C
Rachel Carson was a pioneer in the fight against the excessive use of pesticides.

Carson described a phenomenon known as **bioaccumulation**, whereby a toxic chemical that enters a food chain at a low trophic level becomes more concentrated in organisms higher up the chain, as illustrated in **Figure D**. In bodies of water sprayed with DDT, for example, small amounts of the pesticide were ingested and stored in the nonpolar lipids of aquatic microorganisms. Because these microorganisms serve as food for animals at higher trophic levels, DDT became more concentrated in the body fat of these larger creatures. Predatory birds at the top of the chain accumulated the greatest amounts of DDT. Eventually, the elevated DDT levels affected avian population numbers because the eggshells of affected birds were too thin and fragile to support the growing chick embryos. DDT contributed to the decline of many bird populations and the near extinction of some species of ospreys, hawks, eagles, and falcons. In the early 1970s, the United States and many other countries banned the use of DDT. Within a matter of years, many wildlife species in these countries were able to recover.

CONCEPT CHECK

Why are fish-eating birds more susceptible to DDT contamination compared to plankton-eating fish?

CHECK YOUR ANSWER

A plankton-eating fish gets DDT from all the plankton it eats. This same DDT then enters the bird that eats this fish. But because the bird eats many fish, its exposure to DDT is amplified.

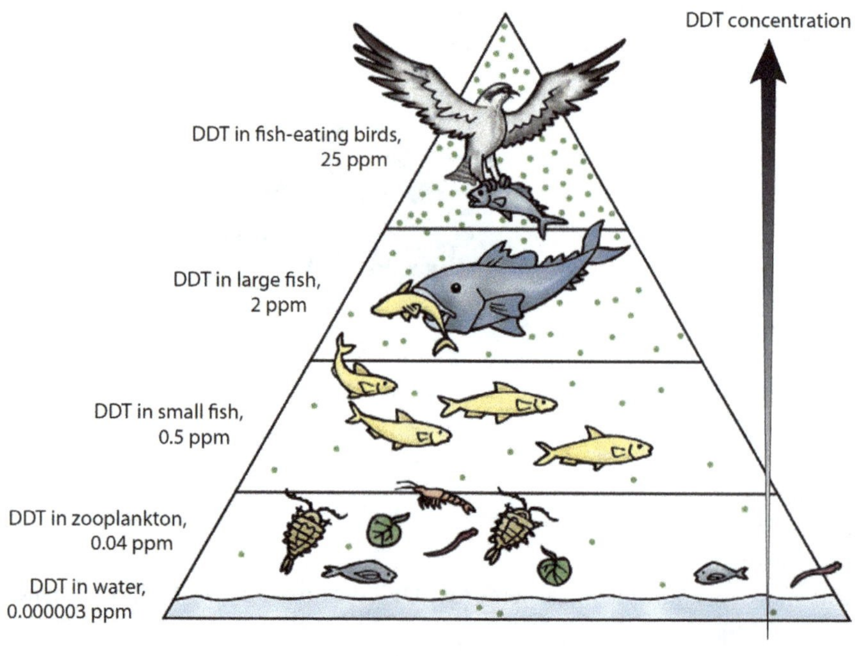

Figure D
The DDT concentration in a food chain can be magnified from 0.000003 parts per million (ppm) as a pollutant in the water to 25 ppm in a bird at the top of the chain.

fyi FOR YOUR INFORMATION

Many nations still rely on DDT as an economical method of controlling insects that carry human diseases. DDT can be most effective in this manner, but not without consequences. In Malaysia, for example, thatched roofs were once sprayed with DDT to kill malaria-carrying mosquitos. Wasps that kill straw-eating moths were also killed. The moths prospered, leading to the destruction of the thatched roofs. Furthermore, cats that ate geckos that ate DDT-laden cockroaches died. With no cats, rats multiplied, as did the bubonic plague-carrying fleas that lived on the rats. Cats were therefore dropped by parachute to remote villages to help prevent plague outbreaks. When it comes to ecosystems, you can never change only one thing.

Alternatives to DDT

Many chlorinated hydrocarbon alternatives to DDT have been developed. One of the earliest substitutes was methoxychlor, as shown in **Figure E**. Methoxychlor has much lower toxicity in most animals and, unlike DDT, is not readily stored in animal fat. Look carefully at the structures of methoxychlor and DDT, and you'll see that they are identical, except that methoxychlor has two ether groups, whereas DDT has two chlorine atoms. Because the structures are nearly identical, they have nearly the same level of toxicity in insects. In higher animals, however, the oxygen atoms facilitate detoxification. Specifically, enzymes in the liver cleave the ether groups to synthesize polar products that are readily excreted through the kidneys. While methoxychlor may be less hazardous than DDT, it nonetheless is a powerful toxic agent to be handled with care and concern.

Organophosphorus compounds and carbamates, in contrast to chlorinated hydrocarbons, readily decompose to water-soluble components and so do not act over extended periods of time. Their immediate toxicity to both insects and animals is much greater than that of chlorinated hydrocarbons, however. Added safety precautions are required during the application of organophosphates and carbamates, especially because of their toxicity to honeybees (**Figure F**).

READING CHECK

Why do organophosphorus and carbamate insecticides have a short duration?

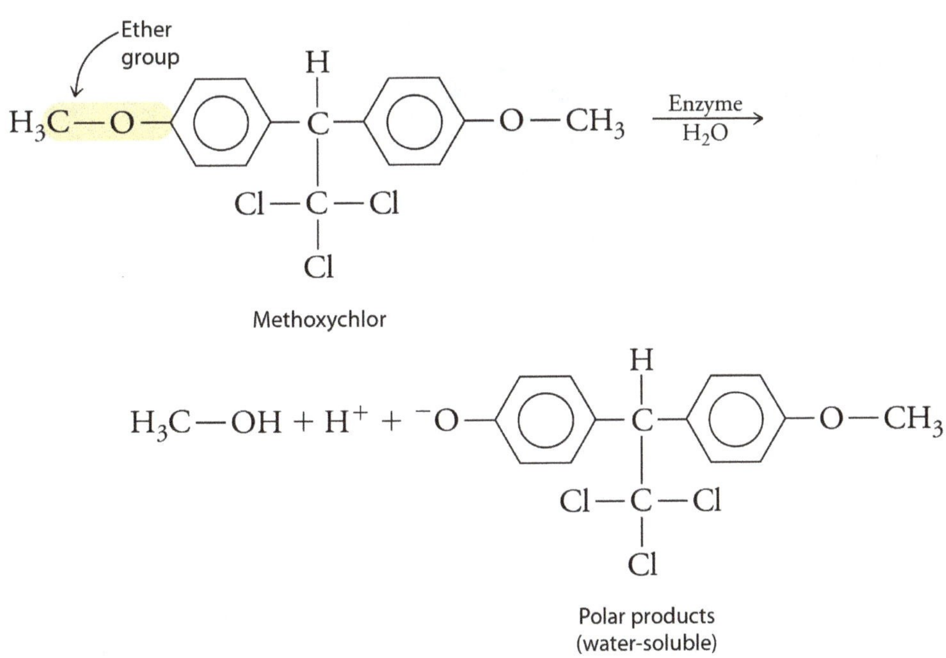

Figure E
Methoxychlor is one of many alternatives to DDT. Enzymes in the liver can cleave the ether groups to produce polar products. Look back at Figure B, and you will see that DDT lacks ether groups.

Figure F
Honeybees do not forage at night. Quick-acting pesticides, such as organo-phosphorus compounds and carbamates, are therefore best applied in the evening. By the time the bees return the next day, these pesticides have lost much of their toxicity.

There are hundreds of organophosphorus and carbamate insecticides in agricultural and household use. Two important examples are malathion, an organophosphorus compound, and carbaryl, a carbamate, both shown in **Figure G**. Malathion kills a variety of insects, such as aphids, leafhoppers, beetles, and spider mites. Carbaryl, like many other carbamates, is relatively selective in the types of insects that it kills.

The most widely used insecticides, however, are the neonicotinoids, such as the compound imidacloprid, shown in **Figure H**. These compounds are derivatives of the compound nicotine, which is a naturally occurring neurotoxin, both to insects and mammals. The neonicotinoids, however, are selectively more toxic to insects than they are to mammals, making them safer for us to handle. These compounds tend to be water-soluble and can be added to irrigation water or even sprayed on the seeds, which grow into plants that then retain enough neonicotinoids to fight off insects. The neonicotinoids are biodegradable, especially when exposed to sunlight. There is strong evidence, however, that these compounds are having a negative impact on bee populations. Notably, they may be responsible, in part, for a phenomenon called colony collapse disorder in which most of the worker bees in a colony disappear.

During the last 60 years, pesticides have benefited society by preventing disease and increasing food production. Our need for pesticides will continue, but greater specificity will certainly be demanded. Furthermore, it is becoming increasingly apparent that the benefits of using pesticides must be considered in the context of potential risks.

To Life!

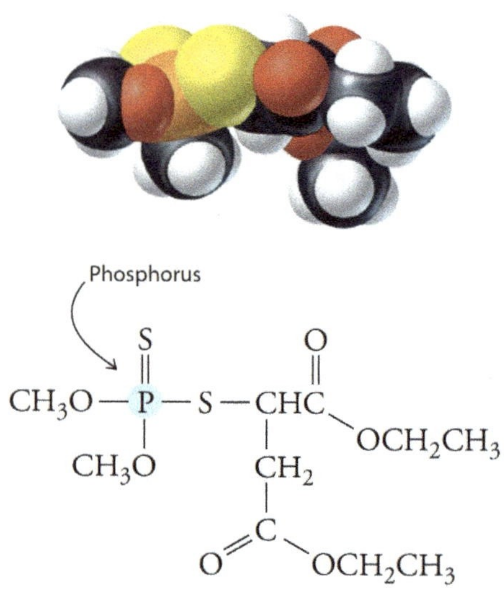

Malathion
(an organophosphorus compound)

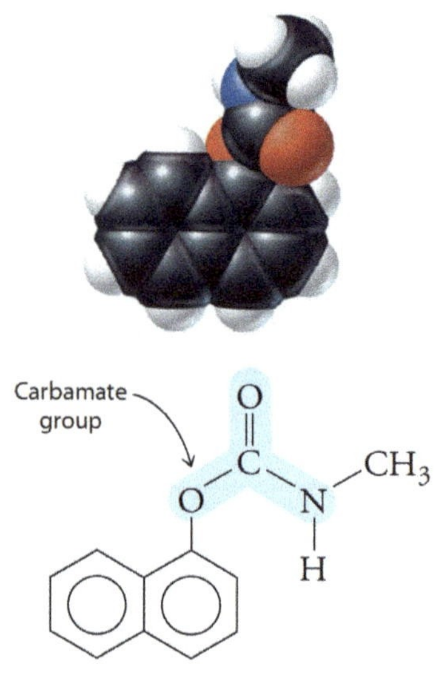

Carbaryl
(a carbamate)

Figure G
The widely used pesticides malathion and carbaryl are short-acting toxic compounds to be handled with much care and concern

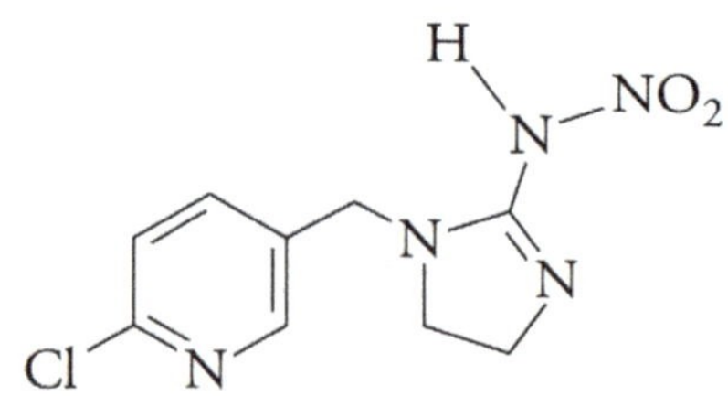

Imidacloprid

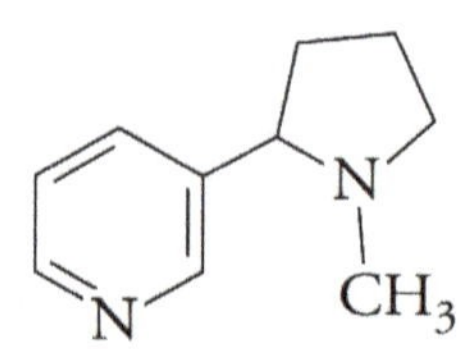

Nicotine

Figure H
Imidacloprid is an example of a neonicotinoid, which is a compound that mimics the insecticidal properties of naturally occurring nicotine.

Chapter 15

Protecting Health

THE MAIN IDEA

Good health comes from good nutrition as well as the protection of the immune system.

15.1 Nutrition, Exercise, and Health
15.2 The Excretory System
15.3 The Innate Immune System
15.4 The Acquired Immune System

We all know that better health arises from adequate nutrition and exercise. But what kinds of nutrients are there, and how can we ensure we receive a proper balance of each? What are vitamins and minerals? How does the body process nutrients and remove wastes? What role does the kidney play? And how exactly does our immune system protect us against bacterial and viral infections? If you cut or burn your finger, why does it swell up and turn red?

15.1 Nutrition, Exercise, and Health?

What does the human body need in the way of nutrition and exercise? What can you do to live a healthier life? After all, health is wealth and never to be taken for granted.

Let's start with your diet. You need a certain number of food Calories to support your activities. Recall that a food Calorie is a unit of energy. Technically, one Calorie (spelled in the upper case) is equal to a kilocalorie, which is 1000 calories (spelled in the lower case). A kilocalorie is the heat energy needed to change the temperature of 1 kg of water by 1°C. These food Calories provide the fuel that your cells use to make ATP. Most people need between 1500 and 3000 Calories a day—the exact number depends on your body size and activity level.

^ Figure 15.1

Tofu is made from soybeans, a plant-based food that provides a complete protein.

You also need certain nutrients that your body can't make on its own. For example, adult humans cannot make 9 of the 20 amino acids needed to build proteins. This is why you need to regularly eat a "complete protein" that contains all the amino acids. Meat is a complete protein, as are beans and rice when eaten together, which is an important part of many vegetarian diets. Soybeans and quinoa are two other plant-based foods that are complete proteins (**Figure 15.1**).

Humans also require a number of vitamins and minerals in small amounts. **Vitamins** are organic molecules that are essential non-protein components of certain enzymes. The body needs many different vitamins, but let's look at just a few. Vitamin C, found in citrus fruits, dark green leafy vegetables, and certain other foods, helps the body resist infections and repair wounds. Vitamin A, which you get from carrots, eggs, fortified milk, and other foods,

READING CHECK

Why is it important to eat foods that contains vitamin A?

is important for proper eye function. Vitamin A enables the eyes to adjust to dim light; insufficient amounts result in night blindness and other problems. The B vitamins are water-soluble and found in leafy greens, dairy products, eggs, beef, and many other sources.

The B vitamins help the body's cells to function properly, allowing you to maintain healthy tissues. Vitamin K, which is found in green leafy vegetables and synthesized by bacteria in the intestines, is essential for blood clotting. Insufficient amounts of vitamin K can result in hemorrhaging. Vitamin D is important for bone growth, immune system function, and other activities. Your body produces some vitamin D when ultraviolet light from the Sun strikes your skin. Vitamin D is not found naturally in many foods, however, and recent studies suggest that many people do not get enough vitamin D.

Minerals are inorganic substances required as components of various body tissues. Important minerals include calcium for bones and teeth, phosphorus for ATP, and iron for hemoglobin. Mineral deficiencies can lead to a variety of health issues.

Beyond calories and nutrients, does it matter what you eat? Yes! And, unfortunately, some of the things people like to eat, such as sweets and fats, aren't very good for them when eaten in excess. Why would humans evolve a preference for such foods? It is likely that people's taste for these foods evolved under a different set of circumstances when these types of food were much less readily available. Their high caloric count made them prized foods to which we became most sensitive.

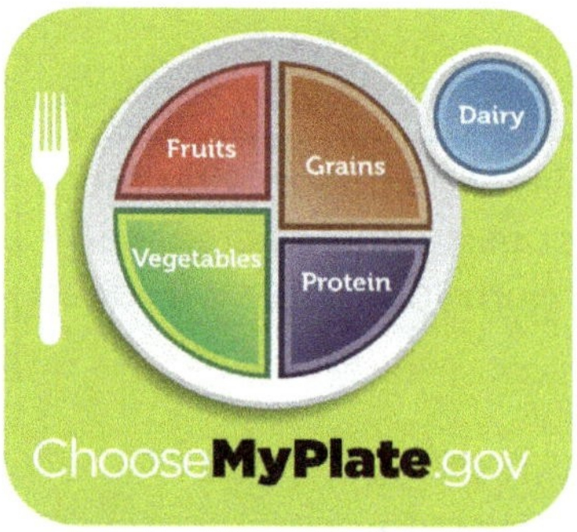

∧ Figure 15.2

The USDA's "My Plate" emphasizes fruits and vegetables, whole grains, and healthy sources of protein.

More than two-thirds of Americans today are either overweight or obese. This has resulted in an increasing number of people affected by metabolic syndrome, a set of characteristics that greatly increases the risk of heart disease, stroke, and type 2 diabetes. *Metabolic syndrome* is associated with two main features: carrying extra weight around the middle and upper parts of the body and insulin resistance, the inability of body tissues to adequately respond to the hormone insulin. Insulin resistance results in higher blood sugar levels as well as increased fat storage. In addition to diet, factors that may also contribute to metabolic syndrome include aging, genetics, and lack of exercise.

So, what makes up a healthy diet? The U.S. Department of Agriculture's "MyPlate" offers some guidelines (Figure 15.2). The plate itself emphasizes the importance of sitting down to a well-planned meal. Half the plate is covered with fruits and vegetables, and another large segment is taken up by whole grains. Other recommendations include obtaining some protein from healthy sources, such as beans and seafood; cutting back on salt, sugar, and solid fats; eating an appropriate number of calories; and exercising.

Exercise is a crucial part of any healthy lifestyle (Figure 15.3). For most of human history, daily life provided people with plenty of physical activity. This was true when most people belonged to hunting and gathering societies and when most people worked on farms. It is still true for a large fraction of the world's population, but not all of it. Desk jobs, cars, television, and the Internet have brought many people too close to a sedentary, couch-potato lifestyle. Many people now have to go out of their way to get the physical activity they need to stay healthy.

What are the benefits of exercise? Regular exercise reduces the risk of heart disease, high blood pressure, colon cancer, breast cancer, osteoporosis, diabetes, obesity, and many other health issues. Exercise improves the performance of the heart and lungs, maintains joints and tendons, and increases muscle mass and bone density. All of this contributes to strength and balance and helps people stay flexible as they age. Exercise is good for mental health too: It helps reduce depression, stress, and anxiety, and even helps you sleep better. There's no doubt that exercise is a good thing.

^ Figure 15.3
Even people with busy lives should make time for exercise, which provides many benefits to the body.

What Are the Odds? Current Major Health Risks

Heart disease is the leading cause of death in the United States, followed by cancer. The 10 leading causes of death in the United States are listed below, along with worldwide data for comparison. This is data from 2019. For some perspective, in 2020, about 400,000 deaths in the United States were from Covid-19, making it the third leading cause of death for that year.

Leading Causes of Death in the United States
(2019 Data from the U.S. Centers for Disease Control and Prevention)

1. Heart disease: 659,041
2. Cancer: 599,601
3. Accidents (unintentional injuries): 173,040
4. Chronic lower respiratory diseases: 156,979
5. Stroke (cerebrovascular diseases): 150,005
6. Alzheimer's disease: 121,499
7. Diabetes: 87,647
8. Nephritis, nephrotic syndrome, and nephrosis: 51,565
9. Influenza and pneumonia: 49,783
10. Intentional self-harm (suicide): 47,511

CONCEPT **CHECK**
What are some guidelines for eating a healthy diet?

CHECK YOUR ANSWER
Eat a lot of fruits and vegetables, choose whole grains, obtain some of your protein from healthy sources such as beans and seafood, and cut back on things like salt, sugar, and solid fats.

More to Explore
To learn more about nutrition, check out the U.S. Department of Agriculture:
https://www.nutrition.gov

15.2 The Excretory System

As cells go about their activities, they generate wastes. We already know what cells do with these wastes—they let them diffuse into the bloodstream. But where do the wastes go from there? How does the body get rid of them? This is the job of the excretory system.

The excretory system, shown in **Figure 15.4**, filters your blood, removing wastes while leaving useful molecules in the blood. The excretory system also controls how much water, sodium, potassium, calcium, and other substances the body keeps. Did you know that your entire blood supply moves through your kidneys 16 times a day? The end result is something we're all familiar with—about 6 cups of urine.

One of the most important wastes found in urine is *urea*. When amino acids are broken down to make ATP, ammonia, a nitrogen-containing waste, is produced. Because ammonia is highly toxic, the liver immediately converts it to urea, a less toxic waste. The liver then releases urea into the bloodstream for excretion.

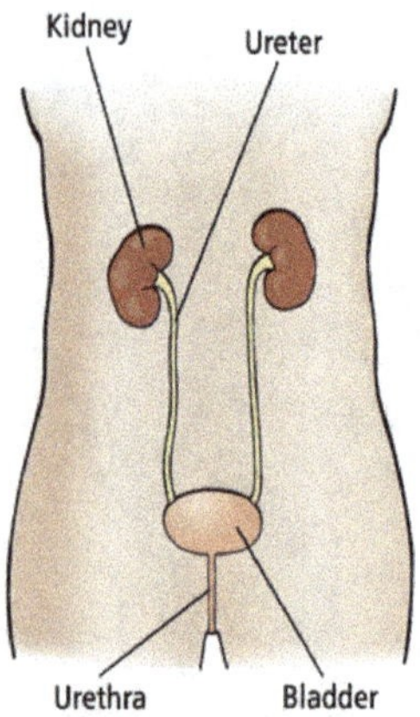

Figure 15.4

The excretory system consists of the kidneys, ureters, bladder, and urethra.

How does excretion happen? The process begins in the **kidneys**, paired excretory organs that filter blood and produce urine. Each kidney contains about a million structures called *nephrons*. The **nephron** is the functional unit of the kidney. What goes into the nephron is more or less blood plasma, and what comes out is urine. Let's look at how this happens.

Each nephron is associated with a cluster of capillaries (**Figure 15.5**). The first part of the nephron, a cup-shaped structure called *Bowman's capsule*, surrounds the capillaries. Blood pressure pushes fluid out of the capillaries and into Bowman's capsule. The fluid that enters Bowman's capsule is called the *filtrate*. At this point, the filtrate is very similar to blood plasma.

From Bowman's capsule, the filtrate enters the *proximal tubule*. The proximal tubule is like a sorting machine. "Good" molecules in the filtrate—such as ions, glucose, vitamins, and amino acids—are transported back into the blood so the body can keep them. "Bad" waste molecules are transported from the blood to the filtrate. The movement of specific molecules into and out of the proximal tubule occurs through active transport, a process that requires energy. This is one of the reasons excretion requires a significant amount of energy.

After moving through the proximal tubule, the filtrate enters the *loop of Henle*, a hairpin-shaped loop. In the loop of Henle, water is absorbed from the filtrate. This helps the body save water. From the loop of Henle, the filtrate moves into the *distal tubule*. In the distal tubule, more wastes are transported into the filtrate.

READING CHECK

What goes into the nephrons that make up your kidneys, and what comes out of them?

Finally, the filtrate moves down the *collecting duct*. In the collecting duct, more water is absorbed from the filtrate. How much water is absorbed depends on whether antidiuretic hormone is present in the bloodstream. If antidiuretic hormone is present, more water is absorbed and kept by the body. As a result, urine becomes more concentrated.

The filtrate—which is now urine—flows from the collecting duct into the *renal pelvis* (see Figure 15.5). The renal pelvis is like a giant funnel that catches the drippings of a million nephrons. From the renal pelvis, urine goes down a tube called the *ureter* to the bladder. The *bladder* is a stretchy sac where urine is temporarily stored. When the bladder is emptied, urine flows down the *urethra* and out of the body.

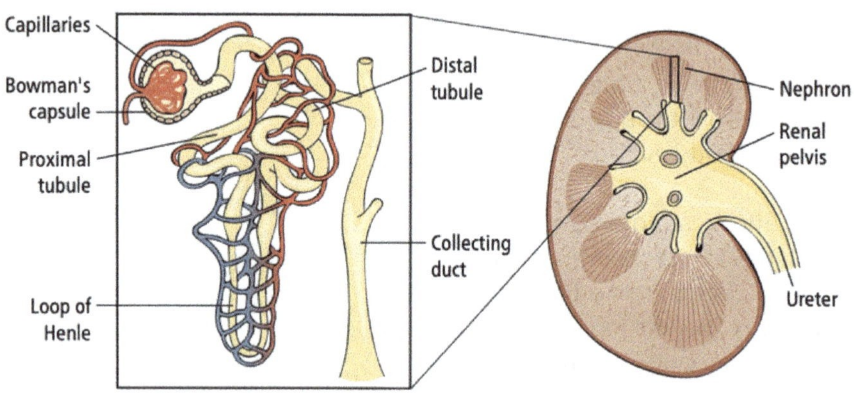

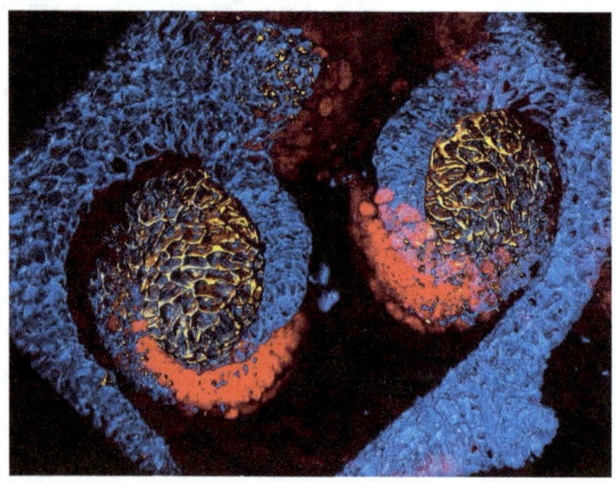

Figure 15.5

The kidney is made up of about a million functional units called nephrons, which help to filter blood plasma of waste products that leave the body as urine. The colorized image on the right shows the capillaries of the nephron bundled together.

CONCEPT **CHECK**

Which organ systems provide the raw materials for making ATP (such as glucose and oxygen)? Which organ systems get rid of the wastes (such as carbon dioxide and urea)?

CHECK YOUR ANSWER

Glucose and other organic molecules that can be broken down to make ATP come from the digestive system, which breaks down the food you eat. Oxygen is brought into the body by the respiratory system. Carbon dioxide is also removed from the body through the work of the respiratory system. The excretory system removes urea and other wastes. The circulatory system is involved every step of the way—it transports raw materials and wastes to their destinations in the body.

More to Explore

You can read more about the excretory system here:

https://www.cliffsnotes.com/study-guides/biology/biology/excretion-and-homeostasis/human-excretory-system

15.3 The Innate Immune System

Without your body's defenses, you would quickly fall prey to the bacteria, viruses, and other **pathogens**—disease-causing organisms—around you. What protects you is your immune system (**Figure 15.6**). Your immune system includes barriers, such as skin, that keep pathogens out. It also includes many different kinds of immune cells, also known as *white blood cells*. White blood cells move throughout your body constantly, looking for pathogens. White blood cells are made in the bone marrow. They then mature in the bone marrow or thymus. Most of your white blood cells are found in your circulatory system, lymphatic system, and spleen.

The *lymphatic system* is an important component of the immune system. The lymphatic system includes lymph vessels that, like blood vessels, travel all over the body. Lymph vessels carry a clear fluid called **lymph**. One function of the lymphatic system is to collect the fluid that leaks out of blood vessels and return it to the circulatory system. Another function of the lymphatic system is to carry white blood cells, which are found in large numbers in lymph. *Lymph nodes* are structures in the lymph vessels where many white blood cells are concentrated. Lymph nodes are found all over the body, including the throat (your tonsils), armpits, and groin. Swollen lymph nodes are a strong sign that the body is fighting an infection.

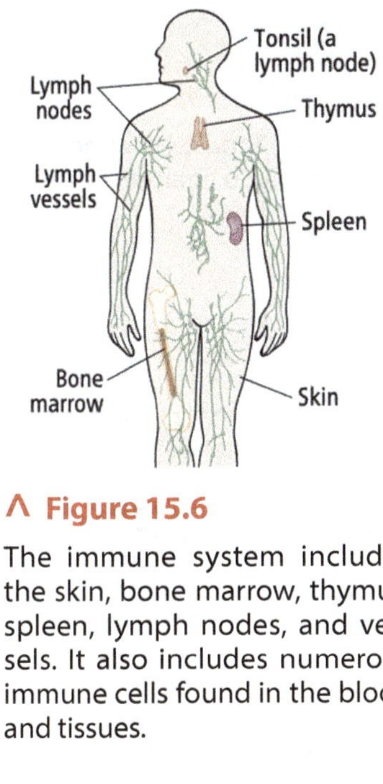

Figure 15.6

The immune system includes the skin, bone marrow, thymus, spleen, lymph nodes, and vessels. It also includes numerous immune cells found in the blood and tissues.

The immune system has two parts, the innate immune system, and the acquired immune system. In the rest of this section, we consider innate immunity. In the next section, we will look at the acquired immune system.

Components of the Innate Immune System

The **innate immune system** includes nonspecific body defenses that work against many different pathogens. These defenses are described as "innate" because they do not have to be activated by exposure to a pathogen—they are all set and ready to go. Three key components of the innate immune system are the skin, the mucous membranes, and innate immune cells.

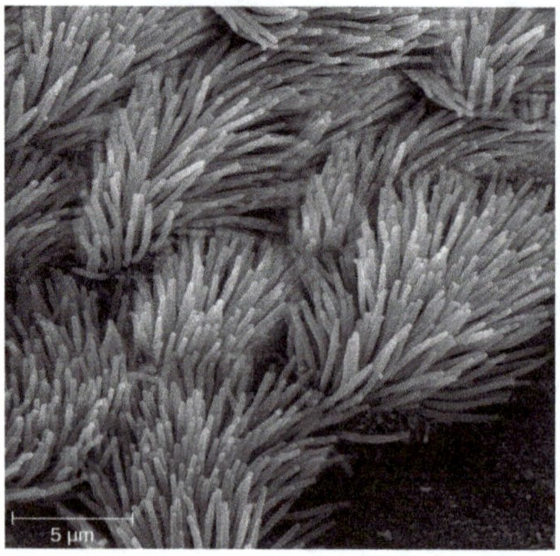

^ **Figure 15.7**

The lining of the respiratory tract includes numerous ciliated cells. The cilia sweep mucus, along with trapped particles or pathogens, up to the pharynx, where it is then swallowed.

Skin

The best way to deal with pathogens is to keep them out of the body in the first place. The skin is a crucial barrier, forming a tough outer layer that is hard to penetrate when intact. In addition, skin cells are shed frequently, making it hard for pathogens to establish a foothold. Hair follicles in the skin also secrete special enzymes and acids that kill bacteria and fungi. In fact, many of your body secretions contain bacteria-killing enzymes—for example, saliva, tears, sweat, and milk.

Mucous Membranes

The inside of your body is lined with *mucous membranes.* Mucous membranes are found inside your nose, mouth, and eyelids, as well as in your respiratory, digestive, urinary, and reproductive tracts. Mucous membranes are not as tough as skin; however, all mucous membranes are covered by a layer of *mucus* that helps trap pathogens. The mucous membranes of the respiratory tract also include cilia that sweep mucus up to the pharynx (**Figure 15.7**). This mucus is then swallowed, and stomach acid kills any pathogens present. Many mucous membranes are also flushed by fluids such as tears, saliva, and urine. These fluids often contain pathogen-killing enzymes. In addition, the constant flushing makes it hard for pathogens to gain a foothold.

READING CHECK

How do innate immune cells protect you against disease?

Innate Immune Cells

What happens when pathogens do get inside the body? Innate immune cells launch an immediate attack (**Figure 15.8**). Innate immune cells have receptors that recognize molecules—usually carbohydrates, proteins, or nucleic acids—found in many different kinds of pathogens (**Figure 15.9**). Because of this, a single innate immune cell is able to respond to many different pathogens. All in all, the cells of the innate immune system include several hundred different receptors. If the innate immune system encounters the same pathogen more than once, its response is the same: The innate immune system retains no memory of pathogens it has encountered in the past.

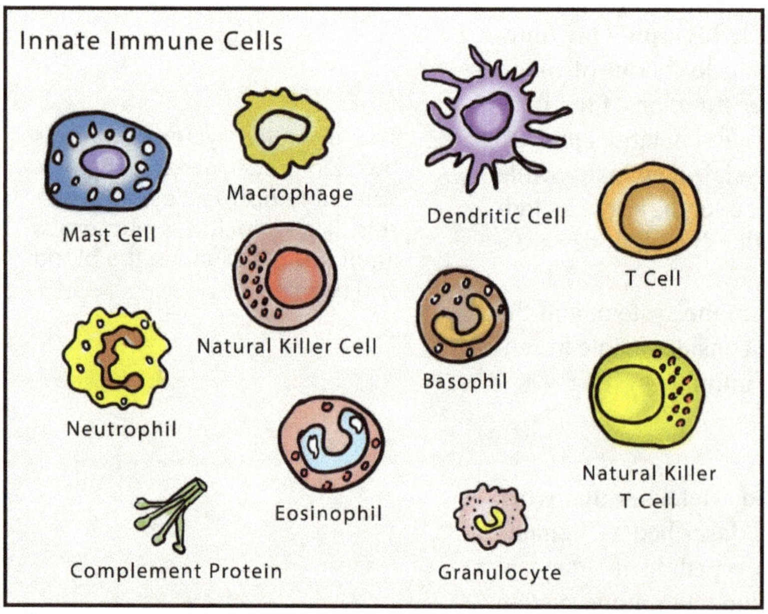

^ **Figure 15.8**

There are a variety of innate immune cells, each with its own specialized function.

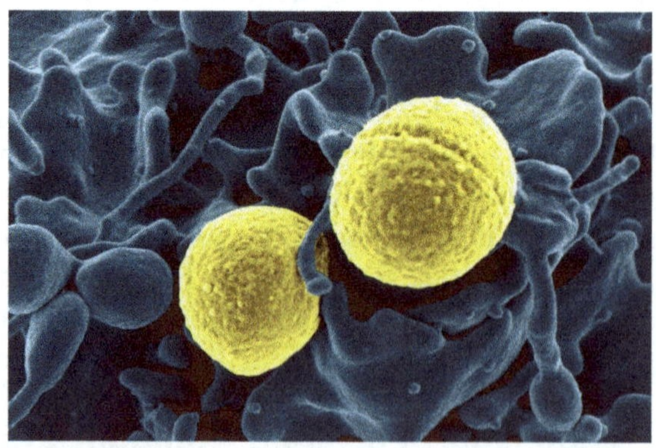

^ **Figure 15.9**

This innate immune cell (colorized as blue) is attacking bacteria (colorized as yellow). The long stringlike structures are extensions of the immune cell's cytoplasm. They are searching the environment for pathogens.

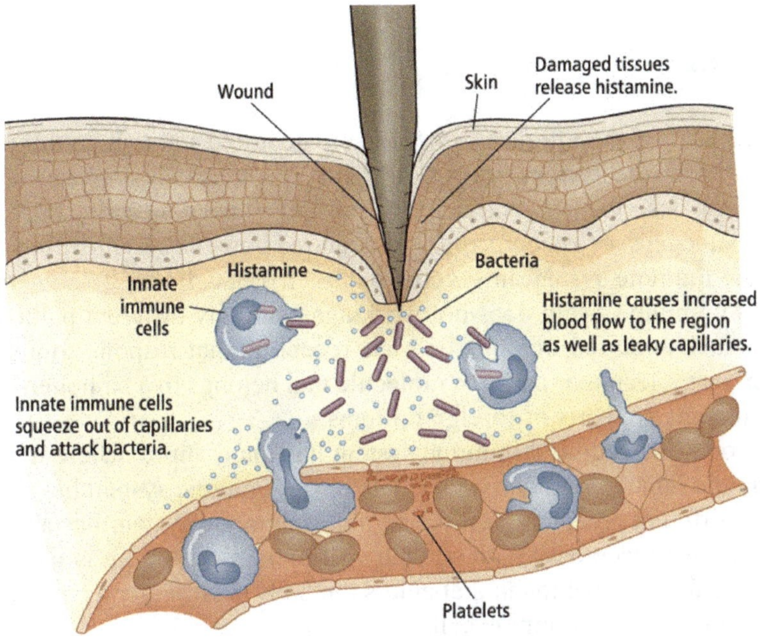

< Figure 15.10

The inflammatory response begins when damaged tissues release histamines, which cause increased blood flow to a wound. Local capillaries become leaky and produce swelling. Innate immune cells squeeze out of the capillaries and attack bacteria or other microorganisms that enter the body.

One of the most important functions of the innate immune system is to initiate the *inflammatory response*, as shown in **Figure 15.10**. You have no doubt experienced the inflammatory response many times—it is what makes your tissues swell and turn red when you cut your finger or scrape your knee. During the inflammatory response, damaged tissues release chemicals called *histamines*. Histamines increase blood flow to the site of the injury and also cause local capillaries to leak fluid.

The extra fluid causes swelling, which helps to isolate the wound from other body tissues. Histamines also attract innate immune cells. These cells squeeze out of the capillaries and attack any pathogens they find. Sometimes, the battle between innate immune cells and pathogens produces a whitish substance called pus. Pus consists of dead bacteria, dead tissue, and dead and living innate immune cells.

CONCEPT CHECK

How does the inflammatory response help defend your body against disease?

CHECK YOUR ANSWER

During the inflammatory response, the tissues surrounding a wound swell, helping to isolate it from the rest of the body. The inflammatory response also brings innate immune cells to the site of injury so that they can attack bacteria or other pathogens.

More to Explore

You can read more about the immune system at the following website:

https://www.niaid.nih.gov/research/immune-system-overview

15.4 The Acquired Immune System

The other part of the immune system is the acquired immune system. The **acquired immune system** is a set of highly specific body defenses. These defenses are described as "acquired" because they become active only after specific pathogens are encountered. The acquired immune system differs from the innate immune system in several ways. First, a cell that belongs to the acquired immune system does not work against many different pathogens. Instead, each acquired immune cell has receptors that respond to a single **antigen**—a molecule or part of a molecule that belongs to a pathogen. Most often, antigens are parts of foreign proteins.

Second, the acquired immune system has many more different receptors than the innate immune system. The innate immune system has several hundred different receptors, but the acquired immune system has about 10 million unique receptors! With so many different receptors, it is just about guaranteed that any pathogen that enters the body will trigger a response by one or more acquired immune cells.

Third, the acquired immune system responds more slowly to pathogens than the innate immune system. Whereas the innate immune system responds immediately, the acquired immune response takes several days to reach full force. This time is needed to make multiple clones of the acquired immune cells that are needed. We will learn more about this process in the following pages.

Fourth, the acquired immune system retains a "memory" of pathogens it has encountered in the past. If it meets the same pathogen again, it reacts more quickly and more aggressively. The innate immune system does not have this kind of memory. **Table 15.1** summarizes the differences between innate and acquired immunity.

TABLE 15.1 Comparison of innate immunity and acquired immunity.

Innate Immune System	Acquired Immune System
Each component can respond to **many different pathogens**	Each component responds to **a single antigen**
Includes **hundreds** of different receptors	Includes about **10 million** different receptors
Has an **immediate** response	Has a **delayed** response that reaches full force several days after the initial exposure to a pathogen
Has **no memory** of past encounters with a pathogen	Has a **memory** of past encounters with a pathogen; has a faster and more aggressive response to repeat exposure

Components of the Acquired Immune System

Let's take a closer look at how acquired immunity works. The acquired immune system has two kinds of cells: B cells and T cells. Both are made in the bone marrow, but each matures in a different organ— in the *bone marrow* for B cells, and in the *thymus* for T cells.

B Cells

B cells react to pathogens in bodily fluids such as blood or lymph. When a B cell encounters a pathogen and binds to an antigen on the pathogen, the cell begins to divide. It makes many copies, or *clones*, of itself (**Figure 15.11**). The time it takes to make these clones is what delays the acquired immune response.

Once the clones are ready, they begin to make and release proteins called *antibodies*. An **antibody** is a large, Y-shaped protein that includes two antigen-binding sites (**Figure 15.12a**). An antibody binds to its antigen in a very specific way, with a fit like a lock and key. A single B cell can make a huge number of antibodies—as many as 2000 every second! These antibodies travel around the body binding to antigens on all the pathogens they encounter.

But what happens after an antibody binds to an antigen? The binding of antibodies interferes with a pathogen's ability to function. For example, bound antibodies can prevent viruses from entering cells and infecting them. Bound antibodies also mark pathogens so that they can be destroyed by other immune cells. Antibodies may also cause pathogens to clump together (see **Figure 15.12b**)—these clumped pathogens become easy targets for other immune cells.

Some of the clones made by B cells are **memory cells**. Memory cells stay in the body for a long time—years or even a lifetime. If the body encounters the same pathogen again, the memory cells initiate a quick, aggressive attack. Pathogens may be eliminated from the body before symptoms even develop. This is why people catch many diseases only once and then are immune for a long time, or even for life.

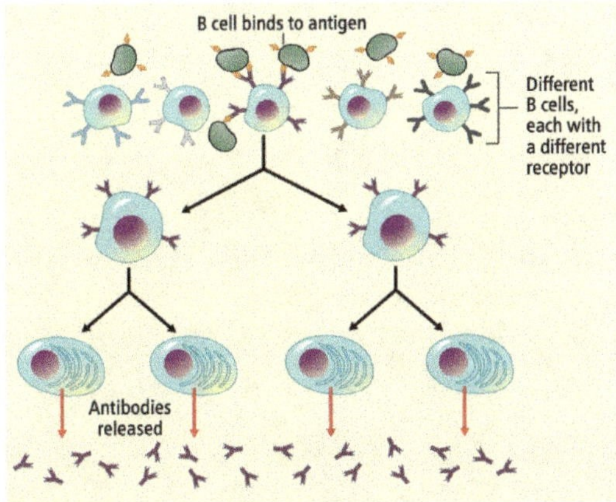

∧ Figure 15.11

When a B cell binds to an antigen, the B cell begins to divide and make clones of itself. Mature clones then make and release large numbers of antibodies.

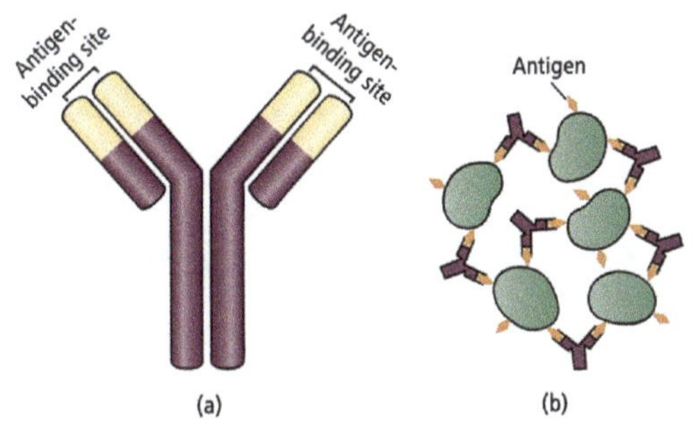

∧ Figure 15.12

(a) Antibodies are large, Y-shaped proteins that include two antigen-binding sites. (b) Because antibodies each have two binding sites, they can cause pathogens (shown here as green ovals) to clump together.

T Cells

T cells attack pathogens that are inside the body's cells as well as the body's own malfunctioning cells. Because viruses enter your cells when they infect your body, T cells are very important in fighting viral infections. T cells are also important in destroying cancer cells or other abnormal body cells.

When a body cell is infected by a pathogen, the cell displays some of the pathogen's foreign proteins on its surface the way a ship might wave a distress flag (**Figure 15.13**). The display of foreign proteins serves as a signal to the body's T cells. A type of T cell called a *helper T cell* binds to

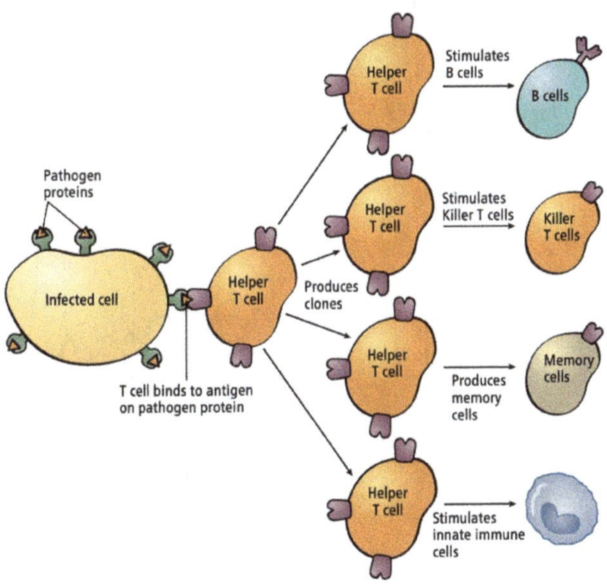

< Figure 15.13

T cells target pathogens that are inside the body's cells. Infected cells display pathogen proteins the way a ship might display a distress flag. A helper T cell binds to an antigen on the pathogen protein and then produces clones that initiate a variety of immune activities.

an antigen on the pathogen protein and begins to divide and produce clones. Helper T cell clones then initiate many different immune activities: They stimulate B cells to produce clones that make antibodies against the pathogen. They also stimulate *killer T cells*, another group of T cells, to divide and produce clones. The killer T cells bind to displayed proteins and then, as their name suggests, kill infected cells. This prevents pathogens from coming out to infect new cells. Helper T cells also stimulate cells of the innate immune system. Finally, helper T cells produce memory cells that remain in the body for a long time and guard against a return of the pathogen.

Vaccines

Vaccines protect you from disease by making use of your acquired immune system's "memory" for pathogens it has encountered in the past. A vaccine exposes your body to a pathogen's antigens, but—this is the key—it does not infect you with the pathogen itself!

Most vaccines contain either dead pathogens or weak strains of a pathogen. A vaccine also may use only part of a pathogen—perhaps a virus's protein coat or a bacterium's flagellum. Your acquired immune system reacts to antigens in the vaccine just as it would react to the real pathogen. That is, the acquired immune system makes antibodies and—most critically—memory cells. If the real pathogen ever shows up, the acquired immune system is ready to attack (**Figure 15.14**).

Vaccines are still the best way we have to fight most viruses. Although many drugs have been developed to attack pathogenic bacteria and fungi, it has been harder to develop drugs that work well against viruses, primarily because viruses are so small and are relatively simple biological agents. This means viruses don't need much to survive, which means they can be difficult to damage. So, don't forget to get your vaccinations!

Diseases of the Immune System

READING CHECK

How does AIDS harm the body?

The immune system normally has no problem recognizing the body's own cells. In *autoimmune diseases*, the immune system identifies certain body cells as foreign and attacks them. *Type 1 diabetes* occurs when the immune system attacks and destroys the insulin-producing cells of the pancreas. *Multiple sclerosis* occurs when immune cells destroy the myelin sheaths surrounding neurons. *Lupus* is a debilitating disease caused by the immune system attacking a wide range of healthy tissues.

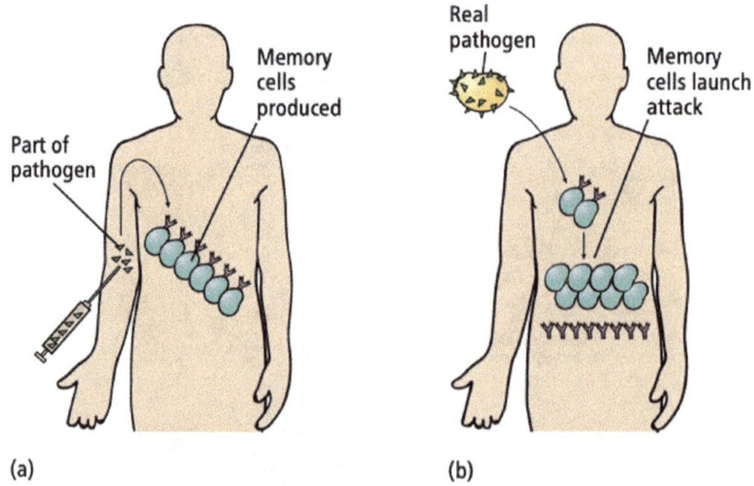

Figure 15.14 >

(a) Vaccines work by introducing dead pathogens, weak pathogens, or parts of a pathogen into the body. The acquired immune system reacts and produces memory cells. (b) If the actual pathogen is encountered later, the memory cells initiate a rapid, aggressive attack.

Acquired immunodeficiency syndrome, or AIDS, is a disease caused by the *human immunodeficiency virus*, HIV. (The "acquired" in the name of the disease refers to the fact that immunodeficiency is acquired during a person's lifetime rather than being inborn.) HIV attacks immune cells, particularly helper T cells (**Figure 15.15**). With these crucial immune cells compromised, infections and cancers that are normally easy to fight off have a chance to overwhelm the body.

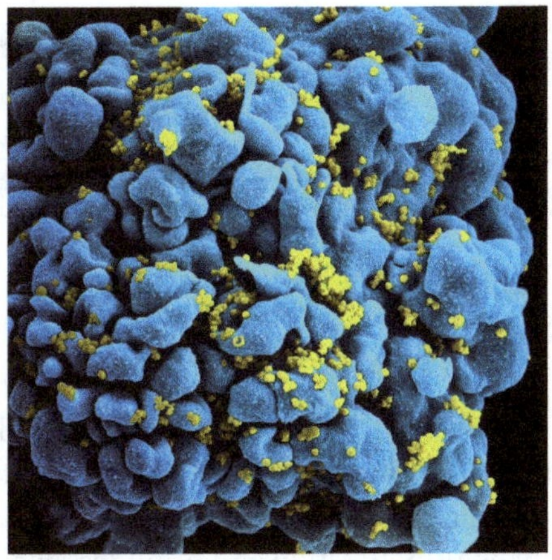

Figure 15.15 >

HIV (colorized as yellow) emerges on the surface of an infected helper T cell (colorized as blue).

CONCEPT **CHECK**

Which immune system cells produce antibodies? How do antibodies help defend your body against pathogens?

CHECK YOUR ANSWER

Antibodies are produced by the B cells of the acquired immune system. Antibodies work by binding to antigens on pathogens. This prevents pathogens from functioning or marks them for destruction by other immune cells. Antibodies may also cause pathogens to clump together so that they become easy targets for immune cells.

More to Explore

To learn more about the acquired immune system:

http://www.biology.arizona.edu/immunology/tutorials/immunology/page3.html

Concept Review

Chapter 15

Detailed Chapter Summary

From their diets, people ideally should obtain Calories to support their activities, a complete protein, and vitamins and minerals in small amounts. Exercise is a crucial part of any healthy lifestyle and provides an extremely wide variety of health benefits. Metabolic syndrome is an increasing health issue and is associated with carrying extra weight as well as insulin resistance.

The excretory system filters blood, removing wastes while leaving useful molecules behind. One important waste is urea, produced when amino acids are broken down to make ATP. The excretory system also controls the amount of water in the body, as well as the levels of different ions and molecules in the blood. The kidneys are made up of functional units called nephrons. Each nephron is associated with a cluster of capillaries. Blood pressure pushes fluid from the capillaries into Bowman's capsule, a structure of the nephron. This fluid, called the filtrate, is initially very similar to blood plasma. From Bowman's capsule, the filtrate moves into the proximal tubule. In the proximal tubule, "good" molecules are transported from the filtrate back into the blood, and waste molecules are transported from the blood to the filtrate. The filtrate then enters the loop of Henle, where water is reabsorbed. From the loop of Henle, the filtrate moves into the distal tubule, where more wastes are transported into the filtrate. Finally, the filtrate moves down the collecting duct. There, more water is reabsorbed if antidiuretic hormone is present. The filtrate is now urine. Urine flows from the collecting duct into the renal pelvis and then to the bladder, where it is stored for later release when convenient.

The immune system protects the body against a variety of pathogens. The lymphatic system includes a network of lymph vessels containing lymph, a clear fluid. The lymphatic system helps return fluids to the circulatory system. Lymph also carries many immune cells. A large number of immune cells are concentrated at lymph nodes in the body.

The innate immune system includes nonspecific body defenses that work against many different pathogens. Skin and mucous membranes provide barriers to the entry of pathogens. Mucous membranes are not as tough as skin, but they are covered with a layer of mucus that helps trap pathogens. Many mucous membranes are also flushed by fluids, such as tears, saliva, etc. Innate immune cells have receptors that recognize molecules found in many different kinds of pathogens. The innate immune system retains no memory of pathogens it has encountered in the past. In the inflammatory response, injury causes the release of molecules called histamines. The release of histamines results in increased blood flow and swelling, as well as the attraction of many innate immune cells to the site of injury.

The acquired immune system includes a set of highly specific body defenses. These are described as "acquired" because they become active only when features of specific pathogens are encountered. Each acquired immune cell has receptors that respond to a single antigen (a molecule or part of a molecule that belongs to a pathogen). All in all, there are over 10 million receptors in the acquired immune system, many more than are found in the innate immune system. The acquired immune system takes longer than the innate immune system to launch an effective response. However, the acquired immune system retains a memory for pathogens encountered in the past, and subsequent responses to the same pathogen are strong and rapid. This memory explains why you catch many diseases only once. It is also the basis for vaccines. B cells are made in the bone marrow and also mature there. B cells are a type of acquired immune cell that reacts to pathogens in body fluids. Activated B cells produce clones that release large numbers of Y-shaped antibodies that bind to antigens on pathogens. The binding can interfere with the function of pathogens, mark them, or cause them to clump together. T cells are made in the bone marrow but mature in the thymus, and they are another type of acquired immune cell. T cells target pathogens inside body cells as well as the body's own malfunctioning cells. T cells mark infected or malfunctioning body cells for destruction and then kill them.

Vaccines rely on the memory cells produced by the acquired immune system to protect us from disease. A vaccine works by exposing someone to a pathogen's antigens, but without infecting them with the pathogen itself. Finally, some diseases of the immune system include autoimmune diseases and AIDS.

Key Terms

Acquired immune system A set of highly specific body defenses that respond to specific features of pathogens.

Antibody A large, Y-shaped protein that binds to a pathogen's antigen, disabling the pathogen or marking it for destruction.

Antigen A molecule or part of a molecule belonging to a pathogen that is recognized by cells of the acquired immune system.

B cells Cells of the acquired immune system that react to pathogens in bodily fluids such as blood or lymph.

Innate immune system A set of nonspecific body defenses that work against many different pathogens.

Kidneys The paired organs of the excretory system that filter blood and produce urine.

Lymph A clear fluid inside lymph vessels that contains large numbers of immune cells.

Memory cells Immune cells made by the acquired immune system that remain in the body for a long time; these cells initiate a rapid, aggressive attack if the same pathogens are encountered again, providing some immunity during repeat exposures to the same pathogens.

Minerals Inorganic substances required as components of various body tissues, such as calcium for bones and teeth and iron for hemoglobin.

Nephron The functional unit of a kidney; it filters waste molecules from blood to produce urine.

Pathogens Disease-causing agents such as bacteria, viruses, or other organisms.

T cells Cells of the acquired immune system that attack pathogens inside the body's cells, such as those infected by viruses, as well as the body's own malfunctioning cells.

Vitamins Organic molecules that are essential non-protein components of certain enzymes; important components of our diet.

Review Questions

Simple Review Questions

15.1 Nutrition, Exercise, and Health

1. Why is it important for you to eat a complete protein regularly?
2. What are some of the important minerals you obtain from your food?
3. What is metabolic syndrome?
4. What are some of the benefits of exercise?

15.2 The Excretory System

5. What is urea?
6. How does fluid move from the circulatory system into a nephron?
7. What happens to the filtrate in the proximal tubule?
8. What is the function of the loop of Henle?

15.3 Body Defenses: The Innate Immune System

9. What are two functions of the lymphatic system?
10. What features of skin make it good at keeping pathogens out of the body?
11. Why is a single innate immune cell able to respond to many different pathogens?

15.4 Body Defenses: The Acquired Immune System

12. What is an antigen?
13. Where in the body are the pathogens to which B cells respond? Where in the body are the pathogens to which T cells respond?
14. What is the function of a memory cell?

Challenging Review Questions

15.1 Nutrition, Exercise, and Health

15. What is the difference between a vitamin and a mineral?
16. What role has evolution played in human food preferences? What effect does this have on us today?
17. What factors make a person more likely to develop metabolic syndrome? What are some of the health risks associated with metabolic syndrome?

15.2 The Excretory System

18. What waste materials are produced during the process of making ATP? What body systems are responsible for removing the wastes from the body?
19. How does the endocrine system interact with the excretory system?
20. What is the difference between elimination (feces) and excretion (urine)? What is the body getting rid of in each case?

15.3 Body Defenses: The Innate Immune System

21. How does the lymphatic system support the work of the circulatory system? How does it support the work of the immune system?
22. How do tears help defend your body against pathogens? Does this explain why people with allergies sometimes get watery eyes?

15.4 Body Defenses: The Acquired Immune System

23. Why is the innate immune system described as "nonspecific"? Why is the acquired immune system described as "specific"?
24. What are some differences between the innate immune system and the acquired immune system?
25. Allergies occur when the immune system is abnormally sensitive to particular substances. Why do people sometimes take antihistamines for their allergies? Can you guess from the word antihistamine what an antihistamine does?
26. If a virus infects your body, will only one type of T cell respond? In other words, does a single virus have exactly one antigen?
27. How does a vaccine protect you from disease?

Apply & Discuss Questions

28. Why do you think people like to eat sweets, fats, and other foods that are not very good for them? Do you think that, over time, our taste for these and other foods might evolve?
29. The liver is an organ that plays important roles in multiple organ systems. What role does the liver play in digestion? What role does it play in excretion?
30. A cell in your body breaks down an amino acid to make ATP and generates a molecule of ammonia, a nitrogen-containing waste. Describe what happens to this ammonia molecule.
31. What do you think explains the placebo effect?
32. The leading causes of death worldwide are listed below. Compare them with the leading causes of death in the United States provided earlier in the chapter.

TABLE 15.3

Leading Causes of Death Globally
(2019 Data from the World Health Organization)

1. Heart disease
2. Stroke
3. Chronic obstructive pulmonary disease
4. Lower respiratory infections
5. Neonatal conditions
6. Trachea, bronchus, lung cancers
7. Alzheimer's disease and other dementias
8. Diarrheal diseases
9. Diabetes mellitus
10. Kidney diseases

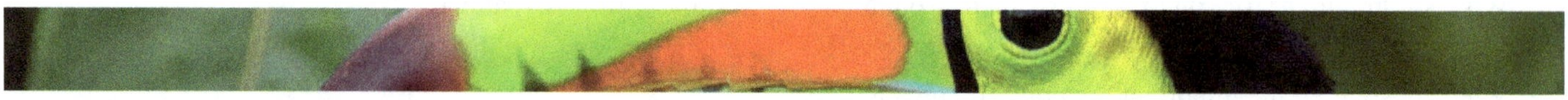

Solutions to Odd-Numbered Questions

Simple Review Solutions

1. Humans cannot make 9 of the 20 amino acids needed to build proteins. This is why you need to eat a "complete protein"—one containing all the amino acids—regularly.
3. Metabolic syndrome is a set of characteristics that greatly increases the risk of heart disease, stroke, and type 2 diabetes. Metabolic syndrome is associated with two main features: carrying extra weight around the middle and upper parts of the body and insulin resistance, the inability of body tissues to adequately respond to the hormone insulin.
5. One of the most important wastes found in urine is urea. When amino acids are broken down to make ATP, ammonia, a nitrogen-containing waste, is produced. Because ammonia is highly toxic, it is immediately converted to urea, a less-toxic waste, by the liver. The liver releases urea back into the bloodstream for excretion.
7. The proximal tubule is like a sorting machine. "Good" molecules in the filtrate—such as ions, glucose, vitamins, and amino acids—are transported back into the blood so that the body can keep them. "Bad" waste molecules are transported from the blood to the filtrate.
9. One function of the lymphatic system is to collect the fluid that leaks out of blood vessels and return it to the circulatory system. Another function of the lymphatic system is to carry white blood cells, which are found in large numbers in lymph.

11. A single innate immune cell is able to respond to many different pathogens. This is because the receptors of innate immune cells recognize molecules (usually carbohydrates, proteins, or nucleic acids) that are found in many different kinds of pathogens.

13. B cells react to pathogens in bodily fluids such as blood or lymph. T cells attack pathogens that are inside the body's cells as well as the body's own malfunctioning cells.

Challenging Review Solutions

15. Vitamins are organic molecules that are essential non-protein components of certain enzymes. Minerals are inorganic substances required as components of various body tissues.

17. People who are overweight are more likely to develop metabolic syndrome. Other factors associated with metabolic syndrome include an unhealthy diet, lack of exercise, aging, and certain genes.

19. The answer is, in many ways! Here are just a few examples: Antidiuretic hormone causes more water to be reabsorbed from the filtrate during excretion. Parathyroid hormone decreases calcium excretion. Mineralocorticoids help regulate water and salt balance in the body by affecting excretion of these substances.

21. The lymphatic system supports the work of the circulatory system by collecting the fluid that leaks out of blood vessels in lymphatic vessels and returning this liquid to the circulatory system. The lymphatic system supports the work of the immune system by carrying large numbers of white blood cells in lymph.

23. The innate immune system is described as nonspecific because its defenses work against many different potential pathogens. The acquired immune system is described as specific because its cells have receptors that respond to a single antigen—a very specific feature of a molecule or part of a molecule belonging to a pathogen.

25. Symptoms of allergies come from the reactions of the innate immune system. As part of the inflammatory response of the innate immune system, tissues produce histamines. Antihistamines, as their name suggests, help to counter these histamines and reduce the redness and swelling that make up the inflammatory response.

27. Vaccines protect you from disease by making use of your acquired immune system's "memory" for pathogens it has encountered in the past. A vaccine exposes your body to a pathogen's antigens, but—this is the key—it does not infect you with the pathogen itself. Most vaccines contain either dead pathogens or weak strains of a pathogen. A vaccine also may use only part of a pathogen—perhaps a virus's protein coat or a bacterium's flagellum. Your acquired immune system reacts to antigens in the vaccine just as it would react to the real pathogen. That is, the acquired immune system makes antibodies and—most critically—memory cells. If the real pathogen ever shows up, the acquired immune system is ready to attack.

Apply & Discuss Solutions

29. The liver makes bile, which is important as an emulsifier for digestion. The liver also converts ammonia to the less toxic urea, an important function related to excretion.

31. That is a very good question that no one fully understands the answer to! It seems that natural endorphin release by the brain may be part of the story, particularly when pain reduction is part of the effect. Clearly, however, there is more to the placebo effect than that. Stay tuned, as the next decade may bring some good answers to this question!

Living Earth Essay

Life on Earth Connections

Topsoil in Trouble

Over the past 100 years, there have been dramatic increases in crop yields. An acre of U.S. farmland in 1900 yielded about 30 bushels of corn. Today, that same acre yields on the order of 130 bushels of corn. This increased efficiency has meant a significant drop in the number of people needed to farm. In the early 1900s, about 33 million people in the United States lived and worked on farms. Today, with a much larger population, only 2 million people are engaged in commercial farming in this country, raising crops and livestock.

Many of the farming methods used to obtain high yields have significant disadvantages. Pesticides and fertilizers, for example, pose certain risks. Pesticides are inherently toxic, and each year thousands of people working in agriculture are poisoned by the mishandling of these dangerous compounds. Fertilizers help plants grow, but major portions of applied fertilizer are washed into streams, rivers, ponds, and lakes, where they upset ecosystems, especially by promoting excessive growth of algae. Fertilizer runoff from fields, illustrated in **Figure A**, can also contaminate drinking water and thus affect human health. An ailment known as blue-baby syndrome, for example, results from drinking water containing high concentrations of nitrate ions, a main ingredient of most fertilizers. Nitrate ions in the bloodstream compete with oxygen for the positively charged iron ions of hemoglobin molecules. This leads to a form of anemia known as methemoglobinemia, to which babies are particularly sensitive. Aside from shortness of breath, one of the major symptoms is a bluish color of the skin.

Poor maintenance of topsoil is also a major concern. Synthetic fertilizers have no organic bulk and do not provide a food source for soil microorganisms and earthworms. Over time, a soil treated with only these fertilizers loses biological activity, which diminishes the soil's fertility. Soils void of organic bulk become chalky and susceptible to wind erosion. Chalky soils lose their capacity to hold water, which means that more applied fertilizer is leached away. Ever-increasing amounts of fertilizer are thus needed.

Over the past 100 years, damaging farming practices have decreased the amount of topsoil in parts of the United

Figure A
The water running off this farm field contains many pesticides and fertilizers that can be harmful to ecosystems and human health.

Figure B
Poor soil conservation practices in the early 1900s contributed to the loss of much topsoil to wind storms thick with dust.

States by as much as 50 percent. During the 1930s, farming practices and drought conditions created giant dust storms, such as the one shown in **Figure B**, that removed major portions of the topsoil in Kansas, Oklahoma, Colorado, and Texas. In one storm, large dust clouds were carried all the way from the Midwest to Washington, D.C., and then into the Atlantic Ocean. Politicians in that city observing the effects of poor soil management right outside their windows quickly passed legislation that created the Soil Erosion Service, which became the National Resources Conservation Service; it continues to this day in its efforts to help protect the nation's topsoil for future generations.

Another limited key resource required for farming is fresh water. In regions where rainfall is insufficient to support large crops, water is either channeled into fields from lakes, rivers, and streams or pumped from the ground. In many areas, groundwater is the primary source of fresh water, but excessive use of groundwater can lead to land subsidence, illustrated in **Figure C**.

Any source of water other than rainwater requires an irrigation method to deliver water to the fields. Flooding is a common method, but it is not efficient because most of the water is lost in runoff and evaporation. Sprinkler systems are an improvement over flooding because they do not cause soil erosion. Such systems also lose large amounts of water, however, because a significant portion of the airborne water evaporates before reaching the ground.

All liquid water on the Earth's surface, no matter how fresh, contains some salts. After irrigation water evaporates from farmland, these salts are left behind, and over time, repeated irrigation causes the salinity of the soil to increase. This process is known as **salinization**, and it leads to a rapid decrease in productivity. To counteract growing soil salinity, farmers flood the land with huge quantities of water. As the water drains into a river, it washes the unwanted salts—along with significant amounts of topsoil—into the

Figure C
The San Joaquin Valley of California has subsided by more than 35 feet since the pumping of groundwater began in the 1920s.

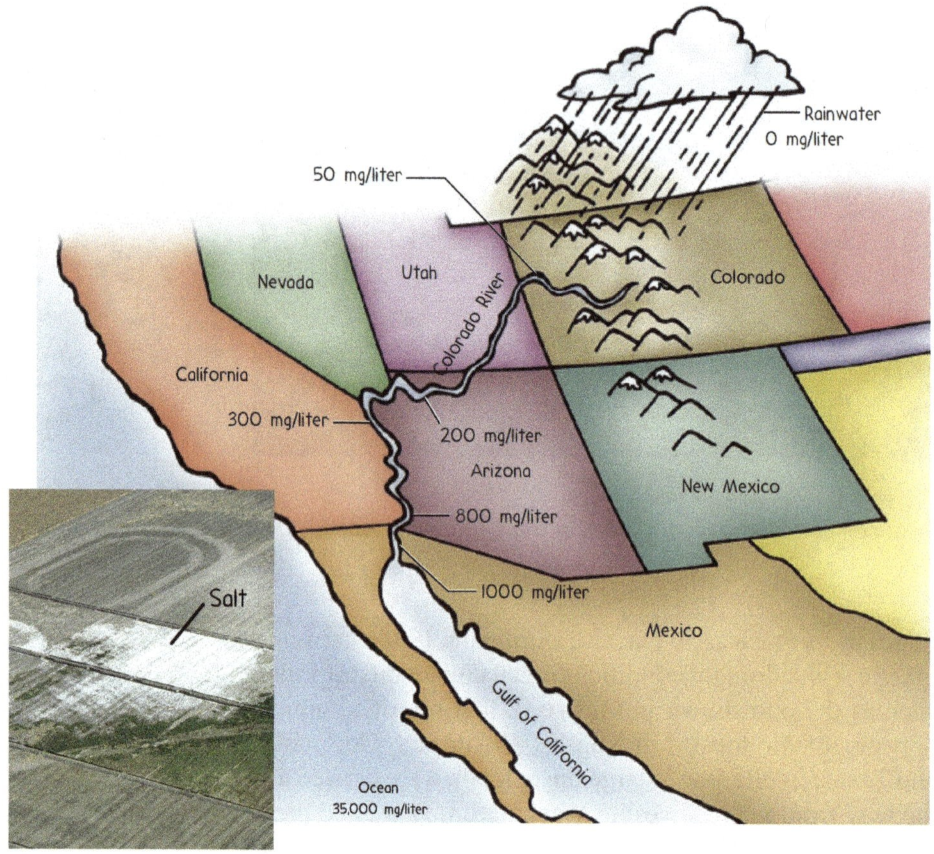

Figure D
As a river flows along, runoff from agricultural fields can add to the river's natural salinity. A typical safe drinking water standard for salt content is 500 milligrams per liter. Agricultural damage occurs when soil salinity reaches a concentration of about 800 milligrams per liter. The inset photo shows salt deposits accumulated in an irrigated field.

river. Thus, a river passing through farmlands gets saltier and saltier as it runs to the sea, as depicted in **Figure D**.

So what's to be done? In a subsequent Living Earth Essay we explore modern-day agricultural practices aimed at optimizing both productivity and sustainability. This includes techniques such as micro-irrigation, organic farming, integrated crop management, integrated pest management, and, importantly, a most promising approach known as regenerative agriculture. To Life!

READING CHECK

What happens to farmland after repeated irrigation?

CONCEPT **CHECK**

Are there any dissolved salts in a mountain stream?

CHECK YOUR ANSWER

Land contains a variety of salts. As water runs over and through the land, these salts dissolve in the water. In general, the farther the water travels, the saltier it becomes. Thus, even water that has traveled only a short distance, which is the case with a mountain stream, contains some salts. So, the answer to this question is a qualified "Yes, but not enough to make the water undrinkable."

FOR YOUR INFORMATION

Those living in the agricultural regions of the arid southwestern United States are most familiar with the accumulation of visible salty sediments, as shown in the inset of Figure D. Agricultural salinization, however, has been a problem since the beginning of agriculture. Archeological evidence shows that the fall of the Sumerian society of ancient Mesopotamia was likely due in part to the accumulation of salts on farmlands irrigated by waters of the Tigris and Euphrates rivers in what is now Iraq. Civilizations along the Nile river in Africa, however, survived through the millennia because, unlike the Tigris and Euphrates, the Nile river floods seasonally, which allows a seasonal flushing of accumulated salts.

Chapter 16

Populations

THE MAIN IDEA

A population is a group of individuals of a single species that live in a specific area.

16.1 Organisms and Their Environments

16.2 Population Growth

16.3 Life History

16.4 Human Population Growth

All living things live somewhere. And wherever they live, they interact with the living and nonliving things around them. In this chapter, we look at ecology, the study of how organisms interact with their environments. We explore how a population of organisms can grow exponentially when resources abound, or logistically when the environment holds a limited carrying capacity. A population can also be described by its life history, which is a description of how long individuals within that population tend to live.

16.1 Organisms and Their Environments

Ecology is the study of how organisms interact with their environments. An organism's environment includes nonliving, or *abiotic*, features, such as temperature, sunlight, rain, rocks, and ponds. An organism's environment also includes *biotic* features—other living things.

Ecology can be studied at many different levels, including those of the individual, population, community, and ecosystem. Studies that focus on individuals may ask how an organism's anatomy, physiology, and behavior help it function in its environment. For example, in terms of anatomy, why do certain plants invest more energy in building root systems, whereas others invest more energy in their stems and leaves? In terms of physiology, does the nutrient content of different seeds explain why finches prefer one type of seed to another? In terms of behavior, how do gulls divide their time between two different feeding strategies: begging for handouts from humans and foraging at sea? Because individual-level studies in ecology focus on adaptation, there is often considerable overlap with evolutionary biology.

^ Figure 16.1

Biologist Pamela Yeh holds an orange-flanked bush robin in Langtang National Park, Nepal. The bird will be measured, weighed, and given a unique identifying leg band before being released.

A **population** is a group of individuals of a single species that live in a specific area. The raccoons in Berkeley, California, are a population. So are the snow leopards of Central Asia and the orange-flanked bush robins of Langtang National Park, Nepal (**Figure 16.1**). At the population level, ecologists are often interested in the size of a population and in how population size changes over time. In addition, population ecologists may study how a specific population uses resources, including the kinds of foods that are eaten and the types of habitats that are used.

(a) (b) (c) (d)

Figure 16.2

Ecologists study different kinds of questions. (a) An individual-level study: How do fireflies use light signals to attract mates? (b) A population-level study: How many giant pandas live in the bamboo forests of central China? Is this number increasing or decreasing? (c) A community-level study: Who eats whom in a grassland in Argentina? (d) An ecosystem-level study: How does an important resource such as water cycle between the biotic and abiotic parts of an ecosystem?

READING CHECK

What is the biological definition of "community"?

A community consists of all the organisms that live in a specific area. For example, the ecological community in an Idaho brush habitat might include sagebrush, prickly pear cactus, insects, squirrels, pronghorns, badgers, lizards, mountain lions, coyotes, and many kinds of microorganisms. At the community level, biologists often look at how species interact with one another. They may study who eats whom in a community or investigate how different species compete for resources.

An **ecosystem** consists of all the organisms that live in a specific area *and* all the abiotic features of their environment. At the ecosystem level, ecologists look at links between the biotic and abiotic worlds. For example, some ecosystem ecologists study the cycling of important resources such as water and carbon between living organisms and Earth. Others look at how energy moves through ecosystems. Additional examples of ecological studies at different levels of organization are shown in **Figure 16.2**.

CONCEPT CHECK

1. An ecologist studies how hyenas and lions compete for food in the African savanna. Is this a study at the population, community, or ecosystem level?
2. An ecologist studies how the number of mountain lions in a certain park has changed over the past decade. Is this a study at the population, community, or ecosystem level?

CHECK YOUR ANSWERS

1. This study examines how two different species interact in a habitat, so it is a community-level study.
2. This study examines a group of individuals of a single species that live in a specific area, so it is a population-level study.

More to Explore

To read more about ecosystems, take a look at the following:
https://education.nationalgeographic.org/resource/ecosystem

16.2 Population Growth

Population ecologists are often interested in how big a population is as well as in how population size changes over time. Sometimes, population ecologists count the total number of individuals in a population. Other times, they prefer to measure the *population density*, the number of organisms per unit area.

Four factors determine how a population's size changes over time: the birth rate, the death rate, the rate of immigration into the population, and the rate of emigration out of the population. Births and immigration increase population size. Deaths and emigration decrease population size. Whether a population is increasing or decreasing in size overall depends on the relative contributions of the four factors. Two models for how populations in the natural world grow are exponential growth and logistic growth.

Exponential Growth

Exponential growth occurs when a population grows at a fixed rate per amount of time (**Figure 16.3**). For example, a population that increases by 10% each year, which means a doubling of population every 9-10 years, grows exponentially. As you can see from the graph in **Figure 16.3a**, a population that is growing exponentially increases in size more and more quickly as time goes on.

What kind of population grows exponentially in the real world? Exponential growth occurs when a population has unlimited resources. In this case, nothing limits population growth, and a bigger and bigger population continues to produce more and more offspring. Although resources are never truly unlimited in the real world, they are sometimes extremely plentiful. For example, organisms may find very plentiful resources when they colonize a new habitat, and their population may begin to grow exponentially. When reindeer were introduced on Saint Paul Island near Alaska, the population grew exponentially for many years (see **Figure 16.3b**). However, exponential growth cannot continue forever. On Saint Paul Island, the reindeer eventually ran out of food, and the population crashed.

Exponential growth is often seen in populations that live in unstable environments. These populations may go through cycles of exponential growth and crash. Good conditions start exponential growth. When resources run out, the population crashes. When good conditions return, the population explodes again, and so on.

READING CHECK

Can a population grow exponentially if the resources it depends upon are in short supply?

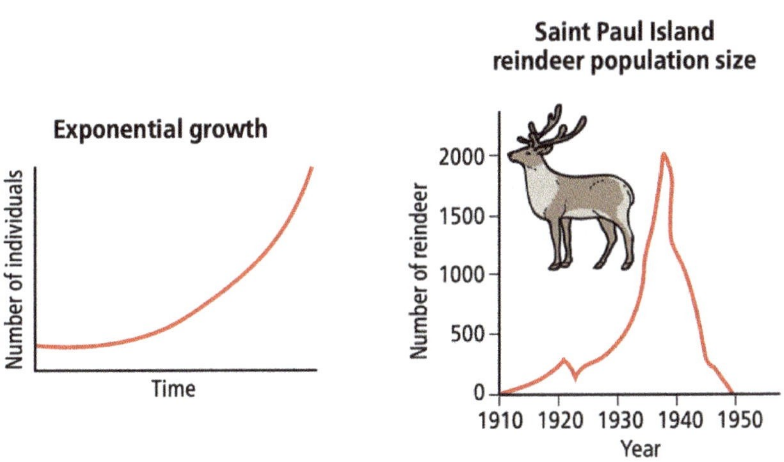

(a) (b)

< **Figure 16.3**

(a) In exponential growth, a population grows at a fixed rate per amount of time. Population size increases faster and faster, as you can see from the way the curve gets steeper and steeper. (b) This graph shows exponential growth followed by a crash in the Saint Paul Island reindeer population.

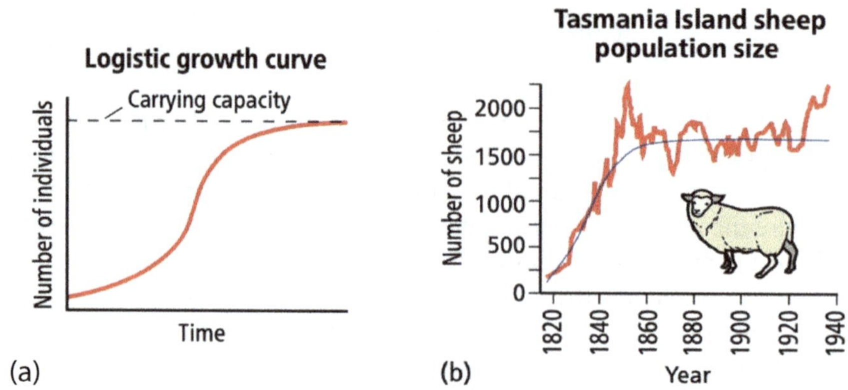

Figure 16.4 >

(a) In logistic growth, population growth slows as the population approaches the carrying capacity of its habitat. (b) This graph shows logistic growth in the Tasmanian sheep population.

Logistic Growth

Logistic growth occurs when population growth slows as the population approaches the habitat's carrying capacity (**Figure 16.4a**). The **carrying capacity** is the maximum number of individuals or the maximum population density that a habitat can support. A certain lake may be able to support 250 trout, for example. Or, a certain grassland may be able to support 35 weasels per square kilometer.

Why would population growth slow as the population approaches carrying capacity? There are many possible reasons. When a population is large, there may be more competition for resources such as food and space. This could make it more difficult to get enough resources to reproduce, causing birth rates to fall. Or, crowded conditions at large population sizes could increase the death rate. In the real world, logistic growth is often seen in stable habitats. Logistic growth in the Tasmania Island sheep population is shown in **Figure 16.4b**.

CONCEPT CHECK

A population of shorebirds occupies a habitat where there are a limited number of safe nesting sites. What kind of population growth do you expect to see in this population?

CHECK YOUR ANSWER

Because only a limited number of nesting sites are available, the population's growth will slow as the population size increases. The population will grow logistically.

More to Explore

You can read more about exponential and logistic population growth here:

https://sciencing.com/difference-exponential-logistic-population-growth-8564881.html

16.3 Life History

How a population grows depends in part on its *life history*—that is, on its schedule of survival and reproduction. This information can be seen in a population's survivorship curve, which shows the proportion of individuals in a population that survive to a given age. Let's look at three different kinds of survivorship curves, shown in **Figure 16.5**.

READING CHECK

What information does a survivorship curve convey?

Type I organisms have low death rates early in life, with most individuals surviving until fairly late in life. Elephants have Type I survivorship—most elephant calves survive to adulthood. Type I organisms typically have large bodies and reach sexual maturity late. They produce a small number of offspring and devote significant resources to each offspring. For example, a female elephant takes about 10 years to reach sexual maturity, gives birth to a calf only every 5 years or so, and provides significant parental care.

Type III organisms have high death rates early in life, with few individuals surviving until late in life. Razor clams have Type III survivorship—most razor clams die early in life, and few survive to adulthood. Type III organisms typically reach sexual maturity early, produce large numbers of offspring, and devote few resources to each. For example, a female razor clam may reach sexual maturity in the first year of life and then produce 100 million eggs, releasing them into the water with no further attention or care. So, whereas elephants produce a small number of "expensive" offspring (expensive in the sense that the parents devote a lot of resources to each offspring), razor clams make many "inexpensive" offspring.

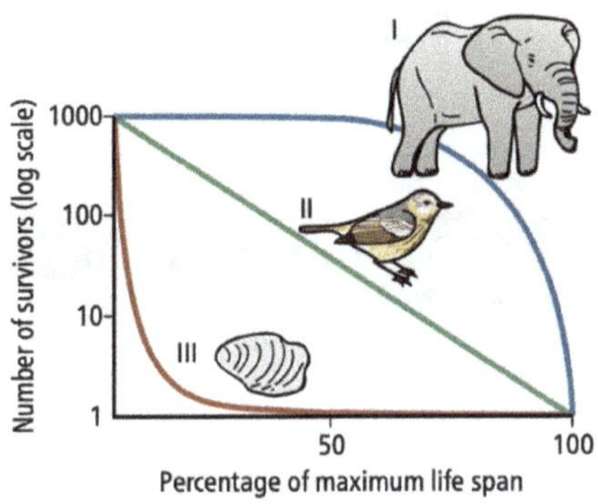

^ Figure 16.5

This figure shows three different survivorship curves. Type I organisms have low death rates early in life—most individuals live to an advanced age. Type II organisms have steady mortality throughout life. Type III organisms have high death rates early in life—few survive to adulthood.

Type II organisms fall between Type I and Type III. Type II organisms experience a steady death rate throughout life—individuals are as likely to die early in life as late in life. Many songbirds show Type II survivorship.

How is life history related to population growth? The production of many "inexpensive" offspring is associated with exponential growth. In the unstable environments of exponential growth, life and death are often chance events. Producing many offspring is adaptive because it becomes more likely that at least some of the offspring will survive if conditions are poor. Also, when conditions are good, there will be many offspring to take advantage of plentiful resources. On the other hand, producing a few "expensive" offspring is associated with logistic population growth. In the stable environments of logistic growth, many populations are at or near carrying capacity. Consequently, an offspring that receives a lot of parental investment is more likely to be able to compete with other members of the population and successfully survive and reproduce. **Table 16.1** summarizes some typical characteristics of Type I and Type III populations.

TABLE 16.1 Type I versus type III populations.

Type I Populations	**Type III Populations**
Large bodies	Small bodies
Reach sexual maturity late	Reach sexual maturity early
Few "expensive" offspring	Many "inexpensive" offspring
Long life expectancy	Short life expectancy
Live in stable environments	Live in unstable environments
Logistic population growth	Exponential population growth

CONCEPT CHECK

What type of life history (I, II, or III) characterizes humans?

CHECK YOUR ANSWER

Humans have a Type I survivorship curve. Most humans survive until late in life. Humans also have most of the other characteristics of Type I populations. For example, we have large bodies, reach sexual maturity late, and produce a small number of "expensive" offspring.

More to Explore

You can read more about survivorship, as well as Type I, Type II, and Type III survivorship here:
https://www.nature.com/scitable/knowledge/library/survivorship-curves-16349555/

16.4 Human Population Growth

More than 8 billion people live on Earth today—the end result of thousands of years of growth (**Figure 16.6a**). Where is the human population headed? Will it crash as resources run out? Or, will population growth slow down and slide toward carrying capacity?

According to United Nations projections, the world human population is expected to increase throughout the 21st century and exceed 11 billion by 2100. However, growth is expected to slow during the second half of the century (**Figure 16.6b**). These projections assume that fertility will continue to decline in high-fertility countries and that death rates will continue to decline everywhere.

We can learn how a population is growing from its *age structure*, the distribution of people's ages within the population. The age structure for the global population at three different points in time can be seen in **Figure 16.7**. In 1950, the age structure diagram resembled a pyramid. This was a sign of a rapidly growing population, where most of the population was young. In 2017, the age structure was less triangular and more uniform. Compared to 1950, the population has aged, and population growth has slowed. The projected age structure for the year 2100 shows that the global population is expected to become older and that growth will slow further.

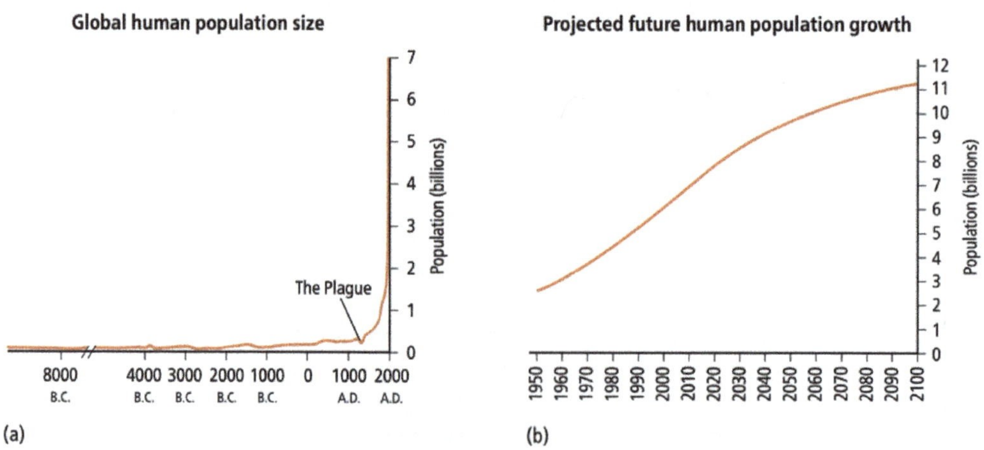

Figure 16.6 > (a) The human population grew exponentially for thousands of years (with the exception of the occasional dip, such as during the bubonic plague in the 14th century). (b) Current projections suggest that population growth may slow considerably in the second half of the 21st century. However, the human population is expected to exceed 11 billion by the year 2100.

Source: United Nations, Department of Economic and Social Affairs, Population Division, *World Population Prospects: The 2017 Revision.*

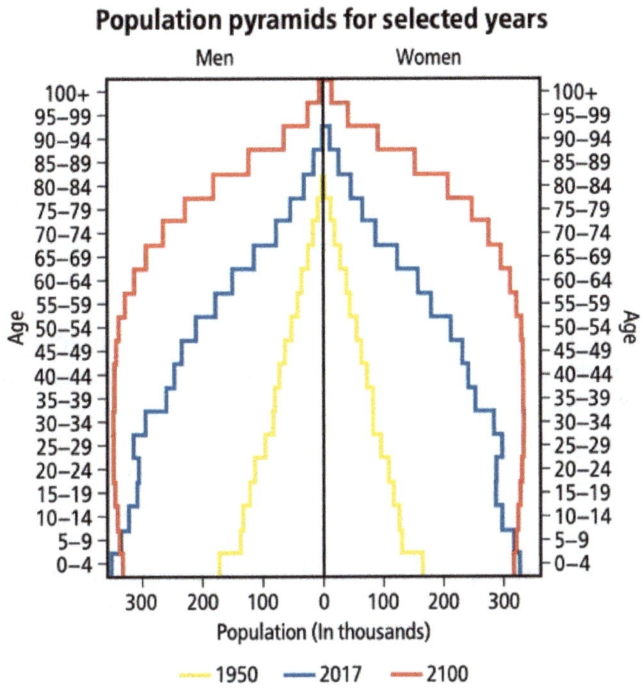

< **Figure 16.7**

Age structures show the distribution of ages within a population. In 1950 (yellow), most of the world's population was young and growing rapidly. By 2017 (blue), the world population had aged, and the rate of growth had slowed. By 2100 (red), the world population is expected to have aged further.

The slowdown in population growth can be explained by a phenomenon called the *demographic transition*—a shift from high birth and death rates to low birth and death rates. The death rate typically decreases first, as a result of medical advances such as better healthcare, improved nutrition, and immunization against childhood diseases. After some time, the birth rate also declines.

Ecological Footprint

Although population growth has slowed, human population size continues to increase, and there are now more humans on Earth than ever before. At the same time, the demand for Earth's resources also continues to increase. An ecological footprint is a measure of how much land and water area a human population needs in order to produce the resources it consumes. The size of this footprint takes into account the population's use of water, soil, energy, food, and other resources. Measuring ecological footprints allows us to understand how resource consumption compares with what Earth can provide. This is an important first step toward the goal of sustainable development.

READING CHECK

What does an ecological footprint tell you about the impact of a population on the environment?

For more than 40 years, the global human population has been in an *ecological overshoot*, consuming more resources each year than Earth can provide. In fact, we would need 1.7 Earths to sustain our current rate of consumption. And, if current trends continue, we will need 2 entire Earths by 2030 (**Figure 16.8**).

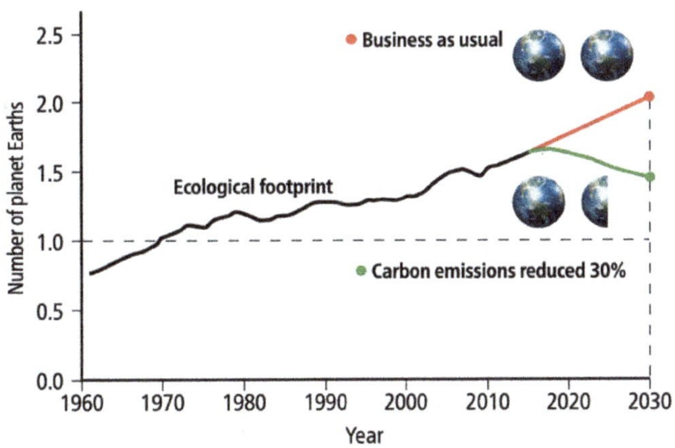

< **Figure 16.8**

This graph shows how our global ecological footprint has increased in the last 60 years. We are now using the resources that would take 1.7 Earths to provide. If present trends continue, we will soon be using the equivalent resources of 2 Earths.

Global overconsumption means that Earth's resources are being quickly depleted. We can already see evidence of this in water shortages, collapsed fisheries, and widespread deforestation. In addition, our overuse of energy, especially fossil fuels, has led to global warming. As Earth's resources begin to run out, it will no longer be possible to support the billions of people who live here. The likely result is catastrophic wars over resources as well as starvation.

Of all the continents in the world, North America has the largest ecological footprint. If everyone in the world consumed as many resources as the average American, we would need the resources of about *5 Earths*. It is vital for all of us to reduce consumption, increase efficiency, reduce energy use, and explore green energy alternatives.

CONCEPT CHECK

What is the age structure of a population, and how has the age structure of the global human population changed from 1950 to today?

CHECK YOUR ANSWER

The age structure of a population describes the distribution of people's ages within the population. From 1950 to today, the age structure of the global human population has changed so that the population is no longer quite as dominated by younger people. There are now relatively fewer children and younger people and relatively more older people.

More to Explore

You can read more about human population growth here:

https://ourworldindata.org/world-population-growth

Concept Review

Chapter 16

Detailed Chapter Summary

Ecology is the study of how organisms interact with their environments. These environments include abiotic features as well as biotic ones. Ecology can be studied at the level of the individual, population, community, and ecosystem. A population is a group of individuals of a single species that live in a specific area. A community consists of all the organisms that live in a specific area. An ecosystem consists of all the organisms that live in a specific area plus all the abiotic features of their environment.

Population ecologists are often interested in population size or population density. Four factors determine how population size changes over time -- birth rate, death rate, immigration rate, and emigration rate. Exponential growth and logistic growth are two models of population growth that are often seen in natural environments. Exponential growth occurs when a population grows at a fixed rate per amount of time. An exponentially growing population increases in size more and more quickly over time. In the real world, exponential growth cannot continue forever. Populations grow exponentially when they have unlimited resources -- or, in the real world, extremely plentiful resources. Exponential growth is often seen (temporarily) in populations that live in unstable environments. These populations may grow exponentially for a while and then crash when resources run out. Logistic growth occurs when population growth slows as the population approaches the habitat's carrying capacity, the maximum population size or density that a habitat can support. Population growth slows at larger population size because of factors such as competition for food or space, disease, and predation. Logistic growth is often seen in stable habitats.

A population's life history describes its schedule of survival and reproduction. A survivorship curve describes the proportion of individuals that survive to a given age. Type I organisms have low death rates early in life, with most individuals surviving until late in life. Type II organisms experience a steady death rate throughout life. Type III organisms have high death rates early in life, with few individuals surviving until late in life. Type I populations are associated with large bodies, late sexual maturity, having few "expensive" offspring, long life expectancy, stable environments, and logistic population growth. Type III populations are associated with the opposite characteristics -- small bodies, early sexual maturity, having many "inexpensive" offspring, short life expectancy, unstable environments, and exponential population growth.

Human population has been growing for thousands of years. There are over 7 billion people on Earth now, and the population is expected to exceed 10 billion by 2100. However, growth is expected to slow in the second half of this century. An age structure diagram shows the distribution of ages within a population. The shape of the age structure diagram reveals how a population is growing -- a pyramid-shaped age structure indicates rapid population growth, whereas a more even structure indicates slower growth. Demographic transition describes a shift from high birth and death rates to low birth and death rates. Many human populations have gone through such a transition. An ecological footprint describes how much land and water area a human population needs to produce the resources it consumes. Globally, we need 1.5 Earths to sustain our current rate of consumption. But if everyone in the world consumed as much as the average American, we would need about 5 Earths!

Key Terms

Carrying capacity The maximum number of individuals or the maximum population density that a habitat can support.

Community All the organisms that live in a specific area.

Ecological footprint The amount of land and water area a human population needs in order to produce the resources it consumes.

Ecology The study of how organisms interact with their environments.

Ecosystem All the organisms that live in a specific area and all the abiotic features of their environment.

Exponential growth A model of population growth in which a population grows at a fixed rate per amount of time.

Logistic growth A model of population growth in which growth slows as the population approaches the habitat's carrying capacity.

Population A group of individuals of a single species that live in a specific area.

Review Questions

Simple Review Questions

16.1 Organisms and Their Environments

1. What is ecology?

2. Explain the difference between the abiotic and biotic features of an organism's environment.

3. What is the difference between a community and an ecosystem?

16.2 Population Growth

4. Describe exponential growth. Under what conditions do organisms grow exponentially?

5. Why do populations that live in unstable environments often grow exponentially and then crash?

6. Describe logistic growth. Under what conditions do populations experience logistic growth?

16.3 Life History

7. What are the differences between Type I, Type II, and Type III survivorship?

8. What type of life history is typical of a top predator?

16.4 Human Population Growth

9. Explain how global human population size is expected to change during the 21st century.

10. What is the age structure of a population? What can you learn from a population's age structure?

11. What is an ecological footprint? What does the ecological footprint tell us about how the global human population is consuming resources today?

12. The human population exceeds 7 billion. The total number of humans who have ever lived, however, is on the order of 110 billion. How have these 110 billion humans not exceeded Earth's resources?

Challenging Review Questions

16.1 Organisms and Their Environments

13. Does a community contain multiple populations? Explain your answer.

14. A scientist studies how water cycles through a desert habitat and how organisms make use of this water. Is this a population-level study, a community-level study, or an ecosystem-level study? Defend your answer.

15. A scientist studies how the number of coyotes in Death Valley National Park, California, has changed over the last decade. Is this a population-level study, a community-level study, or an ecosystem-level study? Defend your answer.

16. A scientist examines how the presence of a nonnative bird species, the starling, affects other species of birds, such as the northern flicker, a kind of woodpecker. The starling (left) and woodpecker (right) in the photo below, for example, fight over the same food and space. Is this a population-level study, a community-level study, or an ecosystem-level study? Defend your answer.

16.2 Population Growth

17. How are exponential growth and logistic growth different? Under what real-world circumstances can populations grow exponentially? When will exponential growth in such a population stop?

18. What factors could cause population growth to slow as population size increases (which is what happens during logistic growth)?

19. The graph below shows the growth of a population of single-celled *Paramecium* in the lab. Did the population grow exponentially or logistically? Can you estimate the carrying capacity of the habitat?

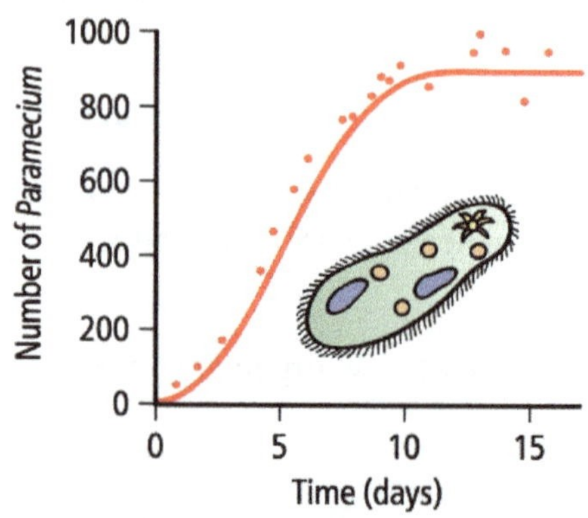

20. Suppose that the carrying capacity of a specific habitat is 50 songbirds. Is it possible for this carrying capacity to change? What could cause the carrying capacity to increase? What could cause the carrying capacity to decrease?

16.3 Life History

21. What is a survivorship curve? What are some characteristics of a population with a Type I survivorship curve?

22. Why is a baby elephant considered an "expensive" offspring?

23. Would you expect a tiger to have a Type I, Type II, or Type III survivorship curve?

24. Name an organism that you might see in your backyard or at the park that has a Type III survivorship curve. What evidence suggests it has a Type III survivorship curve?

16.4 Human Population Growth

25. The graph below shows survivorship curves for males and females in Great Britain and India in 1999. What differences do you see between Great Britain and India in 1999? What might explain these differences?

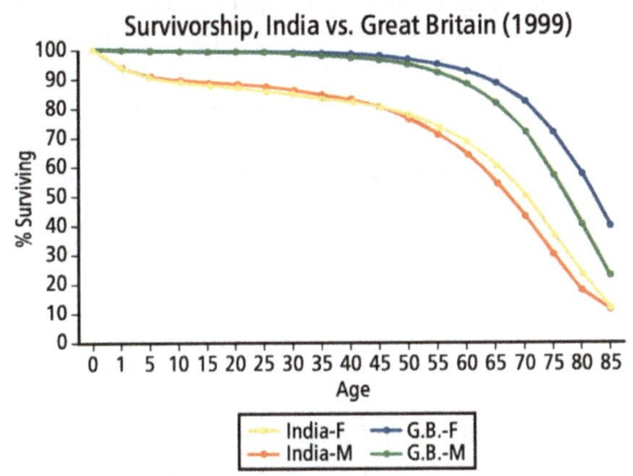

26. The graph below shows survivorship curves for males and females in Great Britain in the 17th century and in 1999. How did survivorship in England change from the 17th century to 1999?

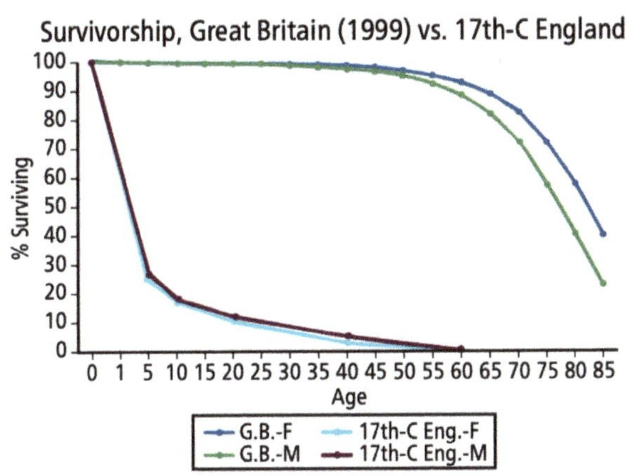

27. All else being equal, will a person who eats a vegetarian diet or a person who eats meat regularly have a larger ecological footprint? Defend your answer.

28. What does it mean that the world human population is in "ecological overshoot"?

Apply & Discuss Questions

29. A habitat's carrying capacity for a population can change over time. For example, a forest may be able to support a certain number of bears, but when the trees are cut down and the land is paved over, the carrying capacity drops (probably to zero). What other factors could cause the carrying capacity of a habitat to change? How has Earth's carrying capacity for humans changed over time? Does technology influence how many people Earth can support?

30. What type of survivorship curve characterizes humans? What other characteristics are associated with this type of survivorship curve? Do humans show many of these other characteristics?

Solutions to Odd-Numbered Questions

Simple Review Solutions

1. Ecology is the study of how organisms interact with their environments.

3. A community consists of all the organisms that live in a specific area. An ecosystem consists of all the organisms that live in a specific area and all the abiotic features of their environment.

5. Good conditions start exponential growth. When resources run out, the population crashes.

7. Type I organisms have low death rates early in life, with most individuals surviving until fairly late in life. Type III organisms have high death rates early in life, with few individuals surviving until late in life. Type II organisms fall between Type I and Type III. Type II organisms experience a steady death rate throughout life-individuals are as likely to die early in life as late in life.

9. According to United Nations projections, the world human population is expected to increase throughout the 21st century and exceed 11 billion by 2100. However, growth is expected to slow during the second half of the century. These projections assume that fertility will continue to decline in high-fertility countries and that death rates continue to decline everywhere.

11. An ecological footprint is a measure of how much land and water area a human population needs in order to produce the resources it consumes. The size of this footprint takes into account the population's use of water, soil, energy, food, and other resources. Measuring ecological footprints allows us to understand how resources are being consumed, and how this compares with what Earth can provide. For more than 40 years, the global human population has been in an ecological overshoot, consuming more resources each year than Earth can provide. In fact, would need 1.7 Earths to sustain our current rate of consumption. And, if current trends continue, we will need 2 entire Earths by 2030.

Challenging Review Solutions

13. A community includes all the organisms in a specific area. A population, on the other hand, is a group of individuals of the same species in a specific area. So yes, a community consists of multiple populations, one for each species found in the community. For example, a community could include a population of ladybugs, a population of aphids, a population of hummingbirds, and many, many other populations.

15. This is a population-level study because it considers a group of individuals of a single species that occupy a given area.

17. A population that is growing exponentially increases in size more and more quickly as time goes on. Under logistic growth, population growth slows as the population approaches the habitat's carrying capacity. What kind of population grows exponentially? In theory, exponential growth occurs when a population has unlimited resources. Although resources are never truly unlimited in the real world, they are sometimes extremely plentiful. For example, organisms may find very plentiful resources when they colonize a new habitat, and their population may begin to grow exponentially.

19. This is an S-shaped logistic curve. Carrying capacity appears to be around 900 individuals per mL.

21. A survivorship curve is a graph that shows the proportion of individuals that survive to a specific age. Type I organisms have low death rates early in life. Most individuals survive till fairly late in life. Populations with Type I survivorship curves typically have large bodies, reach sexual maturity late, have few "expensive" offspring to which they devote considerable resources each, and have a long life expectancy. These populations often live in stable environments and show logistic population growth.

23. A tiger would be expected to have a Type I survivorship curve—it produces expensive offspring that survive reasonably well.

25. There is more infant and early mortality in India than in England in 1999. This is likely because poverty affects more people in India, and more people are unable to get proper medical care.

27. All else being equal, a meat-eater has a larger ecological footprint than a vegetarian. By eating higher on the food chain, a meat-eater ends up using more of Earth's resources. This is because it takes more area to grow the grain to feed animals for consumption (such as cows or chickens) than it would take to supply a person with a vegetarian diet. This goes back to the "10% rule" regarding energy flow through an ecosystem -- 10 percent of the energy available at each level of the food chain moves up to the next level of the food chain.

Apply & Discuss Solutions

29. Many answers are possible. The climate is one obvious variable that can affect carrying capacity. Years with good rainfall may increase carrying capacity for producer populations as well as for populations of species higher up the food chain. Drought may result in a reduced carrying capacity for many populations in the habitat. The introduction of new species into a community can also affect the carrying capacity if this new species competes with or preys on the population of interest. Earth's carrying capacity for humans has increased over time as a result of technological advances such as the origin of agriculture and the green revolution. However, as human activities cause global warming, Earth's carrying capacity for humans may decrease.

Living Earth Essay

Life on Earth Connections

Topsoil Regenerated

Agriculture is the organized use of resources for the production of food. Whether these resources—mainly topsoil and fresh water—will be available for future generations depends on how well we manage them now. We already know from experience that pesticides and fertilizers cannot be applied liberally without threatening both topsoil quality and our supplies of clean groundwater (not to mention our health and that of the planet).

Over the past several decades, there have been strong movements toward developing methods and technologies that can sustain agricultural resources over the long term. Problems associated with irrigation, for example, can be solved by **micro irrigation**, which is any method of delivering water directly to plant roots. Micro irrigation not only prevents topsoil erosion but also minimizes water loss through evaporation, which in turn minimizes the salinization of farmland. One method of micro irrigation is shown in **Figure A**.

Figure A
The micro irrigation method, known as drip irrigation, delivers water through long, narrow strips of punctured plastic tubing. The system provides only as much water as the plants need.

Organic Farming Is Environmentally Friendly

For controlling pests and maintaining fertile soil, the conventional agricultural industry is now looking at the efforts of many small-scale farmers who have demonstrated that significant crop yields can be obtained without pesticides or synthetic fertilizers. This method of farming is known as **organic farming**, where the term *organic* is used to indicate a concern for the environment and a commitment to using only chemicals that occur in nature.

To protect against pests, organic farmers alternate the crops planted on a particular plot of land. Such *crop rotation* works fairly well because different crops are damaged by different pests. A pest that thrives on one season's crop of corn, for example, will do poorly on the next season's crop of alfalfa. Crop rotation also mitigates the depletion of minerals and nutrients. For fertilizer, organic farmers rely on compost, shown in **Figure B**. They also include nitrogen-fixing plants in their crop rotation schedules.

Figure B
Odor-free backyard compost bins are easy to build and maintain. The secret is to optimize the exposure to atmospheric oxygen, which favors aerobic decomposition.

Many claims are made that food produced organically is better for human consumption. Chemically, however, the molecules that plants absorb from synthetic fertilizers are the same as those they absorb from natural fertilizers. If organically grown produce tastes any better or is more nutritious than conventionally grown produce, the reason likely has to do with the genetic strain of the produce or with the greater attention paid to growing conditions, such as the health of the soil and the amount of water made available to plants. But most notable is that organically grown foods are free of residual pesticides, which are certainly not beneficial for human consumption.

Organic farming also tends to be benevolent to the environment. In addition to avoiding the potential runoff of pesticides and fertilizers, organic farming is energy efficient, using only about 40 percent as much energy as conventional farming. A large portion of the energy savings arises because the manufacture of pesticides and fertilizers is energy intensive.

Because much organically grown food, such as that shown in **Figure C**, is grown locally, by purchasing it, you help support local farmers. Ultimately, though, your purchase of organically grown food is a vote in favor of environmentally friendly farming methods.

Figure C
Market forces often result in higher prices for organically grown food. A product can bear this USDA organic seal if at least 95 percent of its ingredients are certified organic.

CONCEPT CHECK
Which are made of organic chemicals, organically grown foods, or conventionally grown foods?

CHECK YOUR ANSWER
Whether grown with natural or synthetic fertilizer, all foods are made of organic chemicals—carbohydrates, lipids, proteins, nucleic acids, and vitamins. Assuming all pesticides can be removed by a thorough rinsing, a conventionally grown carrot may be just as good for you as one grown without the use of synthetic fertilizers and pesticides. The organic in organic farming is a term used to designate a pesticide-free natural method of farming friendly to the environment.

Integrated Crop Management

To meet concerns about sustaining agricultural resources over the long term, groups from industry, government, and academia have identified a whole-farm strategy called **integrated crop management** (ICM). This method of farming involves managing crops profitably with respect for the environment in ways that suit local soil, climatic, and economic conditions. Its aim is to safeguard a farm's natural assets over the long term through the use of practices that avoid waste, enhance energy efficiency, and minimize pollution. ICM is not a rigidly defined form of crop production but rather a dynamic system that adapts and makes sensible use of the latest research, technology, advice, and experience.

One of the more significant aspects of ICM is its emphasis on multi-cropping, which means growing different crops on the same area of land either simultaneously, as shown in **Figure D**, or in rotation from season to season. Like organic farming, multi-cropping achieves significant pest control, and it can be used to improve soil fertility. For example, nitrogen-generating crops such as legumes are a good complement to nitrogen-depleting crops such as corn.

Figure D
Complementary crops such as legumes and corn are grown in alternating strips to enhance soil fertility. The strips follow the contour of the land to minimize erosion from rainwater or irrigation.

An important component of ICM is **integrated pest management** (IPM), one of the aims of which is to reduce the initial severity of pest infestation. This can be accomplished through a number of routes. When a farm is started, for example, only crops that fit the local climate, soil, and topography should be grown. This selectivity makes for crops that are hardy and pest-resistant. Crops should also be rotated as much as possible to reduce pest and weed problems. Another IPM strategy is growing tree crops or hedges either around the perimeter of a farm or interspersed throughout the farm. These trees and hedges provide habitat, cover, and refuge for beneficial insects and such pest predators as spiders, snakes, and birds. As an added benefit,

Figure E
(a) Infrared satellite images reveal much information about our agricultural lands. Shades of red, for example, indicate crop health, browns show how much pesticide has been applied, and black shows areas prone to flooding. (b) In a practice called precision farming, a computer system using data from high-resolution satellite imagery guides a GPS-equipped tractor to pinpoint nutrient and pesticide placement.

the trees and hedges also protect the land from wind erosion. Yet another IPM strategy is to breed and cultivate plants that have a natural resistance to pests. For centuries, this was accomplished by selectively mating plants that showed the greatest resistance. Today, this age-old method is quickly being supplanted by the techniques of genetic engineering.

Another aim of IPM is to minimize the use of pesticides. For example, many farmers now use the global positioning satellite (GPS) system to target precise pesticide applications. Using infrared satellite photography, illustrated in **Figure E**, and careful walk-through assessment of field conditions, farmers can match pesticide blends with crop needs. Computers link application equipment with the GPS satellites, which "beam" pesticide application adjustments every few seconds as a farmer moves across a field. This same technology also works well with the selective delivery of synthetic fertilizers.

Many other methods of pest control can be used in place of chemical pesticides or in combination with them to minimize the need for these agents. Depending on the availability of labor, egg masses or larvae can be handpicked off plants. Instead of using herbicides, weeds can be tilled under. An insect population can also be controlled through various biological approaches, such as by introducing large numbers of sterile insects into a population or by introducing natural predators, as shown in **Figure F**.

Another way to control the proliferation of insects is to modify their behavior through the use of **pheromones**, which are volatile organic molecules that insects release to communicate with one another. Each insect species produces its own set of pheromones, some as warning signals and others as sexual attractants. Sexual pheromones synthesized in the laboratory can be used to lure harmful insects to localized insecticide deposits, reducing the need for spraying an entire field, as depicted in **Figure G**.

Figure F
The almond trees on the right side of the road were decimated by spider mites. Those on the left were protected by the introduction of spider mite predators.

Figure G
Female gypsy moths emit the pheromone disparlure (top) to entice male gypsy moths (bottom left) into mating. The males are so sensitive to this compound that they can detect one molecule in 10^{17} molecules of air. This astounding sensitivity enables them to respond to a female that may be more than 1 kilometer away. However, they can also be tricked into responding to insecticide traps laced with synthetic disparlure (bottom right).

READING CHECK
How are pheromones used to protect crops?

CONCEPT CHECK
Describe Integrated Crop Management.

CHECK YOUR ANSWER
A method of farming that involves managing crops profitably with respect for the environment in ways that suit local soil, climatic, and economic conditions.

FOR YOUR INFORMATION

The outer coat of the naturally occurring rice grain contains bran as well as a number of valuable nutrients. Natural rice is transformed into white rice by milling away this outer coat. But why mill away this outer coat when it is so nutritious? The outer coat also includes fats that, over time, oxidize to make the grain turn rancid. Thus, unprocessed rice, also known as brown rice, is not well suited for long-term storage.

Regenerative Agriculture

The farming techniques described above are part of a movement known as **regenerative agriculture**, which holds that proper agricultural practices can be sustainable as well as help us to recover from past harmful practices. This includes promoting biodiversity, particularly with the use of crop rotation. The plowing of agricultural land should also be minimized while also avoiding the use of synthetic pesticides and fertilizers.

An example of a crop particularly suited to these principles is Kernza® developed by the Land Institute in Salinas, Kansas. This crop is a perennial grain, which means it grows continuously and does not need to be replanted every year (**Figure H**). This, in turn, means that the field need not be plowed routinely, which saves energy as well as protects the soil. Furthermore, the roots grow deep, which means the plants are effective at transporting carbon from the atmosphere into the soil. This is an example of *carbon sequestration*, which is greatly needed to help combat growing levels of atmospheric carbon dioxide, a potent greenhouse gas.

Regenerative agriculture also promotes proper grazing management for livestock. The idea is that animals free to roam across large spans of land help to restore soil fertility, which, in turn, is better able to sequester carbon. This is in contrast to holding livestock within crowded pens or buildings where animal wastes become concentrated and thus toxic, not only to the animals but to the local environment.

While pasture-fed beef is more environmentally sound than factory-fed beef, there remains a deeper problem. The production of beef and other animal products tends to be harmful to the environment no matter how the animals are raised. Livestock, such as cattle, account for about one-third of our total methane emissions, a greenhouse gas up to 80 times more potent than carbon dioxide. Also, the amount of arable pasture land available for livestock is limited. Furthermore, livestock themselves require vast amounts of feed. For example, more than half of the grain grown in the United States is presently consumed by livestock rather than directly by humans (**Figure I**).

(a) (b)

Figure H
(a) Wheat is an annual crop that must be planted each year. (b) Wheat can be hybridized with a wild grass cousin, Thinopyrum, to create a food-bearing perennial crop that grows from one season to the next. As an added benefit, the hybrid grain is low in gluten.

Figure I
Pasture-fed cattle are more sustainable than factory-fed cattle, but neither can be relied upon for feeding the world and helping to maintain a sustainable environment.

The greatest energy efficiency is achieved by eating as low on the food chain as possible. As the world population quickly grows past 8 billion of us, a meat-based diet for all is impossible to achieve. Consider why fossils of *Tyrannosaurus rex* dinosaurs are so rare. Answer: They were meat eaters at the top of the food chain, which necessarily means they were few in number. Both in terms of available calories and also support for sustainable agriculture, we are wise to focus on developing our plant-based diets.

Nature is sophisticated, and if we are to work with nature in a sustainable way, our methods must also be sophisticated. New and improved techniques provide the farmer with a menu of possible actions in response to nature's ever-changing parameters. With each action, however, the farmer must be aware of its potential environmental impact, both locally and globally. In this sense, the human who farms sustainably is not dominating this Living Earth but rather working with it. To Life!

Chapter 17

Communities

THE MAIN IDEA

A community includes all the organisms that live in a specific area.

17.1 Food Webs
17.2 Competition
17.3 Symbiosis
17.4 Invasive Species

Within a community, living things interact in many different ways. Some organisms eat other organisms, whereas others rely on their neighbors for shelter. Individuals of different species may also compete for limited resources, such as food or living space. Why do some eels let tasty shrimp crawl into their mouths without eating them? Why is the orca described as a top predator? How is it wise for humans to eat primarily plants rather than other animals? What is interspecific competition? And how does conflict necessarily follow when the niches of two species overlap? What makes a species new to a habitat an "invasive species"?

17.1 Food Webs

Recall from Chapter 16 that a community consists of all the organisms that live in a specific area. Within communities, species interact in different ways. For example, lions eat zebras, eagles and storks fight over fish, and shrimp clean the mouths of eels. One important set of species interactions is who eats whom. These interactions can be shown in a food chain that contains multiple feeding levels (**Figure 17.1**).

At the bottom of a food chain are producers. A **producer** is an organism that makes organic molecules using inorganic molecules and energy. In almost all communities, the producers are photosynthesizing organisms such as plants.

A **consumer** obtains food by eating other organisms. A *primary consumer* eats producers. Herbivores, such as zebras or rabbits, are primary consumers. A *secondary consumer* eats primary consumers. Carnivores, such as lions and wolves, are secondary consumers. In some communities, there are consumers that eat at even higher levels in the food chain. The species at the top of a food chain is called the *top predator*.
A **decomposer** obtains food by eating dead organic matter. Fungi and bacteria are important decomposers in many communities.

Most ecological communities are complicated enough that food "chains" are more accurately described as **food webs**. **Figure 17.2** shows the food web in the Antarctic aquatic community.

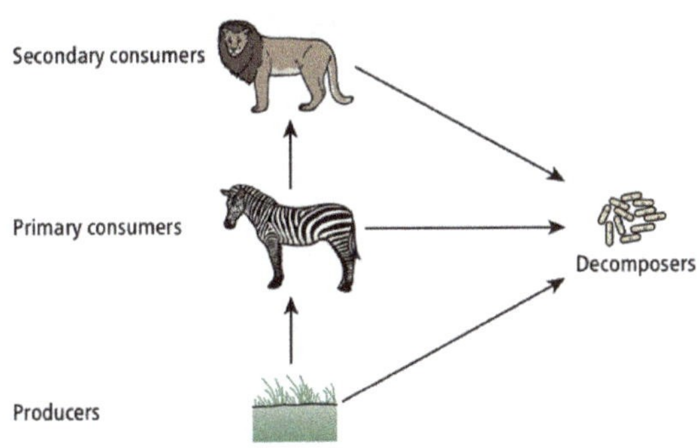

Figure 17.1

This food chain includes producers (grass), primary consumers (zebras), and secondary consumers (lions). Decomposers consume dead organisms at all levels of the food chain.

READING CHECK

Name two types of organisms that are important decomposers in many communities.

Where are humans on the food chain? Most humans are *omnivores* who eat at many levels of the food chain. We eat plants, animals, and even decomposers, such as certain mushrooms. Of course, some humans have a more restricted diet (**Figure 17.3**).

Figure 17.2 >

The Antarctic aquatic food web. Can you find the top predator in this community?

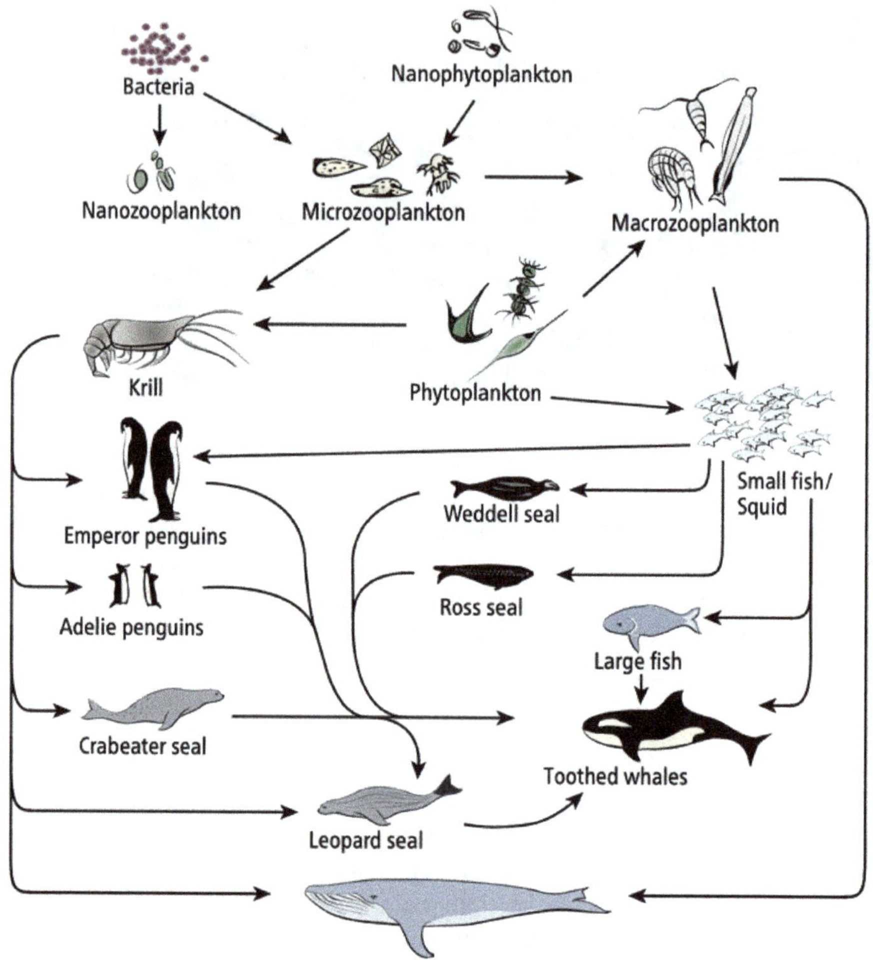

^ Figure 17.3

Most humans eat at many levels of the food chain, but some humans have a more restricted diet.

CONCEPT **CHECK**

Can an organism be a producer and a consumer?

CHECK YOUR ANSWER

Actually, yes. There are some parasitic plants that both photosynthesize and obtain nutrients from their hosts—other plants. This makes them both producers and primary consumers. There are even some predatory plants that eat higher up the food chain by trapping insects. The Venus flytrap, which snaps special leaves together to catch insects, is probably the most famous.

More to Explore

To learn more about food chains, visit the following website:

https://www.nationalgeographic.org/encyclopedia/food-web/

17.2 Competition

Competition between organisms of different species within a community is called **inter-specific competition**. Some instances of inter-specific competition are easy to identify—such as when animals battle over food (Figure 17.4). However, living things don't have to fight each other directly in order to compete. In fact, species within a community compete any time they use the same resource and this resource exists in limited supply. For example, plants growing near each other may compete for limited resources such as sunlight, water, and soil nutrients.

^ Figure 17.4

An eagle and a stork fight over food.

A species' **niche** within a community is the total set of biotic and abiotic resources it uses. This includes the food it eats, the water it drinks, the space it occupies, and any other resource the species uses. So, you can say that two species in a community compete any time their niches overlap.

What is the result of competition? One result of inter-specific competition is evolution. Many organisms have evolved adaptations that enable them to better compete with other species.

For example, some plants that compete for sunlight have evolved large leaves with a large surface area. Others have evolved to grow upward quickly, so that they have a better chance of overshadowing their competitors. Certain plants even release special chemicals that prevent a competitor's seeds from sprouting.

READING CHECK

In biology, what does it mean to say that two species compete with one another?

It is interesting that, although species frequently compete, no two species in a community can have *exactly* the same niche. Otherwise, the species that are better at acquiring and using resources out compete the other species and eventually drive them to extinction. Nonetheless, ecologists have sometimes found species that appear to have identical niches within a community. On closer inspection, however, it has turned out that the niches are not exactly identical.

One of the most famous instances of this occurred in a group of birds now known as MacArthur's warblers. Ecologist Robert MacArthur studied five warbler species living in the coniferous forests of the Northeast United States. All five species lived in the same trees, and all five species ate insects. Did they share the same niche? By watching the warblers closely, MacArthur discovered that there were crucial differences in how the species used resources: Each species of warbler used a different part of the tree,

< Figure 17.5

The five species of MacArthur's warblers initially appeared to occupy the same niche—they all live in the same coniferous forests, and they all eat insects. However, closer inspection revealed that their use of resources differs in significant ways. For example, each species occupies a different part of the tree, as shown here (shaded in green). Each species also has a different way of hunting for insects.

and each had a different way of hunting for insects (**Figure 17.5**). How did MacArthur's warblers come to have such similar, yet nonidentical, niches? Ecologists believe that the niche differences were produced by evolution through natural selection. Individuals of each warbler species were more successful at surviving and reproducing if they competed less with individuals of other species. Over time, this caused the species' niches to diverge.

CONCEPT CHECK

Does inter-specific competition involve members of different species directly fighting over food and other resources?

CHECK YOUR ANSWER

It can, but it doesn't have to. Any time two species within a community use the same resource, and this resource is limited, there is competition.

More to Explore

You can read more about interspecific competition here:

https://bio.libretexts.org/Bookshelves/Introductory_and_General_Biology/Book%3A_Introductory_Biology_(CK-12)/06%3A_Ecology/6.15%3A_Competition

17.3 Symbiosis

In **symbiosis**, individuals of two species live in close association with one another. There are three kinds of symbiosis: parasitism, commensalism, and mutualism.

Parasitism describes a relationship that benefits one member of the interaction and harms the other. Fleas, ticks, and tapeworms are parasites that obtain nutrients from their hosts (**Figure 17.6a**). This relationship benefits the parasite, which gets nutrients and a home, but harms the host. Pathogens, such as many bacteria or viruses, can be considered parasites as well.

Commensalism is a form of symbiosis that benefits one species while having no effect on the other. The remora is a small fish that attaches itself to a shark using a suction cup (**Figure 17.6b**). The remora feeds on small scraps of food left over when the shark eats. It also gets protection from its host. (Would you bother a fish hanging out under a shark?) The remora neither benefits nor harms the shark.

(a) (b) (c)

Figure 17.6

The three types of symbiosis are parasitism, commensalism, and mutualism. (a) Parasitism: A tick feeds on human blood. (b) Commensalism: A remora hitches a ride on a shark. (c) Mutualism: A cleaner shrimp removes parasites from the mouth of an eel.

Mutualism is a form of symbiosis that benefits both species involved. Cleaner shrimp and eels have a mutualistic relationship. The eel opens its mouth, and a shrimp enters and eats the parasites there (**Figure 17.6c**). Both species benefit—the shrimp obtains food, and the eel is freed of its parasites. Sometimes, eels even "wait in line" for a shrimp's services!

Another example of mutualism is mycorrhizae, the close association between fungi and plant roots. The fungi help the plant absorb water and minerals from the soil, and the plant provides nutrients and shelter for the fungi. As another example of mutualism, there are bacteria living in the human gut, commonly referred to as the human microbiome. These bacteria thrive off the food we eat but, at the same time, help us to absorb the remaining broken-down nutrients.

A species that lives in close association with another species evolves in response to that partner. Many parasites, for example, have evolved special adaptations for attaching to their hosts. Of course, hosts also evolve in response to parasites. As we saw, eels have evolved an amazing relationship with cleaner shrimp in order to get rid of their parasites.

READING CHECK

Why is mutualism a "win-win" relationship for the species involved?

CONCEPT CHECK

1. This reed warbler is feeding a large cuckoo chick. Cuckoos lay their eggs in the nests of other species, and the parents then raise the cuckoo chick instead of their own. (The cuckoo mother removes an egg before leaving her own, and the cuckoo chick will often eject the remaining eggs from the nest after it hatches.) What type of symbiotic relationship do these two species have?

2. In some places, mongooses and hornbills (a type of bird) eat together. Both look out for predators while they eat, and each can warn the other of danger. Hornbills are better at detecting predatory birds overhead. Mongooses, with their keen sense of smell, are better at detecting predators on land. What type of symbiosis do the mongoose and hornbill have?

CHECK YOUR ANSWERS

1. This is an example of parasitism. The cuckoo is a parasite that benefits from its relationship with the reed warbler. The reed warbler is harmed.

2. The mongoose and hornbill have a mutualistic relationship—both species benefit from the interaction.

More to Explore

For more on symbiosis, check out the following:

https://www.nationalgeographic.org/article/symbiosis-art-living-together/

17.4 Invasive Species

When the brown tree snake arrived in Guam from its native New Guinea, it found a snake's paradise—plentiful food in the form of forest birds and their eggs and not a single natural enemy in sight. The snake quickly spread through the small island and is now found in startling numbers; in favorable habitats, there can be as many as 5000 individuals per square kilometer (**Figure 17.7**). As the brown tree snake thrives, however, it devastates Guam's unique bird fauna. A dozen species are now extinct, and many others are endangered.

An **invasive species** is a species that has moved from its native habitat to a new area, where it proceeds to do a lot of ecological damage. Although

Figure 17.7 >

Brown tree snake climbing in Guam

READING CHECK

How do most invasive species get introduced into new habitats?

many species were purposely introduced in the past, the introduction of non-native species today is usually accidental, occurring when organisms hitch a ride on a ship or airplane. Many invasive species do a lot of damage to native species by competing with them for resources or by preying on them directly.

Invasive species are responsible for the decline of countless native species worldwide. More than a third of the species listed as endangered under the U.S. Endangered Species Act are threatened wholly or partly because of an invasive species. Some invasive species also have significant economic impact.

The zebra mussel (**Figure 17.8**) is an invasive species that has spread across the eastern United States since its accidental introduction in 1988. Zebra mussels clog water pipes at power plants and water treatment facilities, causing billions of dollars of damage annually. They also threaten native freshwater bivalves by competing with them for food; most of these native species are now listed as endangered.

< Figure 17.8

Invasive zebra mussels crowd out many of the surfaces within the Great Lakes of North America as well as other freshwater aquatic habitats of the world.

CONCEPT **CHECK**

How do invasive species harm native species?

CHECK YOUR ANSWER

Most often by competing with them for food or other resources or by preying on them directly.

More to Explore

You can read more about invasive species at the following websites:

https://www.invasive.org

tps://oceanservice.noaa.gov/facts/invasive.html

https://www.nationalgeographic.org/encyclopedia/invasive-species/

Concept Review

Chapter 17

Detailed Chapter Summary

Species interact in many different ways within communities. A food chain or food web describes who eats whom. The feeding levels in a food chain include producers, various levels of consumers, and decomposers. Producers make organic molecules using inorganic molecules and energy. Consumers eat other organisms. Decomposers eat dead organic matter. Inter-specific competition describes competition between species within a community. Inter-specific competition occurs whenever two species use the same resource, and this resource exists in limited supply.

Inter-specific competition describes competition between individuals of different species within a community. A species' niche describes the total set of biotic and abiotic resources the species uses. No two species can have exactly the same niche. Evolution and adaptation frequently result from inter-specific competition.

Symbiosis occurs when two species live in close association with each other. Parasitism benefits one species and harms the other. Commensalism benefits one species while having no effect on the other. Mutualism benefits both species involved. Species involved in symbiosis typically have evolved in response to their partner.

Invasive species are species that have moved from their native habitat to a new area, where they do a lot of ecological damage. Today, most invasive species are introduced accidentally by humans. Invasive species have damaging effects on communities because they out compete or prey on native species. Invasive species are responsible for the decline of many native species. Some also do considerable economic damage.

Key Terms

Commensalism A form of symbiosis that benefits one species while having no effect on the other.

Consumer An organism that obtains food by eating other organisms.

Decomposer An organism that obtains food by eating dead organic matter.

Food webs Diagrams that show "who eats whom" within a community.

Inter-specific competition Competition between organisms of different species within a community.

Invasive species A species that has moved from its native habitat to a new area, where it does a lot of ecological damage.

Mutualism A form of symbiosis that benefits both species involved.

Niche The total set of biotic and abiotic resources a species uses within a community.

Parasitism A symbiotic relationship that benefits one member of the interaction and harms the other.

Producer An organism that makes organic molecules using inorganic molecules and energy.

Symbiosis A situation in which individuals of two species live in close association with one another, such as is found in parasitism, commensalism, and mutualism.

Review Questions

Simple Review Questions

16.1 Food Webs

1. What is the name for a diagram of "who eats whom" in a community?
2. Explain differences between a producer and a consumer.
3. What is a decomposer? What organisms function as decomposers in most communities?

16.2 Competition

4. What is a niche?
5. Can two species have the exact same niche in a community? Why or why not?

16.3 Symbiosis

6. Define parasitism, and provide some examples.
7. Do species evolve in response to the symbiotic relationships they participate in?

16.4 Invasive Species

8. What is an invasive species?
9. Why are invasive species harmful to native species?

Challenging Review Questions

16.1 Food Webs

10. List three producers, three primary consumers, and three secondary consumers.
11. Think of an example of an organism that eats at multiple levels of a food chain. Is your organism a producer? A consumer? What are the advantages of eating at multiple levels of a food chain?
12. Distinguish between a food chain and a food web.
13. In a habitat, mice eat plant seeds, and owls and lynxes prey on the mice. Draw the simple food chain or food web that shows these relationships. Label the producer(s), primary consumer(s), and secondary consumer(s).

16.2 Competition

14. In a habitat, mice eat plant seeds, and owls and lynxes prey on the mice. What term describes the relationship that exists between two species that eat the same prey, as owls and lynxes do in this example?
15. Explain how two different populations of plants can compete for resources. For what kinds of resources might they compete?
16. What is a niche? Explain the significance of Robert MacArthur's observations of warblers and their niches.
17. How can inter-specific competition result in evolution through natural selection? Use examples in your answer.

16.3 Symbiosis

18. What is an example of a parasite and its host? Explain how the relationship in your example benefits the parasite while harming the host.
19. What is an example of mutualism? Explain how the relationship in your example benefits the two species involved.
20. Some flowering plants rely on insects to carry pollen from male flowers to female flowers. For example, the bee in the photo below is covered with specks of yellow dandelion pollen. Bees are dusted with pollen as they feed on nectar produced by the flowers. Is this an example of parasitism, commensalism, or mutualism?

16.4 Invasive Species

21. Do all species that colonize a new area become invasive species?
22. Why do some species become invasive species in a new habitat?

Apply & Discuss Questions

23. A habitat's carrying capacity for a population can change over time. For example, a forest may be able to support a certain number of bears, but when the trees are cut down and the land is paved over, the carrying capacity drops (probably to zero). What other factors could cause the carrying capacity of a habitat to change? How has Earth's carrying capacity for humans changed over time? Does technology influence how many people Earth can support?
24. What type of survivorship curve characterizes humans? What other characteristics are associated with this type of survivorship curve? Do humans show many of these other characteristics?
25. Two species of salamanders, *Plethodon cinereus* and *Plethodon hoffmani*, overlap in parts of their range in the eastern United States. The two species are very similar in size *except* in places where both species are found. In places where both species are found, *Plethodon cinereus* is smaller than it usually is, and *Plethodon hoffmani* is larger than it usually is. This size difference is related to what the salamanders eat: *Plethodon cinereus* eats smaller prey than it eats in other parts of its range, and *Plethodon hoffmani* eats larger prey than it eats in other parts of its range. How can natural selection due to inter-specific

competition explain the evolution of size and diet in these salamanders?

26. Some acacia trees have evolved a special relationship with certain species of ants. The trees provide food and nesting sites (in the form of hollow thorns, as shown in the photo) for the ants. The ants attack insects and other species that try to eat the acacia, and they may also kill plants that grow in the immediate vicinity. Discuss the species interactions (food chain, competition, symbiosis, etc.) among the species in this community.

Solutions to Odd-Numbered Questions

Simple Review Solutions

1. Food chain or food web.

3. A decomposer obtains food by eating dead organic matter. Fungi and bacteria are important decomposers in many communities.

5. No two species in a community can have exactly the same niche. Otherwise, the species that are better at acquiring and using resources out compete the other species and eventually drive them to extinction.

7. A species that live in close association with another species evolves in response to that partner. Many parasites, for example, have evolved special adaptations for attaching to their hosts. Of course, hosts also evolve in response to parasites. As we saw, eels have evolved an amazing relationship with cleaner shrimp in order to get rid of their parasites.

9. Many invasive species do a lot of damage to native species by competing with them for resources or by preying on them directly.

Challenging Review Solutions

11. An omnivore such as a bear eats both plants and animals. It's both a primary consumer and a secondary consumer. The advantage of eating at multiple levels in a food chain is that there are more potential sources of food.

13. Plants are the producers, mice are the primary consumers, and owls and lynxes are the secondary consumers.

15. Two different populations compete for resources any time they use the same resource, and it is in limited supply. Plants might compete for soil nutrients, water, or sunlight.

17. Inter-specific competition could result in evolution if either (or both) species evolved to be better competitors. For example, if plants were competing for water, they might evolve more extensive root systems. If two plant populations were competing for sunlight, they might evolve larger leaves, or they might evolve to grow very tall to gain a better share of the available light.

19. Many examples are possible. As just one example, humans and dogs have a mutualistic relationship. Dogs get food and protection from humans, and humans gain companionship and love. Depending on the dog, humans may also receive help with herding, hunting, or other activities.

21. No. When an organism is moved to a new place, it might not even survive there. Even if a species can survive in a new area, most species will not become invasive species. Only a few become so successful in their new habitat that they harm native species.

Apply & Discuss Solutions

23. Many answers are possible. The climate is one obvious variable that can affect carrying capacity. Years with good rainfall may increase carrying capacity for producer populations as well as for populations of species higher up the food chain. Drought may result in a reduced carrying capacity for many populations in the habitat. The introduction of new species into a community can also affect the carrying capacity if this new species competes with or preys on the population of interest. Earth's carrying capacity for humans has increased over time as a result of technological advances such as the origin of agriculture and the green revolution. However, as human activities cause global warming, Earth's carrying capacity for humans may decrease.

25. In areas where both species are found, the salamanders experienced inter-specific competition for food. Individuals of each species that happened to differ more in size were more successful because they experienced less competition for food. That is, larger Plethodon hoffmani were more successful at surviving and reproducing because they competed less with Plethodon cinereus. Similarly, smaller Plethodon cinereus were more successful at surviving and reproducing because those individuals competed less with Plethodon hoffmani salamanders. Over time, this led to size evolution in both species.

Living Earth Essay

Life on Earth Connections

Ocean Acidification

The amount of carbon dioxide put into the atmosphere by human activities is growing. Surprisingly, however, the atmospheric concentration of CO_2 is not increasing proportionately. One explanation has to do with the oceans, as described in **Figure A**. When atmospheric CO_2 dissolves in any body of water—a raindrop, a lake, or an ocean—it forms carbonic acid, H_2CO_3. In freshwater, this carbonic acid transforms back to water and carbon dioxide, which is released back into the atmosphere. In the ocean, however, the carbonic acid is quickly neutralized by dissolved alkaline minerals. (The ocean is alkaline, pH = 8.1.) The products of this neutralization eventually end up on the ocean floor as insoluble solids. Thus, carbonic acid neutralization in the ocean prevents CO_2 from being released back into the atmosphere. The ocean, therefore, is an effective carbon dioxide sink—most of the CO_2 that goes in doesn't come out. Pushing more CO_2 into our atmosphere means pushing more of it into our vast oceans. So far, the oceans have been absorbing about one-third of our CO_2 emissions.

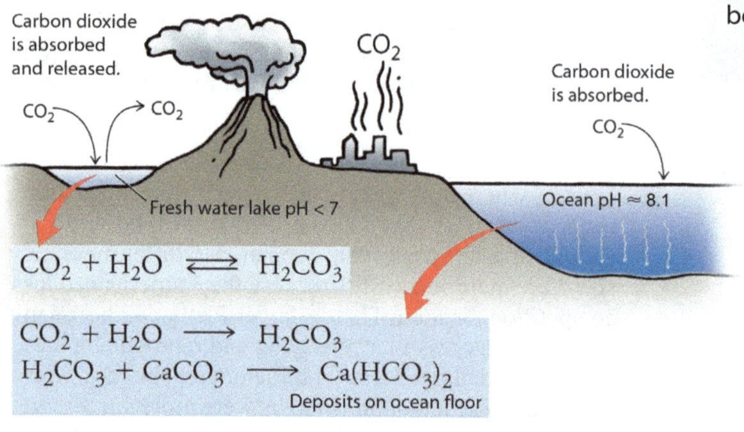

Figure A

Carbon dioxide forms carbonic acid upon entering any body of water. In fresh water, this reaction is reversible, and the carbon dioxide is released back into the atmosphere. In the alkaline ocean, the carbonic acid is neutralized into such compounds as calcium bicarbonate, $Ca(HCO_3)_2$, which precipitate to the ocean floor. As a result, the oceans are a net CO_2 absorber.

The movement of our CO_2 into the oceans, however, comes at a cost. Notice from Figure A that the CO_2-derived carbonic acid, H_2CO_3, reacts with calcium carbonate, $CaCO_3$. The addition of carbon dioxide into our oceans, therefore, has the effect of decreasing the amount of calcium carbonate in the ocean water, as well as other carbonates, such as magnesium carbonate, $MgCO_3$. Coral, shelled organisms, and many other marine species, however, use these carbonates to build and maintain their bodily structures. Importantly, they can only do this when the ocean water is saturated with these minerals, which is currently the case in most surface waters.

CONCEPT **CHECK**

How does adding CO_2 to the oceans cause a decrease in the concentration of carbonate ions, CO_3^{2-}?

CHECK YOUR ANSWER

As per Figure A, the CO_2 reacts with water, H_2O, to form carbonic acid, H_2CO_3, which effectively removes the carbonate ion, CO_3^{2-}, by reacting with it to form the bicarbonate ion, HCO_3^-.

The addition of CO_2 leads to ocean water *unsaturated* with carbonates. In regions where this occurs, the carbonate-based creatures begin to dissolve, which means they perish. This, in turn, can have a significant impact throughout the marine ecosystem, which would affect us as well. Consider this one example: pink salmon off the southern coast of Alaska live on a diet of sea snails, also known as pteropods, which are dependent on carbonate minerals. The destruction of this pteropod population by ocean acidification would also mean an end to the Alaskan pink salmon fishing industry. The immediate reality of this chemistry can be seen by pouring vinegar onto tropical sands, such as described in **Figure B**.

Is the effect of human-produced CO_2 on ocean pH measurable? Absolutely, and it is significant. Over the past 100 years, increases in atmospheric carbon dioxide—primarily due to the burning of fossil fuels—has lowered the average pH of the ocean by about 0.1 pH unit, which is about 1.26 times an increase in acidity. On a geologic timescale, this is lightning fast. The last comparable decrease in ocean pH occurred about 56 million years ago. At that time, major increases in atmospheric carbon dioxide caused the pH of the ocean to decrease by about 0.45 pH units. These changes, however, took place over 5000 years for an average decrease in ocean pH of about 0.01 pH units per century. The result was a huge die-off of marine organisms, which can be seen as a distinct layer of brown sediment within core samplings of the ocean floor (**Figure C**).

READING CHECK

When was the last decrease in ocean pH that would have been comparable to what is taking place today?

We are on track for surpassing the CO_2 event of 56 million years ago at a rate of about ten times as fast. This is indicated on the next page in **Figure D**, which shows the past concentration of atmospheric carbon dioxide. Of note, the most recent levels now exceed 400 ppm, which does not appear to have happened for over the past 800,000 years. You need to go back millions of years to find a comparable level. Our modern times where the human population has quickly eclipsed 8 billion is most unique. We should all understand that our production of copious amounts of carbon dioxide is a two-fold problem. One is the potential for dramatic changes in global climate. The other is a very predictable change in ocean chemistry. Both need to be considered carefully.

To Life!

Figure B

Tropical beaches tend to be made of crunched-up calcium carbonate containing bits and pieces of nearby coral reefs. The acetic acid in vinegar reacts with the calcium carbonate of these sands to produce gaseous carbon dioxide. Pour the vinegar onto these topical sands, and you'll see much bubbling ensues. The oceans and creatures living within the oceans are most sensitive to pH.

Figure C

Sediments that collect on the ocean floor are a physical record of current climate conditions. Because sediments build up over time, deeper sediments are a record of past climate conditions. Core samples from the ocean floor thus help in the study of ancient climates. Shown here is the ocean floor core repository of the Alfred Wegener Institute.

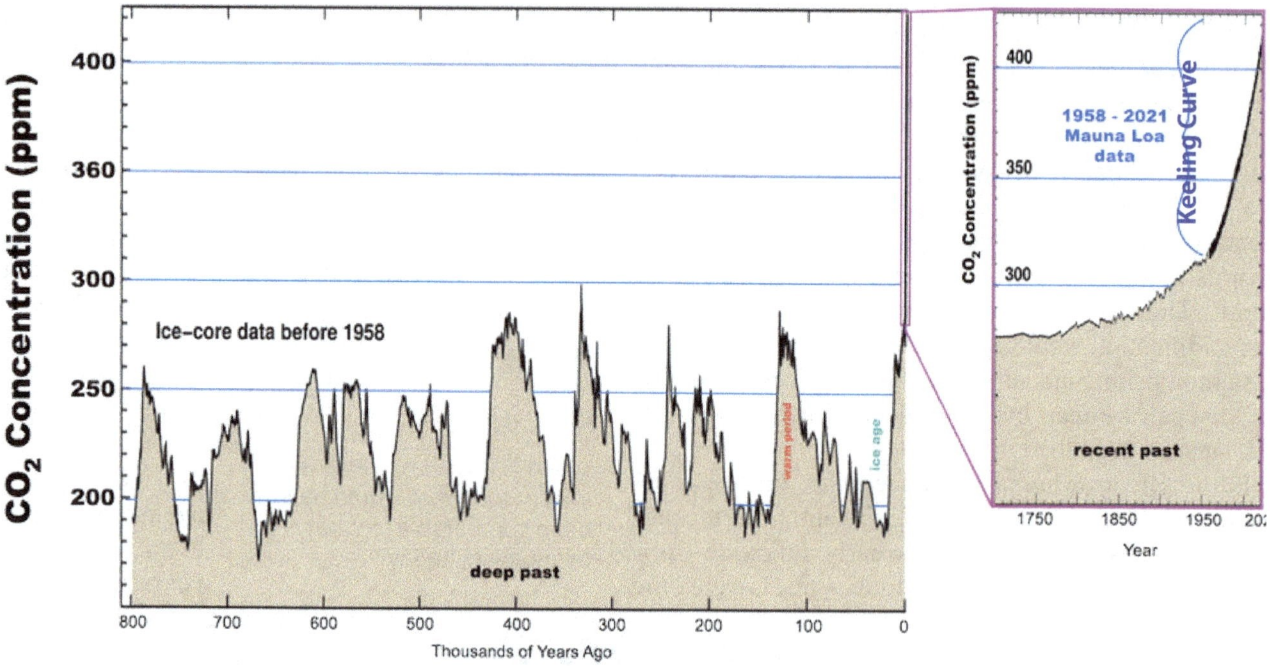

Λ Figure D

These graphs depict the levels of atmospheric carbon dioxide dating back 800,000 years ago as measured by ice core samples, and, more recently, via direct measurements, such as from the Mauna Loa Weather Observatory in Hawaii. Researchers at the Mauna Loa Weather Observatory have recorded increasing concentrations of atmospheric carbon dioxide since they began collecting data in 1958. Data from Mauna Loa forms a graph known as the Keeling curve, after the scientist Charles Keeling, who initiated this project and first noted the trends. (Source: keelingcurve.ucsd.edu)

Think and Discuss

1. How does burning fossil fuels lower the pH of the ocean?

2. How might warmer oceans have the effect of slowing down ocean acidification?

3. Scientists have experimented with ways of enhancing the ocean's ability to absorb atmospheric carbon dioxide. Adding powdered iron to a small plot of the ocean, they found, has the effect of fostering the growth of microorganisms that enhance the rate at which carbon dioxide is absorbed. Might this be a solution to the problem of global climate change? Why or why not?

CONCEPT **CHECK**

Why are atmospheric levels of carbon dioxide not rising as rapidly as might be expected based on the increased output of carbon dioxide resulting from human activities?

CHECK YOUR ANSWER

The ocean absorbs the CO_2 where it is neutralized and is not re-released.

Author Responses to Think and Discuss

1. When fossil fuels burn they react with the oxygen, O_2, in the air to form carbon dioxide, CO_2, and water vapor, H_2O. The carbon dioxide reacts with atmospheric moisture to form carbonic acid, which lowers the pH of rainwater. As the acidic rainwater precipitates into the ocean, it neutralizes alkaline ocean salts, such as calcium carbonate. This, in turn, lowers the pH of the ocean.

2. The warmer the ocean, the lower the solubility of any dissolved gases such as carbon dioxide, CO_2. Less CO_2 would be absorbed, which would slow the rate of acidification.

3. Adding iron to the ocean to enhance its carbon dioxide absorbing power my help to lower the amount of carbon dioxide in the atmosphere. At the same time, however, it might drastically alter the ocean's ecology. For example, the excess carbon dioxide would lower the pH of the ocean. Shelled marine organism would die, which could affect the whole food chain. We could end up trading one problem for another.

Chapter 18

Ecosystems

THE MAIN IDEA

An ecosystem consists of all the organisms that live in a specific area and all the abiotic features of their environment.

18.1 Terrestrial Biomes
18.2 Aquatic Biomes
18.3 Biogeochemical Cycles
18.4 Energy Flow in Ecosystems
18.5 Ecological Succession

Earth is host to numerous ecosystems, both on land and in the water. Examples of land biomes include savannas, tropical forests, and deserts. Examples of aquatic biomes include freshwater lakes and saltwater oceans. What do all ecosystems have in common? How do they differ? How do the living and non-living parts of an ecosystem interact? And what is the general flow of chemicals and energy within not just one ecosystem but all of them combined?

18.1 Terrestrial Biomes

An ecosystem consists of all the organisms that live in a specific area, including all the plants, animals, and microscopic organisms, as well as all the abiotic features of their environment. Earth has many different kinds of ecosystems, and each is occupied by characteristic suites of organisms. These types of ecosystems are referred to as **biomes**. We can divide Earth's biomes into two broad categories: terrestrial and aquatic. We will look at terrestrial biomes in this section and aquatic biomes in the next section.

The land area of Earth is divided into eight major types of ecosystems, known as *terrestrial biomes* (**Figure 18.1**). Each terrestrial biome is occupied by specific types of biological communities and, most importantly, by specific types of plant life.

The same terrestrial biome is often found in completely different parts of the world. For example, there are tropical forests—with their tall trees, dense vegetation, and astounding biodiversity—in the Amazon region of South America, in Southeast Asia, and in eastern and western Africa.

Table 18.1 Weather patterns for various terrestrial biomes.

The type of terrestrial biome found in a certain habitat depends mainly on climate, including factors such as temperature, rainfall, and the degree of seasonal variation. As a result, *latitude* (distance from the equator) and *elevation* (height above sea level) are major influences on the distribution of biomes on Earth. For example, sunlight strikes Earth most directly at the equator, warming the air there and causing it to rise. When the rising air cools, it releases a great deal of moisture as rain. This is why tropical forests, also known as rainforests, are found near the equator. Let's now survey Earth's biomes, which we summarize in Table 18.1.

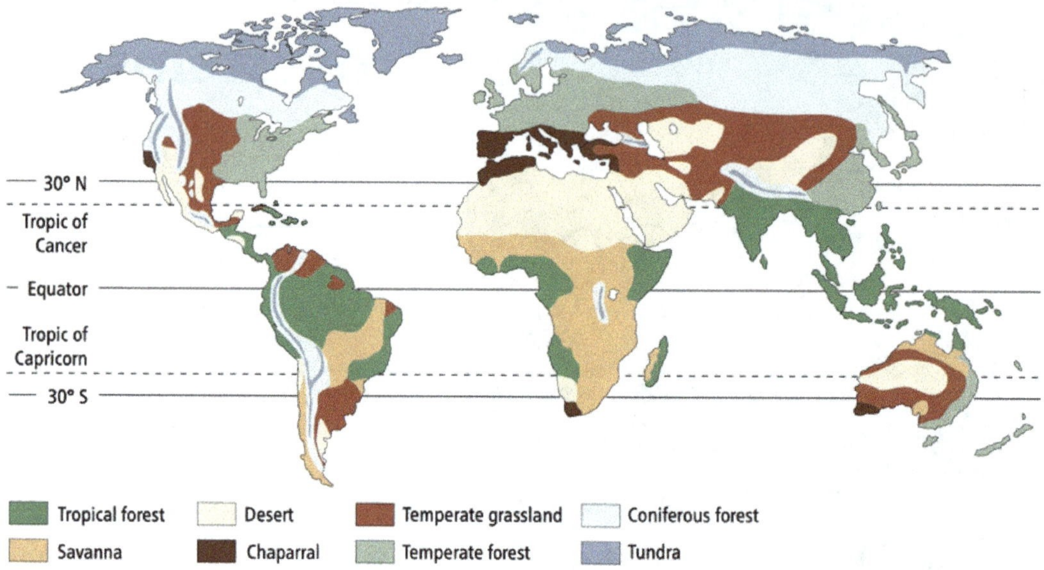

^ Figure 18.1

Earth's terrestrial biomes. Each biome is associated with a specific climate and with specific types of plant life.

Table 18.1 Weather patterns for various terrestrial biomes.

Terrestrial Biome	Winter	Summer	Precipitation (annual in cm)
Tropical forest	none	year-long	wet (200 – 400 cm)
Temperate forest	cold	warm	moderate (75 cm – 150 cm)
Coniferous forest	long	short	snow (40 – 100 cm)
Tundra	extreme	brief	very dry (15 - 25 cm)
Savanna	none	year-long	moderately dry (75 – 90 cm)
Temperate grassland	cold	hot	moderately dry (50 – 90 cm)
Chaparral	mild/rainy	hot and dry	seasonal (40 – 100 cm)
Desert	dry	dry	year-long dry (<50 cm)

Tropical forests (**Figure 18.2a**), sometimes called *rainforests*, are found close to the equator. Temperatures are warm and fairly constant throughout the year. Tropical forests receive between 200 and 400 centimeters of rain a year, distributed between a wet season and a dry season. Tropical forests are famous for their biodiversity—more species are found in this biome than in all other biomes combined. A single square kilometer of tropical forest may contain as many as 100 different species of trees. The tallest trees form a dense canopy, shading the ground. Tropical forests have little leaf litter because all organic material is quickly decomposed. Most of the nutrients present in tropical forests are found in the organisms that live there, and the soil tends to be poor. Tropical forests are rapidly being destroyed worldwide for timber and agriculture. Unfortunately, the poor quality of the soil means that forest that has been cleared for farmland can support crops for only a few years.

(a) (b) (c) (d)

^ Figure 18.2

(a) This tropical forest in Thailand shows the density and diversity of plant life found in all tropical forests. (b) In this temperate forest in New England, the season is autumn, and the trees are shedding their leaves. (c) A coniferous forest in Denali National Park, Alaska. (d) Tundra in Denali National Park, Alaska. Notice the absence of trees in the tundra.

Temperate forests (**Figure 18.2b**) are found in areas that have four distinct seasons, including a warm growing season and a cold winter. Temperate forests receive between 75 and 150 centimeters of rainfall per year. There are usually three or four different species of trees per square kilometer. Some well-known temperate forest trees are elm, oak, beech, and maple. The leaves of temperate forest trees are unable to survive the freezing temperatures of winter. Consequently, these trees are *deciduous*—they drop their leaves in the autumn. Trees recover nutrients from their leaves before they are shed. The soil of temperate forests is fertile, and these forests make good farmland. In fact, many temperate forests have been cut down to make way for agriculture.

Coniferous forests (**Figure 18.2c**), or evergreen forests, are found in areas that have long, cold winters and short summers. Coniferous forests are relatively dry, receiving about 40 to 100 centimeters of precipitation per year. Precipitation falls mainly as snow. As their name suggests, coniferous forests are dominated by conifers such as pine, spruce, and fir. The needle-like leaves of conifers are covered with a thick layer of wax and contain special substances that keep them from freezing during the winter. In coniferous forests, the ground is usually covered with shed needles, and the soil is poor in nutrients. Many coniferous forests are threatened by logging.

Tundra (**Figure 18.2d**) is found in areas that experience extreme cold and little precipitation, averaging about 15 to 25 centimeters per year. One of the defining features of tundra is a layer of *permafrost*, or permanently frozen subsoil, beneath the topsoil. The word tundra comes from the Finnish word tunturia, meaning "treeless plain." Trees cannot survive in tundra because of permafrost and the short growing season. Plant life in the tundra includes low shrubs, lichens, mosses, grasses, and flowers. Plants are often found in close clumps as a defense against cold. The tundra has relatively low biodiversity.

READING CHECK

From where is the word "tundra" derived?

^ Figure 18.3

(a) A savanna in Serengeti National Park, Tanzania. (b) A temperate grassland in South Dakota. (c) Chaparral in California. (d) The Sonoran Desert, Arizona.

Savannas (**Figure 18.3a**) are tropical grasslands with a warm climate and a long dry season. Savannas are covered with grass and occasional scattered trees. Plants that live in the savanna have long roots for coping with drought. Trees also have a thick bark that helps them survive the fires that occur during the dry season. Fires help maintain savannas by preventing tree growth. Living organisms also help maintain savanna habitats—for example, elephants eat and kill trees, and humans burn forests for agriculture. Without this activity, some savannas would grow into tropical forests. Savannas receive a moderate amount of rain, usually between 75 and 100 centimeters each year.

Temperate grasslands (**Figure 18.3b**) are found in areas that have four distinct seasons, including a hot summer and a cold winter. Compared to savannas, temperate grasslands show more temperature variation and receive less rainfall—about 50 to 90 centimeters per year. Several factors prevent trees from taking over the habitat and turning it into temperate forest, including relatively low levels of precipitation, seasonal drought, fire, and grazing. The soil of temperate grasslands is extremely fertile, making for ideal farmland. *Chaparral* (**Figure 18.3c**) is found in places that have mild, rainy winters and hot, dry summers characterized by drought and fire. Chaparral habitats are dominated by small trees and shrubs, many of which have small waxy leaves that help retain moisture. Chaparral plants also have extensive root systems that help them survive the summer drought.

Deserts (**Figure 18.3d**) are habitats that receive very little precipitation, usually less than 50 centimeters per year. Although deserts are usually associated with hot climates, cold deserts also exist. In cold deserts, precipitation may fall as snow. Antarctica, the coldest and driest of the continents, is a desert. Many of the world's largest hot deserts, including the Sahara and

Australian deserts, are found in a band around 30 degrees north latitude and 30 degrees south latitude. Desert plants usually have special adaptations for living in dry conditions, such as extensive root systems and the ability to store water. Desert soils are usually abundant in nutrients. This rich soil, combined with a long growing season and plentiful sunshine, makes deserts productive farmland if water can be brought in through irrigation.

CONCEPT **CHECK**

Name three biomes where seasonal fires play an important role. Why are fires important in these biomes?

CHECK YOUR ANSWER

Fire plays a role in savannas, temperate grasslands, and chaparral. In these habitats, they favor the growth of characteristic plants, in part by preventing the growth of more trees.

More to Explore

To read more about terrestrial biomes, check out the websites below:

https://www.nationalgeographic.org/encyclopedia/biome/

https://earthobservatory.nasa.gov/biome

18.2 Aquatic Biomes

Life likely originated in the oceans, and many species continue to make use of aquatic habitats in both freshwater and saltwater.

Freshwater Habitats

Freshwater habitats include lakes, ponds, and rivers. The still waters of lakes and ponds vary tremendously in size and biodiversity. Nonetheless, all can be divided into four major zones, as illustrated in **Figure 18.4**. Habitats close to the water surface and to shore are part of the *littoral zone*. The littoral zone is relatively warm because of its exposure to sunlight. Organisms that live in the littoral zone include photosynthetic plants and algae, insects, mollusks, crustaceans, fishes, amphibians, ducks, and turtles. The *limnetic zone* includes habitats that are close to the water surface but far from shore. The limnetic zone is occupied mainly by **plankton**, organisms that float in the water rather than swimming actively through it.

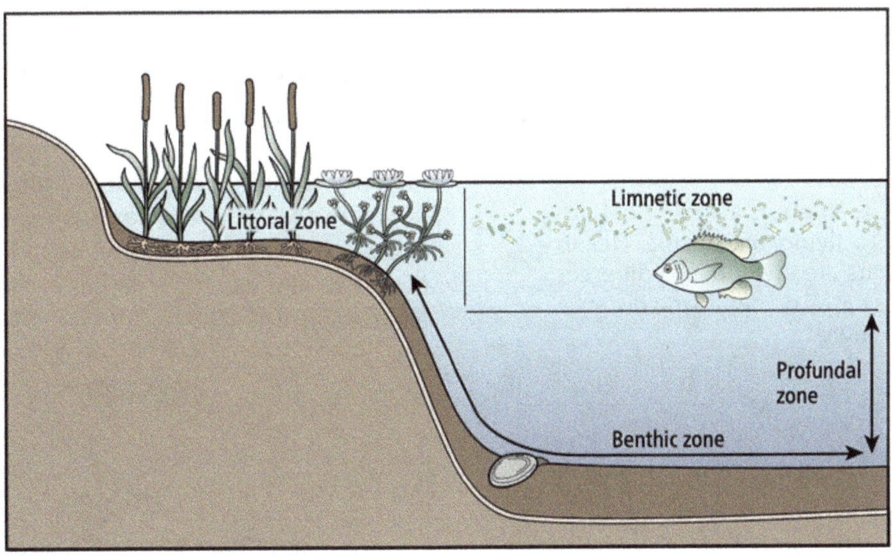

Λ Figure 18.4

Lakes and ponds can be divided into different zones. Each zone is home to distinct kinds of living organisms.

^ Figure 18.5

An estuary in southern Florida.

Phytoplankton are photosynthesizing plankton—they are the main producers in lakes and ponds. Zooplankton are heterotrophic plankton. In some lakes and ponds, fish also live in the limnetic zone.

The *profundal zone* includes the deep-water habitats in ponds and lakes. Most organisms in the profundal zone consume organic debris that drifts down from above. The *benthic zone* describes habitats on the lake bottom. This zone is home to mollusks, worms, microorganisms, and other living things.

Aquatic species that live in the flowing waters of rivers and streams usually have adaptations that keep them from being washed away. Some species have hooks or suckers for attaching to rocks. Others are strong swimmers. As in lakes and ponds, photosynthesizing plankton are the main producers in rivers and streams.

Estuaries are habitats where freshwater rivers flow into oceans. The plants found in estuaries, such as marsh grasses and mangroves (**Figure 18.5**), have adaptations for dealing with changing salinity conditions. Estuaries are also home to many fishes, invertebrates, and birds and are essential nursery habitats for many fishes and invertebrates.

Saltwater Habitats

The oceans offer a wide range of habitats to living organisms, as shown in **Figure 18.6**. The largest habitat, by far, is the pelagic zone, which includes all ocean waters. The pelagic zone is further divided into two parts. The *photic zone* is close to the water's surface and receives enough sunlight to power photosynthesis. The majority of pelagic species are found there. The deeper *aphotic zone* receives little sunlight and is much more limited in food availability and biodiversity.

Pelagic species use one of two modes of locomotion. Plankton floats wherever waters currents take them. Plankton include living things such as diatoms, dinoflagellates, and the larvae of animals such as clams, lobsters, and sea urchins. The phytoplankton, or photosynthetic plankton, are the main producers in oceanic food chains. Nekton swims actively through the water. Fish and sea turtles are examples of nekton. Humans aren't aquatic organisms, of course, but when we do get in the water, we can behave like plankton or nekton (**Figure 18.7**).

^ Figure 18.6

The ocean offers a variety of habitats to living organisms. Ocean habitats are divided into different zones based on depth and distance from shore.

Figure 18.7 >

In the water, humans can behave like nekton or plankton.

< Figure 18.8

Diverse organisms, including starfish, sea anemones, and sea urchins, occupy this tide pool.

Most marine species actually live in the *benthic zone* on the ocean bottom. Benthic species may live on the surface of the ocean bottom, like lobsters, or burrow into the sediment, like clams and worms.

Ocean habitats also can be categorized based on their closeness to shore. The *intertidal zone* is closest to the shore. As the tide moves in and out, the intertidal zone alternates between being submerged underwater and exposed to air. Intertidal species, such as barnacles, sea anemones, and starfish, must be able to deal with exposure to air, temperature fluctuations, and waves (**Figure 18.8**). Special adaptations allow them to live in this environment. Many intertidal species have thick shells or hide in crevices to keep from drying out. In addition, all intertidal species can attach firmly to rocks or other surfaces so that they do not get washed up onto the beach. The *neritic zone* describes underwater marine habitats near the coasts. Coral reefs, the most diverse marine ecosystems on Earth, are found in the neritic zone. Habitats in the neritic zone contain high levels of nutrients that wash into the water from land. Because of this, the neritic zone is richer in life than the regions of the ocean further from land, known as the *oceanic zone*.

READING CHECK

What type of species actively swims through the water?

CONCEPT CHECK

1. Scallops live on shallow seafloors. They move by quickly opening and shutting their shells, shooting jets of water that propel them in spurts. Are scallops pelagic or benthic? Are they nekton or plankton?

2. The flounder has a very unusual feature—one of its eyes migrates from one side of the body to the other side during development. It ends up with two eyes on the same side! What type of aquatic habitat do you think the flounder occupies?

CHECK YOUR ANSWERS

1. Scallops are benthic nekton.

2. Flounders are benthic fish that lie flat on the seafloor. One eye migrates from the "underside" of the body to the "top" side, where it is of more use.

More to Explore

To learn more about aquatic biomes:

https://www.nationalgeographic.org/encyclopedia/aquatic-biome/

https://ucmp.berkeley.edu/exhibits/biomes/freshwater.php

https://ucmp.berkeley.edu/exhibits/biomes/marine.php

18.3 Biogeochemical Cycles

A water molecule in a cell on your cheek once bobbed on the Indian Ocean. A carbon atom on the skin over your knee once floated in the cold air above Antarctica. A nitrogen atom flowing through your veins was once part of Albert Einstein. Parts of you have been all over the world, in places you have never seen with your own eyes, even in important historical figures! How is this possible?

Living organisms are made up of many substances, including water, carbon, nitrogen, phosphorus, sodium, calcium, sulfur, chlorine, and many others. All these substances move around Earth in a series of **biogeochemical cycles**, going back and forth between the tissues of living organisms and the abiotic world. The word *biogeochemical* emphasizes that substances cycle between living organisms (*bio*) and Earth (*geo*)—specifically, Earth's atmosphere, crust, and waters.

As atoms and molecules move through their biogeochemical cycles, they pass through a series of *reservoirs*—that is, places where they stay for a period of time. This is shown in **Figure 18.9**, which shows the water cycle. In this cycle, a water molecule may sit in one reservoir, the ocean, until evaporation moves it to another reservoir, the atmosphere. It may then fall as precipitation into a third reservoir, a lake. The average amount of time a molecule spends in each reservoir—for example, the average amount of time a water molecule spends in the ocean before it moves to another reservoir—is known as its *residence time* in that reservoir. Although every substance used by living organisms has a biogeochemical cycle, we will focus on three of the most important: water, carbon, and nitrogen.

Water

All living things need water. About 98 percent of the water on Earth is found in oceans, rivers, and lakes. The rest—about 2 percent—is found in glaciers, in the polar ice caps, in soil—and in living organisms. The global water cycle, shown in Figure 18.9, describes how water moves around Earth. Water evaporates from the oceans into the atmosphere, is moved around the atmosphere by winds, and then falls back to Earth as rain or snow. Water that falls on land flows back to the oceans through rivers, streams, and groundwater.

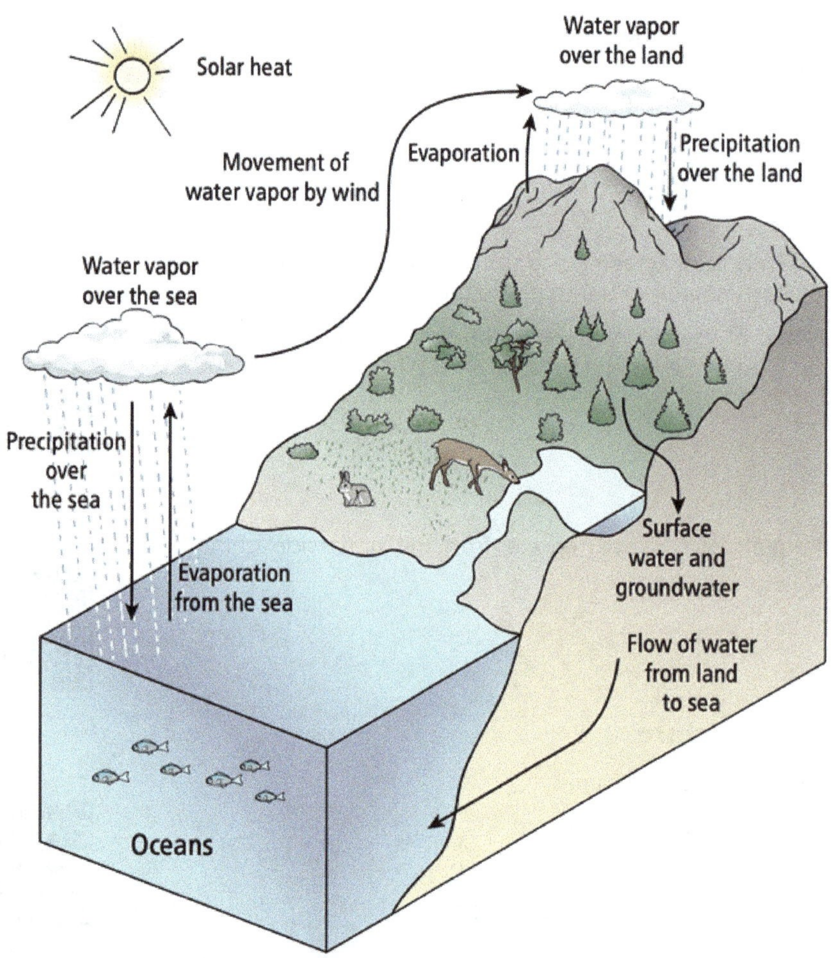

< Figure 18.9

Water cycles through the oceans, atmosphere, and land. Some water enters the bodies of living organisms.

Water enters the biotic world when it is absorbed or swallowed by organisms. Some of this water may pass up the food chain. The rest is returned to the abiotic environment through processes such as respiration, perspiration, excretion, and elimination. Plants also lose water when it evaporates inside leaves and then escapes through the stomata to the atmosphere. Did you know that every time you exhale, you return water to the atmosphere? If you breathe onto a cool mirror, the water vapor in your breath will condense to form a foggy film of liquid water. Thus, with every breath you exhale, you are returning water molecules from the biotic to the abiotic world.

Carbon

Carbon is an essential component of all organic molecules. The carbon cycle is shown in **Figure 18.10**. ≠Much of the inorganic carbon on Earth exists as carbon dioxide and is found either in the atmosphere or dissolved in ocean waters.

Carbon enters the biotic world during photosynthesis, when plants and other producers use carbon dioxide to make the organic molecule glucose. This carbon becomes available to other organisms as it moves up the food chain. Carbon is returned to the abiotic world during cellular respiration, when organisms break down organic molecules (such as glucose) and release carbon dioxide. Carbon is also returned to the abiotic world through the process of combustion, as occurs during a forest fire.

An important part of Earth's carbon supply is found in fossil fuels such as coal and oil. Both coal and oil are formed over long time periods from the remains of dead organisms. This is why they are called "fossil" fuels. Notably, fossil fuels are generally considered biotic because they arose from living organisms.

Human burning of fossil fuels has released so much carbon into the atmosphere that atmospheric carbon dioxide levels are now higher than they have been for 15 million years. Because atmospheric carbon dioxide traps heat on the planet, Earth has warmed noticeably as a result. This warming of Earth is called *global warming*, which is a major issue facing humanity and appropriately addressed in a chemistry or environmental science course.

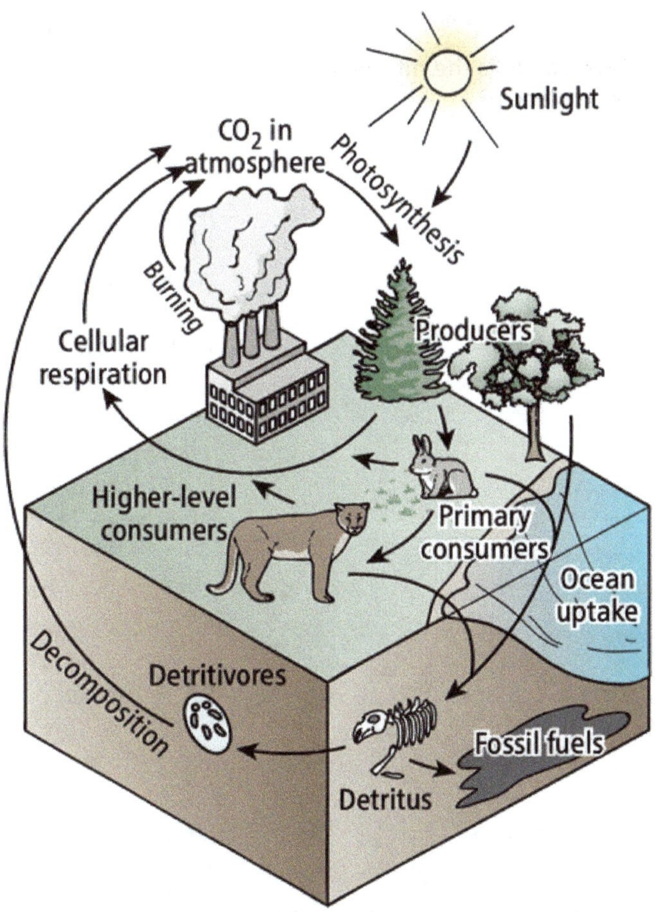

Figure 18.10 >

Carbon is found as carbon dioxide in the atmosphere and oceans. During photosynthesis, producers move carbon into the biotic world. Carbon is returned to the abiotic world as carbon dioxide, a product of cellular respiration.

Figure 18.11 >

Nitrogen is found as nitrogen gas in the atmosphere. Bacteria convert atmospheric nitrogen into forms that can be used by living organisms.

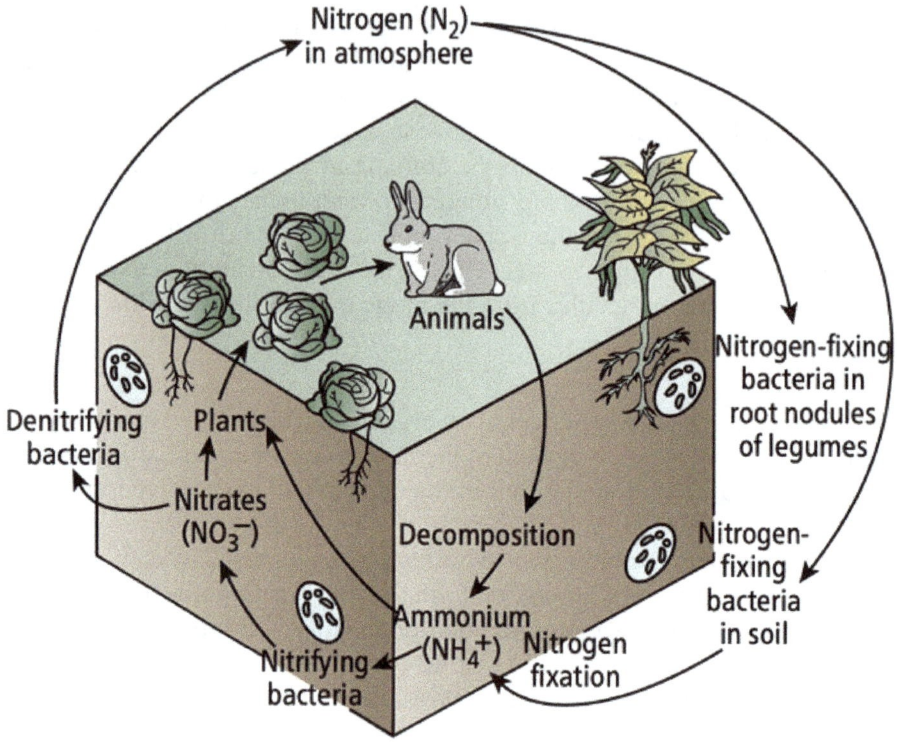

Nitrogen

Nitrogen is an essential component of amino acids, DNA, and many other organic molecules. The nitrogen cycle is shown in **Figure 18.11**. Most of Earth's nitrogen exists as nitrogen gas, N_2, in the atmosphere. However, this nitrogen is not a form of nitrogen that most living things can use. Why not? Because N_2 is an amazingly stable molecule, which means it is rather inert and not easy to transform through chemical or biochemical processes. Certain bacteria, however, have evolved to overcome this barrier.

READING CHECK

How stable is the nitrogen, N_2, molecule?

Nitrogen-fixing bacteria in soil convert nitrogen gas, N_2, into ammonium ions, and then *nitrifying bacteria* convert ammonium ions into nitrates. Plants absorb nitrogen primarily in the form of nitrates, although they make some use of ammonium ions as well. Nitrogen then moves from plants up the food chain. Nitrogen returns to the abiotic environment when it is converted back into nitrogen gas by denitrifying bacteria.

Certain plants, including legumes such as peas, beans, clover, and alfalfa, have evolved a mutualistic relationship with nitrogen-fixing bacteria. These plants shelter the bacteria in special nodules in their roots, and the bacteria provide the plants with nitrogen (**Figure 18.12**). This is why farmers often grow legumes with other plants:

> The excess nitrogen enriches the soil and allows other crops to grow better. These days, however, not all plants rely on bacteria for nitrogen—industrial processes have been invented for making nitrogen-rich fertilizers. This has allowed humans to grow more food and to support larger human populations, but industrial fertilizers also pollute and damage natural environments.

< Figure 18.12

Legumes house symbiotic nitrogen-fixing bacteria in special root nodules like those shown here.

CONCEPT **CHECK**

What living things do you, as a human, rely on to obtain (a) water, (b) carbon, and (c) nitrogen?

CHECK YOUR ANSWERS

a. You don't have to rely on other living things to obtain water—you can drink it directly!

b. All the carbon in your body comes ultimately from plants and other photosynthesizers. This carbon comes to you through the food you eat.

c. The nitrogen in your body also comes through the food you eat. Nitrogen-fixing bacteria and nitrifying bacteria change nitrogen into a form that plants can use, and then you eat the plants (or the animals that eat the plants).

More to Explore

You can learn more about the water cycle here:

https://www.noaa.gov/education/resource-collections/freshwater/water-cycle

Read about the carbon cycle here:

https://www.noaa.gov/education/resource-collections/climate/carbon-cycle

https://earthobservatory.nasa.gov/features/CarbonCycle

And check out this website on the nitrogen cycle:

https://www.visionlearning.com/en/library/Earth-Science/6/The-Nitrogen-Cycle/98

18.4 Energy Flow in Ecosystems

All living organisms need energy to survive, grow, and reproduce. Where does this energy come from? In most ecosystems, energy comes ultimately from the Sun. Sunlight energy enters the biotic world when plants and other organisms use it to build organic molecules during photosynthesis. Photosynthesizing organisms embed about 1% of the sunlight energy that strikes them into organic matter, such as the cellulose in leaves or the starches in potatoes. Globally, this adds up quickly—it is enough to make 170 billion tons of organic material a year!

Biomass is the amount of organic matter in an ecosystem. The rate at which producers build biomass is the ecosystem's **primary productivity**. Different ecosystems vary in their primary productivity—for example, tropical forests, swamps, and coral reefs have very high primary productivity, whereas deserts and tundra have relatively low primary productivity.

Once energy enters an ecosystem as producer biomass (plant material), it goes up the food chain, creating biomass in primary consumers, secondary consumers, and so on. Does all the energy taken in by plants ultimately go into growing zebras, or lions? No. In all ecosystems, only about 10% of the energy at one level of the food chain moves up to the next level. For example, if the producers in one patch of African savanna use 10,000 kcal of energy, you would expect about 1000 kcal to be available to primary consumers (such as zebras), and 100 kcal to be available to secondary consumers (such as lions). This energy flow can be shown in an **energy pyramid**, such as the one in **Figure 18.13**.

Figure 18.13 >

This energy pyramid shows that the amount of available energy decreases as you move up the food chain.

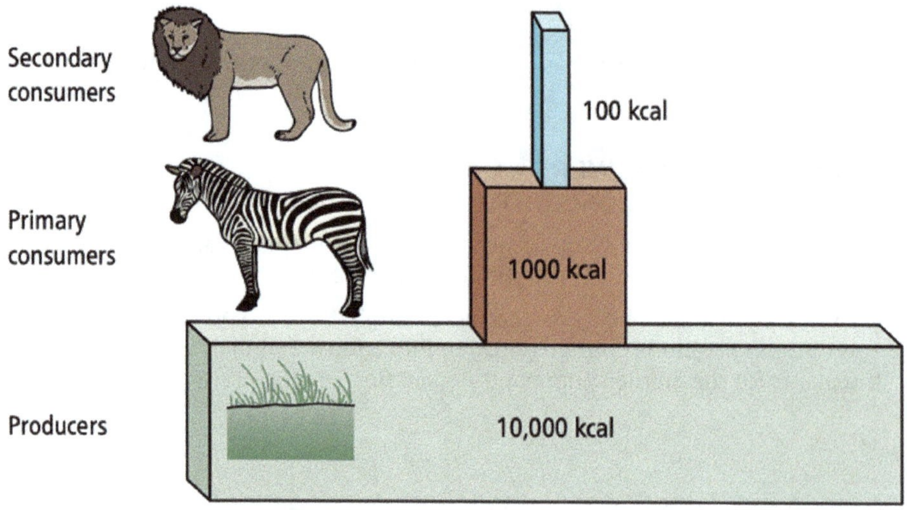

The shape of the pyramid shows the decrease in energy as you go up the food chain.

Why does only 10% of the energy at one level of the food chain move to the next level? What happens to the other 90%? First, some organisms at one level of the food chain are not eaten by organisms at the next level of the food chain. For example, some savanna plants are not eaten by zebras, and some zebras are not eaten by lions. If an organism is not eaten, then the energy in it cannot move up the food chain. Second, not all the energy in an organism's food goes into building biomass—and only the energy that is converted into biomass can move up the food chain. For example, when a zebra eats a plant, some of the energy in the plant is lost in feces and some energy goes into maintenance. Feces contains organic matter that the zebra can't digest. Maintenance is the energy the zebra needs to live—the energy it takes to find food, run, breathe, heat its body, and so on. By the time feces and maintenance have taken their share of energy, only a fraction is left for growth and reproduction—for building new biomass (**Figure 18.14**).

READING CHECK

Why does only 10% of the energy at one level of the food chain move to the next level?

The energy pyramid explains why the higher levels of a food chain have less biomass—why there is more grass than zebras, and more zebras than lions: There is less energy available to organisms that eat higher in the food chain, and fewer of them can be supported. The energy pyramid also explains why the number of levels in a food chain is limited: There just isn't enough energy to support animals that eat at even higher levels. For example, there is no higher-level consumer that feeds exclusively on lions.

^ Figure 18.14

The energy an organism takes in from food goes to three things: feces, maintenance, and growth and reproduction. Only growth and reproduction contribute to building biomass, and only energy that is converted into biomass can move up the food chain.

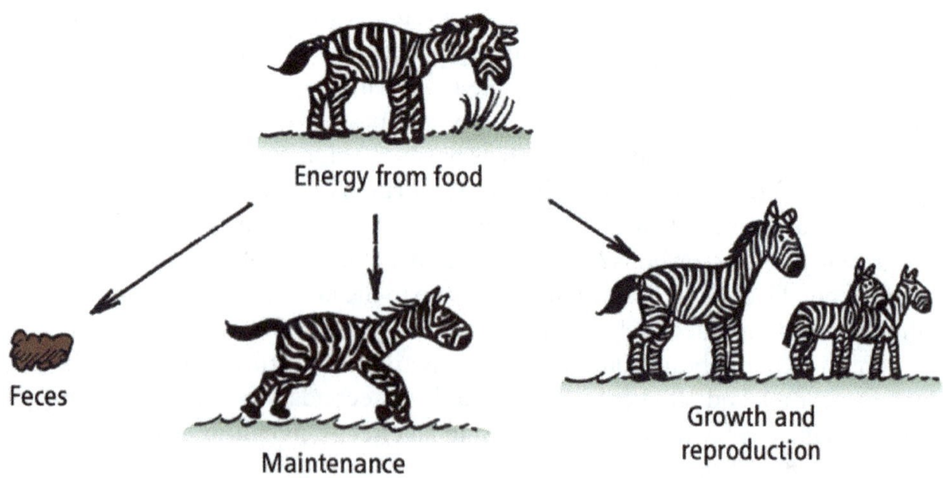

CONCEPT CHECK

1. The producers in a patch of pine forest use about 30,000 kcal of energy. Assuming that 10% of the energy at one level of the food chain moves up to the next level, how much energy is available to the primary consumers in the habitat? To the secondary consumers?

2. Does energy move around Earth in cycles the way water, carbon, and nitrogen do?

CHECK YOUR ANSWERS

1. First, the energy available to primary consumers is = 10%*(30,000) kcal = 3000 kcal.
Then, the energy available to secondary consumers is = 10%*(3,000) kcal = 300 kcal.

2. No, energy flows through ecosystems. Energy comes to Earth from the Sun. A fraction of this energy is captured by producers such as plants. Some of the energy captured by plants is used to build biomass, and some of this energy moves up the food chain. The rest is lost as heat. All living things lose energy to the environment as heat as a result of their activities. All of the energy Earth receives from the Sun is eventually lost to the environment as heat. It can't be recycled.

More to Explore

To read more about energy flow through an ecosystem:

https://www.nationalgeographic.org/topics/resource-library-energy-flow-through-ecosystem/?q=&page=1&per_page=25

18.5 Ecological Succession

In 1883, a huge volcanic explosion tore through the tropical island of Krakatoa. Most of the island was destroyed. The fragment that remained was covered with ash, barren and lifeless. But within three years, a few living organisms, primarily photosynthetic bacteria and ferns, had colonized the island. By 1897, Krakatoa had become a savanna covered with tall grasses. Only a few species of animals were present—almost all of these were flying insects and birds. By 1919, scattered trees could be seen, and by 1931, a forest had developed. As more trees grew, animal diversity increased. Today, Krakatoa is once again a tropical forest.

Krakatoa's history provides an example of ecological succession. **Ecological succession** describes how the community of species living in an ecosystem changes over time. There are two kinds of ecological succession: primary succession and secondary succession.

Primary succession occurs when bare land, devoid of soil, is colonized by successive waves of living organisms. For example, primary succession may begin when new land is formed by volcanic activity or when a glacier's retreat reveals bare rock.

The first species to colonize bare land are known as *pioneer species*. Pioneer species often include photosynthetic microorganisms as well as larger organisms such as lichens and mosses. Pioneer species must be able to survive with few nutrients and little existing organic matter. They also must be able to deal with direct sunlight and the variable temperatures that result from lack of cover. Pioneer species may be succeeded by grasses, shrubs, and finally trees. At each stage of succession, the activities of earlier waves of colonizers build up nutrients and organic matter, allowing later colonizers to thrive. Later colonizers eventually out compete and displace earlier colonizers. Ecological succession ends with a **climax community** that is relatively stable (**Figure 18.15**). During the process of succession, the total biomass

Figure 18.15 >

Primary succession after a glacier's retreat in Alaska: Bare rock was covered by mosses and fireweeds, then shrubs, an alder forest, a spruce forest, and finally, a spruce and hemlock forest. The entire process took place over about 200 years.

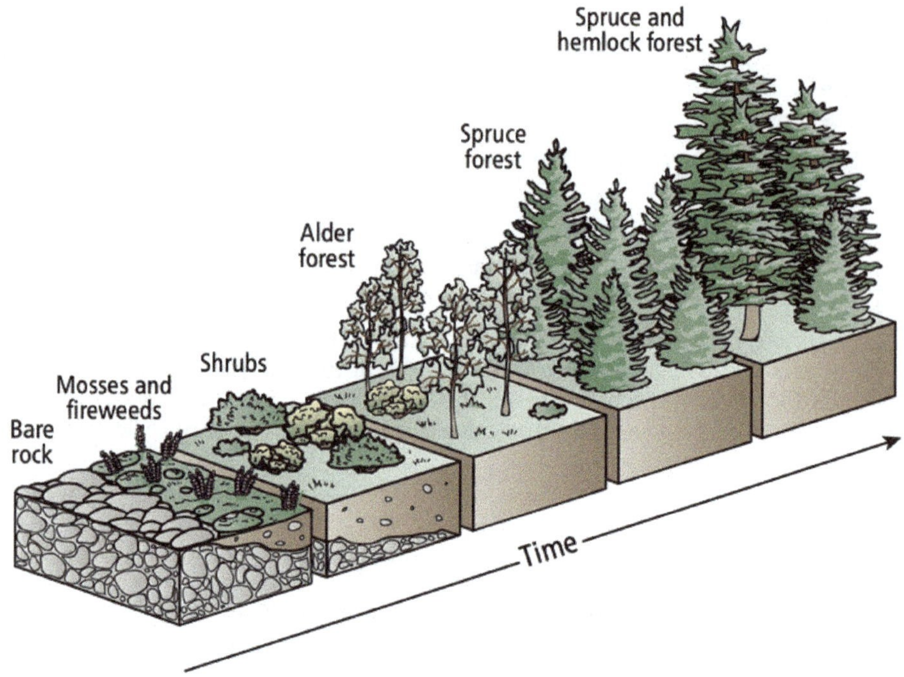

of the ecosystem typically increases, as does the number of species present. *Secondary succession* occurs when a disturbance destroys existing life in a habitat but leaves the soil intact. Secondary succession may begin after a fire or when old farmland is abandoned. Because soil is already present, secondary succession proceeds more quickly than primary succession. As with primary succession, biomass and biodiversity typically increase as secondary succession progresses. For example, during secondary succession on abandoned farmland in the southeastern United States, the number of bird species increased as succession progressed (**Figure 18.16**).

Figure 18.16 >

In secondary succession on abandoned farmland in the southeastern United States, empty fields progressed from grassland to shrubland, pine forest, and hardwood forest over the course of more than 100 years. Greater biodiversity—for example, more species of birds—was seen as succession progressed.

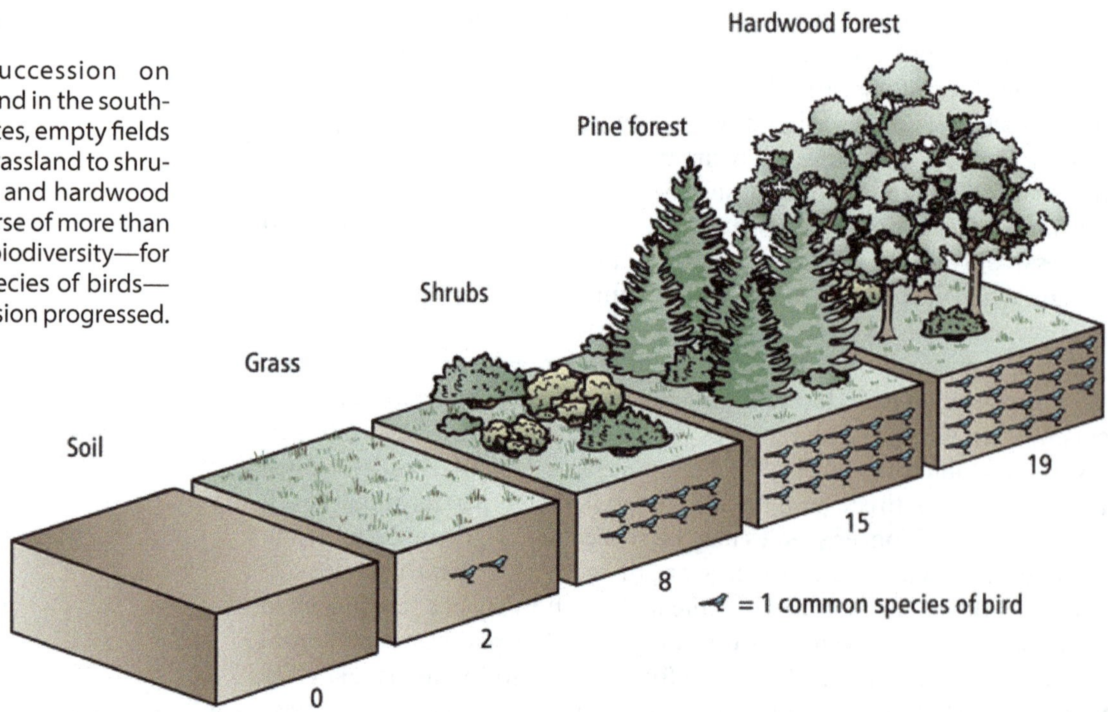

All ecosystems experience change. Most ecosystems never suffer a cataclysm like Krakatoa's, but smaller disturbances, often affecting only a small area of habitat, are common. Fires, floods, and other natural events all damage habitats. Even the fall of a large tree can be a disturbance with significant consequences. In tropical forests, for example, tree falls create holes in the canopy that make sunlight—a rare commodity— suddenly available to organisms in the understory.

According to the *intermediate disturbance hypothesis*, regular disturbances can actually contribute to biodiversity, as long as they are not too extreme. This is because an intermediate level of disturbance guarantees that there will always be habitat at varying stages of recovery, and different species make use of these different habitats. Of course, some disturbances have more lasting effects. Invasive species can devastate ecosystems, as we saw with the brown tree snake in Guam and the zebra mussel in the eastern United States. Human activity can convert one type of ecosystem into another, as when land is cleared for agriculture or for development. Finally, changes in climate, including global warming, can bring big changes to many ecosystems at the same time.

READING CHECK

How can a disturbance of an ecosystem contribute to biodiversity?

CONCEPT CHECK

How might global warming cause ecosystems to change?

CHECK YOUR ANSWER

This is a difficult question that many scientists are trying to answer. As Earth gets warmer, entire ecosystems are likely to shift to areas that have more appropriate climates. Usually, this means higher latitudes (locations farther away from the equator) and higher elevations. However, many organisms have no appropriate habitat to shift to. For example, polar bears already live in the Arctic and cannot search for sea ice farther north. In addition, large areas of Earth have already been developed for human use. Even if there is appropriate new habitat somewhere on a warmer Earth, there may be no way for most organisms to move through human cities or farms to reach it.

More to Explore

To read more about ecological succession, visit the following website:

http://www.countrysideinfo.co.uk/successn/index.htm

Concept Review

Chapter 18

Detailed Chapter Summary

An ecosystem consists of all the organisms that live in a specific area, as well as all the abiotic features of their environment. The major types of ecosystems are known as biomes. Earth's land area is divided into eight major terrestrial biomes. Each biome is characterized by specific types of biological communities -- particularly specific types of plant life. Plant life, in turn, typically depends on climate. For this reason, latitude and elevation are major influences on what biome is found in a particular habitat. The major biomes are: tropical forest, temperate forest, coniferous forest, tundra, savanna, temperate grassland, chaparral, and desert.

Aquatic biomes include freshwater and marine habitats. Lakes are divided into a littoral zone, limnetic zone, and profundal zone. Phytoplankton are the main producers in lakes, ponds, and rivers. Estuaries are habitats where freshwater rivers flow into oceans. Estuaries are characterized by changing salinity. Marine habitats include the pelagic zone and the benthic zone. The pelagic zone describes the water itself and consists of the photic zone near the water surface and the deeper aphotic zone. Pelagic species may be either plankton or nekton. The benthic zone is found on the ocean's bottom. Ocean habitats can also be described based on their closeness to shore. The intertidal zone is alternately underwater and exposed to air. The neritic zone describes coastal marine habitats. The oceanic zone is far from land.

Many of the resources that organisms need cycle between Earth and living things in a series of biogeochemical cycles. The water cycle involves evaporation from the oceans, movement by wind, precipitation, and a return to the oceans in rivers, streams, and groundwater. Organisms swallow or absorb water and lose it through processes such as respiration, perspiration, excretion, elimination, and evaporation.

Carbon enters the biotic world through photosynthesis. Carbon is converted back to inorganic carbon dioxide during cellular respiration, as well as through burning, such as during forest fires. The burning of fossil fuels by humans has added carbon dioxide to the atmosphere and resulted in dangerous global warming.

Nitrogen enters the biotic world after nitrogen-fixing bacteria convert nitrogen gas into ammonium ions. Nitrifying bacteria then convert ammonium ions into nitrates that can be used by plants. Nitrogen returns to the abiotic world when denitrifying bacteria turn it back into nitrogen gas. Legumes have evolved a mutualistic relationship with nitrogen-fixing bacteria that they house in special root nodules.

In nearly all ecosystems on Earth, energy comes ultimately from the Sun. Sunlight energy enters the biotic world when plants and other photosynthetic organisms use it to build organic molecules during photosynthesis. The rate at which producers build biomass is called primary productivity. From the producers, energy moves up the food chain. About 10% of the energy at one level of the food chain moves up to the next level. This percentage is rather low for several reasons. First, not all the organisms at one level of the food chain are eaten by organisms at the next level of the food chain. In addition, energy is lost in feces or goes into maintenance. This leaves only about 10% that contributes to the building of biomass (through growth and reproduction) at the next level of the food chain. The flow of energy through an ecosystem can be diagrammed using an energy pyramid.

Ecological succession describes how the community of species living in an ecosystem changes over time. Primary succession occurs when bare land (without soil) is colonized by successive waves of organisms. Secondary succession begins when existing life in a habitat is destroyed, but soil remains intact. As ecological succession proceeds, the biomass and biodiversity present in an ecosystem typically increase. The intermediate disturbance hypothesis argues that regular disturbances actually contribute to biodiversity, as long as these disturbances are not too extreme.

Key Terms

Biogeochemical cycles The movement of substances such as water, carbon, and nitrogen between living organisms and the abiotic world.

Biomass The amount of organic matter in an ecosystem.

Biome Earth's major types of ecosystems.

Climax community The stable community in which a particular habitat with which ecological succession ends.

Ecological succession Changes over time in the community of species living in an ecosystem .

Energy pyramid A diagram that shows the energy flow through an ecosystem from producers to primary consumers, secondary consumers, and any higher-level consumers.

Nekton Organisms that swim actively through their aquatic habitat.

Plankton Organisms that float passively in their aquatic habitat rather than swimming actively through it.

Primary productivity The rate at which producers build biomass in an ecosystem.

Review Questions

Simple Review Questions

18.1 Terrestrial Biomes

1. What types of living things do scientists use to classify Earth's terrestrial habitats into biomes?
2. What climatic factors help determine the locations of different biomes?
3. Which biome includes more living things than all other biomes combined?
4. Why are fires important in savannas?

18.2 Aquatic Biomes

5. Describe the photic zone of ocean habitats.
6. What are some of the challenges that organisms face in the intertidal zone? What adaptations do intertidal organisms have for dealing with these challenges?

18.3 Biogeochemical Cycles

7. What is a biogeochemical cycle?
8. How does carbon enter the biotic world? How is it returned to the abiotic world by organisms?
9. What role do nitrogen-fixing bacteria and nitrifying bacteria play in the nitrogen cycle?

18.4 Energy Flow in Ecosystems

10. All organisms need energy in order to grow, reproduce, and perform the activities necessary for survival. Where does this energy ultimately come from?
11. How does sunlight energy enter the biotic world?
12. On average, how much of the energy at one level of the food chain becomes available to the next level? What happens to the rest of the energy?

18.5 Ecological Succession

13. How does primary succession differ from secondary succession?
14. Why are the later colonizers of a habitat dependent on earlier waves of colonizers?
15. What usually happens to the total biomass in an ecosystem during succession? Does the number of species present in an ecosystem usually change as succession continues?
16. How can regular disturbances contribute to the biodiversity of a habitat?

Challenging Review Questions

18.1 Terrestrial Biomes

17. What are the major factors that determine what kind of biome is found in a habitat? Does the activity of living organisms affect the type of biome found in an area?
18. Explain why altitude and latitude affect the kind of biome that characterizes a habitat.
19. The map in Figure 18.1 shows the distribution of biomes on Earth. Strips of tundra are seen in both South America and the U.S. mainland. Is this an error? Explain.
20. Do tropical forests tend to provide good farmland when they are cleared? Why or why not?
21. Describe tundra. Why are there no trees in the tundra?

22. Name a biome where regular fires occur. What role does fire play in this biome?

18.2 Aquatic Biomes

23. What is an estuary? What adaptations characterize organisms that live in estuaries?

24. What is an example of plankton? Of nekton?

25. This photo shows a bobbit worm living in the oceans near Indonesia. Does it live in the pelagic zone or the benthic zone? How can you tell?

18.3 Biogeochemical Cycles

26. Describe the water cycle. What are some of the important reservoirs for water on Earth?

27. Was every carbon atom in your body once part of a plant or another producer? Why or why not?

28. Name at least two different processes that return carbon to the atmosphere.

29. How is the carbon cycle relevant to global warming and climate change?

30. Why do legumes grow better in nutrient-poor soil than do many other plants?

31. How has human technology affected the nitrogen cycle?

18.4 Energy Flow in Ecosystems

32. What is primary productivity? What factors could affect primary productivity in a habitat?

33. If you eat a pound of pasta, will you gain a pound of weight? Why not?

34. Why are there more zebras than lions in the African savanna?

18.5 Ecological Succession

35. Why do you think most of the early animal colonizers of Krakatoa were flying insects and birds?

36. Once a habitat is occupied by its climax community, does its species composition continue to change? Why or why not?

37. Could a habitat that received regular mild disturbances be more diverse than one that received no disturbances? How would its biodiversity compare to a habitat that received regular, extreme disturbances?

Apply & Discuss Questions

38. Why can't an ecosystem's energy pyramid be inverted (that is, be upside down)?

39. A single tree can sometimes support many insects and birds. Does this contradict the idea that all ecosystems are characterized by energy pyramids?

40. Different consumers vary in how efficiently they convert food into biomass. Insects use 10% to 40% of the energy they absorb (that is, energy not lost in feces) to build biomass. The rest goes to maintenance. Mammals and birds, on the other hand, use only 1% to 3% of their absorbed energy to build biomass. They use much more energy for maintenance. What accounts for this difference?

41. Would you expect to find more Type I or Type III species among the early colonizers of a habitat? Would you expect to find more Type I or Type III species in the climax community?

Solutions to Odd-Numbered Questions

Simple Review Solutions

1. Each biome is occupied by specific types of biological communities and, particularly, by specific types of plant life.

3. Tropical forests are famous for their biodiversity—more species are found in this biome than in all other biomes combined.

5. The photic zone is close to the water surface and receives plentiful sunlight, enough to power photosynthesis. The majority of pelagic species are found there.

7. Living organisms are made up of many different substances -- water, carbon, nitrogen, phosphorus, sodium, calcium, sulfur, chlorine, and others. All these substances move around Earth in a series of biogeochemical cycles, going back and forth between the tissues of living organisms and the abiotic world. The word biogeochemical emphasizes that substances cycle between living organisms (bio) and Earth (geo) -- specifically, Earth's atmosphere, crust, and waters.

9. Most of Earth's nitrogen exists as nitrogen gas in the atmosphere. However, nitrogen gas is not a form of nitrogen that most living things can use. Bacteria help convert

nitrogen into a usable form. Nitrogen-fixing bacteria in soil convert nitrogen gas to ammonium, and then nitrifying bacteria convert ammonium to nitrates. Plants absorb nitrogen primarily in the form of nitrates, although they make some use of ammonium as well. Nitrogen then moves from plants up the food chain.

11. Sunlight energy enters the biotic world when plants and other organisms use it to build organic molecules during photosynthesis.

13. Primary succession occurs when bare land, devoid of soil, is colonized by successive waves of living organisms. Primary succession may begin when new land is formed by volcanic activity, or when a glacier's retreat reveals bare rock. Secondary succession occurs when a disturbance destroys existing life in a habitat but leaves the soil intact. Secondary succession may begin after a fire or when old farmland is abandoned.

15. During the entire process of succession, the total biomass of the ecosystem typically increases, as does the number of species present.

Challenging Review Solutions

17. The type of biome found in a habitat is determined primarily by climatic variables such as temperature, precipitation, and the presence or absence of distinct seasons. As a result, latitude and altitude are major influences on the distribution of biomes on Earth. Living organisms sometimes do affect the biome found in a particular area. For example, organisms help maintain savanna habitats, preventing them from growing into tropical forest. For example, elephants eat and kill trees, and humans burn forests for cropland.

19. Those strips of tundra occur in high altitudes, where climatic factors such as extreme cold result in tundra.

21. Tundra is found in areas that experience extreme cold and little precipitation. One of the defining features of tundra is a layer of permafrost, or permanently frozen subsoil, beneath the topsoil. Trees cannot survive in tundra because of permafrost and the short growing season.

23. Estuaries are habitats where freshwater rivers flow into oceans. Organisms that live in estuaries, such as marsh grasses and mangroves, need to have adaptations that allow them to live under changing salinity conditions.

25. The bobbit worm lives in the benthic zone. You can see from the photo that it lives on the ocean bottom, rather than in the water itself.

27. Yes. Humans are heterotrophs, and we get all our carbon from the food we eat. Ultimately, this carbon all comes from producers such as plants, diatoms or other oceanic plankton, seaweeds, etc.

29. Global warming and climate change are caused by increasing levels of carbon dioxide in Earth's atmosphere. The carbon cycle describes how excessive amounts of carbon dioxide is released—through the burning of fossil fuels.

31. Humans have affected the nitrogen cycle by inventing industrial processes for making nitrogen-rich fertilizers. This has made larger amounts of nitrogen available for growing crops than could otherwise be achieved. Although this has allowed humans to grow more food and to support larger human populations, fertilizers also pollute and damage natural environments.

33. No, much of the energy in that pound of pasta will go to feces and maintenance (supporting your body's activities).

35. Flying insects and birds had the easiest time getting to the island from the mainland.

37. Yes. According to the intermediate disturbance hypothesis, regular disturbances, if not too extreme, actually contribute to biodiversity because different species make use of different habitats, and periodic disturbances guarantee that there will always be habitat at varying stages of recovery. However, a habitat that received regular, extreme disturbances would probably always be found in the early stages of succession and would probably be less diverse.

Apply & Discuss Solutions

39. Energy pyramids describe the amount of energy at different levels of a food chain. They do not say anything about the number of organisms at each level of the food chain. It makes sense that a single tree, which is very big, and captures significant amounts of sunlight energy, would be able to support many much smaller consumers such as insects and birds.

41. You would expect more Type III organisms among the early colonizers of a habitat - these habitats are unstable and difficult. Producing many offspring is adaptive because it becomes more likely that at least some of these will survive. You would expect more Type I organisms in a climax community - those habitats are more stable. An offspring that receives a lot of parental investment is more likely to be able to compete with other members of the population and successfully survive and reproduce.

Living Earth Essay

Life on Earth Connections

Global Climate Change

The Greenhouse Effect

Park your car with its windows closed in the bright sun and its interior soon becomes quite toasty. The inside of a greenhouse is similarly toasty. This happens because glass is transparent to visible light but not to infrared, as illustrated in **Figure A**. Wavelengths of visible light are shorter than wavelengths of infrared. Visible light wavelengths range from 400 nanometers to 740 nanometers, while infrared wavelengths range from 740 nanometers to a million nanometers. Short-wavelength visible light from the Sun enters your car or a greenhouse and warms various objects—car seats, plants, soil, whatever. The warmed objects then emit infrared light, which cannot escape through the glass, so the infrared energy builds up inside, increasing the temperature.

A similar effect occurs in the Earth's atmosphere, which, like glass, is transparent to visible light emitted by the Sun. The ground absorbs this energy but radiates infrared light. Atmospheric carbon dioxide, water vapor, and other select gases absorb and re-emit much of this infrared energy back to the ground, as **Figure B** illustrates. This process, called the **greenhouse effect**, helps keep the Earth warm. The greenhouse effect is quite desirable, because the Earth's average temperature would be a frigid −18°C without it. Greenhouse warming also occurs on the planet Venus, but to a far greater extent. The atmosphere surrounding Venus is much thicker than the Earth's atmosphere, and its composition is 95 percent carbon dioxide, which brings surface temperatures on Venus to a scorching 450°C—the hottest planet surface temperature of our solar system.

The role of carbon dioxide as a greenhouse gas is well documented. Core samples from polar ice sheets, for example, show a close relationship between atmospheric levels of carbon dioxide and global temperatures over the past 400,000 years, as shown in **Figure C**. Ancient air in bubbles trapped in the ice core, shown in **Figure D**, can be

Figure A
Glass acts as a one-way valve, letting visible light in and preventing infrared energy from exiting.

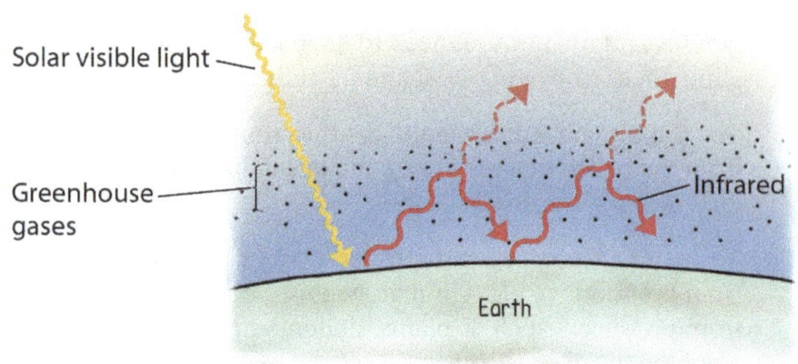

Figure B
The greenhouse effect in the Earth's atmosphere. Visible light from the Sun is absorbed by the ground, which then emits infrared radiation. Carbon dioxide, water vapor, and other greenhouse gases in the atmosphere reflect most of that heat back downward.

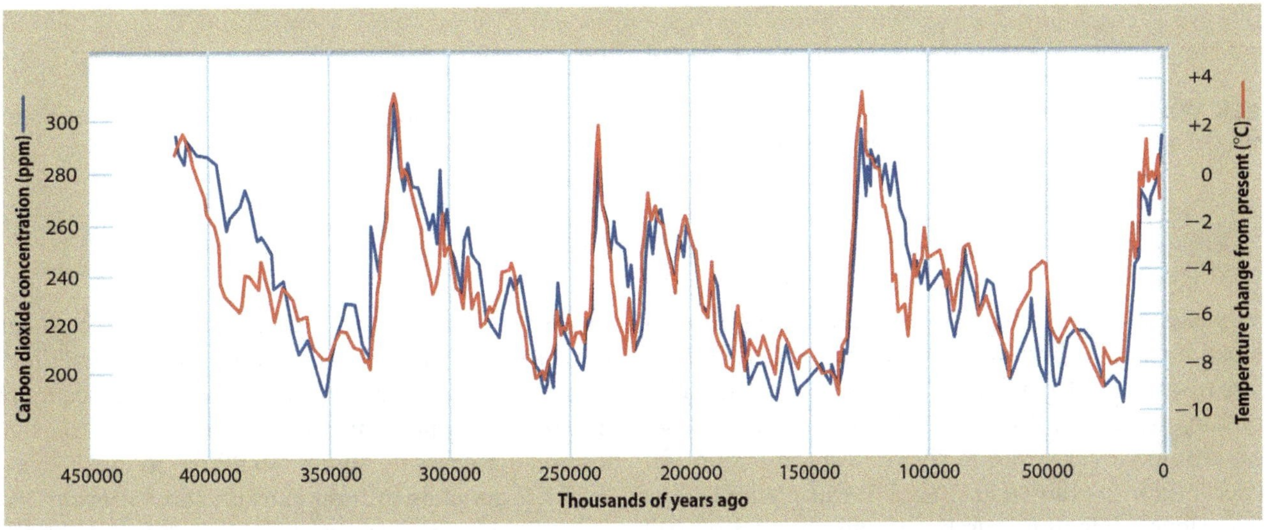

Figure C
Levels of atmospheric carbon dioxide and global temperatures appear to be closely related to each other. Carbon dioxide concentration over the last 450,000 years is shown by the blue line, and temperature is shown by the red line.

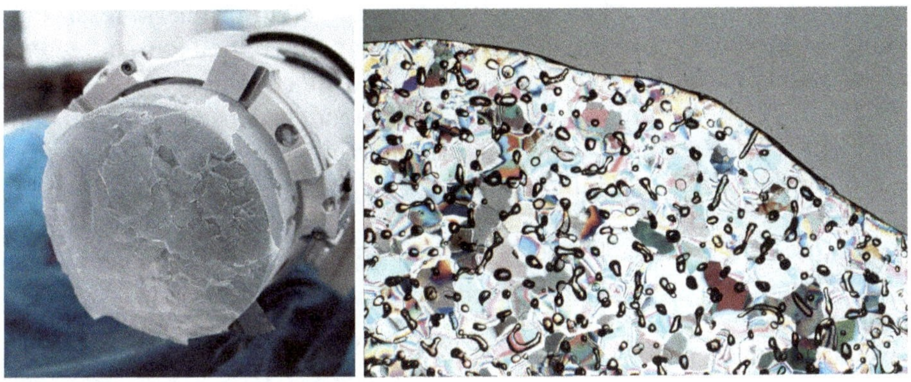

(a) (b)

Figure D
(a) Ice cores reveal information about ancient climates. (b) Crystals of ice photographed in polarized light reveal tiny air bubbles containing ancient air.

sampled directly to measure carbon dioxide levels. The age of the air is a function of the depth of the core. Past global temperatures are determined by measuring the deuterium/hydrogen ratio in the trapped air. When global temperatures are relatively high, the ocean is warmer and larger fractions of water containing deuterium evaporates from the ocean and fall as snow. A high deuterium/hydrogen ratio, therefore, indicates a warmer climate.

READING CHECK

How do we know of the close relationship between carbon dioxide and global temperatures?

FOR YOUR INFORMATION

If we were to hold steady and not increase the current rate at which we put CO_2 into the atmosphere, would the level of atmospheric CO_2 also hold steady? The answer is no because, at our current rate, we already put about twice as much CO_2 into the atmosphere than can be absorbed by plants, soils, and the ocean. To achieve a steady level of atmospheric CO_2, we would need to decrease our emissions by about 80%. It's analogous to a slowly draining bathtub. We're pouring water into this tub faster than it can drain, so the water level naturally rises. We've got to close down the spigot—not just stop opening it up—or the tub will eventually overflow.

CONCEPT CHECK

What does it mean to say that the greenhouse effect is like a oneway valve?

CHECK YOUR ANSWER Both the Earth's atmosphere and glass allow incoming visible waves to pass, but not outgoing infrared waves. As a result, radiant energy is trapped.

Human-Induced Atmospheric Changes

There is exceedingly strong evidence that recent human activities, such as the burning of fossil fuels and deforestation, are responsible for dramatic increases in atmospheric carbon dioxide levels. Prior to the Industrial Revolution, carbon dioxide levels were fairly constant at about 280 parts per million, as shown in **Figure E**. During the 1800s, however, levels began to climb, reaching a level of 300 parts per million in about 1910. Today's level is over 420 parts per million. Interestingly, as can be seen in Figure C, ice samples dating as far back as 400,000 years do not show atmospheric carbon dioxide levels ever exceeding 300 parts per million. In step with these increases, average global temperatures since 1860 have increased by about 0.8°C. (Since 1950, the increase has been about another 0.6°C.) Current estimates are that a doubling of today's atmospheric carbon dioxide levels will increase the average global temperature by an additional 1.5°C to 2.5°C. The Intergovernmental Panel on Climate Change, (IPCC) notes that this needs to remain below 1.5°C to avoid worst-case scenarios.

Carbon dioxide ranks as the number-one gas emitted by human activities. When speaking of atmospheric pollutants such as sulfur dioxide, we talk in terms of millions of tons. The amount of carbon dioxide we pump into the atmosphere. however. is measured in *billions* of tons. as **Figure F** shows. A single tank of gasoline in an automobile produces up to 90 kilograms of carbon dioxide. A jet flying from New York to Los Angeles releases more than 200,000 kilograms (about 300 tons). Above all, our population increases by about 236,000 individuals every day, which is about 86 million individuals every year. In 2012, we passed the milestone of 7 billion humans, and a mere 10 years later, in 2022, we surpassed 8 billion humans, each of us responsible for activities that result in the output of carbon dioxide. A satellite view of North America at night, as shown in **Figure G**, quickly reveals the significant impact we humans are having on a planetary scale.

When direct monitoring of atmospheric carbon dioxide began in 1958, the global atmospheric reservoir of carbon dioxide was about 2,460 billion tons, a figure calculated from the observed concentration of 315 parts per million. By 2020, this amount had grown to 3,280 billion tons, which simple subtraction tells us is an increase of 820 billion tons:

2020 Global atmospheric CO_2:	3,280 billion metric tons
1958 Global atmospheric CO_2:	2,460 billion metric tons
Net Increase:	**820 billion metric tons**

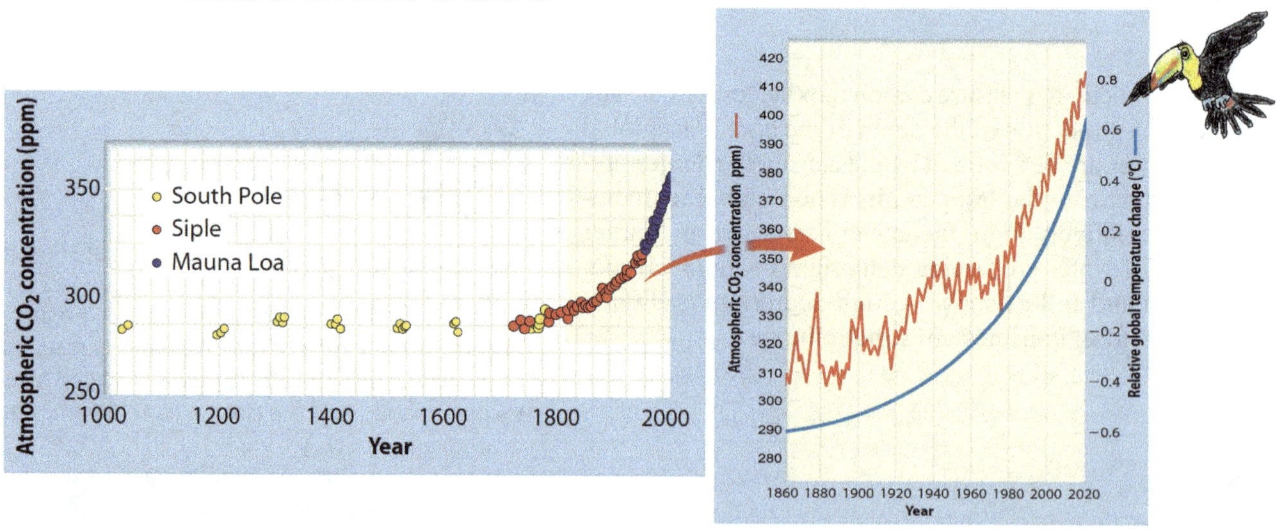

Figure E

Since the Industrial Revolution began in the late 1700s, atmospheric carbon dioxide levels have been increasing at accelerated rates. The yellow and red circles are data from ice core samples, and the purple circles are measurements from the Mauna Loa Observatory. Note: The relative global temperature changes shown in the blue line are relative to the average annual temperature for the years 1950 through 1979.

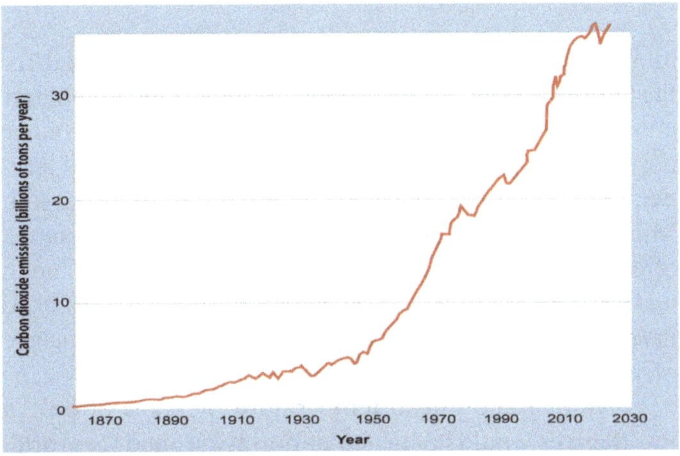

(Source: ESS-DIVE)

Figure F
Carbon dioxide emissions from the burning of fossil fuels have grown dramatically since 1860. Note the dip around 2020 due to the economic burden of the Covid-19 pandemic.

Figure G
Our ability to impact the global environment is clearly indicated by a night-time view of our planet.

Over the same period, humans released on the order of 1700 billion tons of atmospheric carbon dioxide from fossil fuels and other sources. From these data, we can get a feel for nature's ability to absorb carbon dioxide. Even though we pumped out 1700 billion tons of carbon dioxide, the total quantity in the atmosphere went up by only 820 billion tons. Nature is able to reabsorb over half of the carbon dioxide we emit, much of it going into the oceans, which can absorb carbon dioxide because its water is alkaline. Carbon dioxide can also be absorbed by vegetation during photosynthesis, which occurs both in the oceans and the land. It has been shown, for example, that trees grow more rapidly when exposed to higher concentrations of carbon dioxide. The fact that levels of atmospheric carbon dioxide are still going up, however, tells us that we are exceeding nature's absorbing power.

Consider this: according to NASA, more than half of all fossil fuels ever used by humans have been consumed in just the last 20 years. With about 20 percent of the world's population, China ranks first in carbon dioxide emissions and is responsible for about 22 percent of global carbon dioxide emissions. With about 5 percent of the world's population, the United States is the second leading carbon dioxide emitter, producing about 20 percent of the world's total. The third major emitter of carbon dioxide is the European Union, which, with about 6 percent of the world's population, produces about 14 percent. For these industrialized nations, this adds up to about 56 percent of the global carbon dioxide emissions. Developing nations account for the remaining 44 percent, but their sources of carbon dioxide are evenly split between fossil fuels and deforestation.

Deforestation presents multiple threats to atmospheric resources. If the cut wood is used for fuel rather than lumber, burning the fuel releases carbon dioxide into the atmosphere. Whether the wood is used for fuel or for lumber, though, cutting down any forest destroys a net absorber of carbon dioxide. Furthermore, tropical forests have the capacity to evaporate vast volumes of water, which assist in the formation of clouds. The clouds, in turn, keep regions cool by reflecting sunlight and moist by precipitating rain. Farmers who burn down rainforests for farmland are cutting off their future supply of rainwater. When their farms become desert, they are then spurred to burn even more of the rainforest. So far, about 65 percent of all rainforests have been destroyed. At present rates, within a few decades, remaining rainforests will not be able to sustain regional climates, which will leave more than a billion citizens of the rapidly growing communities of South America, Africa, and Indonesia in the midst of arid land.

As their economies and populations continue to grow over the next several decades, developing nations will likely surpass industrial nations in the amounts of carbon dioxide and other pollutants they emit. New energy-efficient technologies that minimize emissions, however, are now available. In a best-case scenario, developing nations will be able to utilize these new technologies while maintaining needed economic growth.

fyi FOR YOUR INFORMATION

Weather is what happens outside on a daily basis. A change in weather is quite common. The climate is different. The climate is the average weather that occurs in a particular region. The climate is rather predictable. It may or may not snow in Vermont on January 26, but the climate of Vermont tells the would-be tourists to bring their snow skis, not their water skis.

Potential Effects of Climate Change

There is consensus among scientists that increased levels of atmospheric carbon dioxide and other greenhouse gases will result in global warming, which will result in changes to global climate systems. How much temperatures may rise, however, is uncertain, as are the potential effects of the temperature increases. This uncertainty is due to the large number of variables that determine global weather patterns. The Sun's intensity, for example, changes over time, as does the ocean's ability to absorb and distribute greenhouse heat. Another variable is the cooling effect of cloud cover, atmospheric dust, aerosols, and ice sheets, which all serve to reflect incoming solar radiation.

A number of mechanisms may ease or even reverse global warming. For example, we may have underestimated the capacity of oceans and plants to absorb carbon dioxide. Greater levels of atmospheric carbon dioxide may simply mean more carbon dioxide in the ocean and more abundant plant life. In addition, warmer global temperatures could mean an increase in cloud cover worldwide and an increase in snowfall in the polar regions. Both of these effects would tend to cool the Earth by increasing the reflection of solar energy. If the cloud cover and snowfall became unusually extensive, continued reflection of solar radiation could even trigger an ice age.

On the other hand, some mechanisms may enhance global warming. Warmer oceans might have a diminished capacity to absorb carbon dioxide because the solubility of carbon dioxide in water decreases with increasing temperature. Also, as the pH of the ocean continues to drop, so will the ability of the ocean to absorb additional carbon dioxide. Rapid climatic changes might destroy vast regions of forests and vegetation, such as through wildfires spurred by drought, meaning those reservoirs for carbon dioxide absorption would no longer exist. Alternatively, more abundant plant life might not be as beneficial for the atmosphere as we had hoped, because although plants absorb carbon dioxide, they also emit other greenhouse gases, such as methane. Importantly, warmer global temperatures might also enhance microbial activity in the soil. Microbes decaying organic matter are a significant source of carbon dioxide in dry soils and a source of methane in wet soils. Furthermore, large quantities of methane locked in Arctic permafrost may also be released as a consequence of warmer terrain.

An average global temperature increase of only a few degrees would not be felt uniformly around the world. Instead, some places would experience wider fluctuations than others. For example, the number of days temperatures exceeding 32°C (90°F) might double in New York City but remain unchanged in Los Angeles. The number of days in polar regions when temperatures rise above 0°C, might double or even triple, causing glaciers and polar ice sheets to melt faster. Melting ice combined with the thermal expansion of ocean waters would lead to an increase in sea levels. Many climatologists project that a global temperature increase of a few degrees over the next 50 to 100 years may raise sea levels by about 1 meter, enough to inundate many coastal regions and displace millions of people.

Small changes in average global temperatures would also change weather patterns. The warming of the equatorial eastern Pacific Ocean during an El Niño, for example, is already known to change local weather patterns throughout the world. If the whole planet were to warm by a few degrees, the impact would be far greater. What is now fertile agricultural land might turn barren, while land now barren might turn fertile. Over the past several decades, for example, average global temperatures have edged upward.

Figure H
Which climate condition might become more prevalent as greenhouse gases continue to increase? The consensus of climate scientists is global warming, as evidenced by recent weather extremes.

In step with this warming trend, the growing seasons of the Great Plains of Canada are up to 2 weeks longer than they were several decades ago. As weather patterns change, one nation's gain may be another nation's loss. Developing nations lacking the resources to make adjustments, however, would be hardest hit.

Science tells us that the potential for human-induced global climate change is real but that the degree of impact is uncertain. Actual quantification will come only from a slow but steady accumulation of evidence. What should be done in the meantime is not a scientific issue but rather a societal one.

CONCEPT **CHECK**

Why are scientists uncertain about the potential effects of global warming?

CHECK YOUR ANSWER

The uncertainty is due to the large number of variables that determine global weather patterns. As the debates continue, bear in mind that the issue is not global warming itself but rather its potential effects.

One societal response to global warming is to adapt to the changes as they occur. Economists argue that large uncertainties in climate projections make it unwise to spend large sums of money trying to avert disasters that may never materialize. Adjusting to immediate changes would be more directed and far less costly. Some measures, however, could be taken now to reduce future difficulties. Irrigation systems, for example, might be made more efficient because even without a major climatic change, such an improvement would make it easier to cope with normal extremes in weather.

A second societal response is to take preventive measures to minimize global warming and the resulting changes in climate. Emissions of greenhouse gases could be kept low by conserving energy and by shifting to fuels containing lower percentages of carbon, such as natural gas or hydrogen. Cars could be powered not by fossil fuels, but by electricity generated from alternative energy sources such as biomass, solar thermal electric generation, wind power, and photovoltaics. Governments may also come to agree on a set of standards for emissions of carbon dioxide and other greenhouse gases. "Polluting rights" may be granted to each nation based on such factors as population and need for economic growth. Consider also the innovative solutions presented in **Figure I**.

The best public policies will be those that yield benefits even in the absence of global warming. A reduction in fossil fuel use, for example, would curb air pollution, acid rain, and the dependence of many countries on foreign oil producers. Developing alternative energy sources, revising water laws, searching for drought-resistant crop strains, and negotiating international agreements are steps that offer widespread benefits. To learn more about greenhouse gases, we recommend you visit Climate.gov and GlobalCarbonAtlas.org.

To Life!

fyi FOR YOUR INFORMATION

Why not just capture the carbon dioxide from the atmosphere and pump it underground? This is a process known as carbon sequestration. The problem here is the vast quantities that would need to be sequestered in order to make much of a difference. Furthermore, the energy source needed to drive sequestration would have to also be non-carbon-based. Despite these challenges, carbon sequestration continues to be an area of intense research. To learn more, visit Climeworks.com.

Figure I
How might seaweed be helpful in fighting global climate change? Certain seaweeds have been found to be high in the chemical bromoform, $CHBr_3$, which, when fed to cattle, cuts their methane, CH_4, emission by over 90 percent. Methane is about 80 times more powerful as a greenhouse gas than carbon dioxide. Total emissions are much less but still significant. About one-third of our methane emissions come from cattle, who release methane as part of their digestion process. Mix the right strain of seaweed into their diet, and this major source of greenhouse gas is greatly reduced.

"Curiosity is a delicate little plant that, aside from stimulation, stands mainly in need of freedom."

—Albert Einstein
(1879 - 1955)

Thank you for your learning efforts!!

Conceptual Academy

Index

A
Absolute time 64
Abyssobrotula galathaea 128
Acetylcholine 280
Acid 30
Acidic 32
Acquired immune system 314
AIDS 317
Actin 279
Action potential 255
Activation energy 77
Active transport 71, 84
Adaptations 162
Adaptive radiations 181
Adrenal glands 273
Age structure 330
Alkaline 32
Allele frequency 176
Alleles 118
Allopatric speciation 180
Alveoli 294
Ambulocetus 184
Amino acids 38
Amniocentesis 133
Amnion 277
Amniotes 235
Amphibians 234
Amygdala 260
Anabolic steroids 273
Anaerobic 82
Anaphase 113, 124
Anaphase I 114
Androgens 273
Animalia 201
Animals 225
Annelids 232
Anterior pituitary 272
Antibiotic resistance 160
Antibiotics 206
Antibodies 315
Anticodon 100, 105
Antidiuretic hormone 272
Antigen 314
Aquaporins 71
Archaea 207
Archaeopteryx 185
Archean 66
Arteries 290
Arterioles 290
Arthritis 278
Arthropods 229
Artificial selection 183
Asexual reproduction 2
Ashfall fossil beds 146
Athlete's foot 225
Atmosphere 16
Atom 17
Atomic mass number 47
Atomic nucleus 18
Atomic symbol 19
ATP 281, 293, 307, 309, 74, 84
Atrioventricular node 290
Australopithecus 190
Autoimmune diseases 316
Autonomic nervous system 253
Autotrophs 205

B
B cells 315
B vitamins 297
B vitamins 308
Bacillus thuringiensis 136, 141
Bacteria 206
Baker, bill 5
Banded iron formations 108
Base 30
Benthic zone 361
Big bang theory 9
Bilayer 58
Bioaccumulation 303, 304
Biodiversity 359
Biogeochemical 362
Biogeography 188
Biogeology 16
Biology 1
Biomass 365
Biomes 355
Biosphere 16
Biotic 325
Bladder 310
Blastocyst 277
Blood pH 293
Blood pressure 291
Blood pressure 291
Blood vessels 290
Bone marrow 278
Brainstem 250
Brown tree snake 369

C
Calcium carbonate 228
Calorie 307
Calvin cycle 80
Cambrian 67
Cambrian explosion 67
Cap 99
Capillaries 290
Carbohydrates 37
Carbon cycle 363
Carbon dioxide 376
Carbon-14 dating 48
Linnaeus, Carolus 201
Carriers 132
Carrying capacity 328
Cartilaginous fishes 233
Catalyst 29
Catalysts 77
Cell 53
Cell cycle 111
Cell membrane 57
Cell wall 52
Cellular respiration 81, 84
Central nervous system 252
Cephalopods 231
Cerebellum 251
Cerebrum 251
Chaparral 358
Charpentier, Emmanuelle 139
Chelicerates 229
Chemical compound 20
Chemical formula 20
Chemical reaction 27
Chemical synapses 257
Chemistry 17
Chemoreception 260
Chemosynthesis 129
Chlorinated hydrocarbons 303
Chlorofluorocarbons 134
Chlorophyll 268
Chlorophyll 79
Chloroplasts 60
Chloroplasts 78
Choanoflagellates 209
Chordates 233
Chorionic villus sampling 133
Chromosome 104
Chromosomes 53
Chromosomes 94
Chyme 297
Ciliates 209
Class 201
Climax community 367
Clones 315
Cloning 136
Cnidarians 226
Codominance 121, 124
Coelacanths 234
Collecting duct 310
Color-blindness 259
Commensalism 346
Competitive inhibition 77, 84
Conclusion 3

Cones 259
Conformations 23
Coniferous forests 357
Conifers 212
Connective tissue 248
Consumer 343
Contamination 141
Continental drift 89
Coral bleaching 227
Corals 226
Cornea 259
Covalent bond 22
Creutzfeldt–Jakob disease 239
Crispr-cas9 139
Crop rotation 338
Crossing over 114
Crustaceans 228
Cuckoos 347
Curare 281
Cyanobacteria 110
Cyanobacteria 66
Cytokinesis 114
Cytoplasm 57
Cytosine 96
Cytoskeleton 57

D
Darwin, Charles 154
Darwin's theory of evolution 175, 188
DDT 303
Decomposer 343
Decomposition 206
Deforestation 377
Demographic transition 331
Deoxygenated 291
Deoxyribonucleic acid 38
Deoxyribonucleic acid 96
Deoxyribose 96
Deserts 358
Deuterium/hydrogen ratio 375
Diabetes 273
Diabetes 316
Diaphragm 295
Diastolic blood pressure 291
Diatoms 208
Diffusion 84
Digestion 296
Dinoflagellates 208
Dinoflagellates 227
Diploid 95
Dipole 24
Dissolving 26
Distal tubule 310
DNA replication 97, 104
Doldrums 222
Dominant allele 118
Dominant traits 117
Dorudon 185
Double helix 104
Doudna, Jennifer A. 139

E
Eardrum 259
Echinoderms 228, 233
Ecological footprint 331
Ecological overshoot 331
Ecological succession 367
Ecology 325
Ecosystem 326
Ectotherm 235
Einstein, albert 362
Einstein, albert 4
El Niño 378
Electric potential 254
Electrical synapse 257
Electron affinity 24
Electron transport 81
Electron transport chain 80
Electron transport chains 81
Electrons 18
Electroreceptive organs 233
Element 18
Elemental formula 19
Elevation 355
Yellow bone marrow 278
Endemic 189
Endocytosis 72, 84
Endotherm 235
Endothermic 28
Endothermic 74
Energy pyramid 365
Enzymes 77
Epididymis 275
Epiglottis 296
Epinephrine 273
Epithelial tissue 248
Esau, Katherine 56
Escherichia coli 135, 52
Esophagus 296
Estrogens 273
Estuaries 360
Eukarya 201, 205
Eukaryotes 201
Eukaryotes 51
Evolution 158
Evolution 176
Evolutionary tree 203
Excitatory signal 258
Excretory system 309
Exercise 308
Exocytosis 72
Exons 99
Exoskeleton 229
Exothermic 28
Exothermic 74
Exponential growth 327
Extremophiles 128

F
Facilitated diffusion 71, 84
$FADH_2$ 81
Family 201
Feedback loop 249
Feedback regulation 249
Fermentation 82
Ferns 212
Fertilization 275
Fertilizers 322
Fibrin 293
Fight or flight 273
Flagella 206
Flagellates 209
Flatworms 229
Flower 213
Flowering plants 213
Fluid mosaic 59
Food webs 343
Fossils 184
Frameshift mutation 103
Freeze 199
Freshwater habitats 359
Frontal lobes 251
Fruit 213
Functional groups 35
Fungi 223
Fungicides 303

G
Gametes 275
Gametophyte 211
Gap junction 73, 84, 257
Gastric juice 297
Gastropods 231
Gastrovascular cavity 226
Gene 93, 104
Gene drives 138
Gene flow 178
Genetic code 100, 105
Genetic drift 177
Genetic engineering 135
Genetic mutation 102, 105
Genetic variation 158, 176
Genetically modified organism 135
Genome 131
Genotype 93, 104
Genus 201
Geosphere 16
Glial cells 252
Global climate change 374
Global warming 363
Glossopteris 89
Glucagon 273
Glucocorticoids 273
Glycolysis 81
Golgi apparatus 60
Grand canyon 65
Great oxygenation event 109
Greenhouse effect 374
Growth hormone 272
Guanine 96

H
Hades 66
Haploid 95
Hearing 259
Heart 289
Helper T cell 315
Heme group 293
Hemoglobin 293
Hemophilia 293
Herbicides 303
Hess, Harry 91
Heterotrophs 205
Heterotrophs 223
Heterozygote 118

Hippocampus 260
Histamines 313
Histones 94
Homeostasis 249
Hominidae 202
Hominids 190
Homo erectus 191
Homo habilis 191
Homo sapiens 68, 190, 202
Homozygote 118
Honeybees 303, 306
Horizontal gene transfer 123, 239
Hormones 251, 271
Horse latitudes 222
Human genome project 131
Human immunodeficiency virus 317
Humus 286
Hybridization 182
Hydrogen bonds 96
Hydronium ion 31
Hydrophilic 58
Hydrophobic 58
Hydrosphere 16
Hydroxide ion 31
Hypothalamus 251
Hypothalamus 272
Hypothesis 4
Hypothesize 3

I
Ice 198
In vitro fertilization 133
Incomplete dominance 121, 124
Inflammatory response 313
Inhibitory signal 258
Innate immune cells 312
Innate immune system 311
Insecticides 303
Insects 231
Insulin 273
Integrated crop management 339
Intermediate disturbance 369
Interneurons 253
Interphase 111
Interspecific competition 345
Intertidal zone 361
Invasive species 347
Ion 21
Ionic bond 22
Iris 259
Isotopes 47

J
Jellyfish 226
Jet streams 222

K
Karyote 51
Keeling curve 354
Kettlewell, Bernard 166
Kidneys 310
Kingdom 201
Krakatoa 367
Krebs cycle 81

L
Lamarck, Jean-Baptiste 154
Large intestine 297
Larva 225
Latitude 355
Law 9
Leaching 286
Libby, Willard F. 48
Light-dependent reactions 78
Light-independent reactions 78
Limnetic zone 359
Linked genes 122, 124
Linnaean system 201
Lipids 37
Liposomes 150
Littoral zone 359
Lizards 235
Logistic growth 328
Loop of Henle 310
Lungfishes 234
Lungs 294
Lupus 316
Lyell, Charles 155
Lymph 311
Lymph nodes 311
Lymphatic system 311
Lysosomes 60

M
Magnetic field 172
Magnetism 172
Magnetoreceptive 173
Malaria 138
Malathion 306
Malthus, Thomas 155
Mammalia 201
Mammals 236
Mariana trench 128
Marsupials 236
Mass extinction 67
Mcclintock, James 5
Meiosis 114
Meiosis I 114
Meiosis II 114
Melatonin 274
Membrane potential 254
Membrane proteins 58
Memory cells 315
Mendel, Gregor 116, 124
Menstrual cycle 273
Mesosaurus 89
Mesozoic 68
Messenger RNA 99
Metabolic syndrome 308
Metaphase 113, 124
Metaphase I 114
Metastasized 134
Methoxychlor 305
Metric system 10
Microbial activity 378
Microcontinents 66
Micrographia 53
Microirrigation 338
Microscope, electron 54
Microscope, light 54
Microvilli 297
Miller, Stanley 150
Mineralocorticoids 273
Minerals 308
Mitochondria 60, 81
Mitosis 111, 114
Mitotic spindle 113
Mixture 25
Mold 223
Molecule 19
Mollusks 228
Mollusks 231
Monotremes 236
Mosses 211
Motor neurons 253
Mucous membranes 312
Multicellular 51
Multiple sclerosis 256, 316
Muscle contraction 280
Muscle tissue 248
Muscles 279
Mutation pressure 177
Mycorrhizae 224
Myelin Sheath 256
Myofibrils 279
Myosin 279
Myplate 308

N
NADPH 79
NASA 152
Nat'l resources conservation 323
Natural selection 156, 158, 176
Neanderthals 202
Negative feedback 249
Nekton 360
Nephron 310
Nervous tissue 248
Neuron 252
Neurotransmitter 257
Neutral 32
Neutrons 18
Niche 345
Nitrogen 364
Nitrogen fixation 268, 269
Nitrogen fixers 268
Nitrogenase 268
Nitrogen-fixing bacteria 364
Noncompetitive inhibition 84
Nonsense mutation 103
Norepinephrine 273
Nuclear transplantation 136
Nucleic acids 38
Nucleus 57

O
Observe 3
Ocean pH 353
Oceanic zone 361
Optic nerve 259
Order 201
Organ 248
Organ of Corti 259
Organ system 248

Organelles 53, 57, 60
Organic chemistry 34
Organic farming 338
Organic farming 338
Osmosis 70
Ovaries 273, 275
Ovulation 33, 275
Oxidation 33
Oxitec 138
Oxygenated 291

P
Pacemaker 289
Yeh, Pamela 325
Pancreas 273
Pangaea 68
Paramecium 111
Parasitism 346
Parasympathetic nervous system 253
Parathyroid 272
Parathyroid hormone 272
Parietal lobes 251
Passive transport 71, 84
Pathogens 311, 312, 315
Pedigree 132, 142
Pelagic species 360
People meet and talk cars 113
Peppered moths 175
Periodic table 18
Peripheral nervous system 252
Peristalsis 297
Permafrost 357
Perseverance 152
Persistence 303
Pesticides 303
Pesticides 322
pH 32
pH of soil 287
Pharynx 294
Phenotype 93, 104
Pheromones 340
Phloem 210
Phospholipids 58
Photic zone 360
Photosynthesis 78, 377
Phyla 201
Pineal gland 274
Pinna 259
Pioneer species 367
Placenta 277
Placentals 236
Plankton 228, 359, 360
Plant-based diets 342
Plants 210
Plasmodesmata 73, 84
Plasmodium 209
Plate tectonics 92
Platelets 293
Pleiotropy 122, 124
Point mutation 102
Polar 24
Polar covalent 24
Pollen 212
Polygenic traits 121, 124
Polyploidy 182
Population 325
Population density 327
Posterior pituitary 272
Postzygotic 179
Predict 3
Prezygotic 179
Primary consumer 343
Primary productivity 365
Primary succession 367
Primates 190, 201
Principle 9
Principle of segregation 118
Prions 239
Producer 343
Profundal zone 360
Progestins 273
Prokaryotes 51
Prolactin 272
Prophase 112, 124
Prophase i 114
Proprioceptors 261
Prostaglandin 261
Prostate gland 276
Protein hormones 271
Proteins 38
Proterozoic 67
Protists 205, 208
Protons 18
Puerto Rican Trench 128
Pupil 259
Pure 25

Q
Quantum theory 9
Question 3

R
Carson, Rachel 304
Radioactive dating 64
Radioactive isotope 47
Radioactive tracing 245
Rainforests 356
Ray-finned fishes 234
Reaction rate 28
Receptor 73
Recessive allele 118
Recessive traits 117
Red blood cells 293
Red bone marrow 278
Reduction 33
Regenerative agriculture 341
Relative time 64
Renal pelvis 310
Reproductive barriers 179
Reptiles 235
Residence time 362
Resting potential 255
Retina 259
Ribonucleic acid 38, 98
Ribosomes 60
Macarthur, Robert 345
Rods 259
Rough endoplasmic reticulum 60
Roundup-ready 141
Roundworms 229

S
Saccharomyces cerevisiae 52
Salinization 323
Saliva 296
Saltwater habitats 360
Sarcomeres 279
Savannas 358
Schleiden, Matthias 54
Schwann, Theodor Schwann 54
Science 3
Scientific method 3, 5
Scientific notation 10
Sea butterfly 5
Seafloor spreading 90
Secondary consumer 343
Secondary succession 368
Seed 212
Seed plants 212
Selective serotonin reuptake inhibitors 258
Semen 276
Seminal vesicles 276
Sensory cells 261
Sensory neurons 252
Sex chromosomes 95
Sex organs 273
Sex-linked traits 122, 124
Sexual reproduction 114
Sexual reproduction 2
Sexual selection 163
Shamloo, Hannah 147
Short carbohydrates 59
Silent spring 304
Simple diffusion 69
Single-nucleotide polymorphisms 132
Sinoatrial node 289
Skeleton 278
Skin 312
Skin cancer 134
Small intestine 297
Smell 260
Smooth endoplasmic reticulum 60
Snowball Earth 67, 110
Soil horizons 286
Solutes 26
Solution 25
Solvent 26
Somatic nervous system 253
Speciation 179
Species 179, 201
Sperm 275
Sphincter 297
Sponges 225
Spores 223
Sporophyte 211
Steroid hormones 271
Stomach 297
Stomata 210
Stop codons 100
Subsoil 286
Substrate 77
Substratum 286

Super continents 67
Superposition 64
Suspension 26
Swim bladder 234
Symbiosis 346
Sympathetic nervous system 253
Sympatric speciation 182
Synapse 257
Système international 10
Systolic blood pressure 291

T
T cells 315
Taste 260
Taste buds 260
Tay-Sachs disease 132
Technology 8
Telophase 113, 124
Telophase I 114
Telophase II 114
Temperate forests 357
Temperate grasslands 358
Tendons 279
Terrestrial biomes 355
Test predictions 3
Testes 273, 275
Testosterone 273
Thalamus 251
The Origin of Species 154
Theory 9
Thermoregulate 164
Thoracic cavity 295
Thymine 96
Thyroid gland 272
Thyroid hormones 272
Tissue 248
Top predator 343
Topsoil 322
Touch 261
Trachea 294
Transcription 99, 104
Transfer RNA 100
Translation 100, 104
Transpiration 246
Transport proteins 71, 84
Tree of life 203
Trisomy 21 115
Tropical forests 356
Tundra 357
Tunicates 233
Type I organisms 329
Type II organisms 329
Type III organisms 329
Tyrannosaurus rex 342

U
Ubuntu 244
Umami 260
Unicellular 51
Uniformitarianism 64
Uniramians 229
United States customary system 10
Upwelling 200
Urea 309
Ureter 310
Urethra 276
Urethra 310
Urey, Harold 150
Uterus 275

V
Vaccines 316
Vacuoles 60
Vascular system 210
Veins 290
Venules 290
Vertebrates 233
Vesicles 60
Vestibular senses 261
Vestigial organs 186
Villi 297
Virchow, Rudolph 54
Virus 238
Vision 259
Vitamin D 308
Vitamin K 297, 308
Vitamins 307
Vocal cords 295
Voorhies, Mike 146

W
Water 198, 362
Water cycle 246, 362
Wegener, Alfred 89
Westerlies 222
White blood cells 293
Wind chill 221
World Health Organization 303

X
Xylem 210

Y
Yellowstone 147

Z
Zebra mussel 348, 369
Zika 138
Zona Pellucida 276
Zooxanthellae 227
Zygote 179

Photo Credits

Chapter 1

CAB01CO	TB studio
Figure 1.2	Ronnie Macdonald
Figure 1.3a	Djpmapleferryman
Figure 1.3b	Christopher Michel
Figure 1.4a	Russ Hopcroft
Figure 1.4b	Jim Baker
Page 8 UNA1	Monkey Business
Figure 1.7	Katya
Page 17 Q15	Ben Goodwyn

Chapter 2

CAB02CO	Bigc Studio
Figure 2.1	Ponor
Figure 2.3	Nick Kellar
Figure 2.7	Nefronus
Figure 2.16	Bob Haarmans
Figure 2.16	Vyacheslav Argenberg
Figure 2.16	Bence Fördős
Figure 2.17	Colin Knowles
Figure 2.20	Sekundemal
Figure 2.21a	Alan Light
Figure 2.21b	John McColgan
Figure 2.30	Dogora Sun
Figure 2.36b	MayStudio
Figure 2.37	Billion Photos
Page 40 UNA1	Warren Garst

Chapter 3

CAB03CO	lHarbingerl
Figure 3.2	Volunteer7
Figure 3.3	Mogana Das Murtey
Figure 3.5a	Kelvinsong
Figure 3.5b	Louisa Howard
Page 65 UNA1	Tuxyso
Page 66 Fig C	Ismoon

Chapter 4

CAB04CO	Judyta Dulnik
Figure 4.3	Animals Network
Figure 4.9a	David M Phillips
Figure 4.9b	Biophoto Associates
Figure 4.16	14 Mostafa&zeyad
Page 89 Fig Bb	Paul Hermans
Page 92 Fig G	M.Bitton

Chapter 5

CAB05CO	Geoff Gallice
Figure 5.1b	Simon Samtleben
Figure 5.2b	AJC1
Figure 5.4b	CalTech
Page 108 Fig A	Arnoldius
Page 108 Fig B	James St. John

Chapter 6

CAB06CO	ParkyJ
Figure 6.1a	500paolo
Figure 6.4	Natalierussell77
Figure 6.18a	Dan-yell
Figure 6.18b	Albarubescens
Page 128 Fig A	Monterey Bay Aquarium
Page 128 Fig B	California Academy of Science

Chapter 7

CAB07CO	vchal
Figure 7.2	Emily Walker
Figure 7.3	Jennifer Yeh
Figure 7.4	Axventura
Figure 7.6	Sam LaRussa
Figure 7.7	A Dombrowski
Figure 7.8	Tom Barros

Figure 7.11a	Bianca Fioretti
Figure 7.11b	Duncan.Hull
Figure 7.12	Thomas Splettstoesser
Figure 7.13	William Murphy
Page 146, Fig A	Mike Voorhies
Page 146, Fig B	James St. John
Page 147, Fig D	Hannah Shamloo
Page 148, Fig E	Hannah Shamloo

Chapter 8

CAB08CO	Wang LiQiang
Figure 8.3a	Arkhipov Sergey
Figure 8.4	Paul Harrison
Figure 8.7	James St. John
Figure 8.10a	Mike's Birds
Figure 8.10b	Putneymark
Figure 8.10c	BB Oros
Figure 8.13	Ben Sale/AJC1
Figure 8.14b	Vittorio Bruno
Figure 8.15a	Andrea Lawardi
Figure 8.15b	HarpersFerryNPS
Figure 8.16	Brian Gratwicke
Page 164 Top	Oliver Koemmerling
Figure 8.18a	Siew Te Wong
Figure 8.18b	Alan Wilson
Page 169 Ex 53	Piccolo Namek
Page 169 Ex 53	Korall
Page 173 Bottom	Laitche
Page 174 Bottom	Dajome 19

Chapter 9

CAB09CO	Anek Sangkamanee
Figure 9.3	Ondrej Prosicky
Figure 9.6a	Mark Mauno
Figure 9.6b	Tony Hisgett
Figure 9.8	Miras Wonderland
Figure 9.9	John Doebley
Figure 9.11a	Wolfgang Sauber
Figure 9.15a	Marko Dimitrijevic
Figure 9.15b	ravas51
Figure 9.16	Forest and Kim Starr
Figure 9.18	Sailko
Figure 9.19	Fährtenleser
Page 195 Ex 28	Steve Gorton
Page 195 Ex 30	Elf

Chapter 10

CAB10CO	Framalicious
Figure 10.5	Janice Carr
Figure 10.7	Hannes Grobe
Figure 10.8	Stef Maruch
Figure 10.9	SmallRex
Figure 10.10	Lyth, O
Figure 10.13	Brian Stansberry
Figure 10.14	Pete Bobb
Figure 10.15b	5snake5
Figure 10.17a	Debi Vort
Figure 10.17b	Bloe Schneid
Figure 10.17c	Andrew Mercer
Page 217 Ex 28	Actinomycetes Society of Japan
Page 217 Ex 37	500px

Chapter 11

CAB11CO	Klaus
Figure 11.1a	dallenstein
Figure 11.1b	Nicole Wolf
Figure 11.3	Ellen Larsson
Figure 11.6	Uwe Nassal
Figure 11.7b	高國維
Figure 11.7c	Ryan Goehrung
Figure 11.8	Mogana Das Murtey
Figure 11.10a	John Turnbull
Figure 11.10b	Don Horne
Figure 11.10c	Peter W. Chen
Figure 11.11	Forest & Kim Starr
Figure 11.12	Albert Kok
Figure 11.13	GlebK
Figure 11.14	Peter Southwood
Figure 11.15	Nick Hobgood
Figure 11.16	Fernando Rodriguez
Figure 11.17a	Mitch Ames
Figure 11.17b	Bruce A.S. Henderson
Figure 11.18	Girish Gowda
Figure 11.19	Mark Pellegrini
Figure 11.20a	Ros Runciman
Figure 11.20b	Jeanlou Justine
Figure 11.21	Steven Penton
Figure 11.22	Silverish Lily
Page 245 Fig B	Tom Harpel
Page 245 Fig C	Tiia Monto
Page 245 Fig C	FungiFriendly

Chapter 12

CAB12CO	NIMEDIA
Figure 12.1b	Bruce Blaus
Figure 12.6	Manuel Schottdorf
Figure 12.8b	Asier Bermejo
Figure 12.11b	OpenStax College
Figure 12.13	Theklan
Figure 12.14b	Mhelga Kolb
Figure 12.15	Shi-Ming Yang
Page 265 Ex 46	Alex
Figure 15.5c	Alandmanson

Chapter 13

CAB13CO	Anne-Marie Keppel
Figure 13.7	Enver Kerem Dirican
Figure 13.8	Designer things
Figure 13.12	Nephron
Figure 13.13a	ScholarJay
Figure 13.13b	Laboratoires Servier
Page 284, Ex 25	Tropen Museum
Page 286 Fig A	Nigel Chadwick

Chapter 14

CAB14CO	TB studio
Figure 14.2	Scientific Animations
Figure 14.5	Pam Belle
Figure 14.6	Fuzis
Figure 14.8	Scientific Animations
Figure 14.9	Cuerdasvocales
Figure 14.15a	Scientific Animations
Figure 14.15b	Ballena Blanca
Page 303 Fig A	David McClenaghan
Page 304 Fig F	Forest Wander

Chapter 15

CAB15CO	Aila Images
Figure 15.1	Dagial
Figure 15.5	eLife-The Journal
Figure 15.15	NIAID

Chapter 16

CAB16CO	Kirill Neiezhmakov
Figure 16.1	Pamela Yeh
Figure 16.2b	Tim Evanson
Figure 16.2d	Bernard Dupont
Page 334 Ex 16	Bernard Spragg
Page 334 Ex 16	Mike's Birds
Page 338 Fig A	Carolina Latorre Canet
Page 338 Fig A	AlvarovegaCLP
Page 338 Fig B	Acabashi
Page 338 Fig B	normanack
Page 339 Fig D	James Brey
Page 340 Fig E	bdk
Page 340 Fig F	Marjorie Hoy
Page 340 Fig G	Ian Alexander
Page 340 Fig G	Albert Herring
Page 341 Fig Ha	Bluemoose
Page 341 Fig Hb	Dehaan
Page 342 Bottom	James Brey

Chapter 17

CAB17CO	Tory Kallman
Figure 17.4	Bruce Bartel
Figure 17.6a	Tomfy
Figure 17.6b	Nicholas Reynolds
Figure 17.6c	Brocken Inaglory
Page 347	Per Harald Olsen
Figure 17.7	Pavel Kirlove
Page 351 Ex 26	Pharaoh han
Page 353 Fig C	Hannes Grobe

Chapter 18

CAB18CO	MakanaCreative
Figure 18.2a	Michael Cory
Figure 18.2b	Siddharth Mallya
Figure 18.2c	Crockerbd
Figure 18.2d	Paxson Woelber
Figure 18.3a	Michelle Maria
Figure 18.3b	USDA NRCS
Figure 18.3c	Krzysztof Ziarnek
Figure 18.3d	Matt Lavin
Figure 18.5	Michelle Reback
Figure 18.8	John Lloyd
Figure 18.12	Terraprima
Page 372 Ex 25	Rickard Zerpe
Page 375 Fig D	Laurent Augustin
Page 375 Fig D	CSIRO
Page 378 Fig H	Thomas Fuhrmann
Page 378 Fig H	Ilya Grigorik
Page 379 Fig I	Jean-Pascal Quod
Page 559 Figure A	Andrei Porzhezhinskii